NELSON CHEMISTRY UNITS 1 & 2

FOR THE AUSTRALIAN CURRICULUM

DEBRA **SMITH**
ANNA **DAVIS**
ANNE **DISNEY**
VERONICA **HAYES**
RACHEL **WHAN**

Australia • Brazil • Japan • Korea • Mexico • Singapore • Spain • United Kingdom • United States

Nelson Chemistry Units 1 & 2 for the Australian Curriculum
1st Edition
Debra Smith
Anna Davis
Anne Disney
Veronica Hayes
Rachel Whan

Publishing editor: Eleanor Gregory
Senior Editor: Kelly Robinson
Editor: Catherine Greenwood
Proofreader: Alexa Cloud
Indexer: Max McMaster
Cover design: Claire Atteia
Text design: Claire Atteia
Art direction: Luana Keays
Cover image: Shutterstock.com/Daniel Sochor
Permissions researcher: Helen Mammides
Production controller: Julie McArthur
Typeset by: MPS Limited
Reprint: Jess Lovell

Any URLs contained in this publication were checked for currency during the production process. Note, however, that the publisher cannot vouch for the ongoing currency of URLs.

Acknowledgements
Seventeen diagrams have been used with permission from Roland Smith, *Conquering Chemistry Preliminary Course* and *HSC Course*, 4th edition, 2004 and 2005 and 2nd edition, 1997, McGraw-Hill, Sydney.

For product information and technology assistance,
in Australia call **1300 790 853;**
in New Zealand call **0800 449 725**

For permission to use material from this text or product, please email
aust.permissions@cengage.com

National Library of Australia Cataloguing-in-Publication Data
Davis, Anna, author.
Nelson chemistry units 1 & 2 for the Australian curriculum
/ Anna Davis, Anne Disney, Veronica
Hayes, Deb Smith, Rachel Whan.

9780170246644 (paperback)
Includes index.
For secondary school age.

Chemistry--Textbooks.
Chemistry--Study and teaching (Secondary)
Chemistry--Problems, exercises, etc

540

Cengage Learning Australia
Level 7, 80 Dorcas Street
South Melbourne, Victoria Australia 3205

Cengage Learning New Zealand
Unit 4B Rosedale Office Park
331 Rosedale Road, Albany, North Shore 0632, NZ

For learning solutions, visit **cengage.com.au**

Printed in China by China Translation & Printing Services.
6 7 8 9 10 11 12 21 20 19 18 17

CONTENTS

Chemistry

Chapter 6: Water and the intermolecular forces 257

Chapter 7: Reactions in aqueous solutions 303

Chapter 8: Gases 341

Chapter 9: Reaction rates 367

Chapter 10: Scientific investigations 397

PREFACE

Nelson Chemistry for the Australian Curriculum Units 1 & 2 has been written to meet the requirements of the Australian Curriculum, Assessment and Reporting Authority (ACARA) Australian Senior Secondary Curriculum – Chemistry.

Nelson Chemistry for the Australian Curriculum Units 1 & 2 is written by classroom teaching experts. They were chosen for their comprehensive knowledge of the chemistry discipline and best teaching practice in chemistry education at secondary and tertiary levels. They have written the text to make it accessible, readable and appealing to students. The text has been written to enable students to meet the A level Achievement Standard. It also allows all students to maximise their learning and results.

The textbook has been written to focus on the human endeavour aspect of chemistry and highlights the relevance of chemistry to students. By providing a contextual base in the form of context chapters, this book provides reasons for developing the understanding and skills in the ACARA Australian Senior Secondary Curriculum – Chemistry.

One aim of structuring the book in this manner is to 'develop students' interest and appreciation in chemistry and its usefulness in helping to explain phenomena and solve problems encountered in their ever-changing world' (ACARA, 2013). It highlights how chemistry relates to each and every thing that can be sensed, from the nanotechnology level to massive complex structures. Students are made aware of how issues such as sustainability, supply of clean drinking water, efficient production and use of energy, acidification of the oceans and climate require the application of chemical knowledge in developing solutions.

Chemistry is frequently called the 'central science' because it incorporates and connects the sciences of biology and physics and serves of great importance in engineering applications and medical fields. This book links chemistry to other disciplines of science by considering examples such as nanotechnology, energy, extraction of metals and production of sustainable materials.

The chemistry chapters explain the understanding and skills that are fundamental to a course of study in chemistry. They deal with how materials form, their composition and properties, how they change and the energy they consume or release through these changes. Theories and models that form the basis of chemical understanding and its use as a predictive tool in investigations and research are emphasised. The experimental nature of chemistry is modelled and students are provided with many opportunities to design and conduct their own experimental research in order to challenge their own understanding of this discipline.

The contextual, methodological approach is designed to ensure students can reach the highest possible standard. The intention has been to ensure all students achieve the level of depth and interest necessary to pursue tertiary studies in chemistry, engineering, technology and other scientific courses.

Each context and chemistry chapter follows a consistent pattern. Relevant learning outcomes from the Science Understanding and Science Inquiry Skills strands appear on the opening page of the chemistry chapters. Science as a Human Endeavour learning outcomes appear on the opening page of the context chapters.

Worked examples, written to connect important ideas and solution strategies, are included throughout the text. Solutions are written in full, including an explanation of the processes used to arrive at the solution. In order to consolidate learning, students are challenged to try similar questions on their own.

Numerous question sets appear at the end of major sections in the book and there is a comprehensive set of review questions at the end of each chapter. All questions have been graded from lower- to higher-order thinking skills: Remembering, Understanding, Applying, Analysing and Reflecting. Numerical answers appear at the end of the book and complete worked answers appear on the teacher website.

Experiments and Investigations demonstrate the high level of importance the authors attach to understanding-by-doing chemistry. These activities introduce, reinforce and enable students to practise Science Inquiry Skills, especially experimental design, data collection, analysis and conclusions. The Scientific Investigations chapter consolidates important investigative concepts and values. It enables students to learn and reflect on their experience as puzzle-solvers and investigators. It is an invaluable tool for students undertaking an Extended Experimental Investigation.

Nelson Chemistry for the Australian Curriculum Units 1 & 2 provides students with a comprehensive study of chemistry that will fully prepare them for a range of different types of assessment and any future studies in the area.

Debra Smith
Series editor

AUTHOR AND REVIEWER TEAMS

Authors

Debra Smith

Debra Smith is an experienced Chemistry teacher and author. Debra has been Head of Science at a number of Queensland high schools and has developed Science curriculum material for Education Queensland schools. She has been President and Treasurer of the Australian Science Teachers' Association. Debra has been involved in writing science curriculum at the local, state and national levels. For her contribution to science education, Debra was awarded the 2010 Prime Minister's Prize for Excellence in Science Teaching in Secondary Schools.

Anna Davis

Anna Davis is the Science Coordinator at Casimir Catholic College, NSW. Anna has extensive experience as a marker and senior marker for the HSC Chemistry examination. She is a Past President of both the Science Teachers' Association of New South Wales and the Australian Science Teachers' Association. Anna has extensive experience presenting professional learning events for science teachers throughout NSW in programming and assessment. Anna was awarded the 2006 Prime Minister's Prize for Excellence in Science Teaching in Secondary Schools.

Anne Disney

Anne Disney is an experienced Chemistry and Physics teacher at a large senior secondary school in Darwin. She works with students in the Centre for Excellence Program in Science and Engineering developing industry-based contextual teaching programs for use in senior science. She has been an exam marker and moderator in South Australia in both Chemistry and Physics for a number of years. Anne has been an active member of the Science Teachers' Association of the NT acting as Vice-President, delegate to the Australian Science Teachers' Association Council, and competitions coordinator.

Veronica Hayes

Veronica Hayes started her career as an industrial chemist. She is now an experienced teacher and Head of Department at St Michael's Grammar School in Victoria. She has taught in a number of schools, is co-director for the Science Drama Awards (STAV) and a Project leader for an ASISTM project (Australian Science Innovation in Science Technology and Maths). She has been a member of the VCAA VCE Chemistry review panel.

Rachel Whan

Rachel Whan studied Veterinary Science at the University of Queensland before completing a Graduate Diploma in Education at the University of Southern Queensland. Since then she has taught Chemistry, Biology, Human Biology, Science, Mathematics and Agriculture in a variety of schools in both Queensland and Western Australia including State, Catholic and Independent schools. She is currently teaching Chemistry, Human Biology and Science at St Brigid's College in Lesmurdie, WA, where she has been the PK–12 Team Leader for Science for the past 7 years. Rachel has been involved in the implementation of state, national and international curriculums across all year levels and has also been involved in the marking and moderation of Chemistry at a state level. Rachel was a co-author of the MYP Science for the International Student IB series of textbooks published by Cengage.

Reviewers

The publisher and authors would like to thank the reviewers for their valuable assistance in developing the manuscript.

Charles Coleiro
Ave Maria College, Vic

Pat Corbin
Lutheran College, Qld

Vicki Cox
Somerset College, Qld

Julie Miles
St Michael's College, Qld

Dr Colleen Spence
Marist College, ACT

Paul Walker
Grace Christian School, WA

Dr Siew Yap
Kingsway Christian College, WA

USING NELSON CHEMISTRY

Nelson Chemistry Units 1 & 2 for the Australian Curriculum provides a contextual approach to the teaching and learning of chemistry. Four relevant contexts have been provided as the first four chapters of the book. These contexts are specifically written to develop the understandings identified in the Science as a Human Endeavour strand. The contexts show how chemistry knowledge can be applied in real-world situations. The chemistry knowledge that links into these contexts has been provided as the next nine chapters. Students will find it easy to navigate between the context and the chemistry content chapters through the use of linking icons. Teachers also have the flexibility of providing their own contexts and still being able to use the chemistry chapters provided in this book.

Science Inquiry Skills have been integrated within the context and chemistry chapters so the students can see the interconnectedness between the theoretical and practical aspects of chemistry.

Each context begins with a concept map that shows the interrelationships between the context, the chemistry content and inquiries and activities. It links to the relevant content of the three strands of the senior Chemistry Australian Curriculum.

Each chapter and context begins with a **Chapter opener**. This page in chemistry chapters presents the Science Understanding content descriptions from the senior Chemistry Australian Curriculum. This page in the context chapters presents the Science as a Human Endeavour content descriptions.

The text has been authored and reviewed by experienced chemistry educators to enable students to achieve the maximum level of achievement of which they are capable. A number of devices have been utilised to improve literacy and understanding. One of these is the use of short sentences and paragraphs. This is coupled with clear and concise explanations and real-world examples. New terms are bolded as they are introduced and appear in an end-of-chapter and an end-of-book glossary.

Important concept

Throughout the text, important ideas, formulas and laws are summarised in **Important concept** boxes.

WORKED EXAMPLE

Step-by-step instructions on how to perform chemical calculations are shown in the **Worked examples**. The logic behind each step is explained. Students are then able to practice these steps by attempting the related problems presented at the end of the worked example.

Chemistry is a practical subject and students need to be given the opportunity to explore and discover through inquiry, which includes practical and research activities. These are presented in three different types of boxes throughout the text.

ACTIVITY

The **Activities** provide the opportunity for short, hands-on tasks to clarify or reinforce a concept. The activities can be performed either individually or in groups.

INVESTIGATION

The **Investigations** allow students to develop, practise and reinforce Science Inquiry Skills. They provide students with the opportunity to design and carry out their own scientific investigation either individually or in a group. Students are prompted to consider ideas for improvement and further investigation to illustrate that investigating involves an ongoing and improving process. Further information on how to conduct a scientific investigation can be found in the Scientific Investigations chapter on page 397.

EXPERIMENT

The **Experiments** reinforce the Science Inquiry Skills strand of the Australian Curriculum. Experiments contain guided instruction on the materials, procedure, collection and analysis of results and discussion.

Risk assessment

The **Risk assessment** table occurs within experiments and investigations. The table highlights the risks to the students and provides suggestions on how to minimise these risks. Teachers are able to supplement this table by adding any further risks specific to the school situation. It also provides a scaffold so that students can begin to conduct their own risk assessments.

Links between Context chapters and Chemistry chapters occur via the **Linking boxes**. One context chapter may be linked to several chemistry chapters, and a chemical concept may be linked to a number of contexts demonstrating application of the same chemistry knowledge in different situations.

At times it is necessary to add an extra piece of information, a curriculum link or an encouraging note. This is done via **Margin notes**.

The Chemistry curriculum draws on knowledge and understanding from across Biology, Earth and Environmental Science, Mathematics and Physics. Knowledge from these areas also provides students with an enriched understanding of chemistry.

Review of student understanding is attained through the **Question sets** throughout the chemistry chapters. Questions are ordered from lower- to higher-order thinking skills. The addition of Reflection questions gives students the opportunity to reflect upon not only what they are learning but why and how they are learning it.

The end-of-chapter review provides:

- a **Summary** or **Checklist** of the important concepts presented within the chapter. This will be a valuable tool when students are revising for tests and exams
- a **Glossary** of all the new terms introduced within the chapter
- **Chapter review questions** (in Chemistry chapters) that review understanding of concepts from the chapter. Questions are ordered from lower- to higher-order thinking skills and include Reflection questions.

Where answers to questions are numerical or only a few words, they are provided in the back of the book.

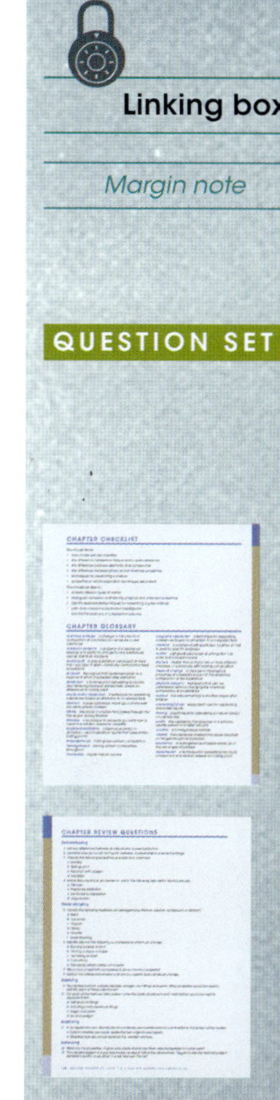

NelsonNet

NelsonNet is your protected portal to the premium digital resources for Nelson textbooks, located at www.nelsonnet.com.au. Once your registration is complete you will have access to an exciting and stimulating digital suite of resources for each chapter that is designed to enhance and reinforce learning.

Each chapter will also be supplemented with the following digital resources.

- **Prior learning worksheet** to revise content from Years 9 and 10 that is a prerequisite to understanding
- **Activity sheets**, including theory and practical exercises
- **Revision sheets** to complete at home to revise class work
- A **review quiz** containing 20 auto-correcting multiple-choice questions at the end of each chapter to review understanding
- **Videos** to assist students in understanding complex concepts. Pages that have videos associated with them are indicated with a blue icon in the footer.
- **Chapter summaries** to support students when revising for tests and exams
- **Links** to websites that contain extra information. These are hotspotted within the ebook and they can also be accessed at http://cac1and2.nelsonnet.com.au.

Please note that complimentary access to NelsonNet and the NelsonNetBook is only available to teachers who use the accompanying student textbook as a core educational resource in their classroom. Contact your sales representative for information about access codes and conditions.

CURRICULUM GRID

		Context				Chemistry									
		1	2	3	4	1	2	3	4	5	6	7	8	9	10
Science as a Human Endeavour	Science is a global enterprise that relies on clear communication, international conventions, peer review and reproducibility (ACSCH009 AND ACSCH048)			✓											
	Development of complex models and/or theories often requires a wide range of evidence from multiple individuals and across disciplines (ACSCH010 AND ACSCH049)			✓	✓										
	Advances in science understanding in one field can influence other areas of science, technology and engineering (ACSCH011 AND ACSCH050)	✓	✓	✓											
	The use of scientific knowledge is influenced by social, economic, cultural and ethical considerations (ACSCH012 AND ACSCH051)	✓	✓	✓	✓										
	The use of scientific knowledge may have beneficial and/or harmful and/or unintended consequences (ACSCH013 AND ACSCH052)	✓		✓	✓										
	Scientific knowledge can enable scientists to offer valid explanations and make reliable predictions (ACSCH014 AND ACSCH053)	✓		✓	✓										
	Scientific knowledge can be used to develop and evaluate projected economic, social and environmental impacts and to design action for sustainability (ACSCH015 AND ACSCH054)		✓	✓											
Science Inquiry Skills	Identify, research and refine questions for investigation; propose hypotheses; and predict possible outcomes (ACSCH001)		✓						✓						
	Identify, research, construct and refine questions for investigation; propose hypotheses; and predict possible outcomes (ACSCH040)			✓	✓						✓	✓		✓	✓
	Design investigations, including the procedure/s to be followed, the materials required, and the type and amount of primary and/or secondary data to be collected; conduct risk assessments; and consider research ethics (ACSCH002 AND ACSCH041)		✓		✓		✓		✓		✓	✓		✓	✓

		Context				Chemistry									
		1	2	3	4	1	2	3	4	5	6	7	8	9	10
Science Inquiry Skills	Conduct investigations, including: – the use of devices to measure temperature change and mass, safely, competently and methodically for the collection of valid and reliable data (ACSCH003)		✓			✓	✓	✓	✓	✓					✓
	– measuring pH and the rate of formation of products, identifying the products of reactions, and testing solubilities, safely, competently and methodically for the collection of valid and reliable data (ACSCH042)				✓						✓	✓	✓	✓	
	Represent data in meaningful and useful ways, including using appropriate graphic representations and correct units and symbols; organise and process data to identify trends, patterns and relationships; identify sources of random and systematic error and estimate their effect on measurement results; and select, synthesise and use evidence to make and justify conclusions (ACSCH004)	✓	✓			✓	✓	✓	✓	✓					
	Represent data in meaningful and useful ways, including using appropriate graphic representations and correct units and symbols; organise and process data to identify trends, patterns and relationships; identify sources of random and systematic error; identify anomalous data; estimate the effect of error on measured results; and select, synthesise and use evidence to make and justify conclusions (ACSCH043)				✓						✓	✓	✓	✓	✓
	Interpret a range of scientific and media texts, and evaluate processes, claims and conclusions by considering the quality of available evidence; and use reasoning to construct scientific arguments (ACSCH005 AND ACSCH044)		✓				✓		✓			✓			✓
	Select, construct and use appropriate representations including chemical symbols and formulae, molecular structural formulae, physical and graphical models of structures, chemical equations and thermochemical equations, to communicate conceptual understanding, solve problems and make predictions (ACSCH006)	✓	✓			✓	✓	✓	✓	✓					

		Context				Chemistry									
		1	2	3	4	1	2	3	4	5	6	7	8	9	10
Science Inquiry Skills	Select, construct and use appropriate representations, including physical and graphical models of molecules, energy profile diagrams, electron dot diagrams, ionic formulae, chemical formulae, chemical equations and phase descriptors for chemical species to communicate conceptual understanding, solve problems and make predictions (ACSCH045)				✓						✓	✓	✓	✓	✓
	Select and use appropriate mathematical representations to solve problems and make predictions, including calculating percentage composition from relative atomic masses and using the mole concept to calculate the mass of reactants and products (ACSCH007)		✓			✓			✓	✓					
	Select and use appropriate mathematical representations to solve problems and make predictions, including using the mole concept to calculate the mass of chemicals and/or volume of a gas (at standard temperature and pressure) involved in a chemical reaction, and using the relationship between the number of moles of solute, concentration and volume of a solution to calculate unknown values (ACSCH046)										✓	✓	✓		✓
	Communicate to specific audiences and for specific purposes using appropriate language, nomenclature, genres and modes, including scientific reports (ACSCH008 AND ACSCH047)	✓	✓		✓	✓	✓	✓	✓	✓	✓	✓	✓	✓	✓

CONTEXT 1
MATTER IN THE UNIVERSE

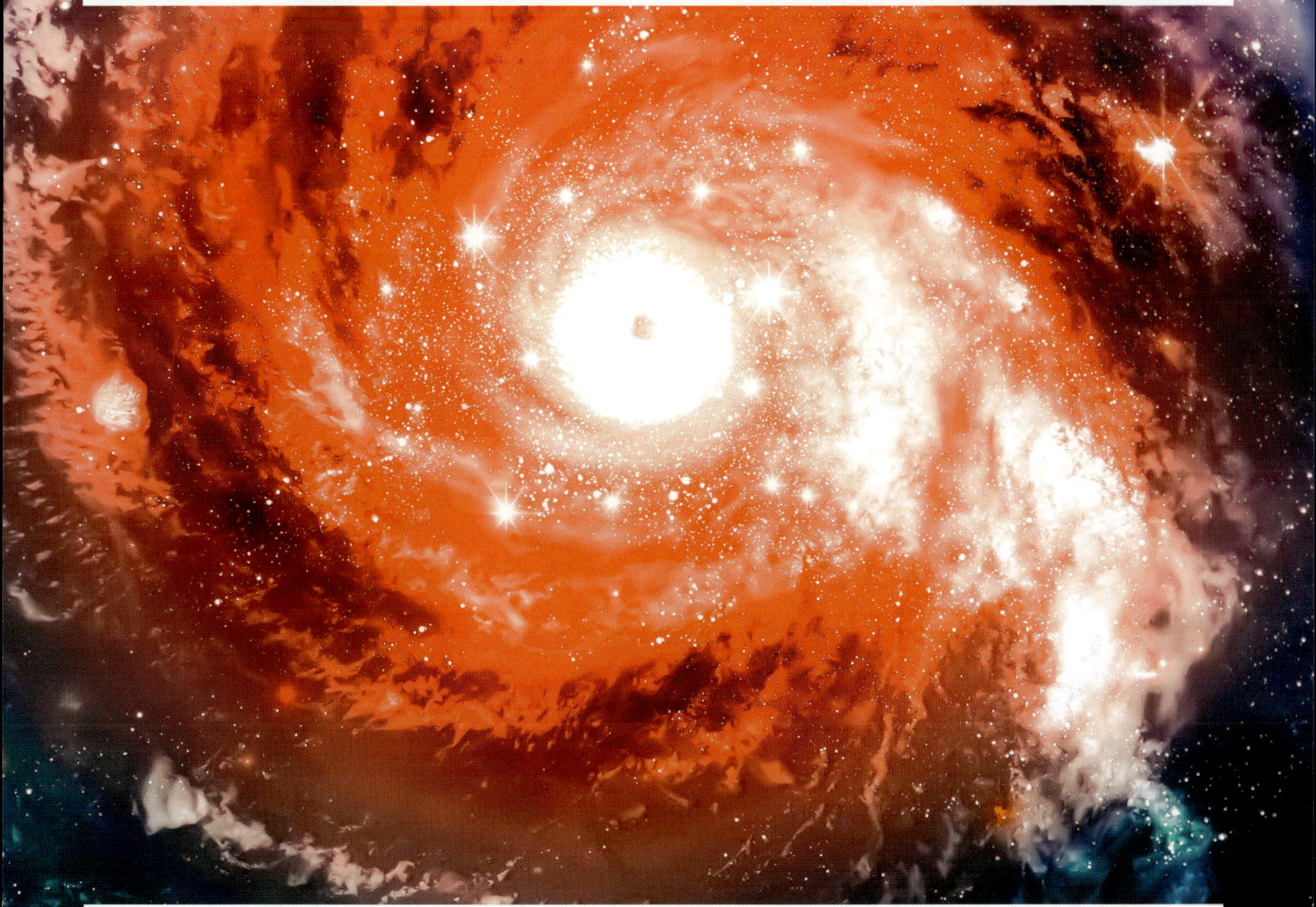

By the end of this chapter you will have covered the following material.

Science as a Human Endeavour

- Advances in science understanding in one field can influence other areas of science, technology and engineering (ACSCH011 and ACSCH050)
- The use of scientific knowledge is influenced by social, economic, cultural and ethical considerations (ACSCH012 and ACSCH051)
- The use of scientific knowledge may have beneficial and/or harmful and/or unintended consequences (ACSCH013 and ACSCH052)
- Scientific knowledge can enable scientists to offer valid explanations and make reliable predictions (ACSCH014 and ACSCH053)

Contextual story – Matter in the universe

- Where elements come from
- Manipulating and using elements
- Elemental isotopes

Chemical understanding

- Atomic structure
- Elemental spectra and spectroscopy
- Periodic table
- Mass spectrometry
- Covalent bonding
- Metallic bonding
- Nanotechnology – elemental carbon
- Isotopes

Investigations

Context 1

- Activities 1.1, 1.2, 1.3, 1.4, 1.5 and 1.6

Chapter 1

- Activity 1.1
- Experiments 1.1 and 1.2

Science as a Human Endeavour (SHE)

- Advances in science understanding in one field can influence other areas of science, technology and engineering (ACSCH011 and ACSCH050)
- The use of scientific knowledge is influenced by social, economic, cultural and ethical considerations (ACSCH012 and ACSCH051)
- The use of scientific knowledge may have beneficial and/or harmful and/or unintended consequences (ACSCH013 and ACSCH052)
- Scientific knowledge can enable scientists to offer valid explanations and make reliable predictions (ACSCH014 and ACSCH053)

Science Understanding (SU)

- Trends in the observable properties of elements are evident in periods and groups in the periodic table (ACSCH016)
- The structure of the periodic table is based on the electron configuration of atoms, and shows trends, including in atomic radii and valencies (ACSCH017)
- Atoms can be modelled as a nucleus surrounded by electrons in distinct energy levels, held together by electrostatic forces of attraction between the nucleus and electrons; atoms can be represented using electron shell diagrams (all electron shells or valence shell only) or electron charge clouds (ACSCH018)
- Flame tests and atomic absorption spectroscopy are analytical techniques that can be used to identify elements; these methods rely on electron transfer between atomic energy levels (ACSCH019)
- The properties of atoms, including their ability to form chemical bonds, are explained by the arrangement of electrons in the atom and in particular the stability of the valence electron shell (ACSCH020)
- Isotopes are atoms of an element with the same number of protons but different numbers of neutrons; different isotopes of elements are represented using atomic symbols (for example, $^{12}_{6}C$, $^{13}_{6}C$) (ACSCH021)
- Isotopes of an element have the same electron configuration and possess similar chemical properties but differing physical properties, including variations in nuclear stability (ACSCH022)
- Mass spectrometry involves the ionisation of substances and generates spectra which can be analysed to determine the isotopic composition of elements (ACSCH023)
- The relative atomic mass of an element is the ratio of the weighted average mass per atom of the naturally occurring form of the element to 1/12 the mass of an atom of carbon-12; relative atomic masses reflect the isotopic composition of the element (ACSCH024)
- Nanomaterials are substances that contain particles in the size range 1–100 nm and have specific properties relating to the size of these particles (ACSCH028)
- The characteristic properties of metals (for example, malleability, thermal conductivity, electrical conductivity) are explained by modelling metallic bonding as a regular arrangement of positive ions (cations) made stable by electrostatic forces of attraction between these ions and the electrons that are free to move within the structure (ACSCH032)
- Elemental carbon exists as a range of allotropes, including graphite, diamond and fullerenes, with significantly different structures and physical properties (ACSCH034)

Science Inquiry Skills (SIS)

- Conduct investigations, including the use of devices to accurately measure temperature change and mass, safely, competently and methodically for the collection of valid and reliable data (ACSCH003)
- Represent data in meaningful and useful ways, including using appropriate graphic representations and correct units and symbols; organise and process data to identify trends, patterns and relationships; identify sources of random and systematic error and estimate their effect on measurement results; and select, synthesise and use evidence to make and justify conclusions (ACSCH004)
- Select, construct and use appropriate representations including chemical symbols and formulae, molecular structural formulae, physical and graphical models of structures, chemical equations and thermochemical equations, to communicate conceptual understanding, solve problems and make predictions (ACSCH006)
- Communicate to specific audiences and for specific purposes using appropriate language, nomenclature, genres and modes, including scientific reports (ACSCH008 and ACSCH047)

Where elements come from

The elements that you see every day, the metals and non-metals, in compounds and mixtures, all had their beginnings in stars. A star is a giant ball of very hot gases held together by gravity. Most stars, including our Sun (Figure 1.1), are mostly hydrogen and helium. A process called stellar nucleosynthesis causes heavier elements to form gradually in larger stars, until we get the elements that lead to the formation of planets such as our Earth. Before we consider the formation of elements, we need to go back further, to the very beginning of the universe, to find out how the basic atomic particles, and the first atoms, formed.

Shutterstock.com/Markus Gann

◀ **Figure 1.1**
This star emits light due to fusion of hydrogen and helium.

To revise atomic structure, refer to Chemistry section 1.1 on page 104.

The Big Bang

Around 13.77 billion years ago, the universe came into being during an event called the **Big Bang**. The current theory states that the universe expanded very rapidly from a very dense and hot state and is still slowly expanding. In the beginning, there were no particles, just a lot of energy. About 10^{-43} seconds after the universe came into being, the first elementary particles formed, and after about 3 minutes, the first protons, neutrons and electrons were formed.

Over the next half a million years, hydrogen and helium nuclei came into existence, and after this the first stars formed and started to undergo fusion. Figure 1.2 on page 4 shows a timeline of the major events of the Big Bang.

Star life cycles

Stars can be many different sizes and temperatures, but all stars undergo nuclear fusion for about 90% of their life. During nuclear fusion, hydrogen is converted into helium inside the core of the star, and a vast amount of energy is released. During this time, the star is referred to as a **main sequence** star, as seen in Figure 1.3 on page 5. At the end of their life, stars start to behave differently, depending on their initial size. Figure 1.3 shows the different possibilities for a star after all its hydrogen has been used up and it is no longer a main sequence star.

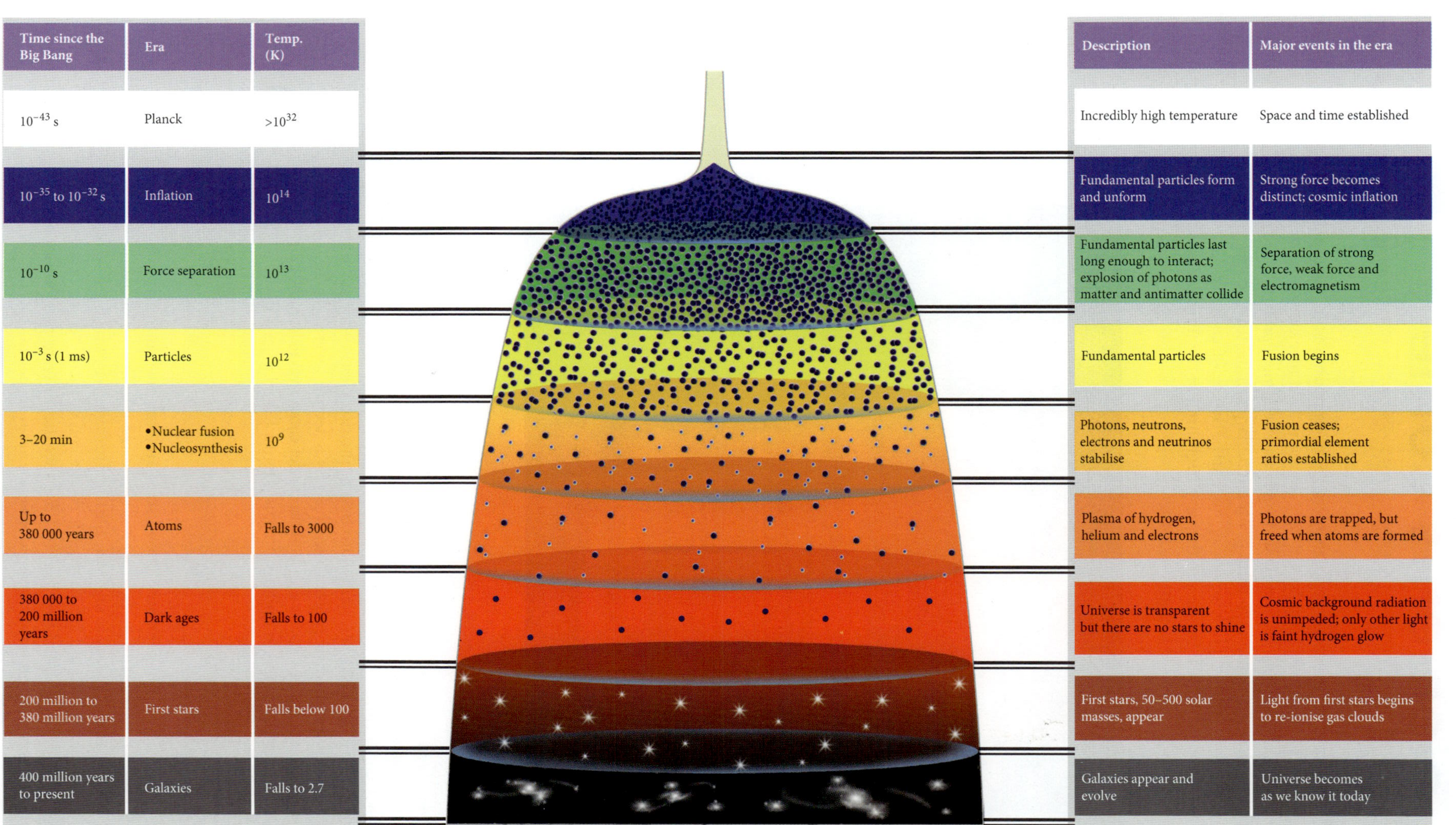

Time since the Big Bang	Era	Temp. (K)	Description	Major events in the era
10^{-43} s	Planck	$>10^{32}$	Incredibly high temperature	Space and time established
10^{-35} to 10^{-32} s	Inflation	10^{14}	Fundamental particles form and unform	Strong force becomes distinct; cosmic inflation
10^{-10} s	Force separation	10^{13}	Fundamental particles last long enough to interact; explosion of photons as matter and antimatter collide	Separation of strong force, weak force and electromagnetism
10^{-3} s (1 ms)	Particles	10^{12}	Fundamental particles	Fusion begins
3–20 min	• Nuclear fusion • Nucleosynthesis	10^{9}	Photons, neutrons, electrons and neutrinos stabilise	Fusion ceases; primordial element ratios established
Up to 380 000 years	Atoms	Falls to 3000	Plasma of hydrogen, helium and electrons	Photons are trapped, but freed when atoms are formed
380 000 to 200 million years	Dark ages	Falls to 100	Universe is transparent but there are no stars to shine	Cosmic background radiation is unimpeded; only other light is faint hydrogen glow
200 million to 380 million years	First stars	Falls below 100	First stars, 50–500 solar masses, appear	Light from first stars begins to re-ionise gas clouds
400 million years to present	Galaxies	Falls to 2.7	Galaxies appear and evolve	Universe becomes as we know it today

Figure 1.2 ▲
The Big Bang: timeline of major events

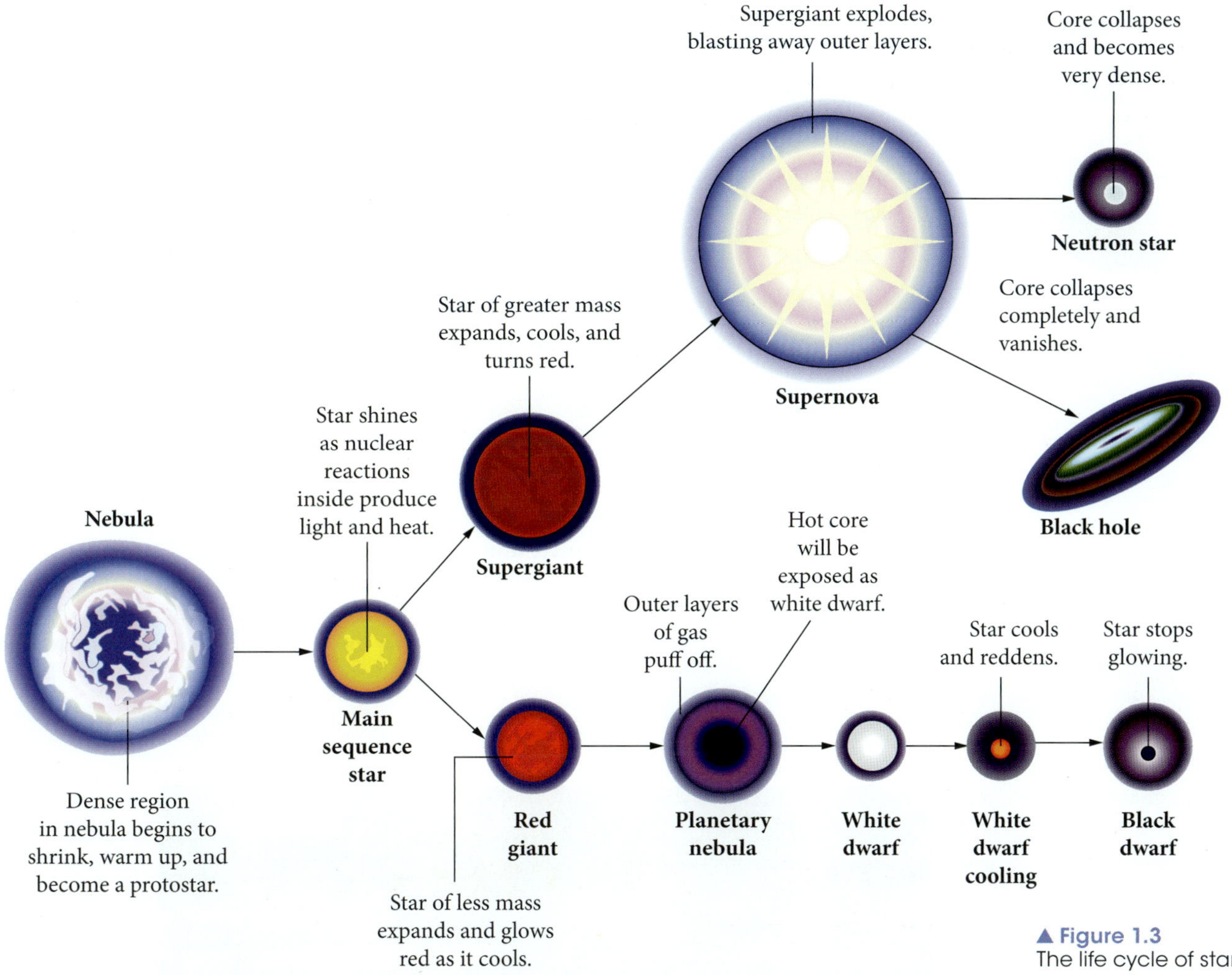

Supergiant explodes, blasting away outer layers.
Core collapses and becomes very dense.
Neutron star
Core collapses completely and vanishes.
Supernova
Star of greater mass expands, cools, and turns red.
Black hole
Star shines as nuclear reactions inside produce light and heat.
Nebula
Supergiant
Hot core will be exposed as white dwarf.
Outer layers of gas puff off.
Star cools and reddens.
Star stops glowing.
Main sequence star
Dense region in nebula begins to shrink, warm up, and become a protostar.
Red giant
Planetary nebula
White dwarf
White dwarf cooling
Black dwarf
Star of less mass expands and glows red as it cools.

▲ **Figure 1.3**
The life cycle of stars

Our Sun is one of the smaller stars in the sky. When all its hydrogen has been used up, in billions of years' time, it will expand and cool, forming a **red giant**. Eventually, it will contract and form a **white dwarf**. Much larger stars form **red supergiants**, also cooler and much larger than the original star. Supergiants eventually explode violently in a huge explosion called a **supernova**, as seen in Figure 1.4. After this explosion, the star may form a dense, fast-spinning **neutron star** or a very dense **black hole**. These larger stars are where the heavier elements formed.

▼ **Figure 1.4**
A supernova, the remains of a massive star

NASA, ESA, J. Hester and A. Loll, Arizona State University

Stellar nucleosynthesis

The core of a massive star becomes extremely hot, much hotter than in smaller stars such as our Sun. The core temperature of larger stars allows the formation of elements heavier than helium. Once certain temperatures are reached, helium begins to fuse to form elements such as lithium, carbon, oxygen and neon. As the star gets denser and hotter, the carbon begins to fuse to form elements such as sulfur and silicon. Silicon fuses to form heavy elements such as scandium, chromium and iron. A massive star at the end of its life looks a little like an onion. It has layers of elements – the lighter elements are on the outside and the heaviest elements are at the core. You can see this in Figure 1.5 on page 6.

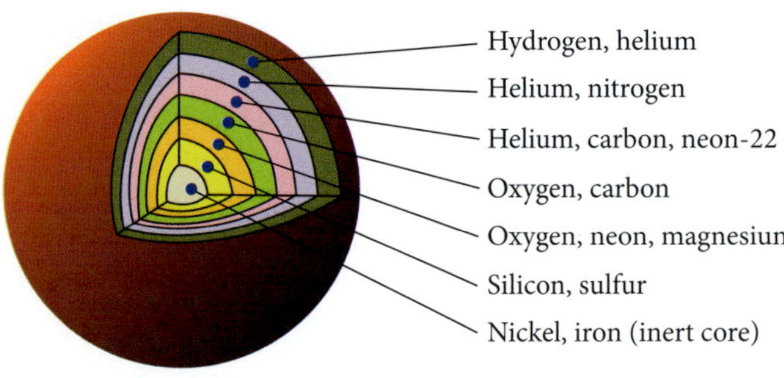

Figure 1.5 ▲
At the end of its life, a very large star has layers of elements.

After iron has formed, the star is unable to continue making heavier elements because the star is not hot enough to cause iron to undergo fusion. When the star reaches this point in its life cycle and it is not undergoing fusion, then it is not producing enough energy to be stable any more. The star will then explode in a supernova. The huge explosion creates a massive amount of energy, which allows the formation of the elements that are heavier than iron. The temperatures that are not possible inside a star exist inside the supernova. The heavier elements can now fuse. All the other natural elements in the periodic table are formed in the supernova explosion. So, without massive stars, the elements that we know and use on Earth today would not exist.

Analysing stars

HOW DO ELEMENTS FORM IN STARS?

Watch this video to see how elements heavier than hydrogen and helium form in stars.

Scientific knowledge can enable scientists to offer valid explanations and make reliable predictions (ACSCH014 and ACSCH053) AC

Even the stars within our solar system are too far away to travel to, so scientists must study them by looking at the light they emit. The most common instrument for analysing light is a **spectroscope**. This device breaks up visible light, and some non-visible light, into its component wavelengths to produce an **emission spectrum**.

Stars are made up of many elements. Each element emits a unique set of wavelengths, so an emission spectrum for a star consists of many lines. This is because the spectrum consists of a mixed pattern of all the elements together. Figure 1.6 shows the spectra of some stars – there are thousands of lines.

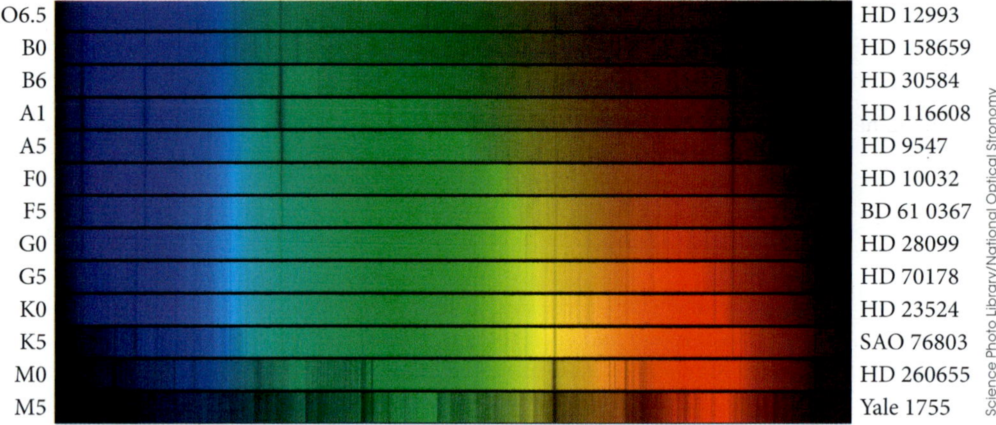

Figure 1.6 ▲
Spectra of 13 stars, each of a different type. The types (from top to bottom) are O6.5, B0, B6, A1, A5, F0, F5, G0, G5, K0, K5, M0 and M5. Our Sun is a G2-type star. The order OBAFGKM ranks the stellar spectral types in order of decreasing surface temperature, while the numbers 0–9 provide subtypes. An O-type would be hotter than 25 000 Kelvin (K) and appear blue. An M-type would appear red and be cooler than 3500K. The dark spectral bands show stellar composition.

To learn why atoms produce unique line spectra, refer to Chemistry section 1.6 on page 113.

To revise elemental spectra, refer to Chemistry section 1.8 on page 123.

Scientists can make a pretty good guess at the most likely elements that are in a star. For example, hydrogen and helium are always present in a star, so scientists look for their emission lines at known wavelengths first. The remaining lines are then compared with known elemental spectra to determine what other elements are present in the stars. For example, if all the known lines for carbon are present, then carbon is present in the star.

This technique is how the element helium was discovered in 1868. Scientists examining the spectrum of the Sun found a line at a wavelength that did not match any known element. Astronomers Jules Janssen and Norman Lockyer independently discovered the line around the same time and are given credit for the discovery of helium. The new element was named helium after Helios, the Greek god of the sun. It was not until 1895 that helium was eventually discovered in Earth's atmosphere.

ACTIVITY 1.1

IDENTIFYING ELEMENTS IN STARS

Aim

To identify the elements present in stars based on their emission spectra

What to do

Use your knowledge about line emission spectra to identify the elements present in star A and star B in Figure 1.7.

What did you discover?

1 Explain how you identified the elements in the two stars.

2 Which of the two stars is more likely to be a massive star? Why?

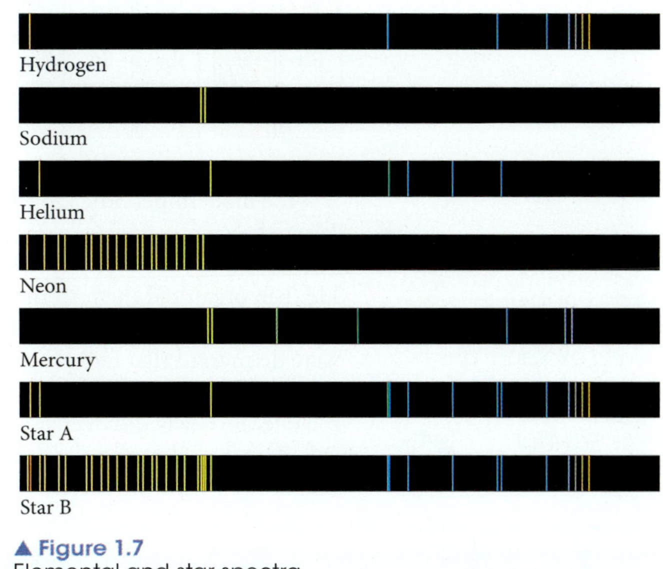

▲ **Figure 1.7**
Elemental and star spectra

Synthetic elements

There are 92 naturally occurring elements that originated in stars and supernovae. The rest of the elements in the periodic table do not exist in nature – they are synthetic. Scientists use a number of different methods to make **synthetic elements**.

A number of other elements have been synthesised by scientists. All of them are unstable because the forces in their nuclei are unbalanced, which causes the nucleus to break down over time. As they break down, the nuclei emit particles and electromagnetic radiation. This process is known as **radioactive decay**. Sometimes, the particles or radiation emitted during radioactive decay can be useful. These will be discussed further on page 22.

To make a new element from an existing element, you need to add extra protons. One way to do this is to bombard an atom with nuclear particles such as protons, helium nuclei, neutrons or heavier atomic nuclei. Some of the new material becomes part of the nucleus of the element, giving it extra protons and forming a new element. Californium, atomic number 98, was formed this way in 1950 – helium ions were fired at the element curium, atomic number 96. The two protons in the helium nucleus added to the curium nucleus, giving it 98 protons and forming a new element.

COSMIC ORIGIN SPECTROGRAPH

Watch this video about the Cosmic Origin Spectrograph installed on the Hubble Space Telescope and see how it is used to find out about the universe.

Organising elements

Scientists organise the elements in the form of the periodic table. Many elements have been known about since prehistoric times, but the scientific concept of elements wasn't devised until around the 17th century when phosphorus as an element was officially discovered. By 1900, there were about 80 recognised elements and the first attempts at organising a system of categorising them really started to take place.

The first attempts at organisation worked around a system proposed by German chemist Johann Dobereiner. In 1817, he noticed that elements that had similar properties could be grouped in groups of three, which he called triads. Dobereiner noted that the groups of three elements had similar properties and their atomic masses showed a specific pattern. Calcium, strontium and

To revise the periodic table, refer to Chemistry sections 1.4 and 1.7 on pages 111 and 119.

Figure 1.8 ▶ Dobereiner's triads on the current periodic table

H							He
Li	Be	B	C	N	O	F	Ne
Na	Mg	Al	Si	P	S	Cl	Ar
K	Ca	Ga	Ge	As	Se	Br	Kr
Rb	Sr	In	Sn	Sb	Te	I	Xe
Cs	Ba	Tl	Pb	Bi	Po	At	Rn

barium were one triad. They all have similar chemical properties and the atomic mass of strontium is almost exactly halfway between that of calcium and barium. Dobereiner noticed the same thing for lithium, sodium and potassium, so he called them another triad. Other triads were phosphorus, arsenic and antimony, and chlorine, bromine and iodine. Figure 1.8 shows where these groups of three elements appear on the current periodic table.

Scientific knowledge can enable scientists to offer valid explanations and make reliable predictions (ACSCH014 and ACSCH053) AC

The modern periodic table was developed by Dmitri Mendeleev and published in 1869. Mendeleev wrote out the names of the elements with their atomic masses and other properties. He laid them out in rows (periods) and columns (groups) by first ordering the elements by mass. He then organised them so that elements with similar properties were in the same group. When he finished, there were gaps in his periodic table. These gaps were later filled as more elements were discovered that had properties and masses that fitted exactly into the gaps. This table has been used with only slight modifications since then. The table has recently been revised to include the newly discovered noble gases, the rare earth elements called the **lanthanoids** (also called lanthanides) and the radioactive **actinoids** (also called actinides). The rare earth elements do not fit in with the periodic nature of the rest of the table so are placed underneath it.

ACTIVITY 1.2

BUILDING YOUR OWN PERIODIC TABLE

Aim

To construct a periodic table based on element properties and masses

You will need

- A set of element cards with information about the elements and their chemical properties, pre-made cards from the Internet, or an interactive periodic table to create your own

What to do

1 If you are making your own cards using the interactive periodic table, decide what properties to include on each card.

Element properties could include:

- name and symbol
- appearance
- state
- atomic number
- atomic mass
- boiling point
- melting point
- valency.

2 Arrange your element cards (either created or printed) in a line in order of increasing atomic mass.

3 Identify where the properties repeat in the line and use this to create a logical set of rows that show trends.

ELEMENT CARDS

Use these sets of element cards with information about the elements.

▶

What did you discover?

1 Compare your table to the periodic table. Make a list of the similarities and differences between the two tables.
2 Examine the properties you looked at when constructing your table and explain why you had differences.
3 Rearrange your cards so they match the periodic table.
   a Where are the metals in the periodic table? Look at the properties of the metals and list those properties that are common.
   b Where are the non-metals in the periodic table? Look at the properties of the non-metals and list those in common.
   c Identify any more trends in properties that you can see. Look down groups (columns) and across periods (rows).

INTERACTIVE PERIODIC TABLE

Use this interactive periodic table to see what elements look like and read their chemical information.

Manipulating and using elements

Most of the substances on Earth are composed of multiple elements in the form of compounds or mixtures, but there are still a lot of elements that are used or studied in their pure form. We analyse materials for particular elements, for commercial, health or forensic reasons. We use elements such as metals in a variety of fields, including construction. A newer application is the use of elements in nanotechnology to create materials with very specific applications.

To understand more about and revise spectroscopy and spectrometry, refer to Chemistry sections 1.8–1.10 on pages 123–131.

Analysing elements

Element analysis can focus on identifying the element present – this is called **qualitative analysis**. But if you know which elements are present, then you can focus on measuring how much of the element is present. This is called **quantitative analysis**. A number of techniques, such as spectroscopy, involve the absorption and emission of light by atoms of elements as electrons move between energy levels.

Heavy metal poisoning

Certain metals are extremely toxic to humans and if ingested in large enough amounts can cause severe illness, DNA damage and even death. Mercury, lead, aluminium, arsenic and cadmium are found in small amounts in humans because of everyday contact with materials and the environment. These elements are sometimes known as heavy metals because they have quite high atomic masses. Some of them can be seen in Figure 1.9.

Lead

Cadmium

Arsenic

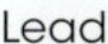

▲ **Figure 1.9**
Some heavy metals: lead, cadmium and arsenic

Mercury is still used in amalgam dental fillings. People consume mercury-containing compounds in fish due to industrial pollutants in the ocean and the effects of biological magnification. This is where small animals consume mercury and a large number of these animals are eaten by larger fish and animals. The amount of mercury in the larger animals is magnified as they consume so many of the smaller animals. Mercury acts as a neurotoxin in humans and can affect development and functioning of the nervous system.

Mercury liquid

Science Photo Library/Cordelia Molloy

Mercury-containing amalgam fillings

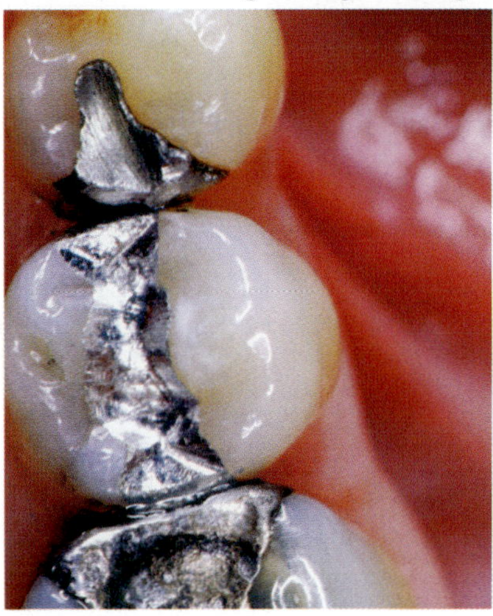

Science Photo Library/CNRI

A mercury thermometer

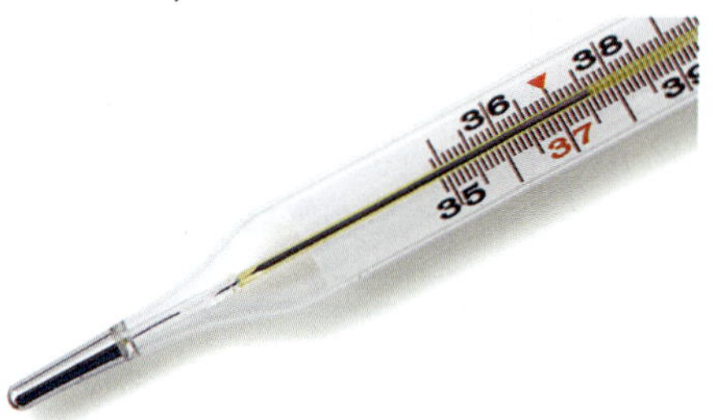

Shutterstock.com/Andrii Gorulko

Figure 1.10 ▲
Mercury has various uses, including in dental fillings and thermometers, but it is toxic.

Lead is used in paints and pigments to give them useful properties. Recently, a large number of toys made in China were recalled. Children in the USA and Europe had become violently ill after playing with toys that had paints and pigments containing lead. Researchers from Greenpeace bought 500 toys in five Chinese cities and conducted tests that showed one in 10 of the toys contained dangerous levels of lead with one containing 1200 times the amount of lead permitted under European safety standards.

Shutterstock.com/Gillian Entress

Alamy/Robert W. Ginn

Figure 1.11 ▲
Lead-based substances

Accumulation of aluminium in the human body can lead to osteoporosis, headaches and problems with liver and kidney function. You come into contact with aluminium through deodorants, drinking water, medications and some processed foods.

Arsenic is a very toxic metal that is less commonly used than the metals discussed so far. It is probably most famous as a method of poison in crime novels. Small amounts of arsenic enter the human body through exposure around industrial sites and to arsenic-treated wood.

Cadmium is a metal that directly damages DNA in the body. Again, you are less likely to come into contact with cadmium but it is used to make batteries and in the mining industry.

Human tissue, blood and urine can be analysed for these toxic metals by a number of different spectroscopic methods. Spectroscopy is a useful analytical method because it generally doesn't destroy materials. Spectroscopy requires only very small amounts of the sample, such as hair, blood and urine, so that most of it can be saved for continued testing or for evidence in court cases.

Scientists have been trying to find the causes for developmental diseases such as autism. One theory is that ingestion of toxic metals could be a contributing cause. Recent studies have shown that children with autism often have significantly higher levels of toxic metals, such as mercury, lead and thallium, in their bodies. Mass spectrometry is used to distinguish heavy metals from elements you expect to find such as carbon, oxygen and hydrogen.

Traditional methods of analysis are often time-consuming and expensive. Spectroscopy allows scientists to quickly analyse information for a vast number of subjects. This increases the chances of finding the causes of problems.

Herbal medications are sold over the counter in a number of countries and are generally considered to be a healthy option for building up the immune system and helping with minor health issues. Regular testing is done on these medications to determine their content. Atomic absorption spectroscopy and X-ray fluorescence spectroscopy testing has found alarming levels of heavy metals, including lead, mercury and arsenic, in some herbal remedies. Sometimes these metals have been added deliberately. Alternative herbal medications popular in India and China combine herbs with metals, minerals and gems. These medications are then sold online all over the world where people have been warned to be aware of what they are taking. Some tests have shown metal levels up to 10 000 times higher than the limits set by health professionals for some tablets.

▼ Figure 1.12
Due to the potential harmful effects of cadmium, some types of battery are now advertised as being free of this metal as well as mercury.

Alamy/Beaconstox

Advances in science understanding in one field can influence other areas of science, technology and engineering (ACSCH011 and ACSCH050)

Forensic testing

Forensic science uses spectroscopy in many ways. A big problem with forensic science is that most analytical methods destroy the evidence. This means that the evidence cannot be subject to further testing and cannot be presented in court. Methods such as spectroscopy that do not destroy the sample are in increasing demand. New methods of analysis are being developed all the time.

American chemist Professor Kenneth Busch developed a method of examining skeletal bones to determine a more accurate time of death, or post-mortem interval. After a person dies, their bones start to lose water and the molecules inside start to decompose. Busch developed a process that uses diffuse reflectance near-infrared spectroscopy. He collected the patterns of spectra from decaying pig bones over a 3-month period and identified a trend as the bones aged. A mathematical model was developed that allowed the post-mortem interval to be determined when bones are analysed. An important feature of this technique is that it uses infrared radiation, which does not destroy the substance during the analysis.

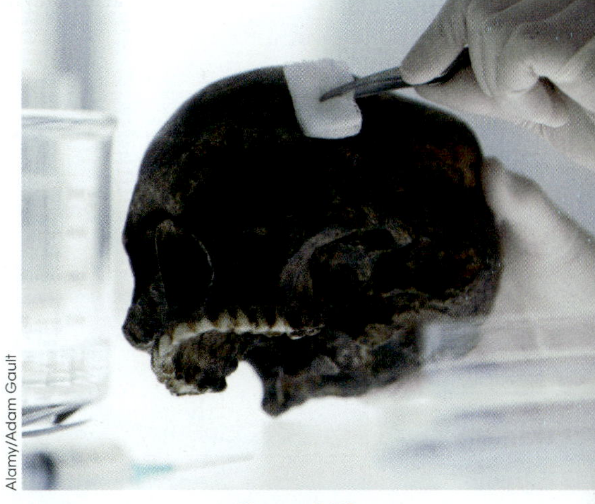

Alamy/Adam Gault

▲ Figure 1.13
Professor Busch's process of diffuse reflectance near-infrared spectroscopy can determine the post-mortem interval of decaying bones such as this skeleton.

To learn about chromatography, refer to Chemistry section 6.9 on page 289.

To revise mass spectrometry, refer to Chemistry section 1.10 on page 131.

In legal cases, documents such as ransom demands, receipts, wills and threatening letters are often involved. Analysis of inks can provide evidence of someone's guilt or innocence. For a long time, the methods involved in analysis destroyed the document, which meant that there was no evidence to display in court and legal cases were often lost. Traditional methods such as **chromatography** require part of the document to be removed, and the more analyses carried out, the more of the document is permanently destroyed.

Scientists in the USA have developed a system that allows immediate analysis of the whole document by mass spectrometry. This is called the DART (direct analysis in real time) system. The DART method removes a microscopic amount of ink from the document, leaving the rest of the document intact. Visually, you would never know that anything was missing. The minute sample of ink is analysed, and a spectrum is produced, which is compared against a very large library of ink samples. Examples of the spectra produced by different inks are shown in Figure 1.14. This method is often used to show when two different pens have been used on a single document to alter a receipt or forge someone's signature.

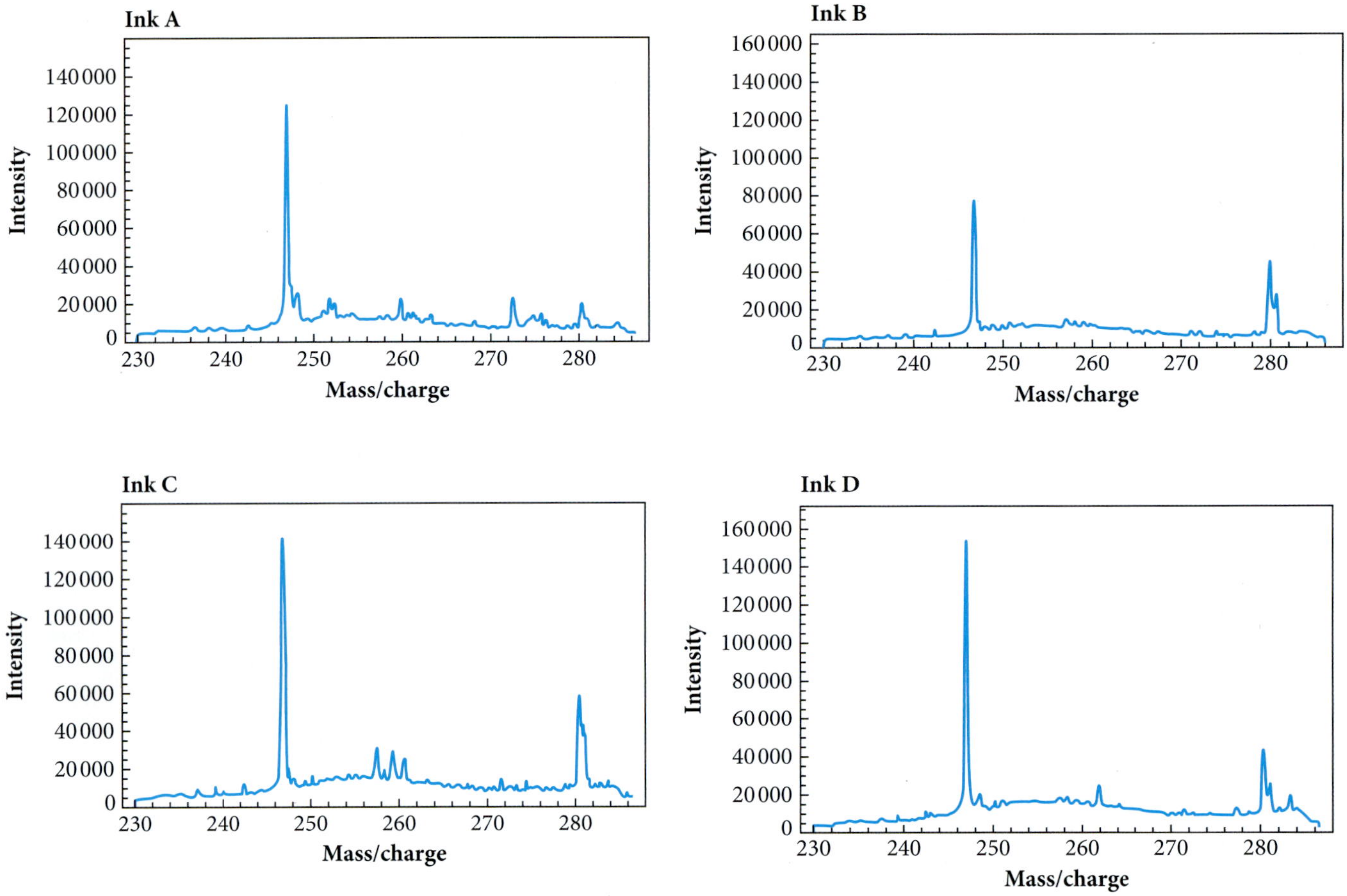

Figure 1.14 ▲
These mass spectra show the different composition of four different inks.

ACTIVITY 1.3

WHODUNIT?

Aim

To identify who forged their signature on a document by analysis of the ink spectra

You will need

- Figure 1.14

What to do

Read the following information carefully. Use Figure 1.14 to help you answer the questions below.

A handwritten will has been found and two siblings are fighting over the contents. The brother claims that he and his sister were each supposed to inherit half their parents' estate. However, this will states that everything is to be given to his sister. The brother suspects that his sister, who found the will, has changed part of it.

You are a forensic chemist assisting the police. You are given the note and the following information.

- Pens A and B were discovered in the parents' house.
- Pens A and C were discovered in the daughter's house.

You take samples of ink from the will and discover that most of the will is written in a pen that has the following characteristics in the spectra.

- There is a peak at a mass/charge value of approximately 247.
- This peak has an intensity value of approximately 80 000.

You also discover that the section of the will that names the daughter as the sole recipient is written in an ink that has the following characteristics in the spectra.

- There is a peak at a mass/charge value of approximately 247.
- This peak has an intensity of approximately 140 000.

What did you discover?

1. Briefly describe the process that you would have used as a forensic chemist to analyse the inks in the document.
2. Describe the information that can be determined from the spectral information given to you in this case.
3. Explain the chemical evidence that links the ink in the pen in the daughter's house to the changed will.

Food analysis

Many elements, such as calcium and iron, are necessary for good health. Other elements, such as heavy metals, can cause severe health problems at high enough concentrations, and should not be in food. Sometimes, food labels can be misleading or not specify all the ingredients contained in the food. Many analytical techniques are used to investigate claims and test foods for safety. The various types of spectroscopy play a major role in this analysis.

Certain foods such as cheese and wine are produced in various areas of the world. Countries and regions are very protective of their food products. For example, technically champagne is produced only in the Champagne region of France, but winemakers around the world have called their sparkling wines 'champagne'. This is no longer allowed and these champagne-style wines must be labelled as 'sparkling wines'. Wines labelled as 'champagne' are often tested to verify that they been produced in France.

The use of scientific knowledge is influenced by social, economic, cultural and ethical considerations (ACSCH012 and ACSCH051) AC

To revise how levels of contaminants are detected and measured, refer to Chemistry section 1.9 on page 127.

Figure 1.15 ▲
Composition of food as listed on the container

Cheese types can also be specific to certain areas; for example, true cheddar cheese comes from the region of Cheddar in the UK. In other cases, some cheeses include the milk of goats or sheep instead of cows, which can cause allergy problems. For this reason, it is important that cheeses are labelled accurately. Testing can determine whether companies are labelling their food products correctly.

There can be serious consequences when products are mislabelled. In some places, there have been problems with so-called beef mince containing other types of meat. It also can indicate substitution with cheaper grades of meat and possible health threats through contamination with untested or unregulated ingredients. People who practice certain religions are not allowed to consume pork. So if pork is present in food, but is not identified on the label, they inadvertently break religious laws. Regular testing of food detects contaminants.

ACTIVITY 1.4

FOOD RECALL

Aim

To devise a process for determining whether a food needs to be recalled due to metal contamination

What to do

Read the following information carefully.

You work for the Food Standards Board in your state and you are given a sample of meat that is suspected of containing two types of metal particles. Under the food standards, mercury is allowable as long as it is below 0.5 parts per million. If cadmium is present in any amount, then the meat must be recalled and banned from sale.

In your group, devise a method for determining whether the food is contaminated past acceptable levels.

1 Using the information on spectroscopic techniques in Chemistry Chapter 1, explain how you would identify whether any cadmium or mercury was present.
2 Using the same information, explain how you would measure the concentration of mercury to determine whether it is below the recommended safe concentration.

In Germany, a project called FreshSCAN has developed an optical sensor that can work through food packaging to get instant feedback about the freshness of meat. No elaborate testing is required and no sample of food is needed for laboratory analysis. Scientists tested the spectroscopic properties of meat in varying stages of freshness and found that the spectra showed patterns that could then be used as a comparison for other samples. As meat ages, chemical changes affect how the atoms and molecules absorb and emit light. This leads to changes in the emitted spectra, as seen in Figure 1.16 on page 15. The benefits of this system are that feedback is instant and unsafe food can be removed immediately. The technology is being examined to see if it can be adapted to other foods by setting up a bank of spectra for foods of different levels of ripeness or spoilage.

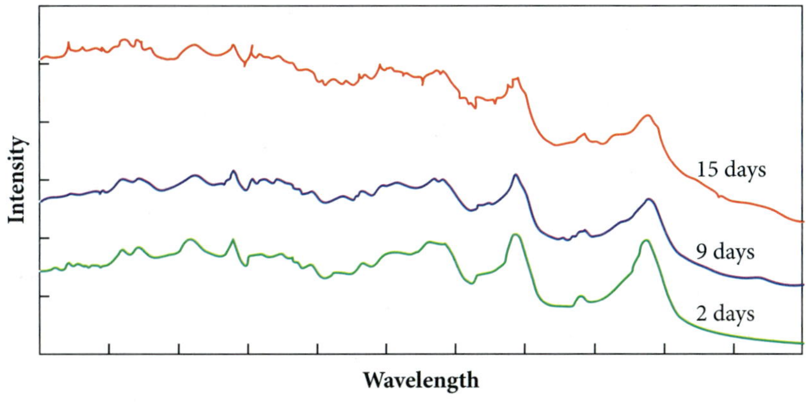

▲ **Figure 1.16**
The spectra of meat change over time, which can be used to determine the age of meat on the supermarket shelf.

Soil pollution is a big problem. Water used for irrigation often contains industrial or agricultural pollutants, which are then transferred to the soil. As vegetables grow, they can take in these pollutants, which are then consumed by humans. By the time the food gets to supermarket shelves, it is too late. In France, a type of spectroscopy is being developed to obtain real-time data about contaminants in vegetables. This process is designed for use in the field. Traditional methods involve time-consuming laboratory work. This system uses a laser pulse to create a plasma sample of a very small amount of the vegetable while it is still growing. Plasma is a gas made up of electrons and ions, perfect for spectroscopic analysis. The plasma then undergoes spectroscopic analysis immediately in the field and the presence of specific metals can be detected. This system primarily analyses for toxic metals such as lead and cadmium, which cause health problems. So it is highly advantageous that food can be analysed before it gets to supermarket shelves and consumer's homes.

◀ **Figure 1.17**
Vegetables can be tested for dangerous levels of contaminants while they are still growing, before they appear on supermarket shelves.

Making useful materials

Almost everything around you is made up of two or more elements. Very few items you come into contact with are made from a pure element. The air you breathe contains elements such as oxygen and nitrogen, but when you breathe, you take in a mixture of air containing lots of different elements and compounds. Very few elements can be used in their pure form.

Advances in science understanding in one field can influence other areas of science, technology and engineering (ACSCH011 and ACSCH050)

To revise electron arrangements in atoms, refer to Chemistry section 1.6 on page 113.

To learn about covalent bonding, refer to Chemistry sections 3.5 and 3.7 on pages 178 and 187.

Silicon

The element silicon is used a lot in technology. Very pure silicon (purity above 99.9%) is used primarily to create solar panels. Silicon is also used in **semiconductors**. A semiconductor is a material with an electrical conductivity somewhere between a conductor and an insulator. When a metal is heated, its conductivity decreases, but when a semiconductor is heated, its conductivity increases.

The silicon that is used in semiconductors has 'nine nines' purity, which means 99.999 999 9% purity. Silicon less pure than this will be inefficient as a semiconductor. Silicon is the best semiconductor because it can withstand the highest temperatures and electrical power running through it and still continue to function. This is important in computers and machinery because they heat up when they are used. A regular conductor will often fail when heat is applied but a semiconductor only gets better at conducting as it gets hotter. Silicon is such a good semiconductor because it has four electrons in its valence shell. When it bonds, it uses covalent bonding between atoms to form large lattice structures called crystals.

In crystal form, silicon is not highly conductive. The structure is slightly altered in a process called doping. This introduces different elements into the crystal structure. A common five-electron element in semiconductors is phosphorus. As seen in Figure 1.19, the phosphorus takes the place of one of the silicon atoms. An extra electron is present with nothing to bond to. This is called an n-type semiconductor.

If the silicon is doped with an element such as boron with only three electrons, then there is a space that is called an electron hole. Free electrons moving around the semiconductor will fill these holes. This p-type semiconductor can be seen in Figure 1.20.

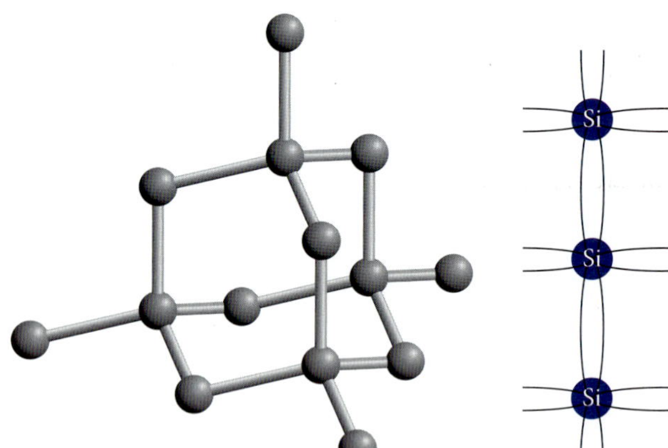

Figure 1.18 ▲ The crystal structure of pure silicon

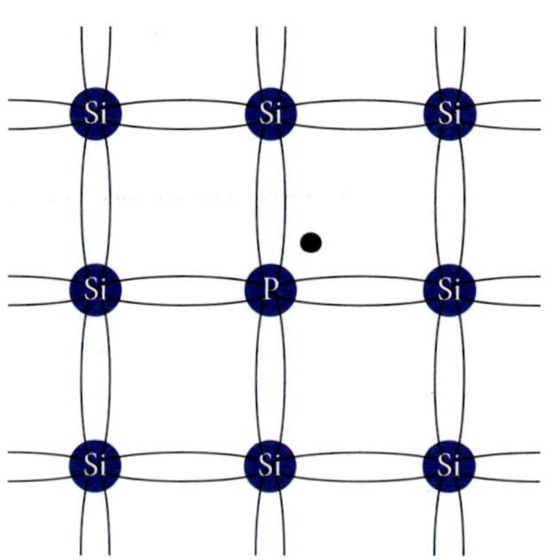

Figure 1.19 ▲ Phosphorus doping of silicon produces an n-type semiconductor.

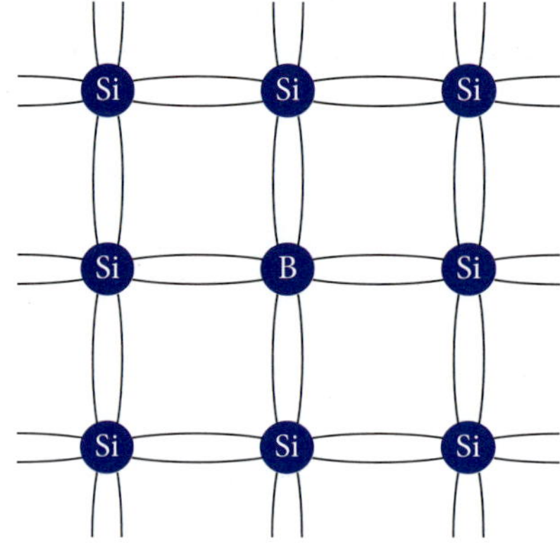

Figure 1.20 ▲ Boron doping of silicon produces a p-type semiconductor.

SILICON IN A SOLAR CELL

Use this interactive activity with animations and explanations to learn how silicon is used inside a solar cell.

When n-type and p-type silicon are put together, the extra electrons from the n-type move across to the p-type to fill the electron holes. The silicon crystal is now a moderate conductor of electricity and that is why it is called a semiconductor. The most common use for silicon in this form is in electronic components, such as transistors, and solar cells.

WOW

Using algae to generate electricity

Scientists are using single-celled algae called **diatoms** to improve the efficiency of solar panels. Specially shaped structures in their silicon shells improve the diatom's use of sunlight in photosynthesis by about 30%.

Scientists grow billions of diatoms in tanks and coat them onto the surface of existing solar cells. The diatoms cover the surface in the most efficient manner for light capture and then overlay their silica shells onto the surface of the solar cell. The living material is removed, leaving a solar panel with increased efficiency. The silica shells that are left behind by the diatoms capture light in hexagonal holes and do not allow it to escape. Tests show that solar panels produced in this manner can triple the amount of electricity produced by traditional panels.

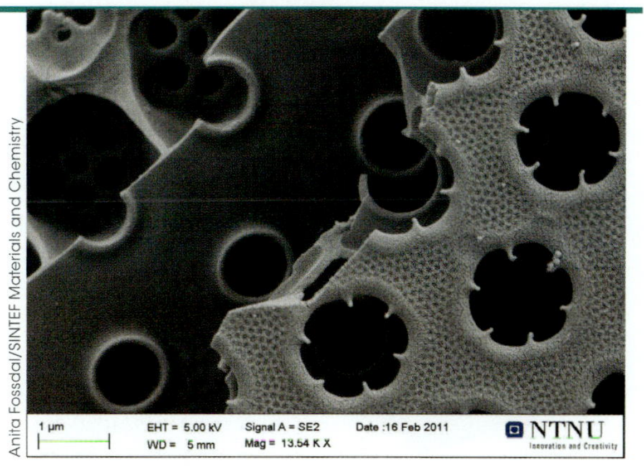

Anita Fossdal/SINTEF Materials and Chemistry

Figure 1.21 ▶ Pores in the silica shell of the diatom *Coscinodiscus wailesii* completely trap sunlight for photosynthesis.

Titanium

The metal titanium has a number of useful properties in its elemental form. Titanium can be alloyed with other metals, such as iron, aluminium, vanadium and molybdenum, to form lightweight metals for use in the aerospace industry.

In its pure form, titanium has the highest strength-to-weight ratio of any metal. This makes it very useful in situations where metals of higher mass are not suitable. Titanium also has excellent corrosion resistance. It will not react with acids unless they are very concentrated. This means that in the environment, titanium will not react with the chemicals it comes into contact with. One of the most well-known uses for titanium is in medical implants as shown in Figure 1.22. It is referred to as a biocompatible material because it is non-toxic and is not rejected by the human body. Its lightweight and anticorrosive properties make it ideal for load-bearing structures such as ball-and-socket joints for hips and shoulders. Titanium pins, plates, bone screws, rods and staples are used throughout the human body to repair and set broken bones or even to replace body parts.

SOLAR PANELS FOR YOUR HOME

Watch this video to see how solar panels work and their advantages in terms of energy production.

To learn about metallic bonding, refer to Chemistry section 3.2 on page 163.

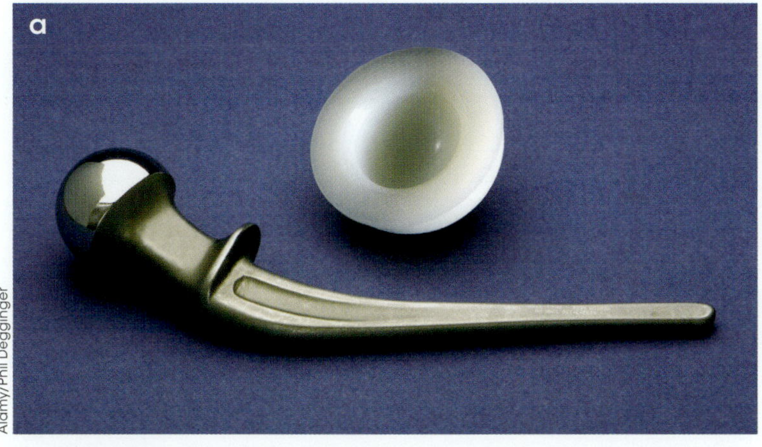

Alamy/Phil Degginger

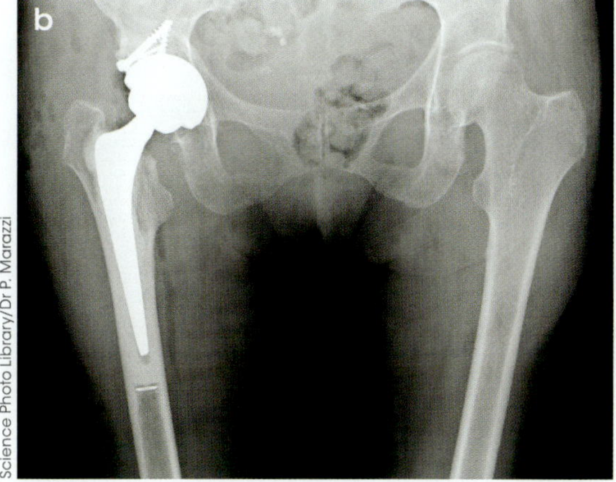

Science Photo Library/Dr P. Marazzi

▲ **Figure 1.22**
A titanium hip joint a) prior to implantation and b) as seen on an X-ray

Alamy/Iain Cooper

Figure 1.23 ▲
Titanium jewellery is becoming popular because of its useful properties.

Titanium jewellery is gaining popularity because it is non-corrosive, lightweight and strong. Also, titanium jewellery is suitable for people with allergies who often react to other metals such as silver and gold.

Elemental gases

Most people are familiar with neon signs, but most of these signs do not involve the element neon. However, these signs all use elemental gases to create coloured light. Figure 1.24 shows a simple neon sign, while Figure 1.25 shows how neon signs are used in a city.

When elements are heated in their gaseous forms, they emit a specific colour of light related to the movement of electrons in the energy levels around the atoms. A 'neon' sign, regardless of what gas is used, works by applying an electric current to a tube of the gas. The current gives the electrons energy and a particular colour of light is emitted. Specific gases are selected to produce particular colours. Many signs incorporate different gas tubes to create a multicoloured effect. Table 1.1 shows the different colours that can be generated with different elements.

Shutterstock.com/FootToo

Figure 1.24 ▲
A simple neon light

Shutterstock.com/Lee Yiu Tung

Figure 1.25
Neon lighting in a city

Table 1.1 Neon sign colours and elements

Element	Colour
Neon	Red
Argon	Lavender-blue
Helium	Pale yellow
Sodium	Yellow
Krypton	Pale green-grey
Xenon	Blue-grey

To revise the emission of light by elements, refer to Chemistry section 1.8 on page 123.

ACTIVITY 1.5

USEFUL PURE METALS

Aim

To investigate the use of a pure metal and communicate this knowledge

What to do

1 Research a metal that is used in its pure form.
2 Find out about the metal's uses and describe them.
3 Relate the uses of the metal to properties such as strength, hardness and conductivity.
4 Create a poster, PowerPoint presentation or short video to communicate your research.

Nanotechnology

The field of **nanotechnology** deals with matter at the atomic and molecular scale. **Nanomaterials** are materials that have particles between 1 and 100 nanometres (nm) in size. A nanometre is one-billionth of a metre. Objects this small cannot be seen with the naked eye or even by light microscopy. For example, a virus particle is around 40 nm while a single strand of DNA is around 2 nm. The diameter of an atom is even smaller than this, around 0.2 nm. For comparison, a single human hair is around 100 000 nm thick. So you can see that nanoparticles are extremely small.

Advances in science understanding in one field can influence other areas of science, technology and engineering (ACSCH011 and ACSCH050)

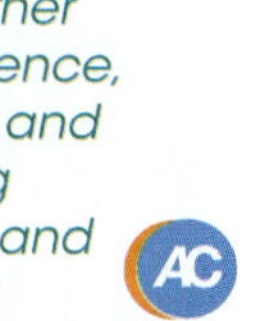

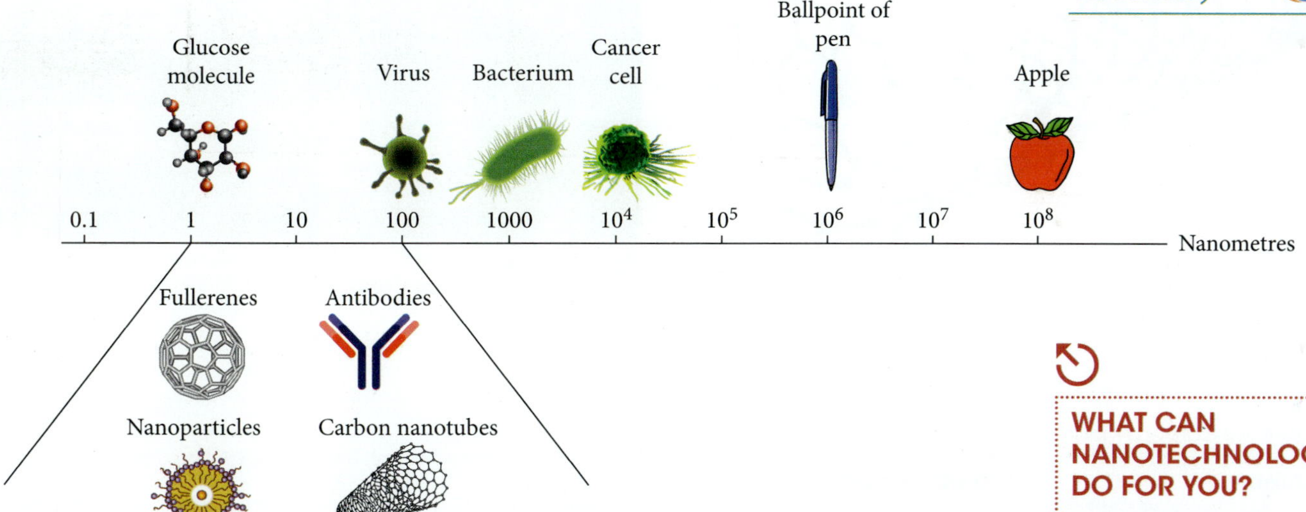

▲ **Figure 1.26**
The size of nanomaterials compared with different objects

WHAT CAN NANOTECHNOLOGY DO FOR YOU?

Use this interactive activity to see how nanotechnology has applications in areas such as computers, medicine, sporting goods and textiles.

Materials of this size have special properties that can be manipulated in order to perform specialised tasks. Applications of nanotechnology are already in use in medicine, materials fabrication, electronics and energy production. Two elements that have shown great promise in nanotechnology are carbon and gold. Carbon nanotubes are used in electronics and optics while gold is used in electron microscopy, electronics and materials science.

To understand more about carbon allotropes, refer to Chemistry sections 1.5 and 3.5 on pages 113 and 178.

Carbon nanotubes

Carbon nanotubes are an allotrope of carbon. An allotrope is a different physical form of an element.

Carbon has four valence electrons so it will normally form four bonds in its compounds. Sometimes, carbon atoms bond to other carbon atoms but form only three bonds. When this happens, the carbons arrange themselves in a hexagonal, flat sheet that is known as

Figure 1.27 ▶
A flat sheet of graphene

Shutterstock.com/Tlex

CARBON NANOTUBES

Watch this video to see how carbon nanotubes are used for a variety of applications.

graphene, as seen in Figure 1.27. Scientists use graphene sheets to form carbon nanotubes (Figure 1.28). Carbon nanotubes are very small, with a diameter as low as 1 nm. They can be as long as a few centimetres, although most are much smaller. In this form, carbon has the highest strength-to-weight ratio of any known material. When carbon nanotubes are constructed with multiple walls, as seen in Figure 1.29, their strength increases further still. Table 1.2 shows a number of projects that are using the strength of carbon nanotubes.

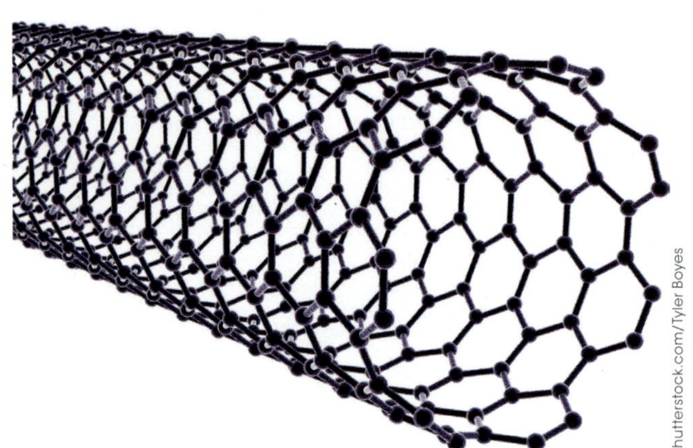
Shutterstock.com/Tyler Boyes

Figure 1.28 ▲
A carbon nanotube

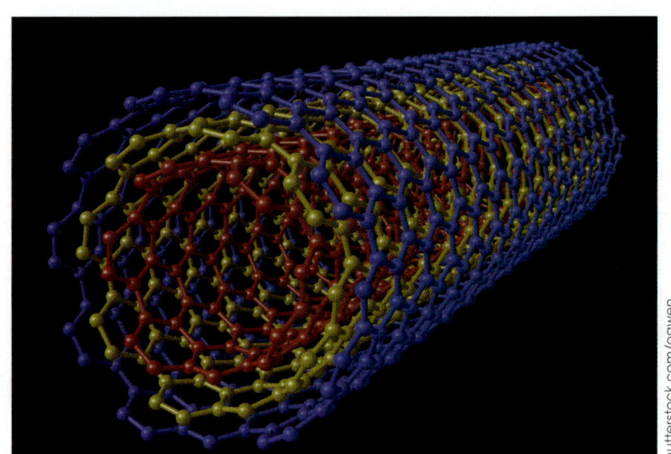
Shutterstock.com/ogwen

Figure 1.29 ▲
A multiwalled carbon nanotube

Table 1.2 Carbon nanotube projects

Area	Project details
Bulletproof armour	Lightweight body armour is being developed that allows the wearer more flexibility than current Kevlar materials. Nanotube armour is also stronger than Kevlar.
Synthetic muscles	Nanotubes are woven in with yarn and filled with wax. These muscles can lift weights 200 times heavier than natural muscle.
Setting broken bones	Nanotubes are used as a scaffold for new bone material to grow around.
Oil spill clean-up	Adding boron to nanotubes gives them sponge-like properties that allow them to absorb many times their weight in oil.
Morphing material	US National Aeronautics and Space Administration (NASA) has developed nanotubes that change shape when a voltage is applied. Controlling the voltage allows particular shapes to be formed.

WOW

Space elevator

Scientists at NASA are working to build a machine that allows objects to travel into space without having to be launched. It is proposed that this space elevator will lift materials into space at about one-fifth of the cost of launching by rocket. So far, no materials have been strong enough to construct the 35 000 km vertical column required to reach an orbiting space station.

Scientists from the University of Cambridge in the UK have combined short carbon nanotubes into long strands that could be the first steps towards a space elevator. The strands are flexible, but extremely strong and, most importantly, very light. All of these properties do not exist in natural materials, and it is only through the use of nanotechnology that a space elevator might one day become a reality. A space elevator could open the door to space tourism and colonisation of other worlds.

Science Photo Library/Christian Darkin

▲ **Figure 1.30**
An artist's impression of a carbon nanotube space elevator

Colloidal gold

Gold is a solid, heavy, dense metal used mostly for jewellery and decorative objects. But gold at a **nanoparticle** size looks and behaves very differently. Gold can exist as a **colloid** in a liquid, usually water, in which particles of gold about 100 nm in size are dispersed. Depending on the size of the particles, colloidal gold can be different colours. The most important property of colloidal gold is its ability to affect light in a predictable, measurable manner. Scientists can look at how light behaves around the particles and use this information in a number of ways.

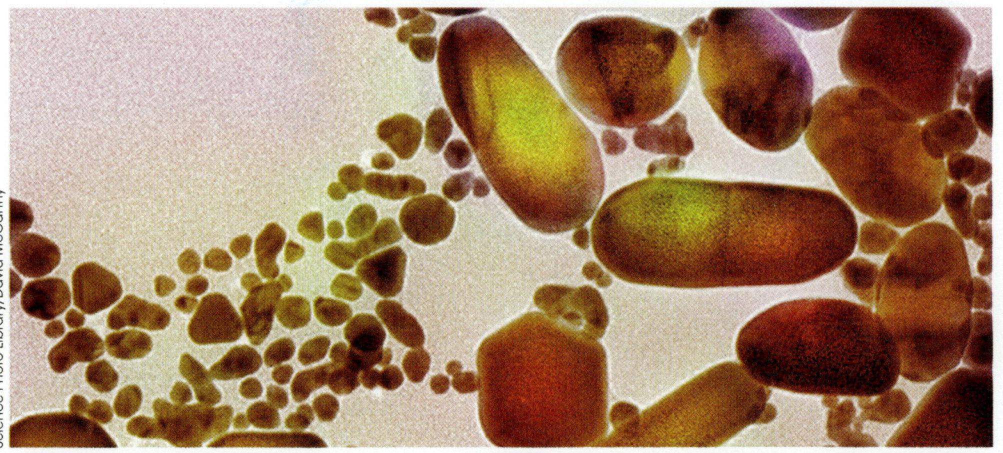

Science Photo Library/David McCarthy

▲ **Figure 1.31**
A scanning electron micrograph of gold nanoparticles, used in cancer research and diagnosis

The use of scientific knowledge may have beneficial and/or harmful and/or unintended consequences (ACSCH013 and ACSCH052) AC

Gold nanoparticles have important applications in medicine. One of their most important uses is as markers. For example, rod-shaped gold nanoparticles called nanorods are being used in cancer diagnosis. The gold nanorods are synthesised so they can attach to the protein markers for a specific cancer, such as breast cancer. After the nanoparticles bind to the cancer cells, scientists examine how the protein–nanorod combination scatters light. This allows for very accurate information about the size, shape and invasiveness of the cancer. A more precise diagnosis leads to better treatment. This method is simpler and uses less expensive equipment than other methods of diagnosis.

EXPLORING NANOTECHNOLOGY THROUGH CONSUMER PRODUCTS

Watch this PowerPoint presentation on nanotechnology, and then explore various consumer products through the information provided.

Gold nanoparticles are also being used to fight cancer, not just detect it. The nanoparticles are synthesised to attach themselves to a particular type of cancer cell. After the gold particles are injected into the body and have attached themselves to the cancer cells, infrared lasers are shone through the body. The nanoparticles convert the otherwise harmless light into heat, which destroys the nanoparticle and any cell that it is attached to. The nanoparticles can also be used as a delivery system for drugs. The nanoparticles seek out the cancer cells, but are also attached to the chemotherapy drugs. These drugs target cancer cells much more effectively when bound to the cell surface with the nanoparticles.

Elemental isotopes

Most elements have at least two isotopes; some are stable and some are unstable. Both types of isotopes can be useful; however, unstable isotopes have particular properties that can be used and manipulated for a variety of purposes.

In an unstable isotope, the forces in the nucleus are unbalanced, so the isotope tends to emit particles and electromagnetic radiation from the nucleus. For example, carbon-12 and carbon-13 are stable and do not undergo radioactive decay. Carbon-14 is unstable and undergoes decay when a neutron in the nucleus converts into a proton. Another product of this conversion is an electron, which is emitted from the nucleus. This process is called beta decay. The electron emitted is known as a beta particle.

To revise isotopes, refer to Chemistry sections 1.2 and 1.3 on pages 108 and 109.

The **half-life** of a radioactive element is the time taken for half of the atoms in a sample to decay to a new element or isotope. The half-lives of these synthetic elements can vary from a few milliseconds to a year. The half-life of a particular element is constant and does not change. For example, an element may have a half-life of 1 minute. This means that after 1 minute, half of the atoms in a sample of this element would have undergone radioactive decay. One minute later, half of the remaining atoms would have decayed. After another minute, half of the remaining atoms would have decayed. If there were 100 atoms to start with, then after 1 minute there would be 50 atoms, after 2 minutes there would be 25 atoms and so on. A typical half-life graph is shown in Figure 1.32. In Activity 1.6 you will simulate the half-life of an element.

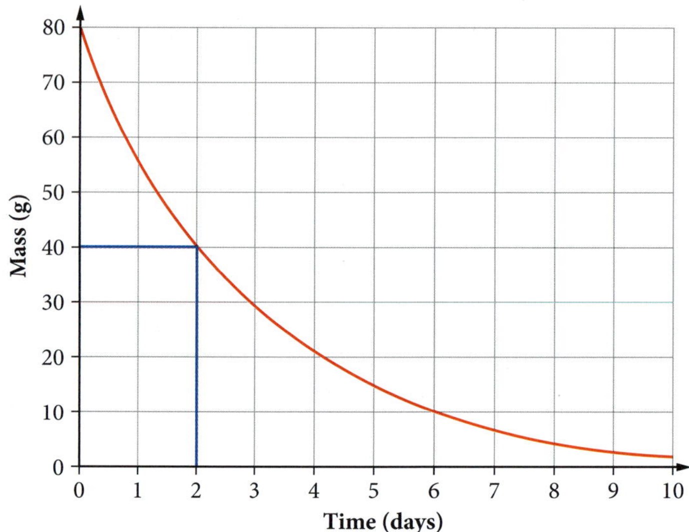

Figure 1.32 ▲
Graph of half-life for the radioactive decay of an element. The half-life is when half of the original sample has decayed, in this case 2 days.

ACTIVITY 1.6

DETERMINING HALF-LIFE OF AN ISOTOPE

Aim

To calculate the half-life of an element

You will need

- 100 M&M'S® or 100 small discs where one side can be easily distinguished from another
- Plastic bag

If you use M&M'S in this activity, perform it outside the laboratory.

What to do

1. Place the 100 M&M'S or discs into a bag and mix them thoroughly.
2. Pour out the M&M'S or discs over a clean piece of paper.
3. Remove the M&M'S that have the logo face down (or the discs with the blank face down). These have 'decayed'.
4. Count the M&M'S that have the logo face up (or the discs with the blank face up). These are still 'radioactive'. Record this number in a results table such as the one below.

Trial number	'Atoms' remaining	Combined class 'atoms' remaining
1		
2		
3		
4		

5. Return the remaining 'radioactive' M&M'S or discs to the bag and mix them thoroughly.
6. Repeat steps 3–6 until you have no more 'radioactive' M&M'S or discs.
7. Record the results of all the class on the board.

What did you discover?

1. On a graph, plot the trial number on the horizontal axis and the 'radioactive' atoms remaining on the vertical axis. Make sure you start your graph at zero. This is your half-life graph.
2. To find the half-life of your M&M'S or discs, draw a horizontal line from a point on the vertical axis that represents half the original number of M&M'S or discs to your graph line. Then draw a vertical line down to the horizontal axis. The place it cuts the horizontal axis is the half-life.
3. How many trials did it take for half the M&M'S or discs to 'decay'?
4. Did half of the M&M'S or discs decay in one trial? Suggest reasons why or why not.

Formation of isotopes

Some radioactive isotopes are made naturally. Carbon-14 is an example of this. When cosmic rays enter the atmosphere, they strike atoms and cause neutrons to be ejected. When a neutron collides with a nitrogen atom, a nuclear transformation occurs and carbon-14 is produced. The carbon-14 then enters the carbon cycle and takes part in normal processes such as photosynthesis, combustion and respiration. Carbon-14 isotopes behave chemically in the same way as carbon-12 because they have the same arrangement of electrons. When the arrangement of electrons is the same, the chemical properties of two substances are the same.

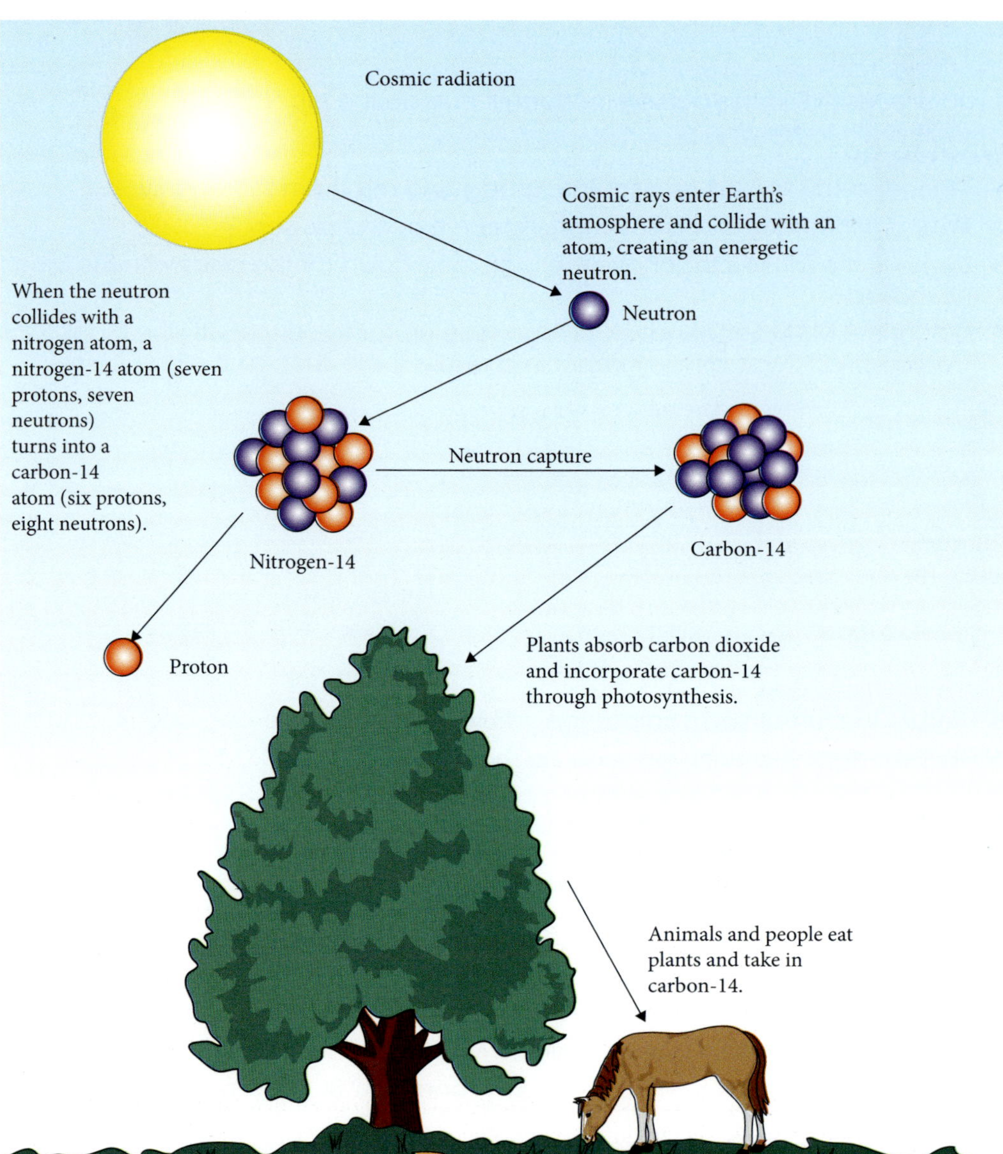

Figure 1.33 ▶ Natural formation of carbon-14 in the atmosphere

Some radioactive isotopes that are formed only in stars do not ever make it to planets such as Earth because they are too short lived. For example, aluminium-26 is found in large amounts in stars, but has never been found naturally on Earth. About 3000 radioactive isotopes that are not found in nature have been made in nuclear reactors and particle accelerators. In a nuclear reactor, large elements undergo fission (splitting) to form smaller elements, which are usually unstable and radioactive. In a particle accelerator, small particles such as protons and electrons are fired at a nucleus to try to make it unstable either by causing it to accept extra particles into the nucleus or by knocking particles out. In both cases, the end result is an unstable nucleus that emits particles and radiation that can be detected and measured.

Uses of isotopes

Both stable and unstable isotopes are very useful in a wide variety of fields. The particles and radiation emitted by an unstable isotope can be detected to find the location of an isotope or measured to determine the quantity present.

Advances in science understanding in one field can influence other areas of science, technology and engineering (ACSCH011 and ACSCH050)

The use of scientific knowledge may have beneficial and/or harmful and/or unintended consequences (ACSCH013 and ACSCH052)

Medical uses

Radioactive isotopes are used in medicine in two ways – to diagnose disease and to treat specific illnesses. Diagnosis of disease uses the radiation that is emitted from an unstable isotope. When an isotope is ingested by a patient, the radiation it emits forms a map of where it ends up in the body. The isotope can be used on its own or be attached to a molecule such as glucose that the body can use.

One of the most common radioactive isotopes used for diagnosis is technetium-99. This isotope has a short half-life of about 6 hours, which means technetium-99 breaks down very quickly and does not remain in the body. It is estimated that every year more than 20 million medical procedures use technetium-99. It has many uses, including imaging organs such as the kidneys, liver, lungs and brain. The isotope attaches to a specific pharmaceutical (a compound used as a medical drug) and is then transported to the required place in the body, where it decays, emitting radiation that is detected and measured. This is usually analysed by computer to produce images to help doctors diagnose their patients, as seen in Figure 1.34.

ALL ABOUT ISOTOPES

This video explains how isotopes are formed and used.

Radioactive isotopes are used in the treatment of diseases, especially cancer. Isotopes that emit radiation are ingested as a drink or in tablets. Cancer cells divide very rapidly so they absorb the radioisotope at high levels. The radiation emitted by the isotope as it decays damages the cells. The most common isotope used in this way is iodine-131. When this isotope is ingested, it concentrates almost exclusively in the thyroid so is used to treat thyroid cancer – the radiation emitted will target cancer cells in that area.

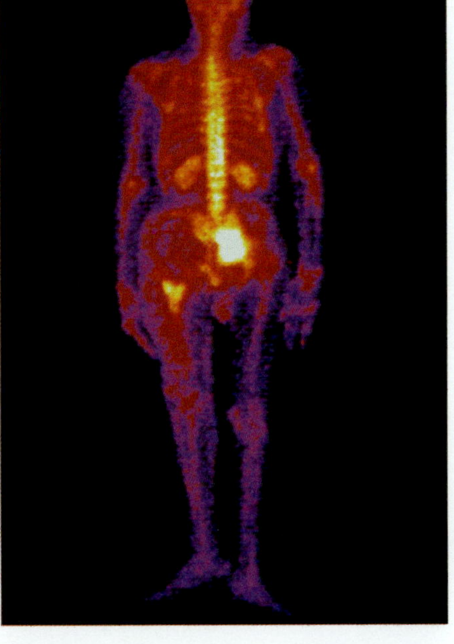

Science Source/GJLP

◀ **Figure 1.34** This image, produced by the radiation emitted by technetium-99, shows metastasised thyroid cancer.

Mineral nutrition

The human body is made up of mostly organic materials containing carbon, hydrogen and oxygen. The presence of other elements such as calcium, sodium, zinc and iron in different forms is important to keep your body functioning. Some people, especially children, have problems absorbing minerals from the food. To diagnose this problem, isotopes are used to track how minerals move through the human body.

When the target group is children, pregnant women or older adults, radioactive isotopes are not recommended because the radiation that is emitted is particularly harmful for these people. Instead, enriched stable isotopes, which do not emit radiation, are used.

Calcium is an essential element for bone and teeth formation. The most common isotope of calcium is calcium-40. The heavier isotope calcium-42 is also usually present in very small amounts in any sample of calcium. Calcium-42 is not radioactive. Sometimes, children do not absorb calcium properly, so doctors measure the uptake of calcium by administering a sample of calcium enriched with calcium-42, which will not harm the child, but is detectable. Doctors measure the amount of the isotope that passes through the child's body and then calculate the amount of calcium absorbed. This allows doctors to identify whether that child has a problem absorbing minerals.

Shutterstock.com/Adam Gregor

Figure 1.35 ▲ People who work with radiation protect themselves against radioactive emissions by wearing gloves, masks and protective suits.

Tagging chemicals in reactions

Chemical reactions inside the human body can be examined by placing isotopic tracers into molecules. For example, a hydrogen-1 atom in a water molecule can be replaced with a hydrogen-2 atom. Hydrogen-2 is known as deuterium or heavy hydrogen. The heavy water molecule behaves in exactly the same way as other water molecules so it is used by the human body in the usual manner. This can then be tracked through the human body and the chemical reactions that occur can be investigated.

Chemical reactions in industry or in a laboratory can be studied by the same methods. Many chemical reactions have been known about for years and even used in industry, but not understood properly. When a specific atom in a reactant is tagged with an isotope, the path that atom takes during a chemical reaction can be followed and recorded. This helps chemists understand the mechanisms of a reaction. When a reaction is understood properly, it can be manipulated. This allows chemists to make more of a particular product, or manipulate a reaction to get a product that may not otherwise form as readily. An example of this is tracking particular proteins during protein synthesis to produce artificial proteins that can then meet specific requirements of the chemist.

Safety of isotopes

Most radioactive isotopes do not pose significant risk because the particles and radiation they emit either are present in very small amounts or do not travel very far in air. Most people would not normally come into contact with isotopes that emit significant levels of dangerous radiation, apart from when undergoing certain medical procedures. People such as medical professionals, who work around radioactive isotopes, take appropriate precautions to limit their exposure through protective materials and secure storage. There are also restrictions on the number of procedures they can carry out each year, which further limits the amount of radiation they are exposed to.

The risk of damage from radioactive isotopes increases with the length of exposure. Radioactive isotopes that are used on or around humans normally have very short half-lives. This means that the total amount of radiation that humans are exposed to is very low as the isotopes decay to minimal amounts very quickly.

Table 1.3 Half-lives of some common radioactive isotopes used in medicine

Isotope	Use	Half-life
Technetium-99	Medical imaging of internal structures	6.01 hours
Iodine-131	Treatment of thyroid problems	8.02 days
Thallium-201	Testing for stress in patients with heart problems	3.04 days
Fluorine-18	Detection of cancer, through attachment to glucose molecules	109.8 minutes

CHAPTER SUMMARY

- Elements up to iron in the periodic table are formed during the life cycles of massive stars. Heavier elements are formed when stars die and explode in a supernova. Scientists can synthesise elements not found naturally on Earth.
- Light from the stars and techniques such as spectroscopy are used to learn about stars. Elements have unique emission spectra, so scientists can determine the elemental composition of a star by examining the light that it emits.
- Scientists organise elements on the periodic table to show patterns and trends. The periodic table was developed after scientists noticed patterns in the physical and chemical properties of elements.
- Scientists analyse materials by different techniques to identify elements present and to measure the quantities of those elements. Analysis of elements can provide information about health problems and the quality of our food, and help solve crimes.
- Elements in their pure form are useful in a wide variety of applications from electronics to medicine. Elements such as silicon, titanium and some elemental gases are used in semiconductors, solar panels, structural materials and signage.
- Nanotechnology deals with the behaviour of materials that has very small particles. Carbon nanotubes have extremely high strength-to-weight properties, which makes them useful in medicine, textiles and materials science. Colloidal gold is used widely in medicine because it can attach to specific cells and be targeted by light sources.
- Isotopes, both stable and unstable, are useful because of their differences from other isotopes of the same element. Isotopes form naturally and artificially when the nucleus of an existing element is changed through fission.
- Radioactive isotopes emit radiation that can be detected and measured. This radiation is used to diagnose and treat illnesses such as cancer. Risk of damage due to isotope exposure is minimised through strategies such as reducing exposure and using isotopes with short half-lives. Stable isotopes that do not emit radiation are used in some procedures involving high-risk groups.

CHAPTER GLOSSARY

actinoids the period of the periodic table that, with the lanthanoids, make up the f block

Big Bang the rapid expansion of matter from a high temperature and density state at the origin of the universe

black hole a region of space with an extremely high gravitational field from which light and matter cannot escape

chromatography techniques used to separate components of aqueous, liquid or gaseous mixtures

colloid a mixture in which tiny clusters of particles are dispersed through another substance; colloids do not settle out due to gravity and the particles are too small to be filtered

diatoms single-celled algae found in water sources

emission spectrum a pattern of bands produced by the emission of light from a source, separated due to the different wavelengths present

half-life the time taken for the radioactive emissions of an isotope to fall to half the original value

lanthanoids the period of the periodic table that, with the actinoids, make up the f block

main sequence stars that are in a state of equilibrium between gravity and the pressure produced by nuclear fusion

nanomaterial a substance that is made of or incorporate nanosized particles, with unique properties

nanoparticle a particle that is on the nanometre (one-billionth of a metre) scale (10^{-9} m)

nanotechnology a branch of science dealing with particles in the range of 1–100 nm

neutron star a celestial body composed of neutrons, with high density and strong magnetic field, formed after the death of a massive star

qualitative analysis analysis that identifies an element or substance but does not make numerical measurements about it, such as mass or concentration

quantitative analysis analysis that measures values such as amount, concentration or volume rather than just identifying the substance

radioactive decay spontaneous disintegration of an atom due to instability in the nucleus, releasing particles or electromagnetic radiation

red giant a star in the late stage of its life cycle, not undergoing hydrogen fusion, with high luminosity but low temperature

red supergiant a massive star that is undergoing helium fusion, with low temperatures and extremely large radii

semiconductor a substance that has little to no electrical resistance when cooled to extremely low temperatures

spectroscope a device used to separate light into its component wavelengths

supernova the explosion of a star at the end of its life cycle, resulting in the ejection of most of the star's mass

synthetic element an element that does not exist in nature, but has been made in a laboratory

white dwarf a planet-sized, dense star formed when a small star ceases nuclear fusion

CONTEXT 2
MATERIALS FOR A PURPOSE

By the end of this chapter you will have covered the following material.

Science as a Human Endeavour

- Advances in science understanding in one field can influence other areas of science, technology and engineering (ACSCH011 and ACSCH050)
- The use of scientific knowledge is influenced by social, economic, cultural and ethical considerations (ACSCH012 and ACSCH051)
- Scientific knowledge can be used to develop and evaluate projected economic, social and environmental impacts and to design action for sustainability (ACSCH015 and ACSCH054)

AC

Contextual story – Materials for a purpose

- History – humans have always tried to make use of the materials around them
- Why most of Earth's materials are not found as elements
- Natural materials
- Other natural materials that are changed to useful materials
- Finding new materials we didn't know existed
- Looking after our resources

Chemical understanding

- Bonding and properties of materials
- Writing formulas
- Pure substances and mixtures
- Chemical and physical properties
- Chemical reactions
- Balancing chemical equations
- Percentage composition
- Ionic bonding and naming
- Reactivity
- Covalent bonding and naming
- Hydrocarbons
- Methods of separation of mixtures
- Exothermic and endothermic reactions
- Mass and moles, basic stoichiometry and law of conservation of mass
- Thermochemistry
- Biofuels and fossil fuels

Investigations

Context 2
- Investigations 2.1 and 2.2

Chapter 2
- Experiment 2.1
- Investigation 2.1

Chapter 3
- Experiments 3.1 and 3.2
- Activity 3.1

Chapter 4
- Investigations 4.1 and 4.2

Chapter 5
- Experiment 5.1

Science as a Human Endeavour (SHE)

- Advances in science understanding in one field can influence other areas of science, technology and engineering (ACSCH011 and ACSCH050)
- The use of scientific knowledge is influenced by social, economic, cultural and ethical considerations (ACSCH012 and ACSCH051)
- Scientific knowledge can be used to develop and evaluate projected economic, social and environmental impacts and to design action for sustainability predictions (ACSCH015 and ACSCH054)

Science Understanding (SU)

- Materials are either pure substances with distinct measurable properties (for example, melting and boiling point, reactivity, strength, density) or mixtures with properties dependent on the identity and relative amounts of the substances that make up the mixture (ACSCH025)
- Differences in the properties of substances in a mixture, such as particle size, solubility, magnetism, density, electrostatic attraction, melting point and boiling point, can be used to separate them (ACSCH026)
- The type of bonding within substances explains their physical properties, including melting and boiling point, conductivity of both electricity and heat, strength and hardness (ACSCH027)
- Chemical bonds are caused by electrostatic attractions that arise because of the sharing or transfer of electrons between participating atoms; the valency is a measure of the number of bonds that an atom can form (ACSCH029)
- Ions are atoms or groups of atoms that are electrically charged due to an imbalance in the number of electrons and protons; ions are represented by formulae which include the number of constituent atoms and the charge of the ion (for example, O^{2-}, SO_4^{2-}) (ACSCH030)
- The properties of ionic compounds (for example, high melting point, brittleness, ability to conduct electricity when liquid or in solution) are explained by modelling ionic bonding as ions arranged in a crystalline lattice structure with forces of attraction between oppositely charged ions (ACSCH031)
- Covalent substances are modelled as molecules or covalent networks that comprise atoms which share electrons, resulting in electrostatic forces of attraction between electrons and the nucleus of more than one atom (ACSCH033)
- Carbon forms hydrocarbon compounds, including alkanes and alkenes, with different chemical properties that are influenced by the nature of the bonding within the molecules (ACSCH035)
- All chemical reactions involve the creation of new substances and associated energy transformations, commonly observable as changes in the temperature of the surroundings and/or the emission of light (ACSCH036)
- Endothermic and exothermic reactions can be explained in terms of the law of conservation of energy and the breaking and reforming of bonds; heat energy released or absorbed can be represented in thermochemical equations (ACSCH037)
- Fuels, including fossil fuels and biofuels, can be compared in terms of their energy output, suitability for purpose, and the nature of products of combustion (ACSCH038)
- A mole is a precisely defined quantity of matter equal to Avogadro's number of particles; the mole concept and the law of conservation of mass can be used to calculate the mass of reactants and products in a chemical reaction (ACSCH039)

Science Inquiry Skills (SIS)

- Identify, research and refine questions for investigation; propose hypotheses; and predict possible outcomes (ACSCH001)
- Design investigations, including the procedure/s to be followed, the materials required, and the type and amount of primary and/or secondary data to be collected; conduct risk assessments; and consider research ethics (ACSCH002 and ACSCH041)
- Conduct investigations, including the use of devices to accurately measure temperature change and mass, safely, competently and methodically for the collection of valid and reliable data (ACSCH003)
- Represent data in meaningful and useful ways, including using appropriate graphic representations and correct units and symbols; organise and process data to identify trends, patterns and relationships; identify sources of random and systematic error and estimate their effect on measurement results; and select, synthesise and use evidence to make and justify conclusions (ACSCH004)
- Interpret a range of scientific and media texts, and evaluate processes, claims and conclusions by considering the quality of available evidence; and use reasoning to construct scientific arguments (ACSCH005 and ACSCH044)
- Select, construct and use appropriate representations including chemical symbols and formulae, molecular structural formulae, physical and graphical models of structures, chemical equations and thermochemical equations, to communicate conceptual understanding, solve problems and make predictions (ACSCH006)
- Select and use appropriate mathematical representations to solve problems and make predictions, including calculating percentage composition from relative atomic masses and using the mole concept to calculate the mass of reactants and products (ACSCH007)
- Communicate to specific audiences and for specific purposes using language, nomenclature, genres and modes, including scientific reports (ACSCH008 and ACSCH047)

Materials

Humans have always been fascinated with the materials around them. We have tried to understand the properties of the materials and use them for specific purposes. Chemistry involves the study of materials – analysing the composition of materials, determining how elements, compounds and mixtures are structured, and exploring the properties of the materials. When we understand the structure and properties of materials, both physical and chemical, we can see the possibilities for their use. Chemistry also investigates the way that substances react to form new materials. These materials may replace other materials because they have properties better suited to a particular use or they may be used for a completely new purpose.

From the earliest civilisations of the Stone Age through the Bronze Age and Iron Age to more recent times, humans have investigated materials. We have utilised the most common and most easily accessible and separated materials, from stone to metals to crude oil. We have found materials that have the properties we need for specific uses. As we have come to better understand the impact of materials and their use, the focus has turned to ways of minimising this impact. We are trying to develop materials that have a reduced **environmental footprint** and develop ways to reuse materials to lessen resource use.

To revise classification and properties of materials, refer to Chemistry sections 2.1 and 2.2 on pages 140 and 143.

To revise mixture separation techniques, refer to Chemistry section 2.3 on page 150.

History – making use of available materials

Stone Age

The Stone Age occurred from the time of the earliest humans to about 3000 BCE. The earliest humans used stone. Initially, they used a piece of stone to smash against another piece of stone to grind ingredients for food. The stones worked like a mortar and pestle (Figure 2.1).

However, over time people realised that different stones have different properties. They found that some stones were very hard, while others could be shaped to produce useful implements and more advanced tools, such as arrowheads for spears that were formed by flaking obsidian (a volcanic rock) or flint.

Shutterstock.com/bonchan

▲ **Figure 2.1**
A mortar and pestle is used for crushing and grinding substances. The pestle grinds the material in the bowl-shaped mortar.

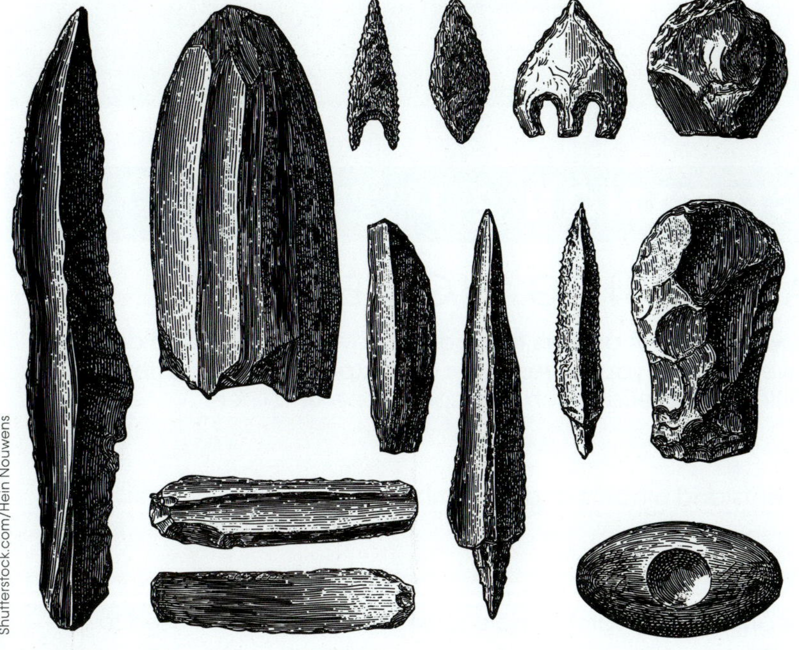

Shutterstock.com/Hein Nouwens

▲ **Figure 2.2**
Stone Age tools

INDIGENOUS CAVE PAINTINGS

See *COSMOS* for an article on Northern Territory cave paintings believed to be the oldest in the world.

However, gradually people found other uses for Earth's resources. They used clay to produce the first pottery vessels for holding food and drink. They exploited different-coloured pigments in different ochres and materials to decorate their pottery. Red ochre was widely used for over 200 000 years. It is derived from the mineral haematite (iron(III) oxide, Fe_2O_3), a compound common in many regions of the world. They also used charcoal (the element, carbon) from their fires for the colour black. Both of these materials are powders and will not adhere to pottery or cave walls. So, they mixed eggs or tree sap with the pigments. The sap and eggs act as binders to enable the coloured powders to bind to the surface.

This mixing of binders and pigments was one of the earliest examples of people mixing materials to change their properties to make them more useful. People used sticks, or sticks with fine grasses attached, for their brushes to apply their pigments to cave walls or pieces of bark. Stone Age people effectively used the materials available to them to create weapons and containers, and for decorating and artwork.

Getty Images/De Agostini/DAGLI ORTI

Figure 2.3 ▶ This Greek pottery vase is from the Late Neolithic period (5300–4500 BCE) in the Stone Age.

INVESTIGATION 2.1

CREATING AND USING PIGMENTS

The earliest people used mixtures of pigments with different binders to produce cave paintings and decorate their bodies for ceremonial purposes. In this investigation, you will research and experiment with different pigments and binders to create artworks on different materials such as paper, stone and linen.

What is your aim?

To identify materials that can be used as pigments and binders and determine which binders are best to use on different materials

What will you need?

Decide on the colours you want to use for your artwork. Research simple chemicals that can be used to produce these pigments.

Find out the types of materials on which the earliest people created their art. Research where you can get samples of these materials.

Research chemicals that the earliest people used to bind pigments to the material on which they produced artwork. Two commonly available examples are water and egg.

What are the risks?

Construct a table similar to the one below. Identify specific risks involved in the investigation and ways that you will manage the risks to avoid injuries or damage to equipment. Ask your teacher to check your risk assessment before you proceed.

What are the risks in doing this investigation?	How can you manage these risks to stay safe?
Pigments may stain your skin.	

How will you carry out your investigation?

Describe a method for conducting this investigation. Clearly indicate which binders will be used with which pigments and which materials will be used for each of the pigment-binder combinations.

What results will you collect?

Make sure that you know which results you will be collecting, and how and when you will be collecting them.

Discuss the best way to record your results. Draw a table or other format in readiness.

Discuss how you will best present your results in a visual way so other people can understand them.

What have you found?

Did you use the same binder to hold the pigments on to all of the materials? If not, justify which binders were most useful for each material.

What do you conclude?

Write a short conclusion to your investigation based on your aim and what you have found.

Ideas for improvement or further investigation

List any ideas on how to improve what you did, or areas that could warrant further investigation.

Bronze Age

The Bronze Age occurred from about 3000 BCE to 1000 BCE. During this time people began to use metals, such as copper and bronze – an **alloy** (mixture of a metal and another element).

Gold was the first metal to be used by ancient civilisations because its colour and lustre made it easy to recognise. Frequently, gold deposits were impure and had a high silver content. Hence, the gold alloy, electrum, was more commonly used. These deposits were rare and, therefore, electrum was used for the jewellery of leaders of these ancient civilisations.

Copper was first used about 5000 BCE. The first copper mineral to be used was malachite, $CuCO_3.Cu(OH)_2$. Initially, these earliest metallurgists processed malachite by placing it in a fire. However, this gave a very low yield. Over time, better techniques were developed, such as using stones to pound the copper ore into smaller pieces, which were then washed and placed into a simple clay kiln. A good air supply and charcoal in the kiln were needed to get a good yield of copper. Once extracted, the copper was reheated so that it could be hammered into shape for jewellery and containers.

To learn about chemical reactions and balancing chemical equations, refer to Chemistry section 4.1 on page 206.

To learn about the relative masses of substances and percentage composition, refer to Chemistry sections 5.1 and 5.2 on pages 230 and 233.

In the fire, and later in clay kilns, two chemical reactions were occurring. The following equations show the reactants and products for these processes. They also indicate the relationship between the various reactants and products.

$$CuCO_3(s) \rightarrow CuO(s) + CO_2(g)$$

$$2CuO(s) + C(s) \rightarrow 2Cu(l) + CO_2(g)$$

As chemistry developed, metallurgists found it useful to determine yields of products through using chemical reactions.

ACTIVITY 2.1

COMPARING COMPOSITION OF COPPER ORES

Aim

To determine the percentage of copper in some copper ores

What to do

1 Determine the percentage of copper in each of the following ores.
   a Malachite, $CuCO_3.Cu(OH)_2$
   b Chalcopyrite, $CuFeS_2$
   c Cuprite, Cu_2O
   d Chalcocite, Cu_2S
   e Tennantite, $Cu_{12}As_4S_{13}$
2 Research and discuss why copper was first extracted from malachite even though it does not have the highest percentage of copper.

Over time, malachite sources were used up and so people began to use copper sulfide ores, such as chalcopyrite, $CuFeS_2$. These sulfide ores were more difficult to smelt and contained more impurities. So, after processing, the final product was rarely pure copper, but was more likely to be an alloy of copper with a small amount of arsenic, tin, zinc, antimony or nickel. These impurities affected the properties of the alloy.

ACTIVITY 2.2

EXTRACTING COPPER FROM CHALCOPYRITE

Aim

To describe the physical and chemical processes required to extract copper from chalcopyrite

What to do

1 a Research the physical and chemical processes used in the Bronze Age for extracting copper from chalcopyrite.
   b Represent these processes using a flowchart.
2 Write balanced chemical equations for each chemical reaction involved in the extraction of copper from chalcopyrite.
3 a Research the processes currently used for the extraction of copper from chalcopyrite.
   b Add these processes to your flowchart.
4 Justify why the process(es) identified in Question **3** could not be used during the Bronze Age.

As time progressed, people extracting the copper realised that by mixing different ores together, they could produce alloys with different compositions and, therefore, different properties – they were practising some of the skills of modern metallurgy.

◀ **Figure 2.4**
a) Bronze Age artefacts in Rustavi Museum, Georgia; b) Bronze ritual food vessel from Zhou Dynasty, China (550–400 BCE)

ACTIVITY 2.3

COMPARING COPPER WITH COPPER ALLOYS

Aim

To compare the properties of copper and copper alloys

What to do

1 Identify the main elements in:
   a bronze.
   b brass.
2 Identify some uses of copper alloys during the Bronze Age.
3 Outline some of the health issues faced by metallurgists during the Bronze Age.
4 Research the properties and uses of copper and at least four copper alloys.

What did you discover?

1 Relate the properties of copper and the copper alloys to their uses.
2 Justify why copper alloys are more commonly used than copper.
3 Justify why there was a time period called the 'Bronze Age' rather than the 'Copper Age'.
4 Explain why the Bronze Age ended.

Advances in science understanding in one field can influence other areas of science, technology and engineering (ACSCH011 and ACSCH050) AC

Iron Age

The Iron Age occurred from about 1000 BCE to 500 CE. Although iron-rich rocks are found all over the world, iron was not commonly extracted until about 1000 BCE. Extraction of iron from iron ores required specialised furnaces, which could achieve much higher temperatures than those used for the extraction of copper. The early metallurgists of Europe and Asia had a better understanding of the processes and conditions required for the production of bronze than they did for the production of iron.

Getty Images/De Agostini

Figure 2.5 ▲ Celtic iron axes and spears, from excavations at Illerup, Denmark

Ancient iron artefacts were very precious because iron is generally found in ores and requires a significant investment of time and energy to extract it. Elemental iron was found at only a few sites where meteors had impacted Earth's surface. Hence, only small amounts were available. Iron's melting point is 1538°C; therefore, it did not melt in the furnaces being used prior to 1000 BCE.

Around 1000 BCE, metallurgists developed processes that enabled them to extract iron from minerals such as siderite ($FeCO_3$). A very hot charcoal fire is required for this chemical process to occur:

$$FeCO_3(s) \rightarrow FeO(s) + CO_2(g)$$

The charcoal burns with oxygen to form carbon monoxide (CO). The CO then reacts with FeO to form iron.

$$FeO(s) + CO(g) \rightarrow Fe(s) + CO_2(g)$$

The iron is then hammered into shape while it is 'red hot' from the flame.

Once people could extract iron and shape it, they used it extensively for the production of weapons since it was harder than bronze.

WOW

Legend of iron smiths

Legend has it that Wayland the Smith from Berkshire, UK, was commissioned by Merlin to make the sword Excalibur for King Arthur.

Why materials are mostly not found as elements

There are now known to be more than 114 elements, with 92 of these elements found in natural materials whereas more than 20 have now been produced using nuclear reactors and particle accelerators.

Elements are pure substances composed of one type of atom. The most stable elements in the periodic table are the noble gases. Except for helium, the atoms of these elements have an electron configuration with eight electrons in their valence shell, the most stable electron configuration. Atoms of other elements react to achieve this stable arrangement of electrons.

Ionic compounds

The simplest ions are formed when elements gain or lose electrons to achieve the same electron configuration as its nearest noble gas. Electrostatic forces of attraction occur between cations (positive ions) and anions (negative ions). The minerals used in the Bronze Age and Iron Age were examples of ionic compounds (Table 2.1). Chemists use chemical formulas to represent the ratio of atoms of each element present in a compound.

To learn about ionic bonding and compounds, refer to Chemistry sections 3.3 and 3.4 on pages 168 and 175.

Table 2.1 Ionic compounds used in the Bronze Age and Iron Age

Mineral	Formula	Ions present
Malachite	$CuCO_3.Cu(OH)_2$	Cu^{2+}, CO_3^{2-}, OH^-
Chalcopyrite	$CuFeS_2$	Cu^{2+}, Fe^{2+}, S^{2-}
Haematite	Fe_2O_3	Fe^{3+}, O^{2-}
Siderite	$FeCO_3$	Fe^{2+}, CO_3^{2-}

Reactivity

To learn about reactivity, refer to Chemistry section 4.2 on page 209.

Elements have different physical properties, such as melting point, density and hardness. They also have different chemical properties; that is, they react in different ways. Many elements react with water and oxygen in the atmosphere to form metallic oxides. This process is called **corrosion**.

Not all metals corrode at the same rate. These metals have different reactivities with oxygen. Iron corrodes quite quickly. Objects made from iron or alloys of iron such as steel corrode quite quickly and need to have a form of protection to minimise the corrosion. Aluminium does not corrode as quickly so is useful in window frames.

Corbis/Gordon Wiltsie/National Geographic Society

▲ **Figure 2.6**
Ships are made from alloys of iron, which corrode easily.

Alamy/Ernst Wrba

▲ **Figure 2.7**
Aluminium window frames do not corrode easily.

However, because most elements are reactive, there are many more compounds naturally present than elements. Minerals are being continuously formed, although it is a very slow process. Sometimes, these processes occur close to Earth's surface where water dissolves rocks, and crystals form over time. Limestone formations in caves are an example of this process. Limestone is mostly calcium carbonate ($CaCO_3$).

iStockphoto/markchentx

▲ **Figure 2.8**
Limestone formations in a cave are produced when water dissolves rocks and crystals of $CaCO_3$ form.

iStockphoto/plasticsteak1

▲ **Figure 2.9**
The granite in these kitchen benchtops formed deep under ground.

Sometimes these processes occur deep under ground where magma comes in contact with rock and melts it. Hence, as these materials react and cool over time, new minerals are formed. Therefore, most natural materials are compounds or mixtures of compounds. Granite has been formed in this way. It is a mixture of quartz, feldspar and other minerals. Granite is polished and used as benchtops in kitchens.

Natural materials

Many materials occur in nature. These are called natural materials. Materials such as wood and stone have been used to make houses. Originally, very little processing was done to materials such as wood. However, over time, more sophisticated structures have been built and the materials used have been protected by stains and varnishes.

123RF/Justin Skinner

Corbis/Frenchie Cristogatin/Arcaid

Figure 2.10 ▲
The wood used to build both this log cabin and the more sophisticated wooden house is a natural material.

Properties

Each material has specific properties, which determine its use. Stone has often been readily available, but it is also generally very dense, which makes it heavy and difficult to move from place to place. Wood, however, is generally much less dense than stone and easier to move. Wood was a common material for building houses – it is strong, withstands rain and wind, and is easy to cut to required lengths and shapes. However, it could not be used for fireplaces because it is flammable. Stone was used for fireplaces and chimneys as it could withstand reasonably high temperatures.

Natural fibres such as cotton and wool were used for clothes and bed coverings.

iStockphoto/jondpatton

Figure 2.11 ▶
Stone is used for fireplaces because it can withstand high temperatures and is not flammable.

iStockphoto/lenta

iStockphoto/Byrdyak

▲ **Figure 2.12**
Cotton and wool are natural materials.

ACTIVITY 2.4

PROPERTIES OF NATURAL MATERIALS

Aim

To identify the properties of natural materials and relate them to their uses

What to do

1 Research the properties of wool and cotton.
2 Discuss why wool and cotton are used for clothes and bed coverings.
3 Identify any limitations of wool and cotton and how this may affect their use for clothes and bed coverings.
4 Choose one other natural material, research its properties and relate these to its uses.

Uses

Some materials that occur in nature can be used with minimal processing. One example is marble, which is formed by the metamorphosis of sedimentary carbonate rocks such as limestone. The many different colours of marble are produced by impurities in the original limestone, as seen in Figure 2.13 on page 40. The stone is cut and polished before it is used for columns and benchtops.

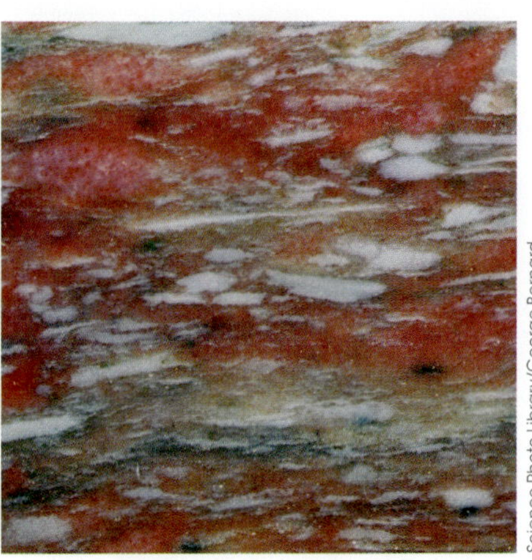

Figure 2.13 ▲
Marble contains impurities, which can make it white-grey, yellow or pink in colour.

To learn about covalent bonding and compounds, refer to Chemistry sections 3.5 and 3.6 on pages 178 and 185.

Covalent compounds

Covalent compounds are pure substances that are formed when atoms of two or more non-metals gain electrons to achieve the stable electron configuration of their nearest noble gas. **Hydrocarbons** (compounds that only contain carbon and hydrogen), such as alkanes and alkenes, are covalent compounds. These compounds are important for fuels.

Natural materials that need processing

Other natural materials, such as crude oil, need more processing before they can be used. Crude oil is a mixture of many different hydrocarbons. Crude oil is classed as a fossil fuel because it is formed by the decomposition of prehistoric living organisms. It takes millions of years to form.

Since crude oil is a mixture of hydrocarbons, it is processed using fractional distillation. Fractional distillation separates substances that have different boiling points. The substance that is most likely to form a vapour is said to be the most **volatile** substance. The substance with the lowest boiling point will evaporate first, then rise up into the fractionating column where it is cooled and collected.

INVESTIGATION 2.2

FRACTIONAL DISTILLATION

Fractional distillation uses the physical property of boiling point to separate mixtures of liquids. In this investigation, you will separate a mixture of ethanol and water because they have different boiling points.

What is your aim?

Write an aim that outlines the purpose of your investigation.

What will you need?

Figure 2.14 indicates some specialist equipment used to separate two liquids.

Research how you will need to modify the equipment to ensure that the mixture is heated gently and safely.

What equipment will you need so that you can identify the liquid that is collected first.

Research how you can test the first liquid to determine its identity.

Research how you can test the second liquid to determine its identity.

What are the risks?

What are the risks in doing this investigation?	How can you manage these risks to stay safe?
Chemicals could splash.	Wear safety glasses and wash your hands at the end of the investigation.
Glassware, especially the flask, will become hot.	Wait for equipment to cool before handling it.
Volatile mixtures could ignite or explode.	Move the Bunsen burner away from the boiling flask periodically to moderate the heat on the flask.
Ethanol is a potentially hazardous organic liquid.	Dispose of ethanol in an organic waste bottle.

In your write-up, add any more risks you can think of, as well as ways to manage them.

How will you carry out your investigation?

1. Figure 2.14 shows a diagram of the equipment required for this investigation.
2. Copy Figure 2.14 and annotate it to indicate the safety precautions that you've taken to ensure that the investigation will be performed safely.
3. Show your annotated diagram to your teacher and get it signed off before commencing your investigation.
4. Clearly describe the test you will perform to identify the first liquid that is collected.
5. Show this description to your teacher and get it signed off before performing the test.

What results will you collect?

1. What information did you record when the first liquid was collected that may assist with identifying this liquid?
2. What information did you record when the second liquid was collected that may assist with identifying this liquid?
3. Describe and record your observations when testing each of the liquids.

What have you found?

1. At what temperature did the first liquid come through the water condenser?
2. What did you notice about the temperature recorded by the thermometer while the first liquid was going through the water condenser?
3. At what temperature did the second liquid come through the water condenser?

What can you conclude?

Justify your identification of each of the liquids with reference to its properties and the tests you performed.

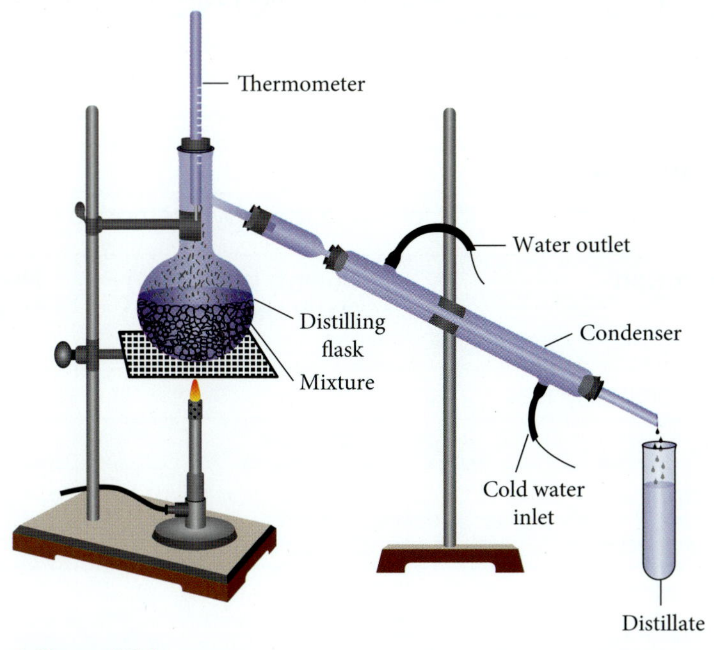

▲ **Figure 2.14**
Distillation apparatus

Oil refineries make the processing of crude oil economically viable and efficient. The crude oil is heated at normal atmospheric pressure to about 400°C in the distillation column, which has multiple exit points for volatile substances. The most volatile substances rise the highest in the column while the less volatile substances condense at lower levels in the column. This enables multiple fractions to be collected at the same time. Crude oil is continuously fed into the distillation column, which allows the process to continue.

Figure 2.15 ▶ Distillation column in an oil refinery

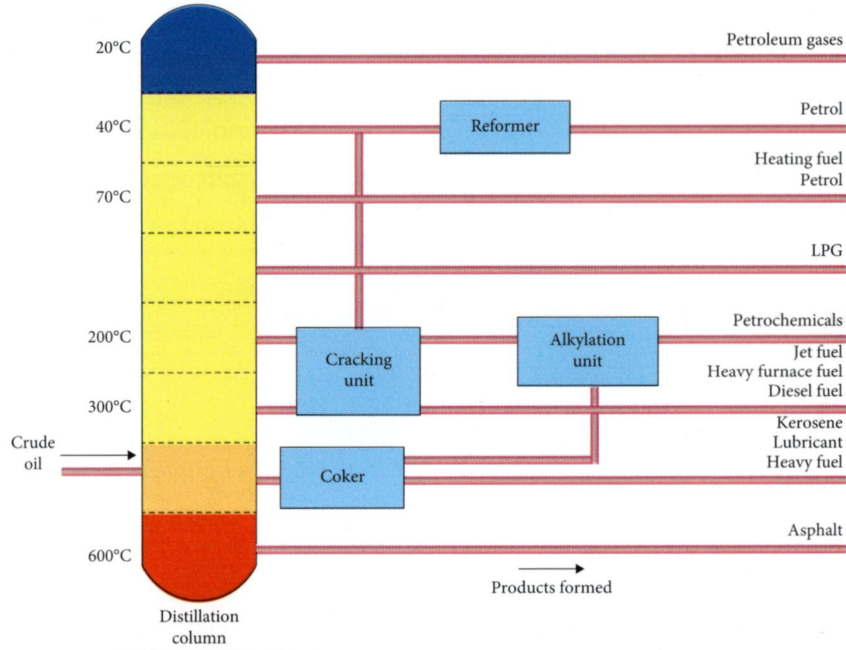

To revise mixtures and methods of separation, refer to Chemistry section 2.3 on page 150.

ACTIVITY 2.5

CRUDE OIL FRACTIONS

Aim

To determine the hydrocarbons present in different crude oil fractions

What to do

Copy and complete the following table by referring to Figure 2.15 and other sources of information.

Fraction	Boiling point (°C) (may be a range)	Number of carbons in chain (may be a range)	Structural formula for one hydrocarbon in this fraction
Petroleum gases			
Jet fuel			
Petrol			
Kerosene			
Heating oil			
Diesel fuel			
Lubricants/waxes			
Asphalt			

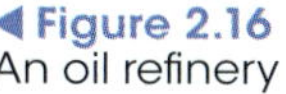

123RF/Dirk Ercken

◀ **Figure 2.16**
An oil refinery

Most fuels are not refined any further. However, some of the heavier fractions are processed further to increase the yields of the lighter fractions, which are the fractions in highest demand.

In Australia, some of the heavier fractions that have carbon chains with 15–25 carbon atoms, undergo **catalytic cracking**. Catalytic cracking is an example of a decomposition reaction where larger molecules are broken down to smaller molecules. This process uses a catalyst, a temperature of about 500°C and a moderately low pressure. The catalysts are **zeolites** (aluminosilicates), which are compounds that contain aluminium, silicon and oxygen. An example of a catalytic cracking reaction is:

$$C_{15}H_{32}(l) \rightarrow 2C_2H_4(g) + C_3H_6(g) + C_8H_{18}(l)$$

pentadecane ethene propene octane

The octane is used for petrol while the ethene and propene are very important starting materials for making plastics.

To learn about hydrocarbons, refer to Chemistry section 3.8 on page 189.

The use of scientific knowledge is influenced by social, economic, cultural and ethical considerations (ACSCH012 and ACSCH051) AC

Fuels

Society uses a number of different fuels for different uses. Common fuels undergo **combustion** reactions to release the energy stored in the chemical bonds within these compounds. The energy is converted into useable forms.

In Australia, coal is the main source of energy for our power stations. Petrol and diesel are commonly used in our cars and propane is the main gas used for our barbecues. Petrol, diesel and propane are all hydrocarbons. These fuels are all classified as fossil fuels.

Hydrocarbons react with oxygen to produce carbon dioxide and water in a process called combustion. These exothermic reactions release large amounts of energy. Below is the combustion reaction for octane:

$$2C_8H_{18}(l) + 25O_2(g) \rightarrow 16CO_2(g) + 18H_2O(l) \qquad \Delta H = -5517 \text{ kJ mol}^{-1}$$

Petrol is a mixture of hydrocarbon chains of 5–12 carbons in length. A common component of petrol is octane (C_8H_{18}). Diesel is a mixture of much longer hydrocarbon chains, 10–24 carbons in length. A common component of diesel is hexadecane, which is commonly called cetane ($C_{16}H_{34}$).

To learn about energy and chemical reactions, refer to Chemistry section 4.3 on page 213.

To learn about moles, refer to Chemistry section 5.3 on page 235.

Figure 2.17 ▶ Diesel and petrol release similar amounts of energy and emit similar amounts of CO_2.

AAP Image/Dave Hunt

Properties and uses

Although petrol and diesel are both mixtures of hydrocarbons, they have slightly different properties since they are composed of different hydrocarbons. The density of diesel is 0.832 kg L^{-1} and the density of petrol is 0.745 kg L^{-1}. Diesel has a higher density than petrol because it consists of longer chain hydrocarbons that have a greater interaction between molecules than in petrol.

Both fuels release similar quantities of heat per kilogram of fuel that is combusted. One kilogram of diesel releases 43.1 MJ while the same mass of petrol releases 43.2 MJ. Carbon dioxide emissions are also important considerations when comparing fuels. Diesel releases 73.25 g CO_2 MJ^{-1} of energy produced, while petrol releases 73.38 g CO_2 MJ^{-1}. Petrol engines require a spark plug to ignite the fuel, whereas diesel engines do not. Diesel engines compress the oxygen in the cylinder to high pressures and temperatures. The diesel is injected into the cylinder and ignites at these temperatures and pressures. Petrol is more flammable, likely to ignite, than diesel. A disadvantage of diesel is that at low temperatures, it becomes quite viscous, which changes its flow rate into the cylinder. This affects the fuel's performance. Additives have been developed that can be added to diesel fuel to overcome this problem. However, diesel engines are generally more fuel-efficient. A Mazda 6 with a diesel engine uses about 6.3 L per 100 km, whereas the same car with a petrol engine uses about 9.7 L per 100 km.

To learn about combustion reactions, refer to Chemistry section 4.4 on page 221.

ACTIVITY 2.6

COMPARING PETROL AND DIESEL

Aim

To compare the properties of petrol and diesel

What to do

A person is trying to decide whether to buy a petrol car or a diesel car. Both cars have a 60 L fuel tank. The cost of diesel is currently 158.9 cents per litre while petrol is 131.9 cents per litre. The motorist travels about 15 000 km per year.

1 Determine which car has the more economical fuel by using the information about distance travelled, fuel cost, density of fuel, fuel economy and energy released per kilogram.
2 Research to determine which car is more environmentally friendly by using the information on the properties of the fuels, such as how much CO_2 they emit per year.
3 Which car would you recommend the person buy, and why?

Other useful natural materials

Australia has large deposits of minerals, and mining is a major industry and a major contributor to Australia's economy. Australia has significant reserves of ores for the metals aluminium, copper, gold, iron, lead, silver, zinc and uranium. These ores must undergo processing to extract these metals.

AUSTRALIAN ATLAS OF MINERALS RESOURCES, MINES AND PROCESSING CENTRES

Create a movie or vodcast about an Australian metal, including location of deposits, processing and uses.

Uranium mining

Uranium is a naturally occurring actinide. Its three main isotopes are uranium-238, uranium-235 and uranium-234, which are radioactive and have long half-lives. Australia has 30–40% of the world's recoverable uranium. There are currently mining operations in South Australia and the Northern Territory and mining started in Western Australia in 2014. Mining of uranium is banned in Queensland. Both exploration and mining of uranium is banned in New South Wales and Victoria, so the extent of deposits in these states is unknown. Currently, Australia exports 10 000–14 000 tonnes each year. Australia only exports to countries that have signed agreements that specify that the uranium will only be used for peaceful purposes in civil nuclear power stations.

Extraction and processing

Extracting uranium from its ore and processing it so that it is suitable for use in a nuclear reactor is a complex process. Some of this process occurs near the mining operation whereas the final processing occurs at a refinery.

The two most common mineral sources of uranium are uraninite and pitchblende. Zircon, thorite and fluorite are generally also present so the extraction process must both isolate and purify the uranium compounds from the raw ore. Uraninite and pitchblende both have the chemical formula UO_2; they also contain the greatest percentage composition of uranium, about 88%. Other minerals, such as carnotite ($K_2(UO_2)_2(VO_4)_2.3H_2O$) have a lower percentage composition of uranium. Hence, a much greater mass of the ore, carnotite, must be processed to produce the same mass of uranium as that produced by pitchblende or uraninite.

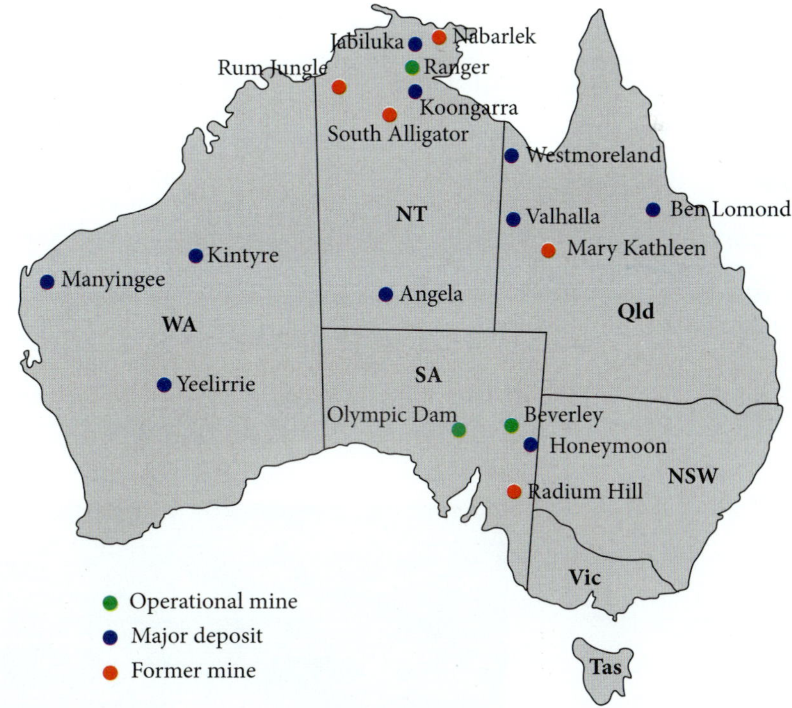

▲ **Figure 2.18** Australian uranium deposits and mines

There are a couple of ways of extracting uranium. The Ranger and Olympic Dam mines in the Northern Territory and South Australia use the method in the following paragraphs. The mine at Beverley in South Australia uses a slightly different method.

The ore is crushed and ground to reduce the particle size. Then it is put through a leaching process to make the uranium salt soluble. Depending on the nature of the ore, this process could use an alkaline leaching agent or an acidic leaching agent. At the Ranger and Olympic Dam mines, an acidic leaching agent, H_2SO_4, will be used to form $UO_2(SO_4)_3^{4-}$ (uranyl sulfate ion). Solvent extraction then both concentrates and purifies the uranium. An organic kerosene solvent containing tertiary amines is added to the uranyl sulfate ion according to the following equation:

$$2(R_3NH)_2SO_4(org) + UO_2\left(SO_4\right)_3^{4-}(aq) \rightarrow (R_3NH)_4UO_2(SO_4)_3(org) + 2SO_4^{2-}(aq)$$

iStockphoto/John Carnemolla

Figure 2.19 ▲
This uranium mine is at Ranger Mine, Northern Territory.

Hence, the uranyl ions are extracted into the organic phase while the impurities stay in the aqueous phase. The organic solvent is removed by evaporation, ammonia is added to neutralise the solution, and the precipitate, ammonium diuranate ($(NH_4)_2U_2O_7$) is formed. This is then heated to form a purified solid, yellow cake, which has the chemical formula U_3O_8 and is the product that Australia exports to Asia, Europe and North America.

The yellow cake is further processed at the refinery by dissolving it in nitric acid to form uranyl nitrate ($UO_2(NO_3)_2.6H_2O$). Uranium nitrate then undergoes a solvent extraction process using an organic kerosene solvent with tributyl phosphate. The uranium is extracted into the organic solvent while the impurities remain in the aqueous phase. The uranium is then washed with nitric acid to remove the kerosene and is then dried to form pure $UO_2(NO_3)_2.6H_2O$. Further heating leads to the production of pure UO_3, which undergoes the following reactions to form uranium hexafluoride (UF_6).

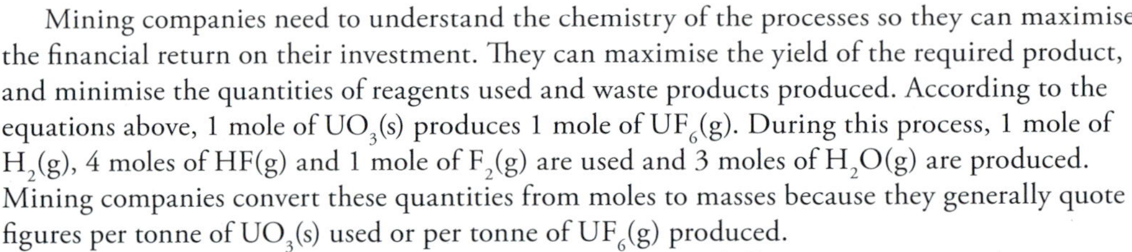

$$UO_3(s) + H_2(g) \rightarrow UO_2(s) + H_2O(g)$$

$$UO_2(s) + 4HF(g) \rightarrow UF_4(s) + 2H_2O(g)$$

$$UF_4(s) + F_2(g) \rightarrow UF_6(g)$$

Mining companies need to understand the chemistry of the processes so they can maximise the financial return on their investment. They can maximise the yield of the required product, and minimise the quantities of reagents used and waste products produced. According to the equations above, 1 mole of $UO_3(s)$ produces 1 mole of $UF_6(g)$. During this process, 1 mole of $H_2(g)$, 4 moles of HF(g) and 1 mole of $F_2(g)$ are used and 3 moles of $H_2O(g)$ are produced. Mining companies convert these quantities from moles to masses because they generally quote figures per tonne of $UO_3(s)$ used or per tonne of $UF_6(g)$ produced.

To learn about mass and moles, stoichiometry and the law of conservation of mass, refer to Chemistry sections 5.4 and 5.6 on pages 239 and 248.

The uranium hexafluoride, UF_6, then undergoes gaseous diffusion, to separate uranium isotopes to produce enriched uranium that can be used in nuclear reactors. This process involves increasing the percentage of uranium-235 from its natural levels of about 0.7% to about 4%. The enriched uranium is then converted to uranium dioxide pellets, UO_2, for use in the nuclear reactors, such as that shown in Figure 2.20.

Corbis/David Veis/epa

Figure 2.20 ▶
The Temelin nuclear power plant in the Czech Republic

ACTIVITY 2.7

PROCESSING URANIUM

Aim

To present information about uranium mining in a flowchart

What to do

1 Present the information about the extraction and refining of uranium in a flowchart. This can be done manually or by using a program such as Gliffy.
2 Identify the physical processes in the extraction and refining of uranium.
3 Identify the chemical processes in the extraction and refining of uranium.

Social, economic and cultural considerations

Uranium mining is a sensitive issue. Australia's uranium mines are located in remote areas near Aboriginal communities. The Ranger mine, near Jabiru (about 260 km east of Darwin) in the Northern Territory, is on land owned by the Mirarr Aboriginal people. It is surrounded by the Kakadu National Park, which is a World Heritage listed area due to its extensive Indigenous art sites and internationally significant wetlands. The Mirarr people have lived on this land for more than 50 000 years. Dreaming tracks cross the land and there are many sacred sites, including the Ranger mineral lease.

The use of scientific knowledge is influenced by social, economic, cultural and ethical considerations (ACSCH012 and ACSCH051) AC

Corbis/Steve Parish/Steve Parish Publishing

Figure 2.21 Wetlands, Yellow Water Lagoon, Kakadu National Park

The Ranger Project Area was exempt from the Aboriginal Land Rights Act so the Aboriginal people did not have a say on whether mining occurred. The government considered the benefits to the Australian economy in terms of jobs generated and export dollars received to outweigh the concerns of the traditional landowners. However, the Mirarr people did have a say on the conditions under which the mining occurred. There was considerable dispute between the Northern Land Council, with whom the Australian Government and the mining company, Energy Resources of Australia (a subsidiary of Rio Tinto Pty Ltd), were negotiating, and the Mirarr people, the traditional owners of the land. The Ranger mine commenced operation in 1980. Over the last 30 years, over 100 000 tonnes of uranium oxide have been produced during open-cut mining operations at Ranger – one of only three uranium mining operations in the world to reach this figure. These operations ceased at the end of 2012. In January 2013, a 14-year battle to renegotiate the original agreement between Energy Resources of Australia (ERA) and the Gundjeihmi Aboriginal Corporation, who acted on behalf of the Mirarr people, was settled. This agreement will lead to ERA paying increased royalties to the Mirarr people. They will also set up a sustainability fund for the Mirarr people. ERA is planning to continue mining at Ranger; however, they are moving to underground mining operations.

Evaluating media and scientific claims

Much has been and continues to be written in the media about the Ranger uranium mine. Most of the articles are about the agreement between the Mirarr people and ERA or environmental issues related to the mining operation, specifically water management.

ACTIVITY 2.8

EVALUATING CLAIMS ABOUT THE RANGER URANIUM MINE

Aim

To evaluate claims made about issues related to the Ranger uranium mine

What to do

1 Find three articles about the Ranger uranium mine from different organisations; for example, the Australian Broadcasting Corporation (ABC), Environment Centre NT, Energy Resources of Australia (ERA) and CSIRO.

2 For each article:

  a identify the claim(s) made.

  b outline the implication(s) for each claim.

  c research and identify any scientific evidence available that may support or refute the claim.

3 For one of the claims that you have researched, present your findings in a format of your choosing; for example, iMovie, Blabberize, podcast, cartoon, PowerPoint.

Finding new materials

Advances in science understanding in one field can influence other areas of science, technology and engineering (ACSCH011 and ACSCH050) AC

Many materials are abundantly available and easy to extract or process. These natural materials have been used for thousands of years. Some materials have only been discovered or synthesised more recently. Other materials are only present in very small quantities so specialised equipment is needed to find and extract them. Other substances are only produced when scientists understand the chemical and physical properties of the material and therefore can see the possibilities in terms of manipulating the material or reacting it with another material to form a new material with different properties.

Rare earth elements

The rare earth elements, more commonly known as the lanthanoids, are elements 58–71 in the periodic table; that is, from cerium to lutetium. The International Union of Pure and Applied Chemistry (IUPAC) classifies 17 elements as rare earth elements – the lanthanoids and lanthanum, yttrium and scandium. Although they are called rare, they are actually reasonably abundant. Thulium, which is the least abundant, is more abundant than gold. However, their deposits are fairly localised; hence, the reference to 'rare'. Originally, they were discovered in Sweden, but there are now known deposits on every continent except Antarctica.

About 2000 tonnes of mine tailings are produced for each tonne of rare earth element mined. The mine tailings contain radioactive thorium. China is currently the world supplier of rare earth elements; it has significantly decreased its production over the last couple of years, which has led to an increase in the price of these elements. Other countries, such as the USA and Australia, are currently determining whether it is economically viable to develop their rare earth deposits and the infrastructure necessary to process these elements.

WOW

Magnets

The magnetic fields of neodymium magnets are 10–50 times stronger than those of iron-based magnets. Therefore, if we used iron-based magnets in our mobile phones, then they would need to be 10–50 times larger to exhibit the same size magnetic field. Imagine how 'chunky' our mobile phones would be!

Extraction

Deposits of rare earth elements are fairly localised and most deposits contain mixtures of rare earth elements, which contain relatively small amounts of individual rare earth elements. This makes their extraction an expensive process.

◀ Figure 2.22
Rare earth elements – properties and uses

The oxides of the rare earth elements all have very similar high melting points. Not only do the oxides have similar properties, but the rare earth elements themselves have similar melting points and they also have the same stable valency, 3+. All of this means that it is extremely difficult to extract rare earth elements.

AN INTRODUCTION TO THE RARE EARTH ELEMENTS

Watch this video from Molycorp Minerals that shows uses for some of the rare earth elements.

Although the ions of the rare earth elements have the same charge, they have different ionic radii since the ratio of protons to electrons is slightly different. This, in turn, leads to different-sized forces of attraction between the nucleus and the valence shell electrons. Generally, as the number of protons increases, the radius of the ion decreases.

The breakthrough technology for extracting rare earth elements was the invention of ion-exchange chromatography. This process allows for the collection of substantial quantities of pure rare earth elements.

During ion-exchange chromatography, the cations of the various rare earth elements bind to the **chelating ligand** (a covalent substance whose molecules can form several bonds with each single metal ion) in the **eluting** solution to different degrees. Hence, they take different amounts of time to be washed from the chromatographic column. The smaller ions bind more tightly to the chelating ligands and, therefore, they move through the chromatographic column in the shorter time.

Uses

THE GLOBAL RACE FOR RARE EARTHS

Watch this video from the Bureau for International Reporting about some of the issues related to mining, supply and uses of rare earth elements.

Rare earth elements have many uses. New hybrid vehicles such as the Toyota Prius use nickel–lanthanum hydride batteries. These batteries are more powerful than the lead–acid batteries used in ordinary vehicles and yet they are also smaller in size. Each Prius contains about 4.5 kg of lanthanum.

Europium was used in cathode ray tube televisions for the red phosphors that produced the vibrant red colour in pictures. This led to the introduction of colour television in the late 1960s and early 1970s. Although these televisions are being phased out and replaced by LED/LCD (light-emitting diode/liquid-crystal display) and plasma televisions, europium is being used in one of the colours that are mixed to produce the white LED-based lights.

However, generally only small quantities of rare earths are added to other materials. Erbium is added in small amounts to optical fibres to amplify the light pulses; this means that the signal maintains its strength and hence can travel greater distances. Rare earth elements are at the forefront of many environmentally friendly technologies.

ACTIVITY 2.9

RARE EARTH ELEMENTS

Aim

To research the properties and uses of a rare earth element

What to do

Choose a rare earth element.

- **a** Identify its properties.
- **b** Describe at least one use and relate it to its properties.
- **c** Outline its abundance.
- **d** Discuss any implications for society if its supply were to be interrupted or terminated.

Looking after our resources

Scientific knowledge can be used to develop and evaluate projected economic, social and environmental impacts and to design action for sustainability (ACSCH015 and ACSCH054) AC

We live in a consumer world. We have more goods and appliances than ever before. All of these goods are packaged. Hence, there is high demand for our resources. Scientists are continually trying to develop new materials or new processes for production of existing materials that have less impact on the environment. Some examples are recycling existing materials and the development of biofuels.

AAP Image/AP/Mark Lennihan

Corbis

▲ **Figure 2.23**
Consumer goods such as smart phones and home cinemas require a lot of resources and energy to produce.

Social, economic and ethical considerations

Society has started to become more aware of the impact of consumerism, not just in terms of environmental effects but also the social and ethical considerations. Society's demand for affordable and luxury goods means that companies are trying to produce them for the cheapest possible prices. This means that many of the manufacturing plants are in countries such as Thailand, China, India and Bangladesh, where wages are low.

Our consumer world has had many positive impacts on our lives. However, many people are concerned about the changing nature of society – for example, people sitting inside watching television and playing computer and video games rather than participating in more active and communal pastimes such as sports.

ABC NEWS BREAKFAST - GET IT RIGHT ON BIN NIGHT

Watch this interview with a spokesperson from Sustainability Victoria on disposal of household waste. It indicates that people 15–30 years of age are least likely to recycle.

ACTIVITY 2.10

ETHICAL CONSIDERATIONS OF CONSUMERISM

Aim

To compare the working and living conditions of workers in Australia and in a country from which we get consumer goods

What to do

1. a Choose a type of manufacturing industry, such as the electronics, car or clothing industry.
   b For workers in that industry in Australia, outline their:
      i working conditions; for example, number of hours per day, any loadings such as sick leave, pay scales, holidays.
      ii living conditions; for example, housing, appliance ownership.
2. a Choose a country from which Australia imports goods in that manufacturing industry.
   b For workers in that industry in that country, outline their:
      i working conditions; for example, number of hours per day, any loadings such as sick leave, pay scales, holidays.
      ii living conditions; for example, housing, appliance ownership.
3. Discuss how our standards of living are affecting the standards of living in other countries.
4. Justify whether people living in Australia should expect a higher standard of living than people in other countries.
5. Create an advertisement to make your peers more aware of the working conditions of workers in other parts of the world.

ACTIVITY 2.11

SOCIAL CONSIDERATIONS OF CONSUMERISM

Aim

To analyse some of the social impacts of living in a material world

What to do

1. Research and justify the positive impacts on society of living in a material world.
2. Research and justify the negative impacts on society of living in a material world.
3. Conduct a class debate on the topic 'People who lived in the 1950s had a better quality of life than people living in the 2010s'.

iStockphoto/Nnehring

iStockphoto/mevans

Figure 2.24 ▲
Changing times: the shape, purpose, materials and technology of cars have changed over time.

Recycling

Materials from the earth are limited resources. As the resource becomes more scarce, the cost increases – it's all about supply and demand. Companies have to weigh up the options of whether it is more economically viable to mine for metals such as aluminium or to recycle existing stock.

Aluminium used to be a very expensive metal, so expensive that the Washington Monument in Washington, DC, USA, was given an aluminium tip to symbolise its value. However, in about 1886, a cheaper method was developed to extract aluminium from bauxite. The cost of aluminium dropped and the metal was used extensively for wrappings, window frames and soft-drink cans.

The extraction of aluminium from bauxite is still a very energy-intensive process. Recycling aluminium only uses 5% of the energy required to mine, extract and purify it from bauxite. To produce 1 kg of aluminium requires:

- 200 MJ of energy to mine, extract and purify it from natural ores, such as bauxite
- 7 MJ of energy from recycled material.

Alamy/National Geographic Image Collection

◀ **Figure 2.25** The Washington Monument was completed in 1885 at a time when aluminium was very expensive to produce.

HOWSTUFFWORKS – RECYCLING ALUMINIUM

Watch this video about how aluminium is recycled.

ACTIVITY 2.12

MINING VS RECYCLING ALUMINIUM

Aim

To compare the chemical and physical processes in the production of aluminium from mining and recycling

What to do

1. a Research the physical and chemical processes involved in extracting aluminium from bauxite, including the reaction conditions.
   b Present the information as a flowchart, either hand drawn or created by using a program such as Gliffy.
2. a Research the physical and chemical processes involved in recycling aluminium.
   b Present the information as a flowchart, either hand drawn or created by using a program such as Gliffy.
3. Refer to your flowcharts to explain why two-thirds of all aluminium mined is still in use today.

Biofuels

There are four main areas of focus in terms of biofuels for cars – ethanol, biodiesel, biogas and green gasoline. Biodiesel can be used with existing diesel engines, although currently in Australia only blends that have 5–20% biodiesel are available. They are labelled as B5 or B20. Green gasoline can be used with existing petrol engines, although this is not currently available in Australia. Compressed biogas can be used in motor vehicles. Although this is also not available in Australia, it is used in Sweden, Switzerland and Germany for cars, trucks and trains. In Australia, only ethanol is currently used as a fuel additive – E10 is a mixture of unleaded fuel and 10% ethanol.

Scientific knowledge can be used to develop and evaluate projected economic, social and environmental impacts and to design action for sustainability (ACSCH015 and ACSCH054) AC

WOW

First diesel engine

Rudolph Diesel, who invented the first diesel engine, designed it to run on peanut oil.

Green gasoline is petrol produced from plants. Petrol is a hydrocarbon – it contains only carbon and hydrogen. Plants contain cellulose, a carbohydrate consisting of many glucose molecules, which contains carbon, hydrogen and oxygen. In petrol, the carbon–hydrogen bonds are broken and carbon–oxygen and hydrogen–oxygen bonds are formed. Glucose, $C_6H_{12}O_6$, already contains some oxygen, so some of the carbon–oxygen and hydrogen–oxygen bonds are already formed; hence, less energy is released when combustion occurs than with a similar-size hydrocarbon such as cyclohexane, C_6H_{12}. Approximately 3 kg of glucose would produce the same amount of energy as 1 kg of cyclohexane.

US scientists at the University of Massachusetts Amherst, Massachusetts, have successfully developed a process for stripping oxygen from glucose to form hydrocarbons similar to cyclohexane. However, the challenge is to make it economically viable on a larger scale. Virent Energy Systems in Madison, Wisconsin, are working on developing this process. The other issue is the source of cellulose, which is currently the waste material from agriculture and forestry, such as sawdust and the stalks left after the cobs of corn have been picked.

The largest supplier of ethanol as a fuel additive in Australia is the Manildra Group. Their source of ethanol is wheat. They process over 1 million tonnes of wheat each year for industrial uses, including the production of ethanol. Their flour mill is located between Orange and Parkes in western New South Wales, about 300 km west of Sydney. Their ethanol plant is located in Nowra, on the south coast of New South Wales, about 160 km south of Sydney.

WOW

Power from the people

In 1895, the city of Exeter, in the UK, used gas produced from sewage to power its street lamps.

Biogas is commonly used to refer to the gas that is produced from the decomposition of organic matter in the absence of oxygen; that is, under anaerobic conditions. The gases that are released are a mixture of methane (60–70%) and carbon dioxide (30–40%) with very small amounts of other gases such as hydrogen, nitrogen and hydrogen sulfide. Since biogas can be combusted to release energy, it can be used as a fuel. The composition of biogas varies with the biodegradeable material used for its production. Biodegradeable materials such as sewage, waste in landfill sites, waste in septic tanks and manure can all be used for the production of biogas. Many of Australia's waste recovery companies have developed pilot plants for biogas production at landfill sites or sewage treatment plants. The biogas at these sites is released into the atmosphere if the site has not been engineered to collect the gas.

Small-scale digestion plants have been set up in over 2 million households in India and several hundreds of thousands of households in Pakistan to produce 'gobar gas', which is produced from cow manure. It provides households with gas for cooking.

In Australia, biodiesel is produced from vegetable oils such as canola oil, animal fats and even recycled vegetable oil. There is also research into using algae as the source of vegetable oils for biodiesel. The advantage of using algae is the potential for high yields per hectare. Algae could produce 100 000 L of oil per hectare, while the next best yield is 5000 L per hectare from palm oil. The other advantage of algae is that prime agricultural land would not be required for its production.

TURNING ALGAE INTO OIL THE NASA WAY

Watch this video on cutting-edge research being carried out on the production of biofuels.

To learn about biofuels, refer to Chemistry chapter 4 'Biofuels' on page 225.

Properties, uses and comparison to fossil fuels

Diesel is a mixture of different hydrocarbons in different proportions; therefore, its properties differ slightly depending on its source, which could be from fossil fuels or the biodiesel sources of vegetable oils and animal fats.

Biodiesel is a safer fuel than diesel because it has a higher **flashpoint** than diesel. The flashpoint is the temperature at which there is enough vapour present to ignite with oxygen when an ignition source is present. Biodiesel is also better for the environment. It is almost carbon neutral since CO_2 is taken out of the atmosphere to produce the plant sources for the biodiesel and returned to the atmosphere when the biodiesel is used. Some extra CO_2 is released into the atmosphere during the production of biodiesel from the vegetable oils and animal fats. However, diesel from fossil fuels does not take CO_2 from the atmosphere since it has taken millions of years to be produced, but it does release CO_2 into the atmosphere during the refinery process and when the diesel is used. Biodiesel also has a lower energy content than diesel from fossil fuels: it releases about 10% less energy.

Ethanol releases more than 30% less CO_2 into the atmosphere than petrol. This means that there are fewer environmental concerns about the use of this fuel. However, ethanol only releases about 60% of the energy that is released by petrol during combustion. This means that considerably more ethanol would be needed to provide the same amount of energy for use in a car. In Australia, E10 is a mixture of unleaded fuel with 10% ethanol. Therefore, more E10 would be used to produce the same energy as unleaded fuel without ethanol added. However, the difference is less than if the vehicle was using ethanol alone.

Biogas is a mixture of gases but the major component is methane. Methane is also available as the major component in natural gas from reserves off the coast of north-west Western Australia, the Otway and Gippsland regions of Victoria and smaller basins in the Northern Territory, Queensland and South Australia. Natural gas and biogas can both be compressed and hence have the potential to be used in vehicles.

The greatest issue for the use of all biofuels is the cost of production when compared to production from fossil fuels. To reduce these costs, at present, ethanol is blended with petrol and biodiesel is blended with diesel.

ACTIVITY 2.13

BIOFUEL PRODUCTION AND USE IN AUSTRALIA

Aim

To describe the production process for one type of biofuel in Australia and compare it to one of the fossil fuels

What to do

1 Choose one biofuel being produced in Australia – biogas, ethanol or biodiesel. Describe how it is produced, including relevant diagrams.
2 Describe some of the properties of your chosen biofuel, either in absolute terms or by comparison with an identified fossil fuel.
3 Develop an advertisement to promote awareness of the availability of your chosen biofuel.

Biomimicry – learning from nature

Scientists are always looking to develop new materials that perform better than existing materials or are used for new purposes. Geckskin is a new material developed by scientists at the University of Massachusetts Amherst. Researchers looked at how geckos were able to cling to very smooth surfaces instead of falling off. Geckskin is a material that adheres very strongly to smooth surfaces such as glass. A 10 cm by 10 cm piece of the material can support the weight of about four people hanging from it. Although it adheres very strongly, it can be removed from the glass very easily and without leaving any residue. Possible uses could be to replace ladders in rescue situations.

Another new material is Shrilk. This biodegradable material has been developed by scientists at the Wyss Institute for Biologically Inspired Engineering at Harvard University, Massachusetts, USA. Shrilk is made from chitosan, a polymer developed from chitin, which is the substance in the exoskeletons of crustaceans and insects, and fibroin, a protein derived from silk. The researchers looked at insects, not just for the materials but also for how to put the materials together. In insects, the chitin and fibroin are layered, so researchers layered the chitosan and fibroin to produce a material that is thin, clear, flexible and strong. A piece of Shrilk is lighter but stronger than a comparably sized piece of aluminium. Due to the availability of chitosan and fibroin, Shrilk is relatively cheap to manufacture and, importantly, it is biodegradable under conditions at landfill sites. The raw materials for making Shrilk have already been approved by the US Food and Drug Administration for use on humans. Potential uses of Shrilk include plastic bags, sutures that would dissolve over time and possibly as a scaffold for regenerating tissue.

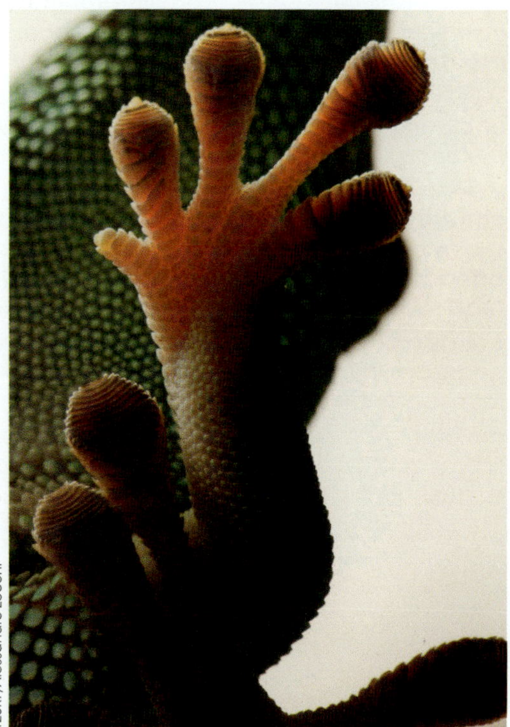
123RF/Alessandro Zocchi

▲ **Figure 2.26**
Gecko feet were the inspiration for a new material called Geckskin, which held up to four people while adhering to a smooth wall.

CHAPTER SUMMARY

- Humans have been looking for materials with specific properties to use for specific purposes for thousands of years. This began with very primitive materials in the Stone Age, and then became more advanced as processes were developed to extract metals from ores during the Bronze Age and Iron Age.
- Some natural materials are useful with minimal processing; for example, marble, wool and cotton. However, many materials need to be processed to be useful; for example, fuels. These materials can undergo physical and chemical processes to make them useful.
- Most elements are very reactive so are found naturally in compounds. The more reactive elements require more technologically advanced methods to extract them from their compounds.
- Mining is a major industry in Australia, which involves large-scale operations of physical and chemical processes to extract the metal from its ore. Uranium mining is one such example. Before a mining operation is established, many economic, social, cultural, environmental and ethical issues need to be considered.
- A deep understanding of the properties of different elements has enabled processes to be developed for isolating specific rare earth elements from ores. This has enabled the development of new technologies, such as stronger magnets and batteries for hybrid vehicles.
- Humans are becoming more conscious of their impact on the environment. Recycling is one way of minimising this impact. Chemists are also developing new materials from more renewable sources; for example, crops being used to produce green gasoline and biodiesel. Renewable fuels must have similar properties and uses as existing fuels from non-renewable sources.
- Chemists are continually developing new materials. They are analysing materials in nature to develop new materials, such as Geckskin and Shrilk.

CHAPTER GLOSSARY

alloy a mixture of two or more elements, one of which must be a metal

catalytic cracking the chemical process of breaking longer-chain hydrocarbons into shorter-chain hydrocarbons

chelating ligand a covalent substance whose molecules can form several bonds with each single metal ion

combustion a reaction with oxygen to form the oxides of each of the elements present; with adequate oxygen, combustion of a hydrocarbon will produce carbon dioxide and water

corrosion a chemical reaction in which a metal degrades in the presence of oxygen and water to form the oxide of the metal

covalent compound a compound composed of atoms of at least two different non-metals chemically combined in definite proportions

eluting separating the parts of a mixture by using the property that they travel through a solvent at different rates

environmental footprint a measure of the impact of humans on Earth's ecosystems and resources

flashpoint the lowest temperature at which a fuel has enough vapour present to ignite in air

hydrocarbon a chemical compound composed only of carbon and hydrogen

volatile can change from a liquid to a gas (vaporise) easily

zeolite an aluminosilicate mineral; i.e. contains Al, Si and O

CONTEXT 3
WATER, THE VITAL SUBSTANCE

By the end of this chapter you will have covered the following material.

Science as a Human Endeavour

- Science is a global enterprise that relies on clear communication, international conventions, peer review and reproducibility (ACSCH009 and ACSCH048)
- Development of complex models and/or theories often requires a wide range of evidence from multiple individuals and across disciplines (ACSCH010 and ACSCH049)
- Advances in science understanding in one field can influence other areas of science, technology and engineering (ACSCH011 and ACSCH050)
- The use of scientific knowledge is influenced by social, economic, cultural and ethical considerations (ACSCH012 and ACSCH051)
- The use of scientific knowledge may have beneficial and/or harmful and/or unintended consequences (ACSCH013 and ACSCH052)
- Scientific knowledge can enable scientists to offer valid explanations and make reliable predictions (ACSCH014 and ACSCH053)
- Scientific knowledge can be used to develop and evaluate projected economic, social and environmental impacts and to design action for sustainability (ACSCH015 and ACSCH054)

AC

Contextual story – Water, the vital substance

- Properties of water
- How water affects the world
- Ice and water
- Sticking together
- Oil and water don't mix
- Learning from nature
- The universal solvent
- Optimising performance for endurance sports
- Blood chemistry
- What is in your water?
- Managing water quality
- Water in crisis

Chemical understanding

- Water and intermolecular forces
- Shapes of molecules
- Uneven electron sharing
- Three types of intermolecular forces
- The importance of bonding
- Solutes
- Measuring solubility and concentration
- Practical applications
- Chromatography
- Acids and bases, pH
- Precipitation reactions

Investigations

Context 3

- Investigation 3.1
- Activities 3.1, 3.2, 3.3, 3.4, 3.5, 3.6, 3.7 and 3.8

Chapter 6

- Activities 6.1, 6.2, 6.3, 6.4, 6.5 and 6.6
- Experiments 6.1, 6.2 and 6.3
- Investigations 6.1 and 6.2

Chapter 7

- Experiments 7.1, 7.2, 7.3 and 7.4
- Investigation 7.1

Science as a Human Endeavour (SHE)

- Science is a global enterprise that relies on clear communication, international conventions, peer review and reproducibility (ACSCH009 and ACSCH048)
- Development of complex models and/or theories often requires a wide range of evidence from multiple individuals and across disciplines (ACSCH010 and ACSCH049)
- Advances in science understanding in one field can influence other areas of science, technology and engineering (ACSCH011 and ACSCH050)
- The use of scientific knowledge is influenced by social, economic, cultural and ethical considerations (ACSCH012 and ACSCH051)
- The use of scientific knowledge may have beneficial and/or harmful and/or unintended consequences (ACSCH013 and ACSCH052)
- Scientific knowledge can enable scientists to offer valid explanations and make reliable predictions (ACSCH014 and ACSCH053)
- Scientific knowledge can be used to develop and evaluate projected economic, social and environmental impacts and to design action for sustainability predictions (ACSCH015 and ACSCH054)

Science Understanding (SU)

- Observable properties, including vapour pressure, melting point, boiling point and solubility, can be explained by considering the nature and strength of intermolecular forces within a substance (ACSCH055)
- The shapes of molecules can be explained and predicted using three-dimensional representations of electrons as charge clouds and using valence shell electron pair repulsion (VSEPR) theory (ACSCH056)
- The polarity of molecules can be explained and predicted using knowledge of molecular shape, understanding of symmetry, and comparison of the electronegativity of elements (ACSCH057)
- The shape and polarity of molecules can be used to explain and predict the nature and strength of intermolecular forces, including dispersion forces, dipole-dipole forces and hydrogen bonding (ACSCH058)
- Data from chromatography techniques (for example, thin layer, gas and high-performance liquid chromatography) can be used to determine the composition and purity of substances; the separation of the components is caused by the variation of strength of the interactions between atoms, molecules or ions in the mobile and stationary phases (ACSCH059)
- The concentration of a solution is defined as the amount of solute divided by the amount of solution; this can be represented in a variety of ways including by the number of moles of the solute per litre of solution (mol L^{-1}) and the mass of the solute per litre of solution (g L^{-1}) (ACSCH063)
- The presence of specific ions in solutions can be identified using analytical techniques based on chemical reactions, including precipitation and acid–base reactions (ACSCH064)
- The solubility of substances in water, including ionic and molecular substances, can be explained by the intermolecular forces between species in the substances and water molecules, and is affected by changes in temperature (ACSCH065)
- The pH scale is used to compare the levels of acidity or alkalinity of aqueous solutions; the pH is dependent on the concentration of hydrogen ions in the solution (ACSCH066)
- Patterns of the reactions of acids and bases (for example, reactions of acids with bases, metals and carbonates) allow products to be predicted from known reactants (ACSCH067)

Science Inquiry Skills (SIS)

- Identify, research, construct and refine questions for investigation; propose hypotheses; and predict possible outcomes (ACSCH040)
- Design investigations, including the procedure/s to be followed, the materials required, and the type and amount of primary and/or secondary data to be collected; conduct risk assessments; and consider research ethics (ACSCH041)
- Conduct investigations, including measuring pH and the rate of formation of products, identifying the products of reactions, and testing solubilities, safely, competently and methodically for the collection of valid and reliable data (ACSCH042)
- Represent data in meaningful and useful ways, including using appropriate graphic representations and correct units and symbols; organise and process data to identify trends, patterns and relationships; identify sources of random and systematic error; identify anomalous data; estimate the effect of error on measured results; and select, synthesise and use evidence to make and justify conclusions (ACSCH043)
- Interpret a range of scientific and media texts, and evaluate processes, claims and conclusions by considering the quality of available evidence; and use reasoning to construct scientific arguments (ACSCH044)
- Select, construct and use appropriate representations, including physical and graphical models of molecules, energy profile diagrams, electron dot diagrams, ionic formulae, chemical formulae, chemical equations and phase descriptors for chemical species to communicate conceptual understanding, solve problems and make predictions (ACSCH045)
- Select and use appropriate mathematical representations to solve problems and make predictions, including using the mole concept to calculate the mass of chemicals and/or volume of a gas (at standard temperature and pressure) involved in a chemical reaction, and using the relationship between the number of moles of solute, concentration and volume of a solution to calculate unknown values (ACSCH046)
- Communicate to specific audiences and for specific purposes using language, nomenclature, genres and modes, including scientific reports (ACSCH047)

Water is a tasteless, odourless liquid, which is essential to our survival. All organisms are at least 60% water. Although water is very common, it has particular properties that make it an unusual material. These properties have enabled life to evolve on Earth and explain why water is vital for life.

The importance of water was recognised by Empedocles (490–430 BCE). He claimed that water was one of the four elements that formed everything that we know. But he never realised the reasons why this substance is so vital for life.

Water has been found not only on Earth but also in black holes, on Mars and on Saturn's moon Titan. But life is only expected to occur where water exists as a liquid.

GOLDILOCKS PLANETS

Visit this website to learn about NASA's search for habitable planets.

Dreamstime/Antartis

Figure 3.1 Earth is a Goldilocks planet. It is not too hot and it is not too cold. Just perfect: it can have liquid water.

WOW

The hunt for Goldilocks planets

Venus is too hot, Mars is too cold and Earth is just right. NASA's astronomers are looking for planets around other suns in the Milky Way that are also just right. The planets have to be in the habitable zone of a solar system – where temperatures allow water to exist as a liquid. The astronomers are looking for planets that may have liquid water. Evidence of planets in other solar systems has only recently been found. Now, hundreds of planets have been found, including a few Earth-size planets. It is these planets that may contain life.

To learn about how molecular shape is determined and how the shape influences the bonding between molecules, refer to Chemistry section 6.2 on page 258.

Development of complex models and/or theories often requires a wide range of evidence from multiple individuals and across disciplines (ACSCH010 and ACSCH049) AC

Properties of water

Water freezes at 0°C and boils at 100°C. It can act as a **solvent** over this entire temperature range, dissolving and transporting materials across a cell or over the oceans. It is the only substance on Earth that exists naturally in its three states of matter: solid, liquid and gas. Unlike other substances, solid water is less dense than the liquid. So ice floats.

The boiling point, density in solid and liquid phases, **surface tension** and its ability to act as a solvent are the unique properties of water. These properties are due to the water molecule's shape, bonding within the molecule and the bonds it forms with other substances. Ultimately it is water's bonding that makes it vital in biological, chemical and physical processes on Earth.

How water affects the world

Figure 3.2 ▶ The tip of the iceberg: why do icebergs float in water?

Scientists study ocean currents when they are investigating the climate and marine environments. Cold, salty water is denser than warm, fresh water. Ocean currents are generated when cold, salty surface water sinks and moves by convection towards the equator along the ocean floor. The less dense warmer water flows in from nearby regions along the surface.

Sailors in the tropics first noticed the stream of freezing cold water near the bottom of the ocean. This cold water must have come from the Arctic. This cold stream served the useful purpose of chilling bottles of wine dunked overboard. The currents may take more than 500 years to move over the planet but they have a dramatic effect on our climate. Oceanographers, meteorologists, software engineers, mathematicians and chemists have developed complex climate models that are used to predict the effects of events such as **El Niño** on Australia's climate.

The El Niño events highlight the link between water and Australia's climate. During an El Niño event, there is a warmer than usual ocean current off the South American coast. This leads to cooler sea surface temperatures across the Pacific. We feel this along the east coast, with drier conditions, more droughts and more severe bushfire seasons. Paradoxically, warmer tropical winds result in more floods in the north of Australia.

The temperature and salt levels of the oceans both affect the currents. A sample of Antarctic water would weigh more than an equal volume of fresh, warm water. Antarctic water is saltier partly because of sea ice. As sea ice forms, dissolved salt is squeezed out. The salt cannot be part of the ice crystal. The expelled salt remains dissolved in the surrounding water. This helps make Antarctic water the densest in the ocean.

OCEAN CURRENTS 1

Visit this website to learn more about ocean currents and how this knowledge is supporting our understanding of climate change.

At the surface, carbon dioxide and oxygen gases dissolve into the ocean. They dissolve better in cold water. Cold water sinks, taking oxygen and carbon dioxide to great depths. This provides a sink for carbon dioxide from our atmosphere and a source of oxygen for the life at these depths. This explains why there is so much life in Antarctic oceans.

Oceans affect the water cycle. The ocean is the main source of the water that is evaporated, cooled, condensed and precipitated. Flowing rivers cause weathering of rocks, washing sediments and dissolved salts into oceans, lakes and underground aquifers. During this cycle, water will go from gas to liquid to solid. Solid water forms the ice that breaks rock cracks open, falls as snow or forms sea ice in the Arctic and Antarctic. The unique properties of water affect these processes and have a dramatic effect on life on Earth.

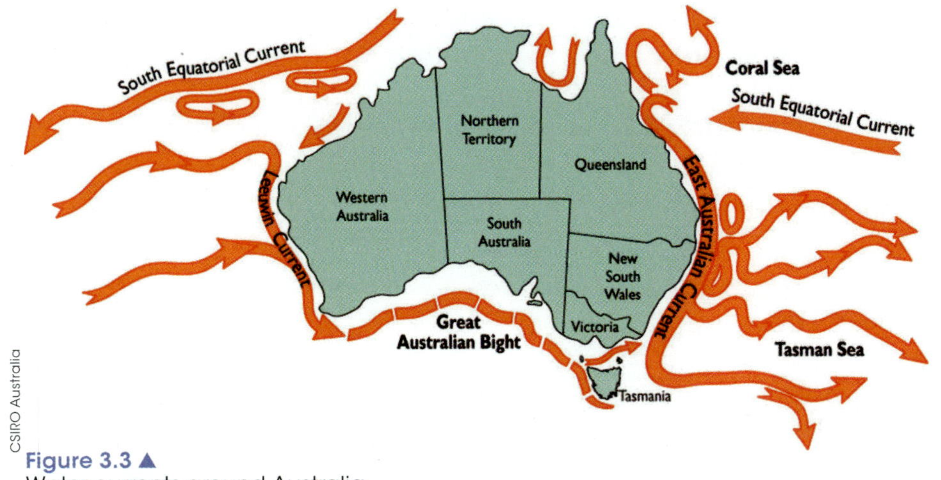

Figure 3.3 ▲
Water currents around Australia

OCEAN CURRENTS 2

Visit this website to learn how ocean currents are affecting the Australian climate and marine environment.

INVESTIGATION 3.1

DENSITY CHANGES IN WATER DUE TO TEMPERATURE AND SALINITY

Two factors affect the density of water: temperature and salinity (the concentration of salt). Density is a measure of mass (g) per unit volume. Ocean currents are formed as the cold salty water sinks underneath the warmer fresher water.

Aim

To design an experiment to determine which factor, temperature or salinity, has the greater effect on the density of water.

You will need

- Accurate thermometer (digital thermometer) that measures from 0°C to 50°C
- Glass measuring cylinder
- Electronic balance
- Source of heating or cooling
- Distilled water
- Salt

You can prepare a range of samples by adding 50 g of table salt to 500 mL of distilled water. This will make the solution equivalent to 100 g of salt per litre of solution. Suggested dilutions are:

- Add 50 mL of distilled water to 50 mL of the solution to make a 50 g L^{-1} sample.
- Add 80 mL of distilled water to 20 mL of the solution to make a 20 g L^{-1} sample.
- Add 90 mL of distilled water to 10 mL of the solution to make a 10 g L^{-1} sample.

What are the risks?

Construct a table similar to the one below. Identify specific risks involved in the investigation and ways that you will manage the risks to avoid injuries or damage to equipment. Ask your teacher to check your risk assessment before you proceed.

What are the risks in doing this investigation?	How can you manage these risks to stay safe?

How will you carry out your investigation?

1 How will you measure the density of each solution?
2 How will you heat or cool your solution?
3 Will you start with a warm solution and cool it down? Or will you prepare fresh solutions each time? Or will you start with a cool solution and warm it up?

How reproducible was your method?

1 What results will you collect?
2 How many samples will you test?
3 How many temperatures will you choose?
4 What range of temperatures will you choose?
5 Will you test fresh water?
6 Collate your results in a table similar to the following and give your table a title.

Salinity	Temperature 1	Temperature 2	Temperature 3

How will you analyse your results?

Will you graph the results? What type of graph will you use: a column graph or a line graph?

What have you found?

1 How does the temperature affect the density of water?
2 How does the salinity affect the density of water?

What do you conclude?

1 From your results, which factor had the biggest effect on the density?
2 Explain how the density changes with temperature and solute levels.

Ideas for improvement

From your results, which method gave the most consistent measure of density?

Taking it further

Some soft drinks contain a large amount of sugar or other additives. Apply your method to determine the effect of sugar levels on density. Was it a similar to result salt levels?

Further extension

How is this knowledge used in the fruit juice, wine, brewing, sugar and honey industries?

Ice and water

Water is the only substance that expands on freezing. As water cools, the molecules become closer together, so the density increases until the temperature approaches 4°C. After this point, the water begins to expand and become less dense. Ice floats. For most other substances, the solid form is denser than the liquid and sinks.

Ice has a definite hexagonal pattern of water molecules. This arises through the interaction between the bonding within and between the water molecules. The molecules are further apart in ice than in liquid water, which is why ice is less dense than liquid water.

If ice did sink, then in the cold regions of the world and during the Ice Ages, the bottom of the lakes and oceans would remain permanently frozen. Never being exposed to the sun, the layer of ice would increase until the whole lake or ocean froze solid. This would kill all life in the oceans and lakes. However, a floating layer of ice insulates the water beneath from freezing. Animals and plants can exist in lakes when the air temperature is below 0°C.

It is difficult to live in near frozen conditions because of the threat of water in cells freezing. With organisms composed mostly of water, the formation of ice can be lethal because an ice crystal can puncture the cell membrane. Fresh water freezes at 0°C and sea water will freeze at −2°C. Frostbite occurs when ice crystals are formed in and around skin cells when a person is exposed to extreme cold. However, life does exist under extremely cold conditions. One reason is that some organisms' cells can produce antifreeze agents. These chemicals, typically salts or proteins, disrupt the ice crystal structure. Some plant cells do not freeze until −40°C. This enables those plants to survive in frozen areas through the winter.

Advances in science understanding in one field can influence other areas of science, technology and engineering (ACSCH011 and ACSCH050) AC

Frozen blood and blood products are used by the armed forces during conflict when haemorrhage is a major cause of death of soldiers. There are limited options for treating blood loss on the battlefield and logistically it is difficult to rely on fresh blood. Frozen blood and products can be stored for years at −80°C. The blood products are not snap-frozen but treated with glycerol and other chemicals to stop ice crystals forming. Within 90 minutes, which includes 30 minutes of thawing, red blood can be treated and ready to use.

By understanding the chemistry that is occurring in the blood, the army has been able to integrate fresh and frozen blood products to save lives on the battlefield. This system could also be used in isolated rural areas of Australia. This is an example of how the understandings developed by chemists are applied in other areas of science, in this case biology and medicine.

To understand more about why ice has a lower density than water, refer to Chemistry section 6.5 on 269.

◀ Figure 3.4
Life under the ice. Ice is less dense than liquid water so floats on top. Colder water is less dense than warmer water so sits above it. The layer of ice insulates the water below it against freezing.

ACTIVITY 3.1

MAKING ICE CREAM

The effect on freezing point of the interaction between solutes and water molecules is not always detrimental. Research on the Internet to find out how ice-cream manufacturers use the effect of solutes on the melting point of water to make ice cream. A useful search term would be 'freezing point depression'.

The use of scientific knowledge is influenced by social, economic, cultural and ethical considerations (ACSCH013 and ACSCH052) AC

IVF: frozen embryos or frozen eggs?

The human egg (oocyte) is the single largest cell in the body. It is harder to freeze than an embryo. During in vitro fertilisation (IVF) procedures, embryos are frozen rather the woman's eggs. If the egg were placed in a freezer, ice crystals would rupture the cell membrane. To avoid this, the water has to be removed from the egg and replaced with a **cryoprotectant** ('antifreeze solution').

Research is being undertaken to successfully freeze eggs. This will allow women to freeze their eggs before treatments such as chemotherapy, which would normally destroy their eggs. Improving the technique may remove some ethical concerns that some people have about frozen embryos. These concerns include what to do with excess embryos that result from IVF treatment, especially if the donor couple cannot decide or can no longer be contacted. Excess embryos can be frozen for later treatments, donated for medical research or to infertile couples, or allowed to thaw and be destroyed. Some couples will only donate the embryo if they can determine the age, race, socioeconomic status or religion of the infertile couple. If the original couple separate, or the man dies, then the status of the embryo is not certain. The legal system is struggling to keep up with the advances in technology.

Freezing eggs also has ethical issues. Fertility drops dramatically after 35 years of age, so some women want to freeze their eggs until they are ready to have children. Is this a medical or a social issue? Women undergoing IVF treatment also need to be aware that the chances of having a healthy carefree pregnancy also decrease with age. Also, the effects of the cryoprotectants on the chromosomes, development and survival rate of the embryo are not completely known.

WOW

Surprise

Scientists are still being surprised by nature. Lake Vostok is the largest subglacial lake in Antarctica and the seventh largest lake in the world. It was first identified in 1950 beneath the Russian Vostok Station, but its existence would not be confirmed until the 1990s. It is a freshwater lake. On 10 January 2013, Russian scientists successfully obtained a core of freshly frozen water from the lake. RNA (ribonucleic acid) was found in the core, which has led to questions about the type of organisms in the lake, which has not been in contact with the atmosphere for millions of years.

Temperatures in Antarctica can get as low as −89°C. Lake Vostok is 500 metres below sea level and the average temperature of the water is −3°C, well below the freezing point of water. It has about 4 km of ice above it, yet it contains liquid water. A similar phenomenon is observed when water is formed under a glacier and under the blades of ice skaters.

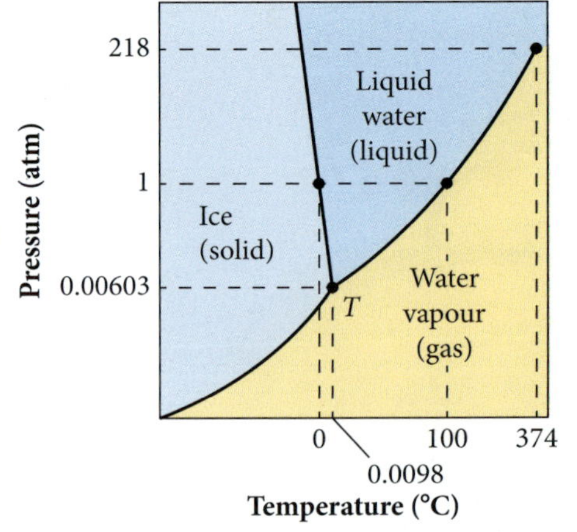

▲ **Figure 3.5**
The state of water depends upon its temperature and pressure, as shown in this triple point (T) graph.

Perfecting the temperature

Water affects all biology on Earth through its effect on heat exchange. The cooling breeze off an ocean can be very welcome. In Perth, the breeze that cools the city after a hot summer day is called the Fremantle Doctor. Water has a high specific heat capacity so it takes a lot of energy or heat to warm the ocean. Large bodies of water have a more constant temperature than land.

The body's reactions are catalysed by proteins called **enzymes**, which operate optimally at about 37°C. If the body gets too hot, then the reactions will stop. Similarly, if the body gets too cold, then the reactions are too slow to sustain life. Water's high specific heat means that cell temperatures usually remain within the narrow range required for life.

One function of blood is to regulate body temperature. As you exercise, blood flow increases and you sweat. Energy is absorbed from the environment when a liquid changes state to a gas. You feel cool as the water in sweat absorbs heat and changes state.

To revise specific heat of capacity and latent heat, refer to Chemistry section 4.3 on page 213.

To understand more about the intermolecular bonds and surface tension in water, refer to Chemistry sections 6.5 and 6.8 on pages 269 and 286.

The effect of extremes of temperature on enzymes will be discussed in Context 4 'Making reactions work for us'.

Advances in science understanding in one field can influence other areas of science, technology and engineering (ACSCH011) and ACSCH050 AC

Sticking together

Water molecules stick together. The high surface tension explains the **viscosity**, or stickiness, of water. It may appear very slight, but this high surface tension, which allows insects to walk on water, has provided the evolutionary pressure for selection of a streamlined shape, and affects the blood and fluids flowing through your body.

Surface tension explains why oil and water do not mix. Water's surface tension is high and the oil molecules cannot break the molecules apart to mix with it.

Oil and water don't mix

In 1989, the Exxon Valdez super tanker crashed into Prince William Sound, Alaska, USA. More than 40 000 tonnes of oil spilt in the pristine environment. The oil spill contaminated over 2000 km of coastline, killing record numbers of sea life: tens of thousands of sea birds, thousands of sea otters, hundreds of seals and more than 20 orca whales. Despite US$2.5 billion

Alamy/Accent Alaska.com

▲ Figure 3.6
The Exxon Valdez oil spill caused damage to the environment in Prince William Sound, Alaska, USA.

being spent on the clean-up, oil still lingers on the coast today. The wild life, such as sea otters and fish, are not recovering as well as hoped. The fishing industry collapsed in 1993.

Oil and most organic compounds are **hydrophobic** ('water-hating') and do not dissolve in water. Substances that dissolve easily in water are called **hydrophilic** ('water-loving'). Hydrophobic substances will form a layer either above or below the water. The oil from the disaster has not been washed away with the water but is found under the surface of some beaches near the intertidal zone.

Chemists can confirm that the oil is from the Exxon Valdez disaster and not from a new deposit spill by identifying components of the oil by very sensitive techniques such as **chromatography**, in particular high-performance liquid chromatography.

SURFACE TENSION IN SPACE

Watch this video to see how the surface tension of water is affected by gravity.

To understand more about solubility, refer to Chemistry section 6.6 on page 275.

The Exxon Valdez accident was depicted in an episode of the Simpsons, Bart after Dark, in which Lisa cleaned the rocks of oil. The ditty 'What do you do with a drunken sailor? Put him in charge of an Exxon tanker' highlighted one of the main causes of the disaster – the captain was drunk.

Mixing it up

Surfactants are molecules that are used to break the surface tension of water. In this way, they can allow oil and water to mix. Surfactants are used in many industries and are a major component of household cleaning agents. Detergents are one type of surfactant that are used to wash away the grease from dirty plates. Surfactants are generally used in oils spills to disperse the oil. In the Exxon Valdez oil spill, weather and tidal factors reduced the effectiveness of the surfactants.

The Exxon Valdez has become the benchmark for other oil spills. Spills are compared to this spill to indicate the relative size of the accident. But it also spurred an industry that aimed to minimise the environmental and economic damage from such incidents. When other disasters like the Gulf of Mexico oil spill occurred in 2010, expertise gained in the Exxon Valdez incident was applied to try to minimise the effect on the local fishing and tourism industry.

Learning from nature

Scientists are mimicking hydrophobic interactions in nature to develop nanomaterials. Nanostructures in the lotus leaf force water into beads so it runs off the leaf. This process is being mimicked in the laboratory to develop self-cleaning surfaces. Nanopatterned surfaces are being developed with nano bumps to generating water-grabbing and water-shedding materials.

Interactions between oil-type molecules and water are the foundation of life. These interactions allow the formation of cell membranes and the folding of protein and nucleic acids. By copying and manipulating these interactions, scientists have produced nanomaterials that self-assemble.

One current area of research is superhydrophobic surfaces. These could lead to better heat transfers in power and desalination plants.

To learn more about surfactants and chromatography, refer to Chemistry section 6.8 on page 286.

Water and biomolecules interact

When some organic compounds such as protein, RNA and DNA crystallise, water is included in the crystals. This is called **water of crystallisation**.

To determine the structure of a protein, scientists often need to get it in crystal form. Up to 50% of a protein crystal could be due to water. To make a protein crystal, a **supersaturated solution** (an unstable **solution** that has additional **solute** dissolved for the particular temperature) of the protein is made, and then the crystal is allowed to slowly precipitate out. It is important that the protein remains folded while the structure is determined. To do this scientists have to control the water content consistently.

NANOTECHNOLOGY

Watch this video of demonstrations of the benefits of nanotechnology and some amazing advances that are just around the corner.

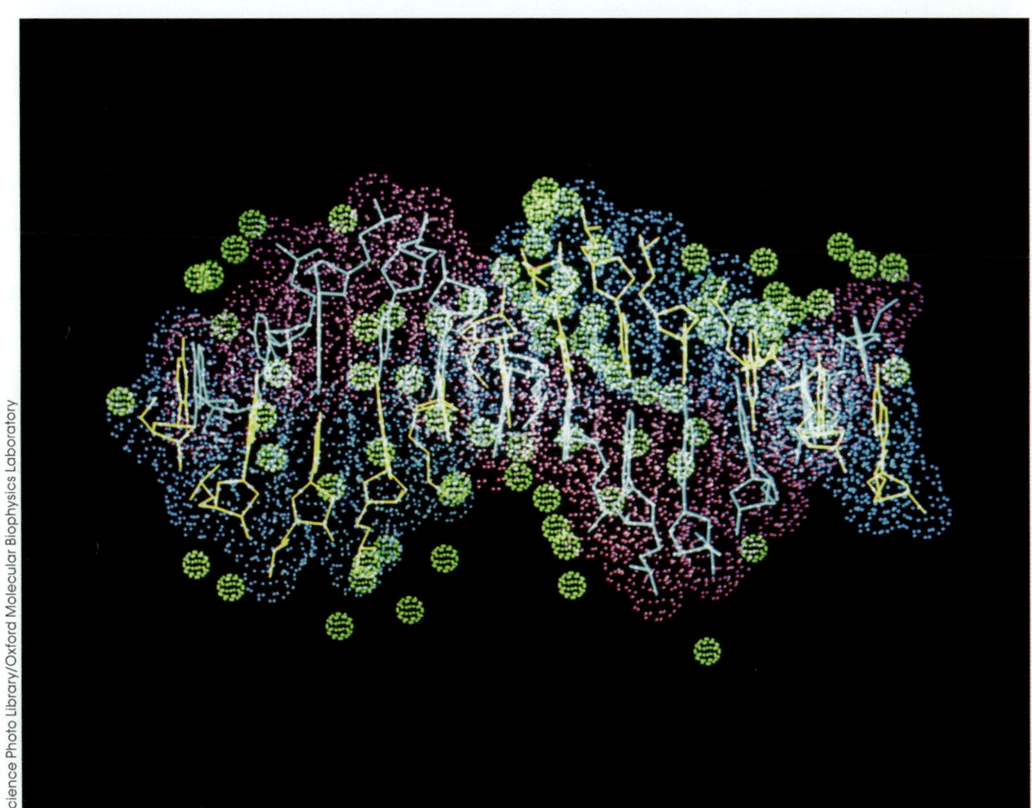
Science Photo Library/Oxford Molecular Biophysics Laboratory

◀ **Figure 3.7**
DNA and water interact. A DNA molecule has a hydrophilic backbone and a hydrophobic inner region of nitrogenous bases (adenine, guanine, thymine and cytosine). The hydrophobic region repels any water (green) between them, which forces the bases towards each other.

To learn about supersaturated solutions and water of crystallisation, refer to Chemistry sections 6.5 and 6.7 on pages 269 and 279.

The different shapes of DNA crystals when water is present

Rosalind Franklin (1920–58) used her understanding of chemistry and her skills in X-ray diffraction to produce a superior image of the DNA molecule. Under controversial circumstances, her co-worker Maurice Wilkins showed the image to James Watson. This was the proof needed to confirm that DNA was a helix. Franklin's analysis had also enabled her to point out errors in Watson and Crick's earlier models.

In 1962, the Nobel Prize for Physiology or Medicine was awarded to Watson, Crick and Wilkins for the discovery of the structure of DNA. Did Franklin miss out because Nobel Prizes are not awarded posthumously? Or was it due to the sexist attitude of the 1940s and 1950s? Franklin had attended one of the only schools that taught Chemistry and Physics to girls. Her work has since been acknowledged.

Development of complex models and/or theories often requires a wide range of evidence from multiple individuals and across disciplines (ACSCH010 and ACSCH049) AC

Manipulating molecules

The shape and structure of molecules will affect how they behave. The shape and structure affects the intermolecular bonds that are formed. By changing or adding different groups to the molecule, the properties of the material changes.

Organic molecules tend to be hydrophobic because they are mostly non-polar. Adding polar groups such as the hydroxyl (OH) group may make the molecule become polar. Polar molecules are hydrophilic.

Methane (CH_4) and ethane (CH_3CH_3) are hydrophobic and form dispersion forces with water. Methanol (CH_3OH) and ethanol (CH_3CH_2OH) can form hydrogen bonds with water and will dissolve easily in water.

In Unit 4 Chemistry, you will examine how chemists change one molecule into another.

The organic functional groups –OH, –C=O and –NH_2 will be covered in detail in Unit 3.

ACTIVITY 3.2

VITAMINS: GOOD OR BAD?

Vitamins are a group of essential nutrients that we only need in small amounts. They are a diverse group of organic compounds. For example, unlike other animals, humans must consume vitamin C. A lack of vitamin C causes scurvy – a deadly disease that used to kill sailors on long voyages. In 1770, Captain James Cook accepted the scientific view that a diet including citrus fruit was essential. He was the first to circumnavigate the world without losing a man to scurvy. An excess of vitamin C is never fatal.

Vitamin A is believed to have caused the death of Antarctic explorer Xavier Mertz in 1913. Short of food Mertz and fellow explorer Douglas Mawson resorted to eating their sledge dogs. Dog livers are high in vitamin A and an excess of this vitamin could have caused his death.

Vitamin A is a non-polar fat-soluble vitamin. It accumulates in the liver and is harder to remove from the body than water-soluble vitamins. This leads to it being toxic in large doses.

1 Copy Figure 3.8. For each vitamin, identify and circle the polar groups.
2 a Why is vitamin C considered polar and water-soluble?
  b Why is vitamin A considered non-polar and fat-soluble?
  c How does this explain why an excess of vitamin C is never fatal but excessive vitamin A can be fatal?
3 Decide whether the other vitamin molecules overall are polar or non-polar. Justify your decisions.
4 On the basis of polarity, classify vitamins B, D and E as likely to be water- or fat-soluble. Research to see if your predictions are correct.

Extension

Research how excess chemicals are removed from the body.

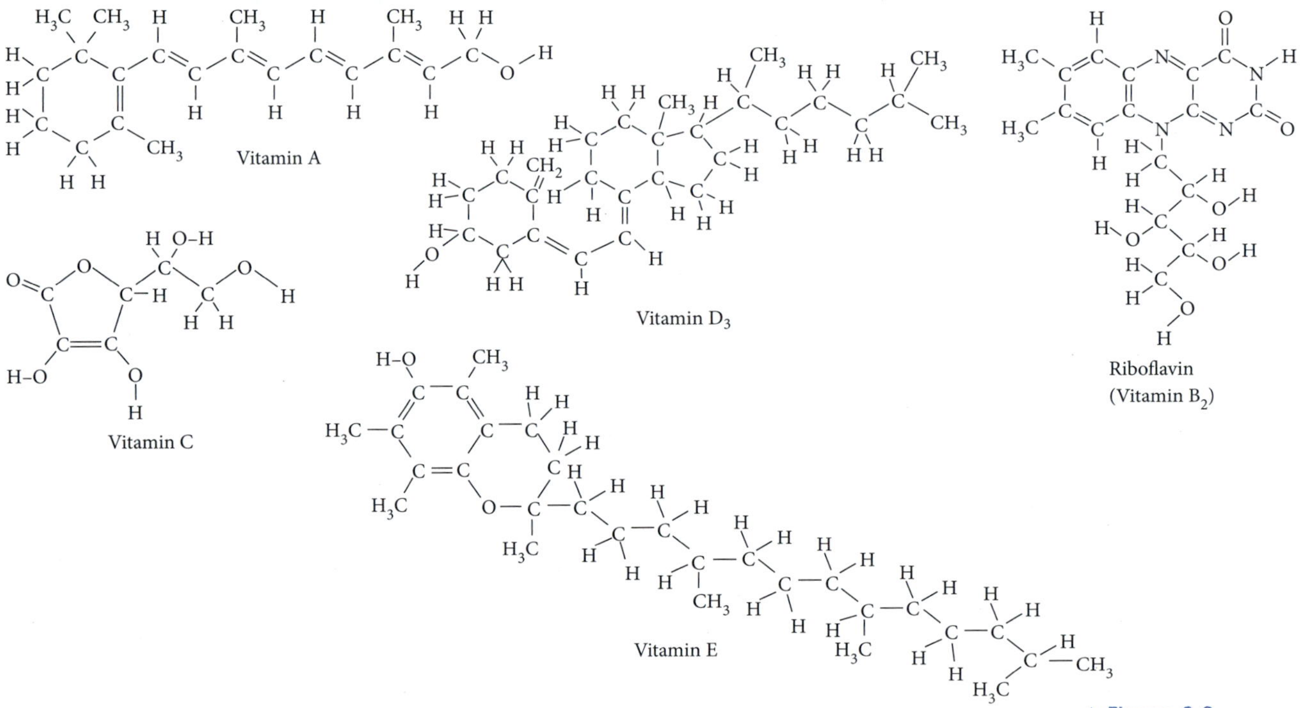

Vitamin A

Vitamin D$_3$

Riboflavin (Vitamin B$_2$)

Vitamin C

Vitamin E

▲ **Figure 3.8** Vitamins A, B, C, D and E

Surfactants and premature babies

Babies born prematurely often die when they struggle for breath. In the 1950s, US paediatrician Dr Mary Ellen was puzzled as to why there was no residual air in the lungs of these babies. It appeared that their lungs could not retain air and collapsed every time the baby breathed out.

Dr Ellen and others that followed discovered the reason behind one of the major causes of death in preterm babies. If babies are born too early they are unable to produce a lung surfactant. The surface water tension in lungs is so high that the alveoli (air sacs) collapse. The treatment is to supply artificial surfactant to the baby. The surfactant breaks the surface tension, allowing the baby to take in a breath of air. This will occur until the baby is able to make its own surfactant, typically in a few days' time. With enough warning, the mother can also be given steroids that stimulate the baby's lungs to start producing the surfactant. It took until 1980 before the first human babies started to benefit from the use of surfactants.

The use of scientific knowledge may have beneficial and/or harmful and/or unintended consequences (ACSCH013 and ACSCH052)

LIQUID BREATHING

Watch this video to see a mouse breathe in liquid, and some future applications.

Around 1989, another treatment started to be trialled – the use of 'liquid breathing'. When babies are born very prematurely, doctors fill their lungs with perfluorocarbons (PFCs). Perfluorocarbons are hydrocarbons in which the hydrogen atoms have all been replaced with fluorine atoms. Perflubron ($C_8F_{17}Br$) (Figure 3.10) is also included in this class of compounds, even though it contains bromine. Perflubron is an excellent medium for carrying respiratory gases. These liquids can carry large amounts of oxygen and carbon dioxide and reduce the surface tension, allowing the baby to breathe.

It was a strange birth for this range of molecules. During the development of the atom bomb (The Manhattan Project 1942–5), chemists tried to develop an inert (unreactive) solvent for the highly reactive uranium chemicals. What they produced was perfluorocarbons. These remarkable liquids are clear, odourless and very dense and can carry 20 times as much oxygen as other fluids. Perfluorocarbons are now also being used as blood substitutes.

◀ **Figure 3.9** This mouse is breathing in a liquid-filled jar, demonstrating liquid breathing.

ACTIVITY 3.3

PERFLUOROCARBONS

The structure of a perfluorocarbon molecule gives remarkable functionality to the material.

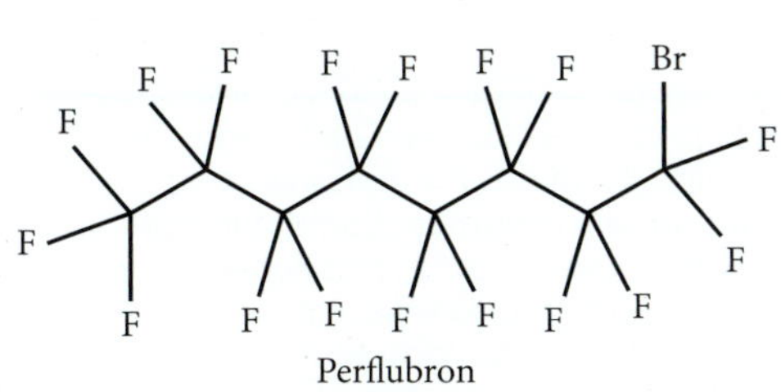

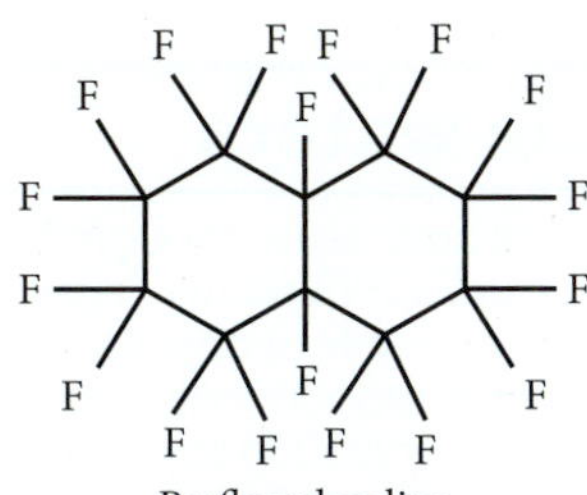

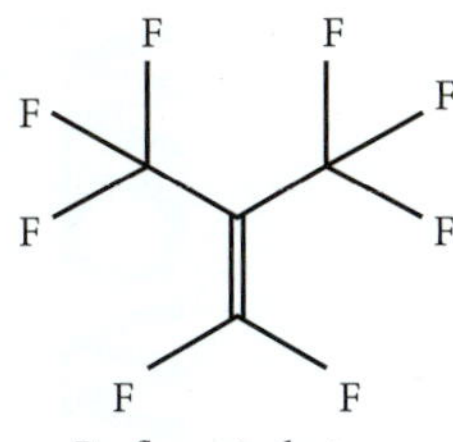

▲ **Figure 3.10** A range of perfluorocarbon molecules

a What type of bonding would you expect between the molecules of a perfluorocarbon?

b Explain why perfluorocarbons are essentially insoluble in water and alcohols and only sparingly soluble in some lipids.

c Would you expect the surface tension of perfluorocarbons to be high or low? Explain your reasoning.

Extension

The solubility of gases in the liquid perfluorocarbon is due to the presence of intermolecular cavities. Explain how this is different from and similar to how gases are dissolved in water.

The universal solvent

Three-quarters of the planet is covered with water, which exists as a liquid between 0°C and 100°C. Water dissolves both organic materials, such as enzymes and amino acids, and inorganic materials, such as salts and minerals. Generally, water has a low viscosity, remaining a liquid rather than becoming a thick slimy soup. Oxygen has to be dissolved in water for our cells to access it. Water plays a key role in the body's hydrolysis reactions in which it is split into H^+ and OH^- ions, which you will probably recognise as the ions involved in acid and base reactions. When carbon dioxide reacts with water, it forms a weak acid, releasing H^+ ions, which play a role in maintaining the cell's environment. Later you will see how rain that contains H^+ ions can dissolve rocks. A substance's solubility in water is determined by intermolecular forces.

In Chemistry Chapter 7, you will learn more about the role of H^+ and OH^- in acid and bases. Acid and bases have key roles in many of the biochemical reactions.

Water's unique ability to act as a universal solvent can explain the chemical processes on Earth and the biological processes in our bodies.

When a substance mixes in water, it may no longer be visible. But it doesn't disappear. The substance may have either reacted with or dissolved in the water. Substances dissolving in water affect the solvent by changing its density, and melting and boiling points.

Soluble salts and soluble organic substances

To learn more about why materials may or may not dissolve in water and these reactions, refer to Chemistry section 6.6 on page 275.

The water in the ocean has a lot more material or solute dissolved in it than fresh water. You can tell this by the taste or because you can float easier in the ocean.

Seawater typically contains six major ions: sodium (Na^+), chloride (Cl^-), sulfate (SO_4^{2-}), magnesium (Mg^{2+}), calcium (Ca^{2+}) and potassium (K^+). River water has lower levels of all salts and a higher proportion of calcium (Ca^{2+}) and hydrogen carbonate (HCO_3^-). Obviously you can't simply get seawater from evaporating river water. The salt levels present are the result of weathering of mostly igneous rocks, erosion, land run-off and the precipitation of different ions onto the floor of the ocean.

Every naturally occurring element is also present in the ocean. Unfortunately, because of pollution, some are at a higher than desired level.

The main ions in your body fluids are sodium (Na^+), potassium (K^+), chloride (Cl^-), calcium (Ca^{2+}), magnesium (Mg^{2+}), bicarbonate (HCO_3^-), phosphate (PO_4^{2-}) and sulfate (SO_4^{2-}). This is very similar to seawater.

WOW

Miracle waters

Mineral waters have been thought to have curative properties for over 2000 years. The common minerals in mineral water are Mg^{2+}, Ca^{2+} and K^+. Mineral waters can also include small amounts of arsenic, cadmium, chromium and cyanide. Some mineral waters even have abnormally high levels of radioactivity owing to the presence of radioactive elements such as uranium, radon and thorium. At one time, rather than a cause for alarm, this radioactivity was considered a bonus. And promoted!

In the 1800s, J. J. Schweppe developed artificial mineral waters. They were very popular due to the poor quality of tap water available. Manufactured mineral waters (without radioactivity) are no longer sold as mineral waters. Legally, mineral waters are defined as waters from an underground source.

The use of scientific knowledge is influenced by social, economic, cultural and ethical considerations (ACSCH012 and ACSCH051) AC

Contradictory gases

In Australia, the soft-drink manufacturing industry is worth over $4 billion. Cans or bottles of Coca-Cola® and Pepsi® are probably found in most homes. These businesses were started by using the research undertaken by Joseph Priestley (1733–1804). Priestley is credited with discovering oxygen and his interest in gases was further advanced in 1767 when he moved next door to a brewery. He started a series of experiments with the brewery gas called 'fixed air'.

Fixed air is carbon dioxide. Being non-polar, carbon dioxide is insoluble in water. Even though carbon dioxide is classed as insoluble, extremely low amounts can be present in solution. Carbon dioxide in water slowly acquires a shell of water molecules. This allows reactions to occur that result in the water becoming acidic. So despite the concentration of carbon dioxide being low, there is a detectable change in the water at the end, as shown by the reactions:

$$CO_2(g) \rightleftharpoons CO_2(aq) \quad CO_2(aq) + H_2O \rightleftharpoons H_2CO_3(aq) \quad H_2CO_3(aq) \rightleftharpoons H^+(aq) + HCO_3^-(aq)$$

This slightly acidic taste is the pleasant taste that Priestley enjoyed. Soda water was invented. He shared his knowledge on how to make soda water. More flavourings were added to make the soft drinks we know today. Manufacturers use high pressures and very low temperatures to dissolve carbon dioxide at about 150 times the concentration in tap water.

To learn about how solubility is expressed, how to calculate the concentration of the solute and how to convert between different units (ppm and % v/v, % w/w and % v/w), refer to Chemistry section 6.7 on page 279.

The use of scientific knowledge may have beneficial and/or harmful and/or unintended consequences (ACSCH013 and ACSCH052) AC

ACTIVITY 3.4

BREATHING AND OXYGEN

Land-based animals breathe oxygen in air, which then dissolves in the water that lines the lungs. Most aquatic organisms must obtain oxygen directly from the water.

Polar substances readily dissolve in water to give a relatively high concentration of solute. Concentrations of about 50 g in 100 g of water are not unusual. Non-polar substances do not readily dissolve; their solubility is usually measured in parts per million (ppm).

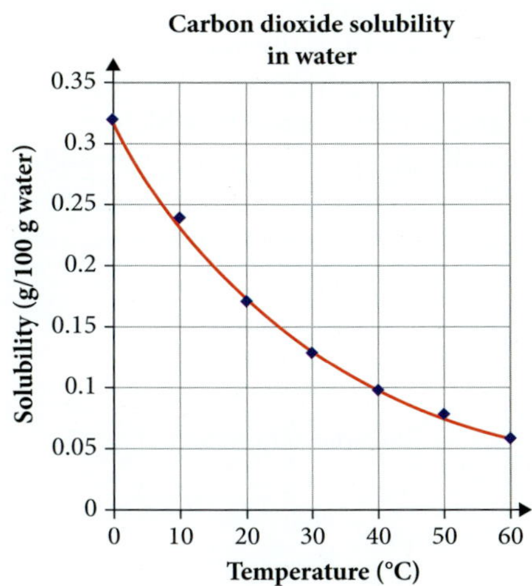

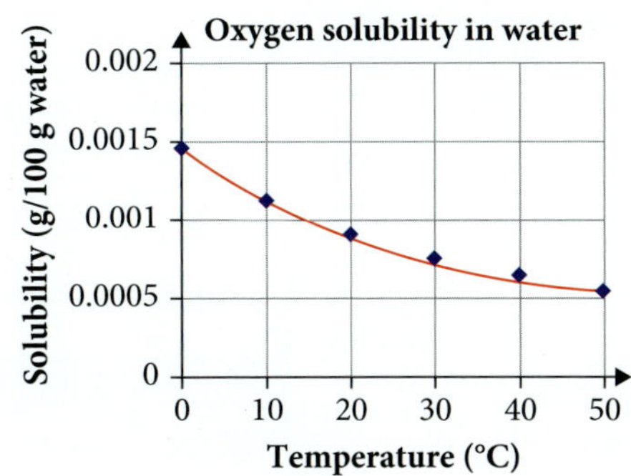

◀ **Figure 3.11** Solubility of carbon dioxide and oxygen in water

1 Examine Figure 3.11.

a At what temperature is most oxygen dissolved in water?

b Compare the relative amounts of oxygen and carbon dioxide that can dissolve in water at a certain temperature. Propose a reason why this occurs.

c What is the relationship between temperature and oxygen solubility?

d Determine the amount of oxygen and carbon dioxide that dissolves in water at 37°C.

2 Colder regions of the ocean support more life than warmer regions. What do these facts indicate about the consequences of global warming heating the oceans?

To learn about the unit moles per litre that chemists use as the unit of concentration, refer to Chemistry section 7.2 on page 312.

Optimising performance for endurance sports

Water acts as a solvent in our bodies. Our body is about 70% water and each cell contains a lot of ions in solution. An important task of chemists is determining what substances are present in water and aqueous solutions and their concentrations. This is because the concentration of a substance is important.

The ions formed from the minerals sodium and potassium are important for transmitting electrical impulses along our nerves. Together with calcium ions, sodium and potassium ions are important for muscle contractions. Their salts are called **electrolytes** and a solution made by dissolving salts is called an electrolytic solution. Salt ions disperse when dissolved in water and the solution is then able to conduct electricity.

When you sweat, you lose these electrolytes, particularly potassium and sodium ions. It is important to maintain salt and water levels during exercise and sport. But it becomes more important when hot and when you are exercising for a long time. If you sweat and lose even 2% of your body weight, your blood volume decreases, which will cause your heart to pump harder. This reduces an athlete's performance.

When exercising for a long time, pure water is not enough to replace electrolytes. Sport drinks are designed to cater for this need. If they are labelled as **isotonic**, they claim to have the same salt, sugar or electrolyte level as the body.

Some drinks are **hypotonic** because the salt levels are lower than the body's salt levels. But if you consume these sport drinks when you are not exercising and are fully hydrated, you may be drinking a **hypertonic** solution. A hypertonic solution is one in which the salt level is higher than the surrounding fluid. This may actually make it harder for your body to absorb the water.

The difference between hypotonic, isotonic and hypertonic solutions is the concentration of solutes. You can dilute a hypertonic solution to make an isotonic solution.

▼ Figure 3.12
Isotonic, hypotonic and hypertonic solutions

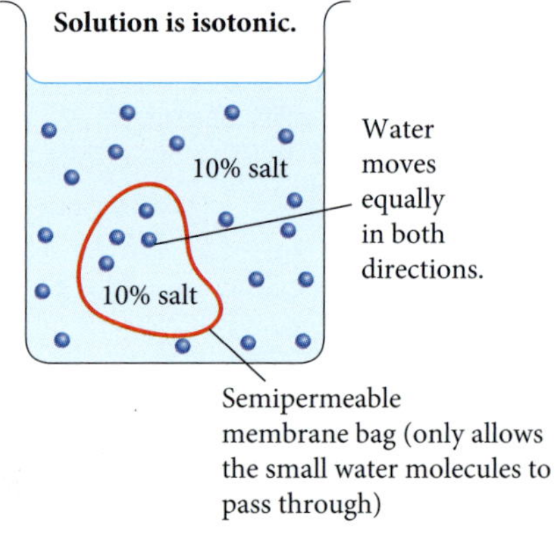

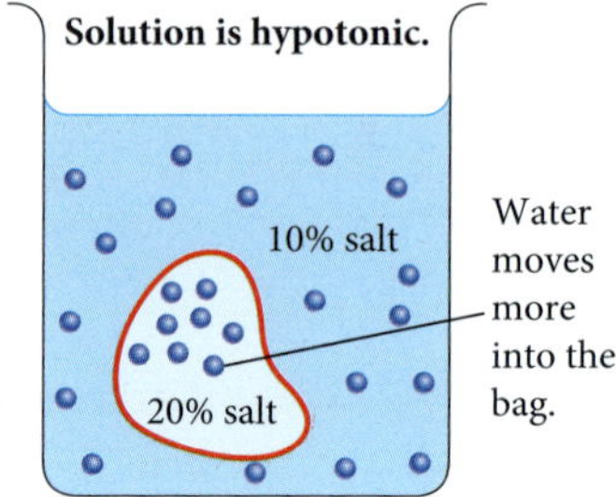

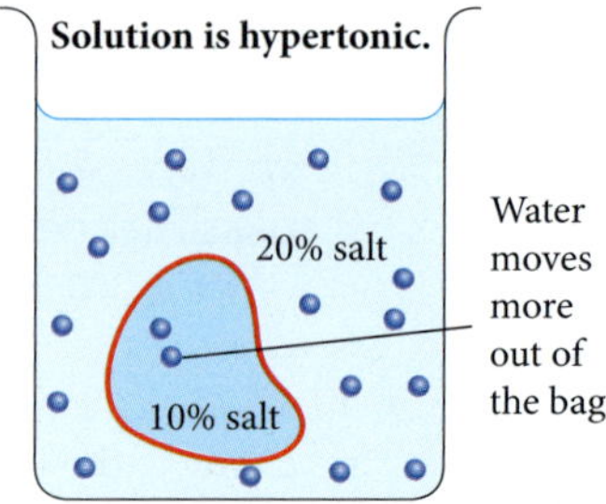

Photo by Brice Oxenham

Figure 3.13
During endurance events it is important to maintain hydration and sugar levels. Cycling events such as the Murray to Moyne, in which teams ride 520 km over 2 days, are a challenge for most people. This cyclist, Samantha, has the added complication of having to also balance her own blood sugar levels due to her type-1 diabetes.

WOW

Water can be fatal

Drinking too much water too quickly can cause water intoxication and is potentially fatal. With too much water, the electrolytes in the body are diluted. Fatalities are rare and have occurred during water drinking contests, when trying to rehydrate after excessive exercise or when attempting to counteract the effects of drugs such as Ecstasy. In 17th- and 18th-century France, forcing people to drink water was a legal method of torture handed down by the French courts.

Scientific knowledge can be used to develop and evaluate projected economic, social and environmental impacts and to design action for sustainability (ACSCH015 and ACSCH054) AC

Blood chemistry

Blood is a complex **suspension** of plasma (fluid component) and cells. It is more viscous than water due to the dissolved salts and the biomolecules and cells being transported through the body. Polar substances can directly dissolve into the plasma. Non-polar substances such as many hormones have complex carrier proteins to transport them along the blood vessels. Blood is responsible for the delivery of oxygen and nutrients to all parts of the body, temperature regulation and waste removal.

Another role of blood is to control the pH of the body despite all the chemicals and reactions occurring. Blood has to maintain a pH of 7.4. There are three chemical buffer systems for this: the carbonate/carbonic acid buffer, the phosphate buffer and the buffering of plasma proteins. The carbonate (CO_3^{2-})/carbonic acid (H_2CO_3) buffer is the most important.

To learn about acids, bases, pH and their reactions, refer to Chemistry section 7.4 on page 320.

Blood monitoring

One vital nutrient transported by blood is sugar. The main blood sugar is glucose. Sugar levels are typically controlled between 79 and 110 mg per 100 mL of blood. In diabetes, the sugar levels in blood are not fully controlled by the body. People with diabetes have to inject a hormone (insulin) to decrease sugar levels. It can be fatal to have too low a level of sugar (hypoglycaemia). In the short term, low sugar levels are far more dangerous than high sugar levels (hyperglycaemia). The long-term health issues facing some people with diabetes are thought to be due to the high sugar levels. Blood becomes more viscous with excess sugar levels. This changes the microcirculation, particularly near the kidneys, liver and heart.

Diabetes Australia runs a national research program administered through the Diabetes Australia Research Trust (DART). Research is being undertaken in many areas from investigating different types of insulin to better monitoring and injection methods. Gone are the days when one injection of insulin was given and the person had to limit what they could eat. Now insulin can be long acting, short acting or designed for use during pregnancy. Insulin can be taken as a tablet, injected via a pen or continuously pumped in. This research is providing better health outcomes for people with diabetes.

There are many reasons why the level of particular chemicals in the blood is important. Using the known healthy range for particular substances, doctors can diagnose a health condition.

To learn more about how substances are detected by chromatography, refer to Chemistry section 6.9 on page 289.

ACTIVITY 3.5

BLUE BLOOD

Oxygen-rich blood is red. Blue blood indicates deoxygenated blood. When deprived of oxygen, cells will die. Permanent damage may occur within 3 minutes.

Use Figure 3.11 (page 71) to determine the amount of oxygen that can be dissolved at 37°C. This level is too low to sustain life. Unlike carbon dioxide, oxygen does not react with water, so it is transported around the body by a molecule called haemoglobin.

Conduct an Internet search to determine how haemoglobin works and then complete one of the following tasks.

- Compare haemoglobin to other respiratory pigments, such as myoglobin or foetal haemoglobin.
- Compare haemoglobin to the range of respiratory pigments used by some invertebrates, the true 'blue bloods'.

Copper – the Jekyll and Hyde element

Copper is an essential trace element. The recommended minimum level dietary intake for infants is around 200 μg/day. Copper is toxic, so maximum levels in drinking water are around 2 ppm. Determining the copper levels in blood can indicate if too much or too little copper is present for good health.

In Chapter 4, you learnt about atomic absorption spectroscopy (AAS). The first few atomic absorption spectrometers built by Alan Walsh in 1954 were used to solve medical treatment problems. The value of this technique can be seen when sick infants are suspected of having copper poisoning. Infants do not have much blood, so a technique that only requires a small amount of blood for analysis is very useful. AAS can confirm, with minimal harm, high copper levels in an infant's blood.

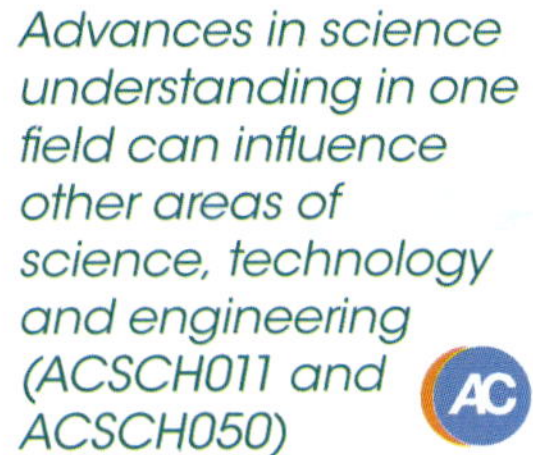

Hot water systems are typically made of copper. Copper dissolves more readily in hot water than in cold water. If baby formula is made up with hot water, it can contain high enough levels of copper ions to cause copper poisoning and even death. The infant's small size and their reliance on baby formula as their sole source of nutrients makes this an important issue.

You should not use hot water for cooking or drinking. Always let the cold water run for 30–60 seconds, particularly if you are preparing baby formula.

What is in your water?

To revise atomic absorption spectroscopy, refer to Chemistry section 1.9 on page 127.

Australia is the driest continent, so the supply of water is an important issue. In a drought region, sources of water such as bore water may be investigated. In other areas after a flood, the quality of the water can be cause for concern. We need to have a clean pure source of water for our health, but not all water sources are pure. Scientists use different methods of analysing water depending on what they are looking for; for example, salt, organic compounds or gases. Some of these methods are listed in Table 3.3.

Table 3.3 The different solutes present in water require different techniques to detect them

Solute	Technique	Key points	Relevant chapter
Salts	Range of precipitation reactions	Need to know solubility rules and common precipitates	Chemistry Chapter 7
	AAS	Metal ions detected to ppm	Chemistry Chapter 1
Organic compounds	Chromatography	Relies on the relative solubility of the different substances	Chemistry Chapter 6
Gases	Gas chromatography	Relies on the relative solubility of the different substances	Chemistry Chapter 6

Heavy metals

One of the main concerns with water supplies is the presence of heavy metals. These are metal or metalloid elements that are at least five times denser than water. The most common heavy metals of concern are arsenic, cadmium, iron, lead and mercury.

Heavy metals can enter the body in food, water, or air or by absorption through the skin. Some heavy metals such as copper, iron and zinc are essential to life at low levels and can be found in many food items. Heavy metals become toxic when they accumulate in the soft tissues of the body. Acute poisoning normally occurs after inhalation. The presence of heavy metals in water can slowly accumulate over time to dangerous levels, causing major health issues.

Science is a global enterprise that relies on clear communication, international conventions, peer review and reproducibility (ACSCH009 and ACSCH048) AC

To learn about how precipitation reactions can be used to detect metal ions in solution, refer to Chemistry section 7.1 on page 304.

To learn how dilutions affect concentration, refer to Chemistry section 7.3 on page 317.

Ouch ouch disease

The first documented case of mass cadmium poisoning occurred in Japan. Itai-itai (ouch ouch) disease was first diagnosed in the 1950s although it had been noticed in 1912. This disease is named after the painful cry of the patients. Cadmium is of no benefit to the body. Even at low levels of exposure, cadmium will accumulate in and poison soft tissues such as the kidneys and liver. At higher levels, the kidneys start to fail and cannot remove acid from the blood. The side effect of this is the development of gout, a painful condition. High levels of cadmium cause bones to soften and fractures become more common.

In Japan, the exposure to cadmium was due to mining along the river that was the source of water for the rice fields. Although mining for other metals had occurred along the river for centuries, increased demand during World Wars I and II caused the outbreaks. The worst symptoms occurred in mothers who had borne many children. Cadmium enters the body in the same way as zinc. In pregnancy, there is an increase demand for zinc. The body absorbs more zinc (which means correspondingly higher levels of cadmium).

ACTIVITY 3.6

WHAT TO DO?

You work for the EPA (Environmental Protection Authority) and have just been advised that water contaminated with cadmium has been released into a local lake.

A sample of the industrial wastewater contains 3000 $\mu g\ g^{-1}$ cadmium ions and every 100 L of this waste is diluted with 100 L of pure water before it is released.

- **a** What is the concentration of cadmium ions in the new, diluted, solution?
- **b** A total of 500 L of this dilute solution was released into a lake that contains 1×10^6 L of water. If there was no cadmium in the lake initially, what approximate concentration would it have after the solution was added?
- **c** The cadmium concentration in the unpolluted water is generally less than 1 $\mu g\ L^{-1}$. How does this compare to the lake after diluted wastewater containing cadmium is added?
- **d** The World Health Organization has reported the cadmium uptake by humans in a smoker's house to be 2–4 μg per day. How many glasses (300 mL) of the lake water would you need to drink per day to reach this amount?
- **e** What would your advice be to the town council, the manager of the site? Would you drink the water and would it be any different if you were pregnant?

Science is a global enterprise that relies on clear communication, international conventions, peer review and reproducibility (ACSCH009 and ACSCH048) AC

Fluoridated water

Fluoride, in the form of sodium fluoride (NaF), has been added to the water of most Australian states for decades to reduce tooth decay. Fluoride has been added at a level of 1 ppm. This is similar to the level in seawater, while lake water has typical levels around 0.5 ppm. At these low concentrations, fluoride significantly reduces the number of dental cavities. However, when the natural level in the water is already high, dental fluorosis (discoloration or staining of teeth) can occur. With excessive exposure to fluoride, a more severe and painful disease of the bones and joints, skeletal fluorosis, may occur. Excessive fluoride levels are generally due to industrial pollution, very high levels in local water or volcanic activity. It can occur when the water supply originates from certain rocks in parts of India, Africa and the Middle East. In some areas, most of the water supplies contain above the recommended levels, sometimes as high as 45 mg L^{-1}. Some bores in Kenya have values between 1640 and 2880 mg L^{-1}.

While some people have expressed concern about the fluoridation of their water, others are concerned about the increased consumption of bottled water as it does not have the recommended levels of fluoride.

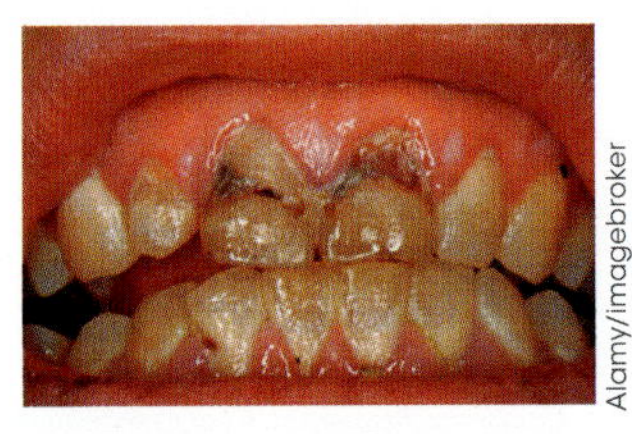
Alamy/imagebroker

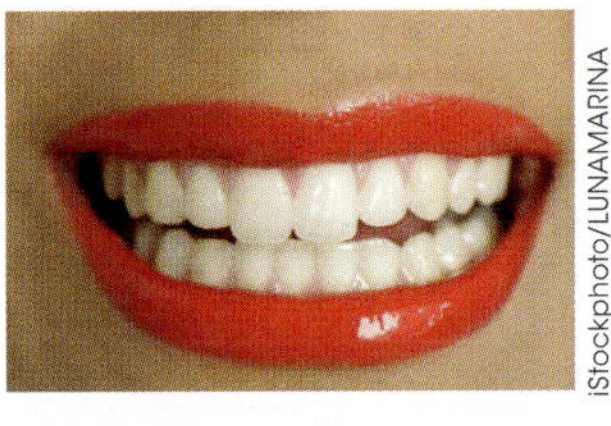
iStockphoto/LUNAMARINA

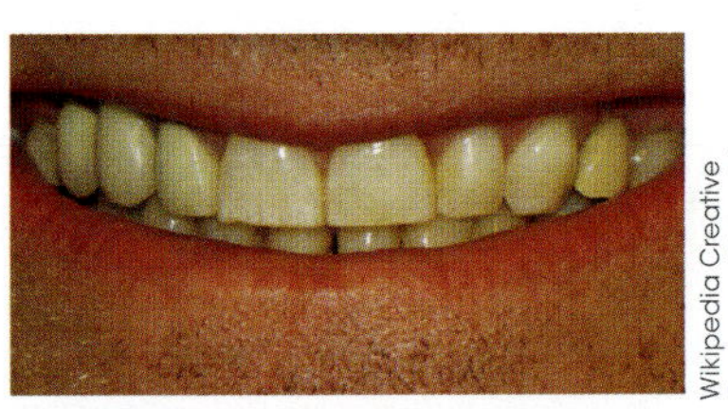
Wikipedia Creative Commons/Dozenist

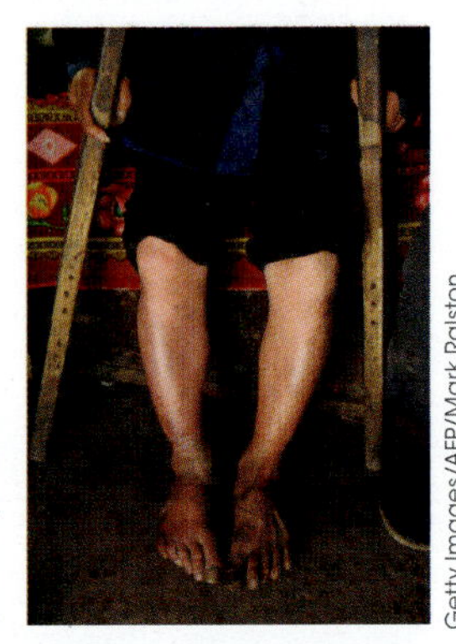
Getty Images/AFP/Mark Ralston

<1 ppm	1 ppm	Mild dental fluorosis	Skeletal fluorosis
Dental caries	Adult 1.5–4 mg/day	>3 mg/day (for a child) Exposure during teeth development when around 1–4 years old	>20 mg/day (over years)

▲ **Figure 3.14**
The relevance of fluoride in water to health was first noticed in native-born residents of Colorado Springs, Colorado, USA. Cosmetic stains on teeth were found to be caused by high levels of fluoride in the local water supply. People with these stains also had an unusually high resistance to dental cavities.

ACTIVITY 3.7

BABY TEETH AND GOOD HABITS

Children's toothpaste has a lower amount of fluoride ions than adult toothpaste.

a Explain why fluoride is added to most toothpastes.

b Explain why children's toothpaste has lower levels of fluoride ions.

c What other measure would you recommend for children?

Managing water quality

Australia has guidelines to ensure the quality of the drinking water. These are determined in collaboration with National Health and Medical Research Council and the Natural Resource Management Ministerial Council. These guidelines help manage the water supplies and protect the safety of the community.

Scientists are constantly developing more accurate and reliable measurements of the substances in the waterways. Metals are detected at very low levels by AAS. Sometimes, chemists can use knowledge about precipitation reactions to work out what is present.

Detecting organics in water

Coal and coal seam gas (CSG) is natural gas (methane) and is found in coal deposits. CSG is formed from plant matter that has been under extremely high pressure for millions of years. CSG is used in the same way as any other form of natural gas, such as for cooking or generating electricity. CSG is plentiful in Australia.

CSG is extracted by drilling a well into a coal seam and pumping in large volumes of water and sand at high pressure. This fractures the coal seam and releases the gas. The sand deposits in the fractures to prevent them closing when pumping pressure ceases.

CSG can have environmental and economic benefits in a low-carbon future. Not everyone agrees that this industry should develop. The debate over CSG has united farmers and environmentalists. Environmentalists are concerned about damage to the landscape while rural community members may also see a threat to their way of life.

One concern is the amount of water that is used to extract the methane. There is a risk that the contamination of aquifers may occur because of:

- chemicals injected through the fracking process (in which rocks are fractured by injecting chemicals under high pressure)
- chemicals released from the coal seam
- removed water being replaced by water flowing in from other aquifers into the coal seam.

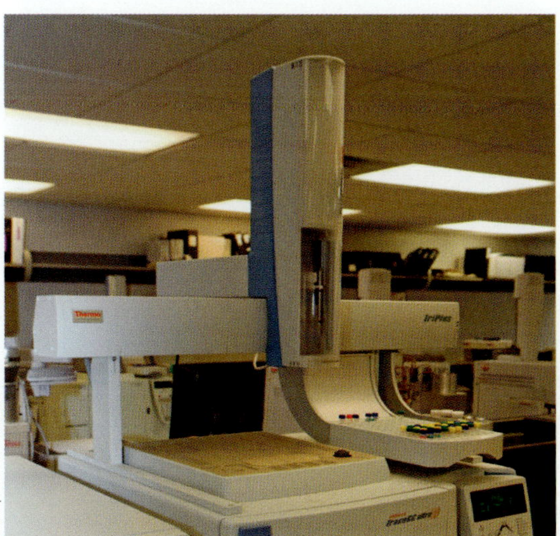

Alamy/Mikael Karlsson

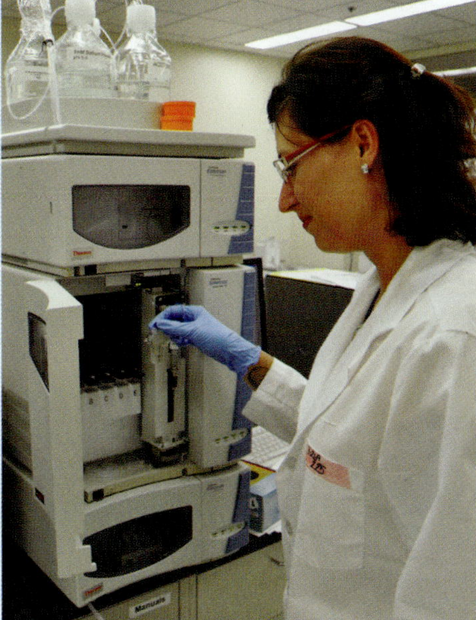

Science Source

▲ **Figure 3.15**
Chromatography instruments: gas chromatography and high-performance liquid chromatography are analytical techniques used to separate organic substances.

Science is a global enterprise that relies on clear communication, international conventions, peer review and reproducibility (ACSCH009 and ACSCH048)

DRINKING WATER GUIDELINES

Visit this website to read the latest drinking water guidelines in Australia.

To learn about how to identify unknown ions in solution, refer to Chemistry section 7.7 on page 332.

COAL SEAM GAS

Visit this website to learn more about the issues associated with coal seam gas.

Scientific knowledge can be used to develop and evaluate projected economic, social and environmental impacts and to design action for sustainability (ACSCH015 and ACSCH054)

To learn about chromatography techniques, refer to Chemistry section 6.9 on page 289.

Sensitive techniques such as gas chromatography and high-performance liquid chromatography can be used to detect very small amounts of possible pollutants. These techniques rely on surface interactions to separate substances on the basis of their relative differences in attraction between a solid and a mobile phase.

SAFE WATER SUPPLIES

Water treatments are vital for securing safe water supplies. To learn how it is done, take one of Seqwater's (Queensland) virtual tours.

Water in crisis?

Australia is a dry continent with a growing population. This makes water availability an important sustainability issue. Rivers, lakes and ocean may be vastly different but the threats to them are remarkably similar. Some water supply issues are due to the European settlers clearing the land and not understanding Australia's unique environment. Climate change is adding another dimension to the problem. The availability of reliable fresh supplies of water is something that governments have to consider. Many cities around Australia have water treatment plants that remove the salts from seawater or highly saline water to produce drinking water. In Australia, a variety of techniques are used to produce the energy required for these processes, such as wind farms.

ACTIVITY 3.8

SALINITY

1 Salinity is a large problem in particular regions of Australia. The Murray River runs along the border of Victoria and New South Wales. It is the source of Adelaide's water.
   a Find out two main sources of the salt in the Murray River.
   b What could cause the variation in salinity observed in Figure 3.16?

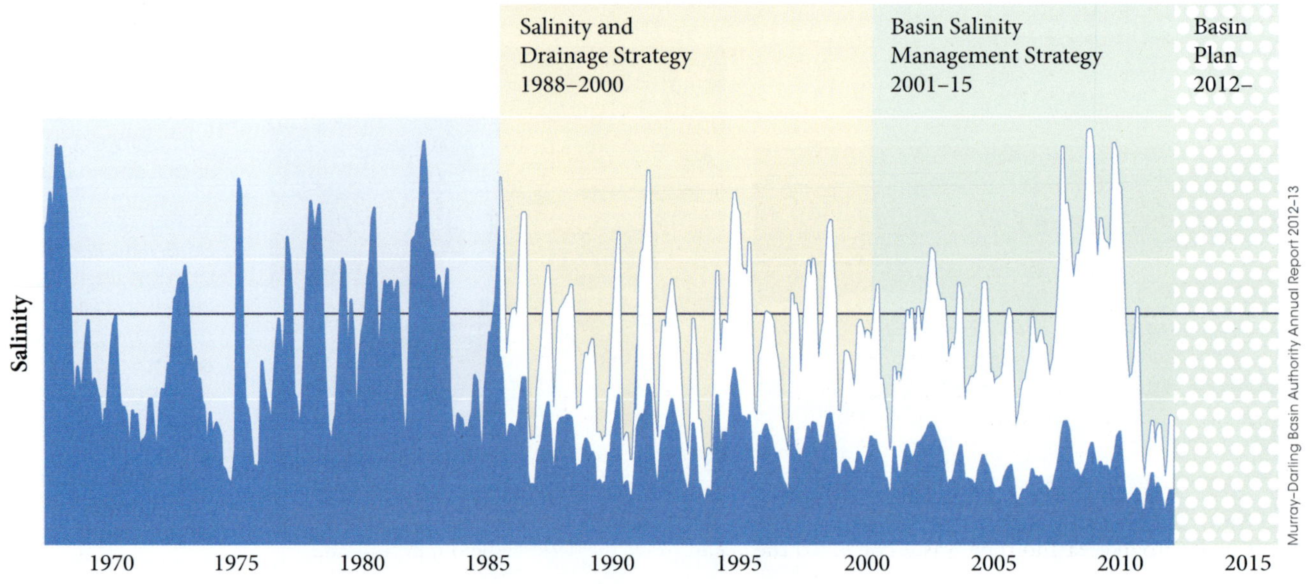

▲ **Figure 3.16**
The effects of salinity management in the Murray–Darling Basin: average salinity levels in the Murray River at Morgan, South Australia, 1970–2010

c Why is it important to examine water systems over time?

d Name some of the interventions that have been implemented to decrease salinity levels.

e What would be some consequences of not intervening?

f Adelaide is down river from Morgan. Explain how the graph could have changed if the testing was done in Adelaide.

g What other options could be available to the cities to guarantee water supply?

h Besides the supply of water to towns and cities, what problems arise from high salinity in the rivers?

2 Table 3.5 lists typical calcium carbonate levels of various waters around Australia. Calcium carbonate results in **hard water**, which is water that soap won't lather in and leaves lime scale in kettles.

a Which city has the softest water?

b Which has the hardest water?

c Suggest why the composition of water varies from city to city.

Table 3.5 Calcium carbonate levels in potable (drinkable) water in major cities in Australia

Major city	$CaCO_3$ concentration (ppm)
Adelaide	134–148
Brisbane	100
Canberra	40
Darwin	31
Hobart	5.8–34.4
Melbourne	10–26
Perth	29–226
Sydney	39.4–60.1

Extension

In Singapore, drinking water is produced from recycled sewage. Research to find out why we don't recycle water in Australia.

Great Barrier Reef

The Great Barrier Reef is the world's largest reef and it is visible from outer space (Figure 3.17). This area of more than 900 islands stretches for more than 2600 km. It has been on the UNESCO World Heritage List since 1981. The reef has a very high biodiversity, being home to the dugong, six species of turtles, more than 300 species of hard, reef-building corals, 400 species of sponges, 240 species of birds and more than 1500 species of fish.

Scientific knowledge can be used to develop and evaluate projected economic, social and environmental impacts and to design action for sustainability (ACSCH015 and ACSCH054) AC

Major developments along the coast of Queensland, such as farming, aquaculture and urban development, have increased the run-off to the reef. Stormwater is water that runs off after rain falls in an urban area and then washes down to the ocean. It carries with it the oil, petrol, dog droppings, litter and sediments that were on the streets. Rain in rural areas may include livestock waste, fertilisers and pesticides. Approximately 25% of the land drains directly into the reef. This run off affects the **turbidity** of the water. Turbidity is a measure of how much the suspended particles reduce the amount of light that penetrates the water. Increased turbidity reduces photosynthesis and blocks fish's gills.

Run-off from the land consists of soluble fertiliser ions such as nitrate (NO_3^-) and phosphate (PO_4^{3-}). These ions are used by the single-celled, plant-like organisms called **phytoplankton**;

Getty Images/Universal Images Group

Figure 3.17 ▶ The Great Barrier Reef is so large that it is visible from space.

an increased supply of nitrates and phosphates can cause algae blooms or **eutrophication**. Algal blooms block the light to the organisms such as corals or the sea grass beds in the Hinchinbrook Channel near Townsville. This reduces the food source for some vulnerable species such as the dugong and green turtles.

Run-off increases the amount of organic matter and heavy metals entering the waters. Microbes will feed on this matter, consuming oxygen in the water. There is an increase in the biochemical oxygen demand (BOD) in the water, which deprives other aquatic organisms such as fish of oxygen. Mining in the area has led to increased levels of cadmium, mercury, copper, lead, nickel, zinc and arsenic. Near the Great Barrier Reef catchment area, coal and shale oil are mined north of Gladstone, silica is mined near Cape Flattery and magnesia is mined north of Rockhampton. With increased mining, there is always a demand for bigger ports to bring the ships in and out of the area.

Ocean acidification due to high levels of carbon dioxide is a major threat. Higher acidity levels mean less carbonate is available for the coral and they are less able to build their skeletons and repair themselves when damaged. A recent study showed a 14% decrease in coral growth since 1990 due to increased levels of carbon dioxide. This was the most significant decrease in coral growth in the last 400 years.

The value of the Great Barrier Reef was estimated to be over $51 billion by the Oxford Economic Group in 2010. The agricultural production is estimated to be worth $2.3 billion per year, fishing $500 million, tourism $4.3 billion and recreation $500 million. What would you spend to save it?

CHAPTER SUMMARY

- Water has unique properties that are vital to the biological, chemical and physical processes on Earth. Water exists naturally in three states: solid (ice), liquid (oceans, lakes and rivers) and gas (water vapour).
- Strong hydrogen bonding in water accounts for all the properties that relate to temperature. A relatively large amount of energy is required to break the bonds between the molecules of water to change state from liquid to gas (high latent heat) or to increase the average temperature of the water (high specific heat).
- Evaporation of water has a cooling effect: bodies of water have less extreme temperature ranges than land; and water is liquid over a large temperature range. These phenomena enable organisms to survive on Earth.
- The cohesive nature of the water molecules explains surface tension. Understanding this allowed techniques such as chromatography to be developed and the use of surfactants in the medical, detergent and food industries.
- Ice floats because of the regular crystal structure adopted upon freezing. This protects the life under the ice in the lake from freezing and alters the density of ocean waters.
- Water is the universal solvent, partly because it is liquid from 0°C to 100°C. It can readily dissolve polar substances, such as salts and polar molecules. Salt levels change the temperature at which water freezes, allowing some cells to survive extreme temperatures. Salinity affects the density of the water, which, in turn, affects the convection currents of the ocean.
- Sometimes, there are detrimental effects if the concentration of solutes is too low or too high. This knowledge is used in medical diagnosis and in determining water quality. Techniques such as chromatography can determine the amounts of solute present.
- Understanding water chemistry and quality enables the environmental, social and economic impact of human activity to be developed and evaluated, such as how higher carbon dioxide levels affect the acidification of ocean ecosystems, which in turn affect industries such as fishing and ecotourism.

CHAPTER GLOSSARY

chromatography techniques used to separate components of aqueous, liquid or gaseous mixtures

cyroprotectant a chemical that is used to stop biological tissues from freezing

El Niño a warm ocean current that develops every few years along the coast of Ecuador and Peru

electrolyte a charged solute that allows the solution to conduct electricity

enzyme a protein molecule that catalyses a specific type of reaction by lowering the energy of activation

eutrophication a process in which an additional amount of nutrients leads to excessive growth of plants including microscopic plant-like phytoplankton

hard water water that contains high levels of Ca^{2+} and Mg^{2+} ions, which interferes with the action of soaps

hydrophilic 'water-loving', a particle with polar regions that bond to water

hydrophobic 'water-hating', a particle with mostly non-polar regions that do not bond with water

hypertonic concentration of solutes is higher than another solution

hypotonic concentration of solutes is lower than another solution

isotonic concentration of solutes is the same in both solutions

ocean acidification lowering of pH of the ocean due to the reaction of carbon dioxide with water molecules

phytoplankton microscopic plant-like organisms found at the surface of oceans, seas and lakes

solute the substance that is dissolved or the smaller component of a solution

solution a homogeneous mixture that is formed when a solute dissolves in a solvent

solvent the substance in which the solute dissolves or the greater part of a solution

supersaturated solution an unstable solution that has more solute than possible at that temperature; it can be formed from dissolving solute into the solution at a higher temperature and then allowing it to cool. The extra solute is still in solution but can readily be crystallised

surfactant a chemical that lowers the surface tension of a liquid

surface tension the force that arises from the attraction of the surface molecules to the bulk of the material

suspension a cloudy, heterogeneous mixture containing solid particles that will eventually settle out

turbidity cloudiness of water due to suspended microscopic particles. High turbidity reduces how far light penetrates into water

viscosity the ability of a fluid to resist flow; honey has high viscosity, water has low viscosity

water of crystallisation the water molecules that are bonded inside crystals; there is an exact ratio between the water molecules and either the salt or the polar molecule because of the number of ion–dipole or hydrogen bonds present between the two

CONTEXT 4 MAKING REACTIONS WORK FOR US

By the end of this chapter you will have covered the following material.

Science as a Human Endeavour

- Development of complex models and/or theories often requires a wide range of evidence from multiple individuals across disciplines (ACSCH010 and ACSCH049)
- The use of scientific knowledge is influenced by social, economic, cultural and ethical considerations (ACSCH012 and ACSCH051)
- The use of scientific knowledge may have beneficial and/or harmful and/or unintended consequences (ACSCH013 and ACSCH052)
- Scientific knowledge can enable scientists to offer valid explanations and make reliable predictions (ACSCH014 and ACSCH053)

AC

Contextual story – Making reactions work for us

- The rates of reactions can be altered.
- The rates of reactions in our bodies depend on concentration, temperature, surface area and catalysts.
- We can control the rates of reactions in and around our home and in industry.

Chemical understanding

- Reaction rates
- The collision theory
- The factors that affect the rate of a reaction:
  - Concentration
  - Pressure
  - Temperature
  - Surface area
  - Catalyst
- Kinetic theory of gases
- Calculations involving gases

Investigations

Context 4

- Activities 4.1, 4.2, 4.3, 4.4, 4.5 and 4.6
- Experiments 4.1, 4.2 and 4.3
- Investigation 4.1

Chapter 9

- Activities 9.1, 9.2, 9.3, 9.4, 9.5, 9.6, 9.7, 9.8, 9.9 and 9.10
- Experiments 9.1, 9.2, 9.3, 9.4, 9.5 and 9.6
- Investigation 9.1

Science as a Human Endeavour (SHE)

- Development of complex models and/or theories often requires a wide range of evidence from multiple individuals and across disciplines (ACSCH010 and ACSCH049)
- The use of scientific knowledge is influenced by social, economic, cultural and ethical considerations (ACSCH012 and ACSCH051)
- The use of scientific knowledge may have beneficial and/or harmful and/or unintended consequences (ACSCH013 and ACSCH052)
- Scientific knowledge can enable scientists to offer valid explanations and make reliable predictions (ACSCH014 and ACSCH053)

Science Understanding (SU)

- The behaviour of gases, including the qualitative relationships between pressure, temperature and volume, can be explained using kinetic theory (ACSCH060)
- Varying the conditions present during chemical reactions can affect the rate of the reaction and in some cases the identity of the products (ACSCH068)
- The rate of chemical reactions can be quantified by measuring the rate of formation of products or the depletion of reactants (ACSCH069)
- Collision theory can be used to explain and predict the effect of concentration, temperature, pressure and surface area on the rate of chemical reactions by considering the structure of the reactants and the energy of particles (ACSCH070)
- The activation energy is the minimum energy required for a chemical reaction to occur and is related to the strength of the existing chemical bonds; the magnitude of the activation energy influences the rate of a chemical reaction (ACSCH071)
- Energy profile diagrams can be used to represent the enthalpy changes and activation energy associated with a chemical reaction (ACSCH072)
- Catalysts, including enzymes and metal nanoparticles, affect the rate of certain reactions by providing an alternative reaction pathway with a reduced activation energy, hence increasing the proportion of collisions that lead to a chemical change (ACSCH073)

Science Inquiry Skills (SIS)

- Identify, research, construct and refine questions for investigation; propose hypotheses; and predict possible outcomes (ACSCH040)
- Design investigations, including the procedure/s to be followed, the materials required, and the type and amount of primary and/or secondary data to be collected; conduct risk assessments; and consider research ethics (ACSCH002 and ACSCH041)
- Conduct investigations, including measuring pH and the rate of formation of products, identifying the products of reactions and testing solubilities, safely, competently and methodically for the collection of valid and reliable data (ACSCH042)
- Represent data in meaningful and useful ways, including using appropriate graphic representations and correct units and symbols; organise and process data to identify trends, patterns and relationships; identify sources of random and systematic error, identify anamolous data and estimate their effect on measurement results; and select, synthesise and use evidence to make and justify conclusions (ACSCH043)
- Select, construct and use appropriate representations, including physical and graphical models of molecules, energy profile diagrams, electron dot diagrams, ionic formulae, chemical formulae, chemical equations and phase descriptors for chemical species to communicate conceptual understanding, solve problems and make predictions (ACSCH045)
- Communicate to specific audiences and for specific purposes using language, nomenclature, genres and modes, including scientific reports (ACSCH008 and ACSCH047)

Reactions

Imagine if you were in a car crash and the air bag inflated too slowly. Or if the metal frame of a building rusted as soon as it was erected. Or if the food we eat took weeks to digest. Many reactions affect our everyday lives; however, it is not only what reactions happen, but how quickly they happen that is important. As our understanding of chemical reactions has increased, we have been able to understand why reactions occur at the rate that they do. This has allowed chemists to manipulate reactions so they occur at the rate that we need.

Corbis/Yva Momatiuk & John Eastcott/Minden Pictures

▲ Figure 4.1
Rust has destroyed this building. Rusting is a slow reaction.

Reactions in our bodies

The human body is an amazing system that relies on many chemical reactions to make it work efficiently. These reactions need to occur at the correct **rate** to keep us healthy. Otherwise, many diseases can occur.

WOW

Maple syrup urine?

One in 180 000 people lack the enzyme that breaks down branched-chain amino acids, including leucine, isoleucine and valine. This leads to a build-up of these amino acids and their toxic by-products. People with this genetic disease produce a sweet-smelling urine, similar to the smell of maple syrup, hence the name Maple Syrup Urine disease. Affected individuals need to eat a modified diet that is reduced in these amino acids to prevent loss of appetite, vomiting, lethargy, seizures, coma and possible death.

To learn about how reaction rates are measured, refer to Chemistry section 9.1 on page 368.

To learn about collision theory, refer to Chemistry section 9.2 on page 374.

COLLISION THEORY

View this animation and change the conditions to see what happens in a collision where there is not enough energy, the correct energy and orientations, and the correct energy but the wrong orientation.

Biochemistry

The study of chemical processes in living organisms is known as **biochemistry**. This is a large area combining chemistry and biology. It involves the study of chemicals in plants and animals and the reactions that they are involved in. This ever-expanding field of science links genetics, metabolism, hormones, transport across cell membranes, immunity and many diseases.

Biochemistry includes all aspects of the chemical reaction – the reactants, the products, the process and the rate. All these factors are important in understanding the way our bodies work. This chapter focuses on the rate of the reactions; however, rate is related to these other factors.

To understand why some reactions occur quickly, and others slowly, we must understand what happens during a chemical reaction. **Collision theory** is a model that allows us to explain what happens during a chemical reaction and the factors that affect it.

Enzymes

Chemical reactions occur every day in your body to keep it healthy. Many of these reactions rely on **enzymes** to allow the reaction to occur at an adequate rate. An enzyme is a biological catalyst. This means that it is found in living organisms and, being a catalyst, can speed up a chemical reaction without being consumed.

For more information about catalysts, refer to Chemistry section 9.3 on page 379.

Enzymes are usually large protein molecules. Enzymes are highly specific and act on a particular reaction, type of reaction or bond type by making it easier for the reaction to occur. The enzyme has an **active site** with a shape that matches the reactant particles. When the enzyme and reactants (known as the **substrates**) bind, a change occurs which allows the reactants to form the products in a way that lowers the activation energy. Because less energy is needed for a successful collision, more particles have enough energy and therefore more collisions will result in a chemical reaction.

Our digestive system provides the nutrients we need for growth, repair and energy. The food we eat is largely made up of **carbohydrates**, **proteins** and **fats**. These are large organic molecules that are made up of smaller units (Table 4.1). The digestive system breaks down the large molecules into smaller molecules that can be absorbed into the bloodstream. In the blood, the molecules are transported to where they are needed. However, if the large molecules cannot be broken down, they cannot be absorbed and therefore the food cannot be used by the body.

HOW ENZYMES WORK

View this animation of models of various ways enzymes work.

Table 4.1 Nutrients in food

Nutrient	Small units that can be absorbed
Proteins	Amino acids
Carbohydrates	Simple sugars; for example, glucose
Fats	Glycerol and fatty acids

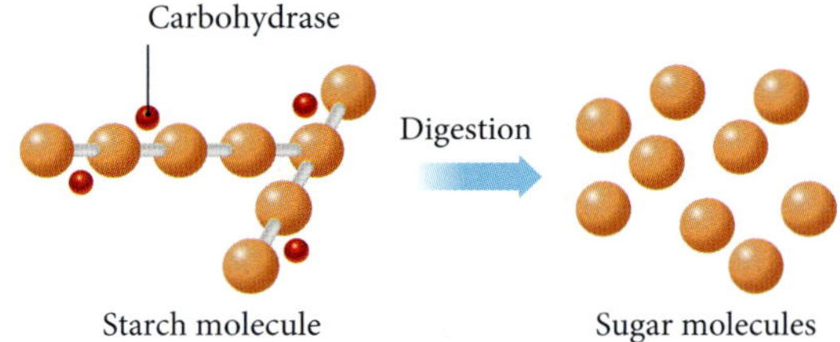

Figure 4.2 ▶ Enzymes help break complex molecules into smaller, simpler molecules that can be absorbed. Carbohydrase acts on starch molecules to produce sugar molecules.

Enzymes can be specific for a particular reaction, breaking a particular bond in a particular molecule (Figure 4.3). This means that the larger molecules are split into smaller sections. Upon binding to the enzyme at the active site, the substrate is changed in such a way as to allow the bond to be broken with less energy. The products formed are then released and the enzyme is available for further reactions. Table 4.2 on page 87 lists some features of some digestive enzymes.

Figure 4.3 ▶ A model of an enzyme

1. The substrate for the enzyme is the molecule sucrose. It is made of glucose and fructose bonded together.
2. The substrate binds to the enzyme, forming an enzyme–substrate complex.
3. The glucose–fructose bond in the sucrose molecule breaks.
4. Glucose and fructose are released as products and the enzyme is free to bind another molecule of substrate (sucrose).

Bond

Active site

Enzyme

H_2O

Glucose

Fructose

You will learn more about proteins, carbohydrates and fats in Units 3 & 4.

Table 4.2 Digestive enzymes

Enzyme	Location	Reactant	Product
Amylase	Saliva in the mouth Pancreas releases it into the small intestine	Starch (a large carbohydrate)	Maltose
Protease (pepsin)	Stomach	Proteins	Smaller proteins
Protease (trypsin)	Pancreas releases it into the small intestine	Proteins	Peptides and amino acids
Lipase	Pancreas releases it into the small intestine	Fats	Fatty acids and glycerol
Peptidase	Small intestine	Peptides (small sections of protein)	Amino acids
Sucrase	Small intestine	Sucrose (a small carbohydrate)	Glucose and fructose
Maltase	Small intestine	Maltose (a small carbohydrate)	Glucose
Lactase	Small intestine	Lactose (a small carbohydrate)	Glucose and galactose

There are thousands of different enzymes in the human body controlling the wide range of reactions that are occurring. These include:

- carbonic anhydrase, which catalyses the reaction between carbon dioxide and water to form carbonic acid, allowing carbon dioxide to travel through the blood:
  $CO_2(g) + H_2O(l) \rightarrow H_2CO_3(aq)$
- DNA polymerase, which is involved in DNA replication
- dehydrogenase, an important enzyme in the first step of cellular respiration in mitochondria to produce energy in our bodies
- cyclin dependent kinases, which control cell division
- tyrosinase, which triggers melanin synthesis and is important in determining skin colour.

The activity or effectiveness of enzymes can be affected by the conditions or environment in which they are found. These conditions include:

- concentration – the higher the enzyme concentration, the more substrate that can bind to an enzyme at a given time and the greater the reaction rate
- temperature – enzymes are proteins, and as such many are **denatured** above about 45°C. Above this temperature, the structure of the protein changes, changing the shape of the active site. This permanently inactivates the enzyme as the substrate is unable to bind to the active site
- pH – different enzymes require different pH values for optimum activity; for example, the pH of the stomach is much less than the pH of the rest of the digestive system, so only specific enzymes will be activated in the stomach
- **co-enzymes** – ions or non-protein molecules such as vitamins that change the shape of the active site in order for the enzyme to function.

WOW

Coenzyme Q10

Coenzyme Q10 is a naturally occurring enzyme that is found in all cells. However, it is more abundant in the hardest working cells such as heart muscle cells. Its functions include acting as an antioxidant and improving the efficiency of energy production. Research is being done into the effects of taking coenzyme Q10 supplements on conditions such as heart disease, cancer and Parkinson's disease.

EXPERIMENT 4.1

EFFECT OF pH AND TEMPERATURE ON THE ENZYME COMPLEX RENNET

Rennet is a complex of enzymes naturally found in the stomachs of mammals. It helps the coagulation of milk – separating it into a solid and a liquid component. It is especially important in the digestion of milk in young animals.

Rennet is also utilised in milk desserts in the form of a setting agent known as junket. The enzymes help the milk set, forming a soft solid.

Aim

To investigate the effect of pH and temperature on the effectiveness of the enzymes in rennet

Materials

- 1 packet of junket (This may vary depending on the quantity that can be made – you will need enough for 65 mL of milk)
- 100 mL of milk
- Distilled water
- 1 mL of 1 mol L^{-1} hydrochloric acid (HCl)
- 1 mL of 1 mol L^{-1} acetic acid (CH_3COOH)
- 1 mL of 1 mol L^{-1} sodium acetate ($NaCH_3COO$)
- 1 mL of 1 mol L^{-1} sodium hydroxide (NaOH)
- 13 test tubes
- Test tube rack
- Water baths at 10°C, 20°C, 30°C, 40°C, 50°C, 60°C, 70°C and 80°C
- 10 mL measuring cylinder
- Thermometer
- Stopwatch
- pH meter

What are the risks in doing this experiment?	How can you manage these risks to stay safe?
Chemicals could splash in your eyes.	Wear safety glasses at all times.
1 mol L^{-1} HCl is corrosive to skin and clothes.	Wear gloves and aprons.
1 mol L^{-1} NaOH is corrosive to skin and clothes.	Wear gloves and aprons.

In your write-up, add any more risks you can think of, as well as ways to manage them.

Procedure

Part A: Effect of temperature

1. Place 5 mL of milk into each of eight test tubes.
2. Place one test tube in each water bath and leave it until the milk has reached the desired temperature.
3. Dissolve the junket in 15 mL of water. (You will need to keep some of this mixture for part B.)
4. Place 1 mL of junket into the first test tube, mix and record the temperature inside the test tube, and start the stopwatch immediately.
5. Stop the stopwatch when the milk has become firm. Record the time taken to set.
6. Repeat steps 4 and 5 for each of the milks at the other temperatures.

Part B: Effect of pH

7. Place 5 mL of milk into each of four test tubes.
8. Record the temperature of the milk – these should all be the same.
9. Into test tube 1, place 1 mL of distilled water.
10. Into test tube 2, place 1 mL of 1 mol L^{-1} HCl.
11. Into test tube 3, place 1 mL of 1 mol L^{-1} CH_3COOH.

▶

12 Into test tube 4, place 1 mL of 1 mol L^{-1} NaOH.

13 Into test tube 5, place 1 mL of 1 mol L^{-1} $NaCH_3COO$.

14 Use the pH meter to test the pH of the milk in each test tube. Ensure that the meters are cleaned thoroughly between test tubes. Record the pH.

15 Add 1 mL of the junket solution from part A to test tube 1 and time how long it takes the milk to become firm. Record the time taken.

16 Repeat step 15 for test tubes 2–5.

Results

Construct a results table like this in your notebook. Record your data in the table.

Variable	Test tube	Temperature (°C)	pH	Time for milk to become firm (min)
Temperature	1			
	2			
	3			
	4			
	5			
	6			
	7			
	8			
pH	1 (distilled water)			
	2 (HCl)			
	3 (CH_3COOH)			
	4 (NaOH)			
	5 ($NaCH_3COO$)			

Analysis of results

1 Draw separate graphs for parts A and B.

2 What happened to the time taken for milk to become firm as the temperature increased?

3 What happened to the time taken for the milk to become firm as the pH increased?

4 Do these results support your understanding of the effect of temperature and pH on enzymes?

Discussion

1 What aspects of the experiment increased the accuracy and reliability of the results?

2 What were the major difficulties in obtaining accurate results during the experiment?

3 How could the experiment be improved?

Conclusion

1 What conclusion can you make about the effect of temperature on the activity of enzymes in junket?

2 What conclusion can you make about the effect of pH on the activity of the enzymes in junket?

Taking it further

1 In what other ways could you investigate the action of enzymes?

2 Research the role of rennet (or rennin). Explain how your findings about the effect of temperature and pH may play a role in ensuring the enzyme's effectiveness.

iStockphoto/suprun

Figure 4.4 ▲ Pineapples contain bromelain – an enzyme that is proteolytic (breaks down proteins).

Enzymes in our food

Many of our foods come from plants and animals, which contain enzymes. These enzymes are still present in the food we eat.

Pineapples contain the enzyme bromelain. This is a **proteolytic enzyme**, meaning it breaks down proteins. Proteins are made up of many **amino acids** joined together into a very large molecule. Although there are only 20 amino acids, they can combine in many different combinations to produce many different proteins. Proteolytic enzymes, also known as proteases, are able to break the bonds between the amino acids, thereby breaking up the protein. The breakdown of proteins can occur without the presence of proteases, but at a much slower rate.

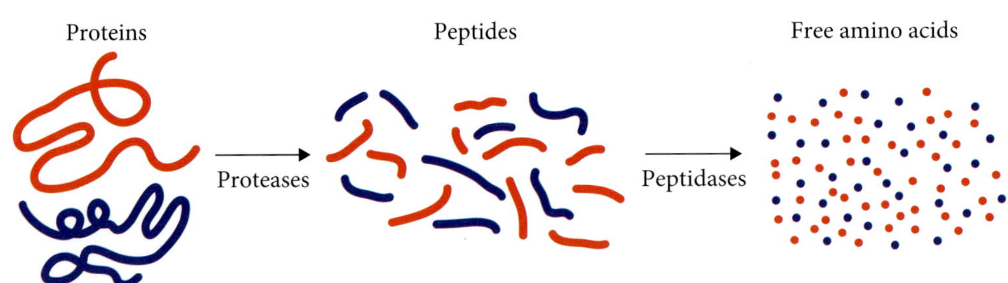

Figure 4.5 ▶ Proteins are broken down into smaller sections called peptides, which are further broken down into individual amino acids.

EXPERIMENT 4.2

ENZYMES IN PINEAPPLE

Pineapples contain the enzyme bromelain, which breaks down the bonds between the amino acids in proteins.

Jelly contains the protein gelatin, which makes jelly set. The enzymes in pineapple are able to break this protein down, causing the jelly to become a liquid.

Aim

To observe the action of the enzyme bromelain in pineapples

Materials

- Fresh pineapple, sliced and cut into quarters
- Jelly crystals
- 2 × 250 mL beakers
- Spatula
- Hot and cold water

What are the risks in doing this experiment?	How can you manage these risks to stay safe?
Chemicals could splash in your eyes.	Wear safety glasses at all times.

In your write-up, add any more risks you can think of, as well as ways to manage them.

Procedure

1 Follow the instructions on the jelly packet to make the jelly.
2 Divide the jelly between the two 250 mL beakers.
3 Refrigerate the jelly until it is nearly, but not completely, set.

4 Place a piece of fresh pineapple on the top of the jelly in one of the beakers.
5 Refrigerate the beakers overnight.
6 Record your observations of the jelly in each of the beakers.

Results

Record your results in an appropriate table. Include an appropriate title for your table.

Analysis of results

1 In which beaker did the jelly remain firm?
2 In which beaker did the jelly turn into a liquid?
3 Use collision theory to provide an explanation for how the enzyme in fresh pineapple is able to speed up the breakdown of proteins in the jelly.

Discussion

In what ways could the experiment be improved to improve the accuracy and reliability of the data collected?

Conclusion

What conclusion can you make about the activity of the enzymes in pineapple?

Taking it further

How could you test the effect of different treatments on the activity of the enzymes in pineapple; for example, temperature and pH?

ACTIVITY 4.1

TENDERISING STEAK

What makes some steaks tender, and other steaks tough?

What you will need

- Computer with Internet access

What to do

1 Read the first two pages of the article located at the weblink 'Tender steak'.
2 Explain what it is that makes meat tough, and how it is made tender.
3 List the different methods of tenderising steak that are mentioned in the article, and explain how each method works.
4 Use your knowledge of digestive enzymes to explain how the steaks are tenderised by enzymes.
5 Use collision theory to explain why meat can be tenderised quickly using pineapple, but it takes days if it is just left out of the fridge.

TENDER STEAK

Read the article to find out what makes steak tough and how to make it tender.

Energy production

Our bodies require energy for just about everything – for our brains to work, for our muscles to maintain posture, for our heart to pump, for movement and to maintain our body temperature.

Cellular respiration is a series of chemical reactions occurring in cells that produces energy from the food we eat and oxygen we breathe. More specifically, the reactants are glucose and oxygen while the products are carbon dioxide, water and energy.

Word equation: Glucose + oxygen → carbon dioxide + water + energy

Chemical equation: $C_6H_{12}O_6(aq) + 6O_2(g) \rightarrow 6CO_2(g) + 6H_2O(l)$ + energy

To learn about what affects rate of reaction, refer to Chemistry section 9.3 on page 379.

It is important that this reaction occurs at a rate that matches the needs of the body. The body can ensure this happens by controlling the concentration of the reactants – oxygen and glucose. The higher the concentration of the reactants, the greater the chance that the reactant particles will collide. A greater number of collisions mean a greater number of successful collisions and therefore reaction rate increases.

To learn about gas pressure and amount of gases in the atmosphere, refer to Chemistry section 8.1 on page 342.

After skipping a meal, you often feel tired and lethargic. This is due to a reduced blood glucose concentration, lowering the rate of cellular metabolism and therefore energy production. People who climb mountains often feel very light-headed, nauseous, tired and lethargic and have trouble thinking clearly. This is due to the reduced oxygen concentration in the air at higher altitudes. This follows through to a reduction in the blood oxygen concentration, rate of cellular metabolism and therefore energy production.

WOW

Altitude sickness

Atmospheric pressure decreases the higher up a mountain you travel. This means that there are fewer oxygen molecules available – at 3600 m above sea level there are only 60% of the number of oxygen molecules per breath. This reduction in oxygen along with the lower atmospheric pressure can lead to altitude sickness if you don't acclimatise your body as you travel up the mountain.

There are three types of altitude sickness – acute mountain sickness, high-altitude pulmonary oedema and high-altitude cerebral oedema. The symptoms range from headaches, nausea, shortness of breath to decreased coordination, tightness of the chest, marked fatigue, confusion, decreased consciousness and coma. Many of these symptoms are a direct result of a lack of oxygen and the body trying to increase its oxygen intake.

To learn about the effect of concentration on reaction rates, refer to Chemistry section 9.3 on page 379.

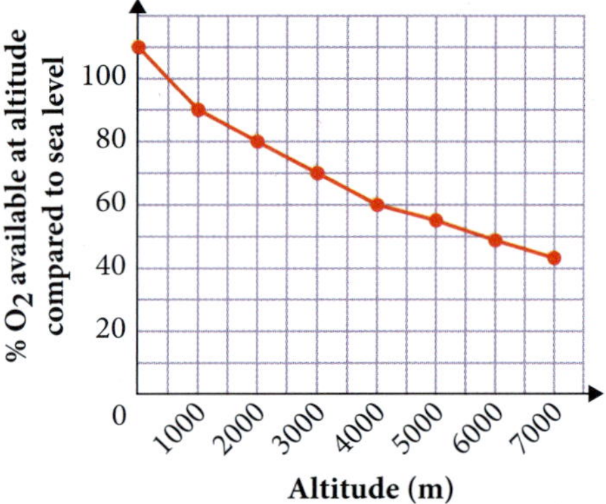

Figure 4.6 ▶ Oxygen availability at different altitudes

Homeostasis involves complex processes that control our internal environment, including oxygen and glucose concentrations. The rate of breathing controls the amount of oxygen entering our blood and which is transported to cells. Blood glucose concentration is controlled by the hormones insulin and glucagon from the pancreas. These concentrations are also influenced by chemical equilibrium, which will be considered in Units 3 & 4.

Homeostasis

Homeostasis is an important area of study in Biology and Human Biology. It considers the various mechanisms that are involved in maintaining a constant internal environment. This is to keep factors such as temperature, pH, gas concentrations, blood glucose and water within the optimum limits so that our bodies function correctly. In many instances, this is directly linked to chemical reactions being able to occur at the required rates.

Why we chew our food

iStockphoto/jsemeniuk

◀ **Figure 4.7** By cutting the food into smaller pieces, a greater surface area is available for reaction.

In most cases, foods contain large, complex molecules. For your body to use these, they must be broken down into small, simple molecules that can be absorbed and then used for the different processes in the body. A lot has to happen to food before your body can use it.

The process of digestion starts in the mouth. Food must be chewed thoroughly to make it easy to swallow. The second reason that you need to chew food is to speed up its digestion. By breaking up the food into smaller pieces, chewing increases the surface area of the food that is exposed to the digestive chemicals. This means that more of the food can be digested at the same time, decreasing the time it takes for the meal to be digested overall.

iStockphoto/Forzua

Chemical digestion can only occur when the digestive juices meet the food particles. These particles must collide so that the reaction can take place. In a solid, this can only occur on the surface because the interior particles are not able to collide with the digestive juices. When the food is broken up into smaller pieces, the particles on the inside become exposed, and are able to react.

EXPERIMENT 4.3

WHY WE CHEW OUR FOOD

Chewing food breaks up large pieces into smaller pieces. This is known as physical digestion and has the function of increasing the surface area of the food. This allows more of the food to be exposed to digestive enzymes and so will speed up chemical digestion.

In this experiment, agar, with phenolphthalein, will represent the food. The phenolphthalein is an indicator that will initially be pink (neutral pH). In an acidic solution, this indicator will become colourless. This will allow you to measure how much of the 'food' has undergone chemical digestion.

Aim

To investigate the effect of surface area on the rate of digestion

Materials

- Agar, with phenolphthalein, cubes: 16 × 1 cm sides, 4 × 2 cm sides, 1 × 4 cm sides
- Vinegar (volume will depend on size of containers)
- 3 containers deep enough for cubes to be submerged in liquid
- Ruler
- Tongs
- Paper towel

What are the risks in doing this experiment?	How can you manage these risks to stay safe?
Chemicals could splash in your eyes.	Wear safety glasses at all times.

In your write-up, add any more risks you can think of, as well as ways to manage them. ▶

Procedure

1 Place all the 1 cm cubes into the first container, ensuring that they are not touching each other.
2 Pour vinegar over the cubes until they are covered.
3 Leave the cubes in the acid for 20 minutes.
4 Use tongs to carefully remove the cubes from the acid and use the paper towel to pat them dry.
5 Cut each cube in half and measure how far from the edge the colour changes from colourless to pink.
6 Repeat steps 1–5 for the other sized cubes.

Results

Record your results in an appropriate table. Remember to include an informative title for the table and appropriate headings on your columns.

Analysis of results

1 Calculate the percentage of each cube that has been 'digested'.
2 Calculate the total surface area for each size of cube. (Remember that surface area is the total area of all the faces of the cube.)
3 Graph your results of total surface area and the percentage digested in an appropriate graph.
4 What is the relationship between the surface area and the percentage digested?
5 Do these results support your understanding of the effect of surface area on the rate of a reaction?
6 Were there any anomalies in your results? If there were, what possible explanations are there for these anomalies?

Discussion

1 Why were there different numbers of cubes for the different sizes?
2 How does chewing (or cutting the agar up) increase the surface area?
3 What errors did you encounter during the experiment?
4 How could the experiment be improved to make it more accurate or reliable?

Conclusion

What conclusion can you make about the effect of surface area on the rate of digestion of food?

Taking it further

How could you further investigate the effect of surface area on the rate of reactions?

To learn about the effect of surface area on the rate of reactions, refer to Chemistry section 9.3 on page 379.

Another benefit of chewing is that it mixes the food with saliva, which starts the process of chemical digestion. Saliva contains enzymes, including amylase and lipase. Amylase is an enzyme for the breakdown of starch, a large carbohydrate molecule, into smaller carbohydrate molecules. Lipase, also an enzyme, is responsible for the breakdown of fats into smaller molecules. As with all enzymes, amylase and lipase allow these reactions to occur at a rate fast enough to be effective.

Hypothermia

Normally, the human body maintains a core temperature of about 37°C. However, in some situations the temperature may drop. **Hypothermia** occurs if the core body temperature drops below 35°C. The first sign of hypothermia is uncontrolled shivering as the body tries to produce heat. Other signs include drowsiness, loss of coordination, lethargy and difficulty breathing. If the core body temperature falls below 32°C, the person may slip into a coma. As the temperature continues to fall, the heart and lungs stop working effectively. This means that the cells do not get enough oxygen, and death may occur.

When the temperature falls, the rate of the reactions occurring in the body also decreases. This is because temperature is a measure of average **kinetic energy**. When the temperature decreases, the number of particles with enough energy for a successful collision decreases.

With fewer successful collisions, the reaction rate decreases. Kinetic energy is related to the velocity, or speed, of the particles. The particles are moving slower, so they will collide less frequently. This also decreases the number of successful collisions and therefore the reaction rate declines.

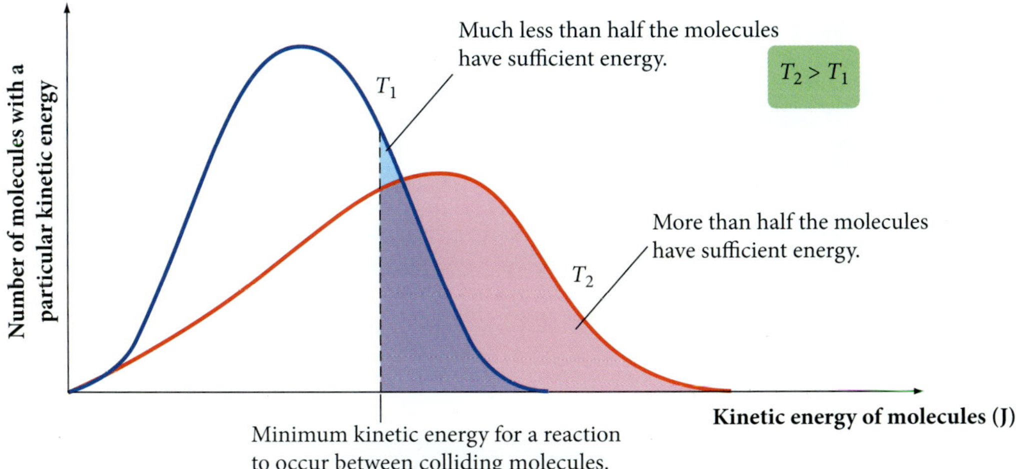

▲ **Figure 4.8**
The number of molecules with enough energy for a successful collision at different temperatures (*T*)

To learn about the effect of temperature on the rate of reactions, refer to Chemistry section 9.3 on page 379.

To learn about the relationship between kinetic energy and temperature of gases, refer to Chemistry section 8.2 on page 346.

The use of scientific knowledge may have beneficial and/or harmful and/or unintended consequences (ACSCH013 and ACSCH052) AC

Scientific knowledge can enable scientists to offer valid explanations and make reliable predictions (ACSCH014 and ACSCH053) AC

Hypothermia decreases the rate of the reactions occurring in the body, and therefore the body is unable to work efficiently. This includes cellular respiration, so the cells are not able to produce the energy needed to function properly.

Reactions around the home

There are many chemical reactions that we depend on to be able to carry out our everyday lives in our homes. Being able to control the rate of these reactions is vital for the efficiency of these reactions as well as our safety.

ACTIVITY 4.2

REACTIONS WHERE WE LIVE

What to do

1 Look around the inside and outside of your home.
2 List all the chemical reactions that take place that you can think of. Remember, a chemical reaction involves the production of a new substance.
3 For each reaction, identify the speed that we want the reaction to occur at (fast, moderate or slow).
4 Try to identify how we manipulate the reaction for it to occur at this desired rate.
5 Share your findings with others in your class.

What did you discover?

1 How reliant are we on chemical reactions in our everyday lives?
2 How are we able to control these reactions to our advantage?

123RF/Moodboard

Figure 4.9 ▲
Relaxing by a camp fire – but it's really chemistry!

Making fires

Before we had reverse-cycle air-conditioners, wood fires were often used to heat our homes. Today, they are still used in some locations for household heating, but are also popular for camping. These fires use the **combustion** reaction of wood, which produces heat and light due to the energy being released.

It is not always easy to get a fire started. To build a fire, you start with kindling, small pieces of wood that are usually ignited by burning some paper. Then, you add other pieces of wood of gradually increasing thickness. Once the fire is burning well, larger logs of wood are added so that they will continue burning slowly. This procedure is not random. It is all related to the rate of the reaction determined by the surface area available for the reaction.

ACTIVITY 4.3

FIRE AS A REACTION

What to do

1 Work through the interactive on the weblink 'On fire' to gain an understanding about what is happening in a fire.

2 a What are the reactants for a wood fire?

b Why is kindling used to start a fire?

c Why would large logs be used to keep a fire burning overnight?

d Hospitals often have cylinders of oxygen gas. Why would it be very important to stop anyone smoking near the cylinders?

3 Go to the weblink 'How fire extinguishers work'. Read the four pages about how fire extinguishers work.

4 Use what you have learnt about fire extinguishers and your knowledge about collision theory and reaction rates to explain how a fire extinguisher can stop a fire.

What did you discover?

1 How is fire an example of a chemical reaction?

2 How can the rate of the reaction be manipulated in a fire?

3 How can our knowledge of reaction rates help us to extinguish fires?

ON FIRE

Learn more about combustion by exploring this virtual fire lab.

HOW FIRE EXTINGUISHERS WORK

Visit this website and read the article about fire extinguishers.

The use of scientific knowledge may have beneficial and/or harmful and/or unintended consequences (ACSCH013 and AACSCH052) AC

Preserving food

Have you ever taken the milk out of the fridge to make a milkshake, only to find that it has gone off? Or left a sandwich in your bag, only to find it weeks later by the smell? There are a variety of reasons that food will spoil. Methods have been developed to slow this degradation so that our food lasts longer.

Food spoilage happens when there is a change in the smell, look, taste or feel of the food that makes it inedible. There are two main reasons that food spoils: enzymes catalyse the breakdown of the food molecules; and **micro-organisms**, generally bacteria or fungi, cause the food to break down. In either case, if we can slow the breakdown of the food, we can make the food last longer.

Some **food preservation** methods are:

- freezing – keeping the food at temperatures below 0°C
- refrigeration – keeping the food at reduced temperatures, usually about 4°C
- canning – processing the food and then sealing it in an airtight container
- pickling – storing or cooking the food in a solution such as brine or vinegar that inhibits the growth of micro-organisms
- vacuum-packing – storing the food in a vacuum container.

Each of these methods utilises our understanding of chemical reactions and the conditions required by micro-organisms. By changing the conditions, we are able to slow the rate of the degradation reaction or microbial growth. This will mean that the food will take longer to spoil.

Scientific knowledge can enable scientists to offer valid explanations and make reliable predictions (ACSCH014 and ACSCH053) AC

ACTIVITY 4.4

FOOD PRESERVATION

Aim

To apply an understanding of reaction rates to understand how various food preservation methods work

You will need

- Computer with Internet access

What to do

1. Research what happens when food spoils. You may wish to use the information at the weblink 'Food spoilage' as a starting point.
2. Identify what the reactants are in the degradation of food. Be general; you do not need to be too specific.
3. Research each of the common methods of food preservation, finding out what is done and how it works. You may wish to use the information in the weblinks 'Methods of food preservation' and 'Food preservation' as a starting point.
4. Based on your research, identify which methods work by reducing microbial growth and which methods work by slowing the chemical reactions. Some methods may come under both classifications.
5. For each method that works by slowing the chemical reaction, identify which factor affects reaction rates.
6. Use your knowledge of collision theory to explain how each of the methods that affect the chemical reaction rate work.

What did you discover?

Explain how our understanding of the rates of reactions is able to help us to preserve food.

FOOD SPOILAGE

Visit this website to look at food spoilage in more detail.

METHODS OF FOOD PRESERVATION

Visit this website to learn more about commercial methods of preserving food.

FOOD PRESERVATION

Visit this website to learn more about preserving food.

Food poisoning

When appropriate food handling and preservation methods are not used, food may become contaminated with harmful micro-organisms. These include bacteria such as *Salmonella*, *Escherichia coli* and *Listeria*, viruses and toxins, which can cause food poisoning. Symptoms of food poisoning include nausea, vomiting, diarrhoea, stomach cramps, fever and headaches. In severe cases, it can cause long-term problems or even death. Each year, 5.4 million people in Australia suffer from food poisoning – something very dangerous that could be prevented!

INVESTIGATION 4.1

INVESTIGATING FOOD PRESERVATION

Many different methods are used to make food last longer. But which one is the most effective? Or how should the method be implemented to be the most effective? Your task is to consider an aspect of food preservation and design and conduct an investigation to further develop your understanding of food preservation and its relation to reaction rates.

What is your aim?

You may choose to compare methods of food preservation or choose one method and compare variations within that method. Your aim should be related to the effectiveness of food preservation of the different methods or variations.

What will you need?

Carefully consider what you will need to conduct a fair and thorough test. You also need to consider what you will need to maintain safety throughout your investigation.

What are the risks?

Construct a table similar to the one below. Identify specific risks involved in the investigation and ways that you will manage the risks to avoid injuries or damage to equipment. Ask your teacher to check your risk assessment before you proceed.

What are the risks in doing this investigation?	How can you manage these risks to stay safe?

How will you carry out your investigation?

Develop a detailed procedure for your investigation. Consider how you will change your independent variable and how you will measure your dependent variable. Ensure that all the other variables are controlled to make your investigation valid. You may wish to carry out a practice test to ensure that your method is going to be effective. Remember that a good method should detail exactly what you did to obtain your results.

What results will you collect?

Consider what data you will need to collect, how many trials you will need to conduct and the number of variations in your independent variable you will need.

How will you analyse your results?

What will you need to do with your data to form a conclusion?

What have you found?

What did your results show? Can you explain this by using your knowledge of food preservation? Are there any anomalies in your data? What are some possible explanations for these anomalies?

What can you conclude?

What conclusion can you make from your investigation?

Ideas for improvement

How could you improve your investigation to allow you to achieve your aim more accurately or thoroughly?

Taking it further

In what ways could you further investigate food preservation methods?

Batteries going flat

A battery uses a chemical reaction to produce electricity. There are many different batteries but, in all types, electrons are transferred from one reactant to another. This movement of electrons allows the production of electricity.

When a battery 'goes flat', the rate of the chemical reaction is too slow to be able to produce enough electricity.

The use of scientific knowledge may have beneficial and/or harmful and/or unintended consequences (ACSCH013 and ACSCH052) AC

The reaction that occurs in batteries is called a redox reaction. You will learn more about redox reactions in Units 3 & 4.

ACTIVITY 4.5

REACTIONS IN BATTERIES

Aim

To understand that batteries utilise a chemical reaction and that the rate of the reaction is related to batteries going flat

You will need

- Computer with Internet access

What to do

1 Choose one type of battery: zinc–carbon, alkaline, lithium ion or lead acid.
2 Conduct some research to find out the reactants that are involved in the reaction utilised in your chosen battery.
3 Explain what will happen to the concentration of the reactants as the battery is used.
4 Explain what you think has happened when a battery goes flat.

Extension

Some types of batteries are rechargeable. Find out what happens to the concentration of the reactants when these batteries are recharged.

What did you discover?

Explain how our understanding of the rates of reactions helps us understand how batteries work.

The recharging of batteries is associated with reversible reactions. You will learn about reversible reactions in Units 3 & 4.

The use of scientific knowledge is influenced by social, economic, cultural and ethical considerations (ACSCH012 and ACSCH051)

THE HABER PROCESS

View this animation to understand how the Haber process occurs.

The Haber process – reactions in industry

Early in the 20th century, a shortage of naturally occurring nitrogen fertilisers led to the development of a method of producing ammonia. A German chemist, Fritz Haber, refined the conditions of the reaction to maximise the yield of ammonia in satisfactory conditions. This is known as the **Haber process**. One major factor that he needed to consider was the reaction rate. If the rate was too slow, not enough ammonia would be produced in a given time.

During the Haber process, nitrogen gas is combined with hydrogen gas in a 3:1 ratio. These reactants undergo the following exothermic reaction.

$$3H_2(g) + N_2(g) \rightarrow 2NH_3(g)\ \Delta H = -92.4\ \text{kJ mol}^{-1}$$

This can be represented by an energy profile diagram, as seen in Figure 4.10.

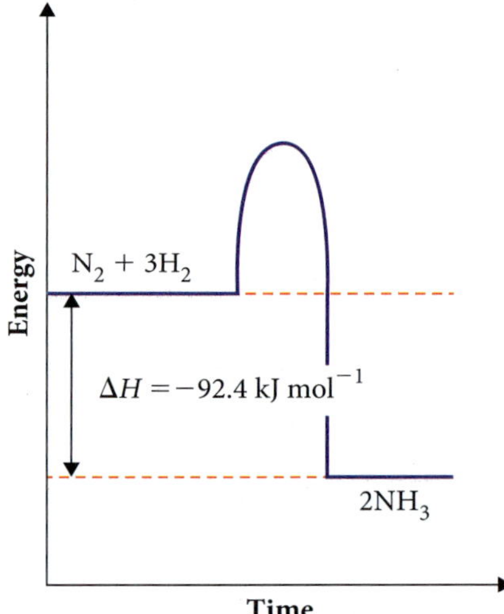

Figure 4.10 ▶ Energy profile diagram for the Haber process

Fastest production of ammonia

You have previously learnt about the factors that may affect the rate of a reaction. We can use this knowledge to understand the conditions that will allow the fastest production of ammonia in the Haber process.

ACTIVITY 4.6

RATE OF THE HABER PROCESS

Aim

To use your understanding of collision theory and the factors that affect the rate of a reaction to identify the conditions that would produce ammonia at the fastest rate

You will need

- Computer with Internet access

What to do

1 Explain how temperature could be used to increase the rate of the reaction for the Haber process. What factors do you think may limit the temperature that could be used?

2 Explain how the surface area is increased by the reactants being gases. How does this influence the rate of the reaction?

3 Explain how the pressure of the system will affect the concentration of the reactants. Would a high or low pressure increase the rate of the production of ammonia? What may limit the extent to which pressure can be manipulated?

4 Research which catalysts are used in the Haber process and how they are supplied in the system. How are the catalysts used to ensure they are the most efficient?

What did you discover?

Present your findings in a concept map to clearly demonstrate your understanding of how each factor can be manipulated to increase the rate of production of ammonia in the Haber process. For each, use collision theory to explain how each factor is able to affect the rate of the reaction.

Conditions used by industry

For the industrial production of ammonia using the Haber process, a temperature of 400–450°C and pressure of 200 atm (20 Mpa) is used. Yet these conditions do not produce the maximum rate of reaction. Why would the manufacturers use conditions that produce ammonia at a slower rate than what is possible?

There are many factors to consider when deciding on the conditions for a reaction. One is the **yield** of the reaction. This means how much of the reactants become products.

Another factor to consider is the cost. It is very expensive to have very high temperatures or pressure. There is a point where the revenue from the extra ammonia produced is outweighed by the cost of maintaining the high temperature and pressure.

When considering the conditions for a reaction, both the rate of the reaction and the yield of the reaction need to be considered. The rate of the reaction is discussed in Units 1 & 2, while the yield will be covered during your study of equilibrium in Units 3 & 4.

To learn about quantities of gases and how to calculate these, refer to Chemistry sections 8.4 and 8.5 on pages 355 and 361.

CHAPTER SUMMARY

- Collision theory allows us to explain the requirements of a reaction and understand factors affecting the rate of a reaction. We are able to alter the rate of reactions by changing the concentration or pressure of reactants, temperature, surface area or presence of catalysts.
- Organic catalysts are known as enzymes. They are usually made of proteins and are specific for a certain reaction. Enzymes are affected by concentration, coenzymes, temperature and pH.
- Cellular respiration is a chemical reaction that occurs in the mitochondria of cells to produce energy. The rate of cellular respiration is dependent on the concentration of oxygen and glucose.
- Digestion provides the body with the required nutrients for survival. Chewing food helps break the food up into smaller pieces, thereby increasing the surface area available for chemical reactions. This increases the rate of digestion.
- The normal human body temperature is approximately 37°C. If the temperature falls too low, then the rate of the reactions in the body decreases to a dangerous level.
- Fires are an example of a chemical reaction. The rate of the reaction for a fire can be manipulated by chopping the wood into smaller pieces or controlling the amount of oxygen available.
- Food spoilage is the breakdown of the food, causing it to change its taste, texture or smell. Food spoilage occurs when micro-organisms or enzymes cause a chemical reaction that degrades the food. Food preservation techniques utilise different methods of killing micro-organisms, slowing the growth of micro-organisms or slowing the chemical reaction.
- Batteries provide electrical energy through a chemical reaction. When the concentration of reactants is low, the rate of the chemical reaction is not fast enough to produce sufficient electricity and the battery is 'flat'.
- The Haber process is an industrial process that produces ammonia from nitrogen and hydrogen gases. The Haber process uses various techniques to increase the rate of the reaction to produce ammonia, including increasing the temperature, increasing the pressure and using a catalyst. However, other factors must also be considered in the Haber process such as cost, safety and yield.

CHAPTER GLOSSARY

active site the section of an enzyme where the substrate(s) binds and undergoes a chemical reaction

amino acid a small organic molecule that combines to form proteins

biochemistry the study of chemicals and their reactions in living organisms

carbohydrate an organic molecule made up of one to many molecules of simple sugars

cellular respiration the reactions that occur in cells to convert the energy in nutrients into energy that can be used; many organisms rely on glucose and oxygen to react to produce carbon dioxide, water and energy

co-enzyme a non-protein chemical that binds to an enzyme and is necessary for the function of the enzyme

collision theory a theory that explains the rate of reaction at a molecular level; it states that, for a reaction, the particles must collide with sufficient energy and the correct orientation

combustion a reaction with oxygen to form the oxides of each of the elements present; with adequate oxygen, combustion of a hydrocarbon will produce carbon dioxide and water

denature altering the chemical structure so that the original properties are lost

enzyme a protein molecule that catalyses a specific type of reaction by lowering the energy of activation

fat an organic molecule that is insoluble in water; many are made up of fatty acids joined to glycerol

food preservation methods of slowing the degradation or spoiling of food

food spoilage negative changes in food causing changes in taste, smell and feel

Haber process an industrial process of ammonia production from nitrogen and hydrogen gases

homeostasis processes that maintain the stable internal environment of an organism within limits

hypothermia the condition of a reduced body temperature (below 35°C), lower than needed for normal metabolism

kinetic energy the energy of movement

micro-organism a very small organism that can be seen only with the use of a microscope such as a bacterium, virus, or protozoan

protein a large organic molecule made up of many amino acids joined together

proteolytic enzyme an enzyme that breaks proteins into smaller lengths

rate how much one quantity changes with respect to another quantity

substrate the chemical that fits into the active site of an enzyme

yield the amount of product produced during a reaction

CHAPTER 1 ATOMS AND ELEMENTS

By the end of this chapter you will have covered the following material.

Science Understanding

- Trends in the observable properties of elements are evident in periods and groups in the periodic table (ACSCH016)
- The structure of the periodic table is based on the electron configuration of atoms, and shows trends, including in atomic radii and valencies (ACSCH017)
- Atoms can be modelled as a nucleus surrounded by electrons in distinct energy levels, held together by electrostatic forces of attraction between the nucleus and electrons; atoms can be represented using electron shell diagrams (all electron shells or valence shell only) or electron charge clouds (ACSCH018)
- Flame tests and atomic absorption spectroscopy are analytical techniques that can be used to identify elements; these methods rely on electron transfer between atomic energy levels (ACSCH019)
- The properties of atoms, including their ability to form chemical bonds, are explained by the arrangement of electrons in the atom and in particular by the stability of the valence electron shell (ACSCH020)
- Isotopes are atoms of an element with the same number of protons but different numbers of neutrons; different isotopes of elements are represented using atomic symbols (for example, ${}^{12}_{6}C$, ${}^{13}_{6}C$) (ACSCH021)
- Isotopes of an element have the same electron configuration and possess similar chemical properties but have different physical properties, including variations in nuclear stability (ACSCH022)
- Mass spectrometry involves the ionisation of substances and generates spectra which can be analysed to determine the isotopic composition of elements (ACSCH023)
- The relative atomic mass of an element is the ratio of the weighted average mass per atom of the naturally occurring form of the element to $\frac{1}{12}$ the mass of an atom of carbon-12; relative atomic masses reflect the isotopic composition of the element (ACSCH024)

1.1 The atom

Our current understanding of the **atom** and its structure has been developed as a result of a number of important experiments over a long period of time. Our knowledge of atomic structure allows us to make predictions about the properties and reactions of different elements and compounds.

Atomic particles

An atom contains three subatomic particles – the **proton**, the **neutron** and the **electron**. The properties of each of the particles are shown in Table C1.1.

Table C1.1 Symbols, charges and masses of atomic particles

Particle	Symbol	Charge	Relative mass
Proton	p	+1	1
Neutron	n	0	1
Electron	e	–1	$\frac{1}{1800}$ (approx.)

ATOMS: THE SPACE BETWEEN

Watch this video that explains the structure of the atom.

The **nucleus** of an atom is a very small region of space that contains all the protons and neutrons. The nucleus contains most of the mass of the atom, but occupies only a small volume of the atom. Surrounding the nucleus are the electrons. Electrons are very small, move extremely fast, and are spread out over a relatively large distance. This creates an electron cloud around the nucleus, as illustrated in Figure C1.1. Despite the large amount of space they cover, electron contribute almost no mass to the atom. If the atom were the size of the MCG, including the grandstands, then each electron would be the size of a pea somewhere in the whole ground, while the nucleus would be no bigger than a peach in the centre of the playing field.

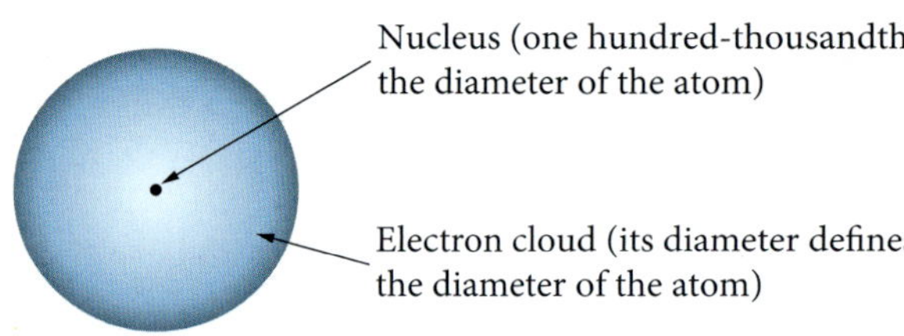

Figure C1.1 ▶ Structure of an atom, showing a nucleus and an electron cloud

RUTHERFORD'S EXPERIMENT

Watch this video explanation and follow the instructions to see how Rutherford's gold foil experiment was conducted.

The current model of the atom was initially theorised by Ernest Rutherford in the early 1900s after an experiment in which he fired **alpha particles** (particles containing two protons and two neutrons) at a thin sheet of gold foil. Earlier models of the atom suggested that all the particles would be evenly spread throughout the atom. It was thus expected that the alpha particles would pass through the gold foil with little or no deflection.

However, Rutherford found that a small number of the alpha particles experienced a significant deflection as though they had struck something large. This did not support his hypothesis and led to Rutherford theorising that most of the mass of the atom was in a structure in the centre. This structure is the nucleus, which contains all the protons and neutrons. This has been verified and further investigated, particularly in recent years with the aid of advanced technological tools.

ACTIVITY 1.1

GOLD FOIL EXPERIMENT

Aim

To simulate Rutherford's gold foil experiment

You will need

- Hula hoop
- Golf ball
- Ping-pong balls
- String

What to do

1. Work in pairs.
2. Tie the golf ball onto the hula hoop so it hangs approximately in the middle of the hula hoop when it is upright.
3. One student is to close their eyes and lightly toss a ping-pong ball 30 times randomly through the hula hoop.
4. The other student is to record how many times the ping-pong ball hits the golf ball.

What did you discover?

1. Explain what is represented by the hula hoop, the golf ball and the ping-pong ball in relation to Rutherford's experiment.
2. Compare the number of times the golf ball was hit by the ping-pong ball with the number of times it was missed. Explain whether this matches what Rutherford found in his experiment.
3. Explain how the results of Rutherford's experiment might lead you to the conclusion he made.

Atomic representation

The number of protons in the nucleus defines the type of atom. A substance made of only a single type of atom is called an **element**. Elements can be represented by symbols. These symbols are found in the **periodic table** along with the element name. Some elements are represented by the first letter of their names – hydrogen is H, carbon is C. The symbols of other elements have two letters – chlorine is Cl, aluminium is Al. The first letter of the symbol is always a capital letter and the second letter is always lower case. So, for example, the symbol for chlorine is Cl, not CL.

Sometimes elements are represented by letters that do not occur in their names. These symbols are based on the element's Latin, Greek or German name. For example, the symbol for gold is Au, and the symbol for potassium is K. The Latin word for gold is *aureum*; hence, its symbol is Au. The Latin word *kalium* means potash, an original source of potassium; hence its symbol is K.

Element names are determined by the International Union of Pure and Applied Chemistry (IUPAC). An element can be named after a mythological concept, mineral, place, country, property or scientist. Many of the elements have been known for a long time, but some elements have only recently been created in laboratories.

Only 92 of the elements identified in the periodic table occur naturally on Earth. All the elements in the periodic table with atomic numbers of 93 or higher have been synthesised by scientists. Many **synthetic elements** have been made and named in the last 60 years. All of the synthetic elements exist for only a short period of time after formation because they decay quickly.

IUPAC approves names suggested by scientists, companies or universities that first synthesise the element. For example, the element berkelium was named after the city of Berkeley in California, where it was made. New elements are given a temporary name based on the Latin and Greek for their atomic number until a formal name is approved. For example, element 117 is currently called ununseptium.

The composition of elements can be represented by $^{A}_{Z}X$. This gives information about the number of protons, neutrons and electrons. This information can be seen in Figure C1.2.

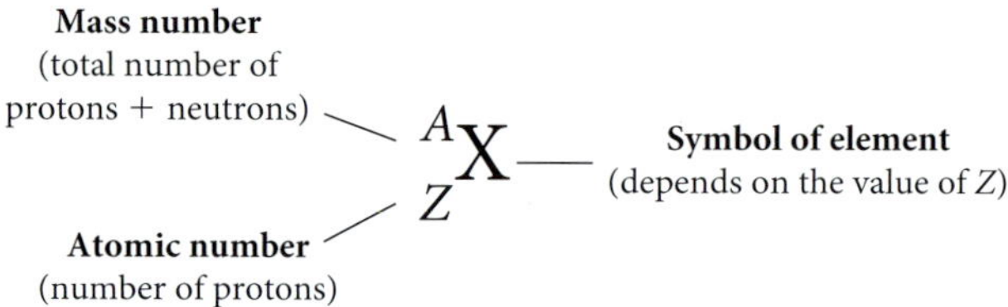

Figure C1.2 ▶ Atoms can be represented by a symbol that includes the atomic number and mass number.

The **atomic number**, Z, represents the number of protons in the atom. This is the number that defines an atom. An atom of carbon will always have six protons, even if the number of neutrons or electrons changes. In an atom, which is uncharged, the number of electrons and protons are equal, so the atomic number will also give the number of electrons in the atom. On a periodic table, the atomic number is the smaller number inside the box, often located above the element symbol.

The **mass number**, A, is the total number of protons and neutrons in the atom. To find the number of neutrons in an atom, subtract the atomic number, Z, from the mass number, A.

$$\text{Number of neutrons} = \text{mass number} - \text{atomic number}$$

To learn about how elements were created, refer to Context 1 'Matter in the universe'.

On a periodic table, the mass number is the larger number inside the box and may not be a whole number. For example, carbon has a mass number of 12.01 in the periodic table. This number represents the average mass of all the forms of the element that exist. When using the mass number to calculate the number of neutrons, round it to the nearest whole number.

The information about mass and atomic number found in the periodic table is illustrated in Figure C1.3.

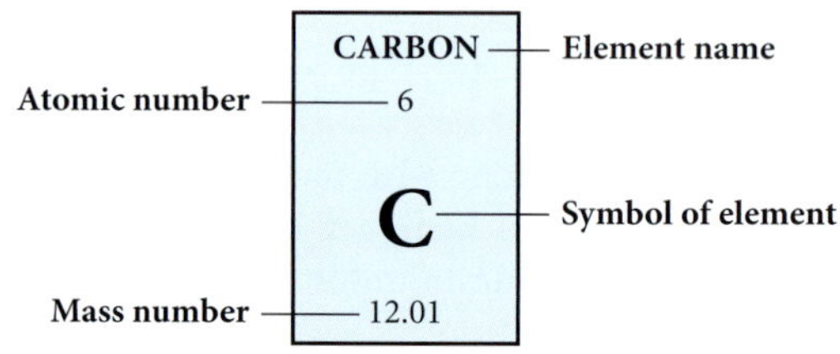

Figure C1.3 ▶ How atomic information is displayed in the periodic table

The calculations that determine the mass numbers shown in the periodic table will be explained in section 1.3.

WORKED EXAMPLE 1.1

Sodium is represented by $^{23}_{11}Na$. Determine the number of protons, neutrons and electrons in sodium.

Answer

Sodium has 11 protons, 12 neutrons and 11 electrons.

Logic

The atomic number of sodium is 11. This means that it has 11 protons.

Its mass number is 23, so sodium has 23 − 11 = 12 neutrons.

The atom is neutral, so there is the same number of electrons as protons.

▶

Try these yourself

a Determine the number of protons, neutrons and electrons of:

- **i** $^{35}_{17}Cl$
- **ii** $^{48}_{22}Ti$
- **iii** $^{27}_{13}Al$
- **iv** $^{137}_{56}Ba$

b Use correct atomic representation to write the element:

- **i** zinc, which has 30 protons and 35 neutrons.
- **ii** phosphorus, which has 15 protons and 16 neutrons.
- **iii** copper, which has 29 electrons and a mass number of 64.
- **iv** lead, which has a mass number of 207 and 82 protons.

What holds an atom together?

In an atom, the nucleus has an overall positive charge. The electrons around the nucleus are negatively charged. The attraction between the positive nucleus and the negative electrons keeps the atom together. This is called **electrostatic attraction**.

An atom can be modelled as a nucleus surrounded by electrons, held together by electrostatic forces of attraction between the nucleus and electrons.

BUILD AN ATOM

Build atoms by adding protons, neutrons and electrons to an atom. 'Fix' an unstable atom by adding protons or neutrons.

How a nucleus stays together

The nucleus of an atom is very stable despite being composed only of positive protons and neutral neutrons. So how do positively charged particles stay together when they should be repelling each other and causing the nucleus to break apart?

Electrostatic repulsion is a force that occurs between particles with the same charge. Repulsive forces act between protons in the nucleus. This repulsive force is illustrated in Figure C1.4a.

Another type of force is present in the nucleus, called the **strong nuclear force**. It occurs between all particles in the nucleus, regardless of charge. This force is attractive, so there is attraction between neutrons and neutrons, neutrons and protons, and even protons and protons. The force is also short range – it only works between particles that are very close to each other. The action of the strong nuclear force is shown in Figure C1.4b.

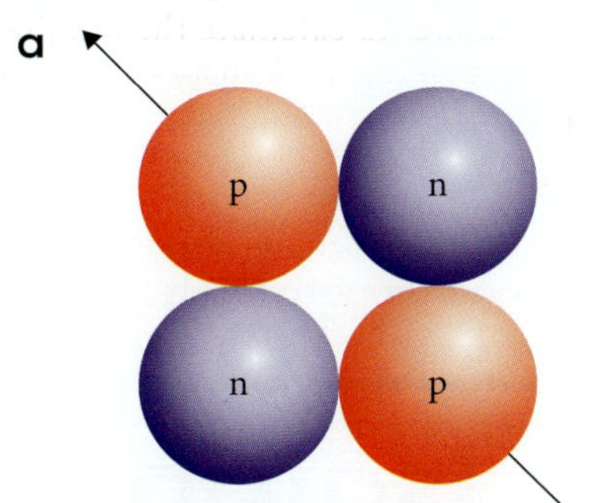

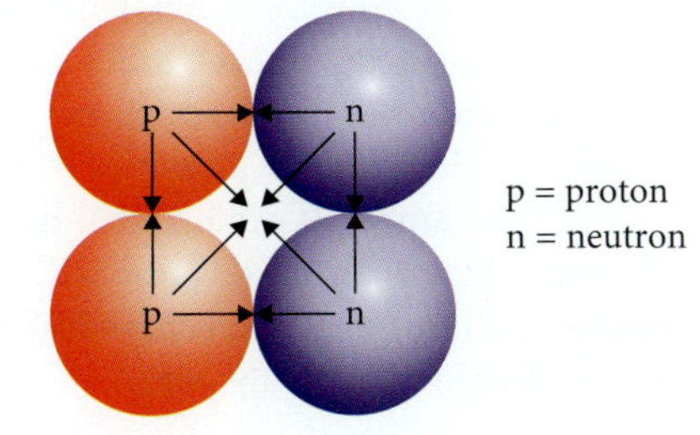

Figure C1.4 a) Repulsive forces act between protons in a nucleus. b) The attractive strong nuclear force acts between all particles in the nucleus.

In a stable nucleus, the electrostatic repulsive forces and the short-range attractive nuclear forces are balanced. If the forces are unbalanced, then the nucleus will be unstable and decay over time. This will be discussed later in this chapter.

QUESTION SET 1.1

Remembering

1 Name the three particles in an atom and state the charge of each particle.

2 Describe where each of the three particles is found in the atom.

Understanding

3 Explain the balance of forces that exists in a nucleus that makes it stable.

Applying

4 Name the following elements and calculate the number of protons, neutrons and electrons.

a $^{19}_{9}F$

b $^{80}_{35}Br$

5 Use a periodic table to help you represent the following as $^{A}_{Z}X$.

a An atom of aluminium with a mass number of 27

b An atom with 4 protons and a mass number of 9

6 Copy and complete the following table. You will need to extract information from the periodic table for some elements.

Element	Atomic number	Mass number	Number of protons	Number of neutrons	Number of electrons
Hydrogen	1	1			
Magnesium			12	12	
Boron	5			6	
Chlorine		35			17
Nickel		59		31	

Analysing

7 Element 113 does not occur naturally. It was synthesised in 2012 by Japanese researchers. It was unstable so it decayed in less than a second. Propose the reason for the rapid decay of this element.

1.2 Isotopes

WHY ARE ATOMS STABLE? AND HOW DO WE MAKE NEW ELEMENTS?

Watch this video to learn how an atom stays stable and how we use this knowledge to synthesis new elements.

Isotopes are different forms of the same element. They are the same element because the number of protons is the same. However, they have different numbers of neutrons, which makes them isotopes of each other. Isotopes of different elements can be naturally occurring or synthesised for a specific purpose. They show some similarities and some differences in their properties. Their relative abundance on Earth is different.

It is the number of protons that defines what an element is. Carbon has a number of isotopes, all containing six protons. Carbon-12 ($^{12}_{6}C$) has six neutrons, while carbon-14 ($^{14}_{6}C$) has eight neutrons. Different elements have different numbers of naturally occurring isotopes and some, such as fluorine, have only one.

Isotopes are atoms of an element with the same number of protons but different number of neutrons.

Isotopes of the same element will have very similar **chemical properties**. Chemical properties relate to how an element participates in chemical reactions. When an atom reacts in a chemical reaction, its behaviour is due to the arrangement and number of electrons. Isotopes have the same number and arrangement of electrons so their chemical properties are similar.

Isotopes of the same element can have different **physical properties**. These properties are features, such as colour, density and mass, that you can observe or measure. Properties can vary because isotopes have slightly different masses due to the different numbers of neutrons. For example, the masses of identical amounts of helium-3 atoms and helium-4 atoms will be different.

Stable and unstable isotopes

Some isotopes of elements are stable because the attractive and repulsive forces in the nucleus are balanced. Other isotopes have unstable nuclei in which the forces are not balanced. In some nuclei, the repulsive forces are stronger than the attractive forces. In other nuclei, the attractive forces are stronger than the repulsive forces. In both of the cases where the forces are not balanced, an unstable nucleus forms and the nucleus will undergo **radioactive decay** to become stable. During radioactive decay, high-energy particles or radiation is emitted, which can be used for such purposes as radioactive dating and medical diagnosis and treatments.

To see formation and applications of isotopes, refer to Context 1, 'Matter in the universe', page 22.

Isotopes of an element have similar chemical properties but different physical properties, including variations in nuclear stability.

1.3 Relative atomic mass

An atom is extremely small, so it is difficult to measure the mass of one atom. Instead, we use **relative atomic mass** A_r, which compares the mass of an atom to the mass of another atom. We talk about atoms having a mass, and this is shown in the periodic table; however, it is not an actual measured mass, but a comparison that scientists have developed.

Scientists arbitrarily picked carbon-12 as the standard for comparison to calculate relative atomic mass. Hydrogen and oxygen were considered and even used for a time, but carbon was selected for ease of use in experiments and because it is a very common element on Earth. The relative atomic mass of an atom is the mass of the atom relative to the mass of a carbon-12 atom. The mass of a carbon atom is considered to be exactly 12 and all others are compared to this.

Magnesium has twice the mass of carbon so has a mass of $2 \times 12 = 24$. Hydrogen is one-twelfth the mass of carbon so has a mass of $\frac{1}{12} \times 12 = 1$. A visual representation of atomic mass comparisons is shown in Figure C1.5.

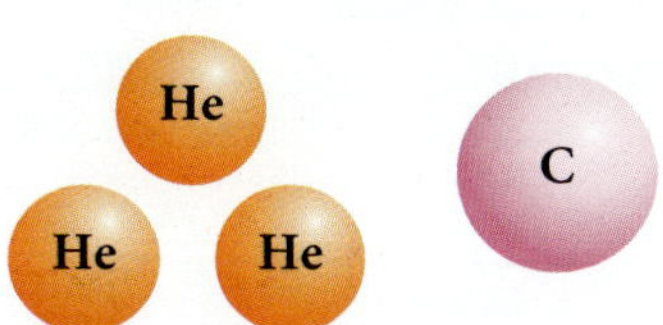

Three helium atoms have the same mass as one carbon atom. Therefore, the relative atomic mass of helium is $\frac{1}{3} \times 12 = 4$.

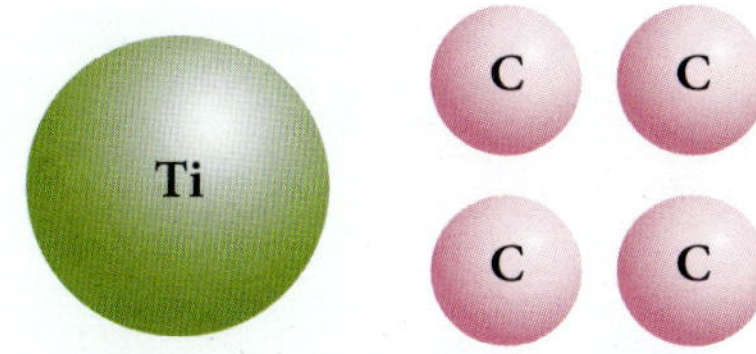

One titanium atom has the same mass as four carbon atoms. Therefore, the relative atomic mass of titanium is $4 \times 12 = 48$.

◀ **Figure C1.5** Relative atomic mass

Isotopes of the same element have different atomic masses. They also have different abundances on Earth, which means they are found in different amounts. For example, two of carbon's isotopes are carbon-12 and carbon-13. On Earth, 98.9% of all carbon is carbon-12 and only 1.1% is carbon-13. Other isotopes of carbon, such as carbon-14, are present in such low levels that they are not considered here.

The mass and abundance of the naturally occurring isotopes of an element are used to calculate the mass numbers shown in the periodic table. For the two isotopes of carbon, the equation is:

$$\text{Mass number} = \frac{(\text{abundance percentage} \times \text{atomic mass}) + (\text{abundance percentage} \times \text{atomic mass})}{100}$$

Relative atomic mass numbers are not whole numbers. It may not seem to make sense that a mass number could be something like 12.01 – it is not possible to have 0.01 of a proton or a neutron. The mass number of carbon is calculated by factoring in the abundance and the mass of each isotope.

$$\frac{(98.9 \times 12) + (1.1 \times 13)}{100} = 12.01$$

So the atomic mass of carbon does not really include 0.01 of a proton or a neutron. It is simply an average of the isotope masses that exist on Earth.

Worked example 1.2 shows you how to calculate the atomic mass seen in the periodic table for any element.

The relative atomic mass of an element is the ration of the weighted average mass per atom of the naturally occurring form and reflects the isotopic composition of the element.

WORKED EXAMPLE 1.2

Chlorine has two isotopes with relative atomic masses of 35 and 37. The percentages of each isotope present on Earth are 75% chlorine-35 and 25% chlorine-37. Calculate the overall relative atomic mass of chlorine.

Answer	Logic
$\frac{(\text{abundance percentage} \times \text{atomic mass}) + (\text{abundance percentage} \times \text{atomic mass})}{100}$	Write out the correct formula to use.
$\frac{(75 \times 35) + (25 \times 37)}{100} = 35.5\%$	Insert amounts into the formula and calculate the answer.

Try these yourself

a Lithium has two isotopes: lithium-6 with an abundance of 7.59% and lithium-7 with an abundance of 92.41%. Calculate the relative atomic mass of lithium.

b Magnesium has three isotopes: magnesium-24 with an abundance of 78.99%, magnesium-25 at 10.00% and magnesium-26 at 11.01%. Calculate the relative atomic mass of magnesium.

QUESTION SET 1.2

Remembering

1 Define 'isotope'.

2 Explain 'relative atomic mass'.

Understanding

3 Explain the difference in structure between the isotopes nitrogen-14 and nitrogen-13.

4 Explain how the relative atomic mass shown in the periodic table is calculated.

5 Explain why isotopes of the same element have the same chemical properties.

6 Explain why some isotopes are stable and some are unstable.

7 Use your understanding of isotopes to explain how they might be useful.

Applying

8 An element has a mass three times that of carbon. Calculate its relative atomic mass.

9 Copper has two isotopes: copper-63 with an abundance of 69.17% and copper-65 with an abundance of 30.83%. Calculate the relative atomic mass of copper.

10 Vanadium (V) has two naturally occurring isotopes, vanadium-50 and vanadium-51. Given that the mass number for vanadium in the periodic table is 50.94, predict which isotope is more abundant and explain your reasoning.

1.4 Periodic table

The periodic table is the chemist's method of arranging all the elements in a systematic way so that trends can be seen. The periodic table arranges the elements in order of increasing atomic number. Trends in atomic radius, metallic character and electronegativity can be determined by examining the periodic table.

History of the periodic table

The first periodic table was developed by Dmitri Mendeleev in 1869 and was the first table to be arranged in the current structure. In his table, the elements were arranged by increasing atomic weight in such a way that trends in columns, rows and across the whole table could be seen. Mendeleev left gaps for elements not discovered at the time and was able to predict their mass and properties from the trends shown in his periodic table.

To learn about organising elements, refer to Context 1, 'Matter in the universe', page 7.

Arrangement of the periodic table

The periodic table is arranged in a series of horizontal rows called **periods** and vertical columns called **groups**. The elements are in order of their atomic number from left to right in each row. A simple periodic table showing groups and periods is seen in Figure C1.6.

Groups

Periods

▲ **Figure C1.6**
Groups and periods on the periodic table

Elements with similar properties are found in vertical groups numbered 1–18. Group 2 includes the elements beryllium, magnesium, calcium and strontium, while group 15 includes nitrogen, phosphorus and arsenic.

Some of these groups have names. These are shown in Figure C1.7. Group 1 is often referred to as the **alkali metals** and includes lithium, sodium, potassium and rubidium. In the middle of the periodic table are the **transition elements** in groups 3–12.

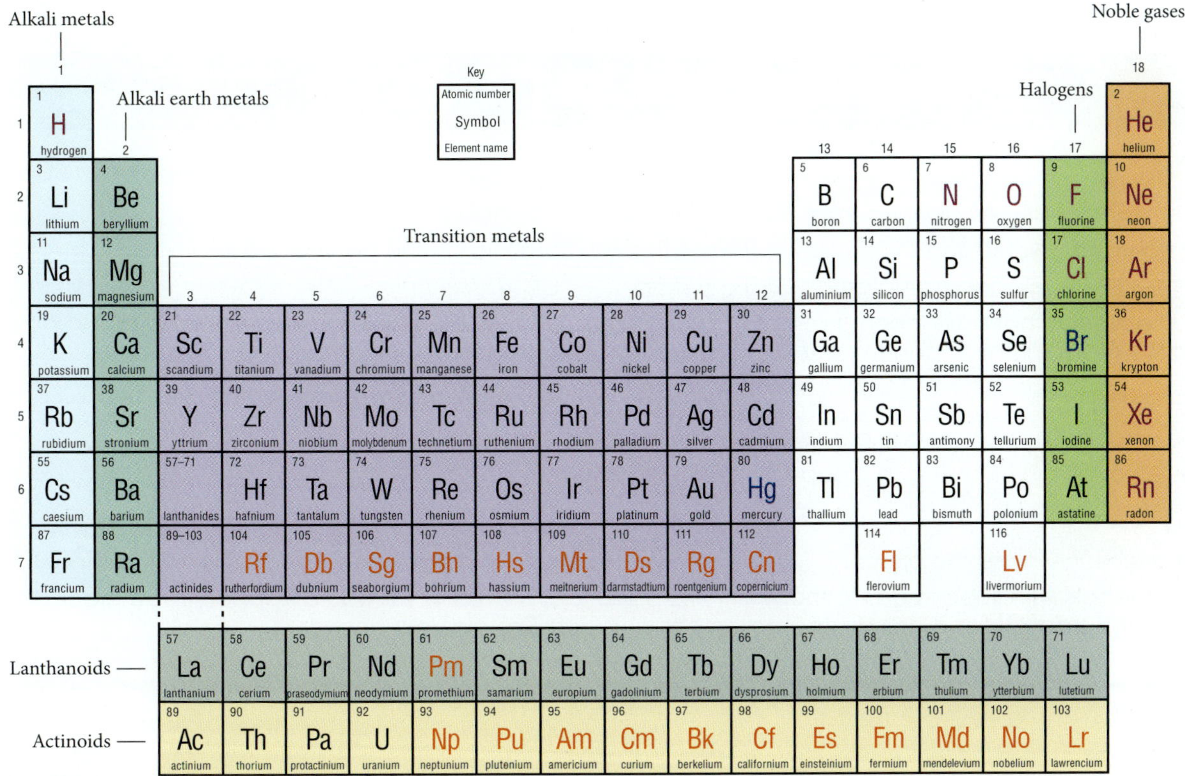

▲ **Figure C1.7**
Common names of some groups and blocks of the periodic table

Each of the elements in the same group shows similar chemical properties. For example, when reacted with water, each of the group 1 elements will form a metal hydroxide and hydrogen gas. This reaction with water becomes more violent as you go down the group. On contact with water, lithium reacts vigorously, sodium reacts a little more violently, while rubidium and caesium explode violently. This is an example of a trend in the groups. The **reactivity** of the elements in group 1 increases down the group.

The periods on the table are given numbers 1–7. Period 1 contains only hydrogen and helium; all other periods contain at least eight elements. The elements in a period display very different properties and chemical reactivity.

The periodic table can also help you determine whether an element is a metal, a non-metal or a metalloid. A **metalloid** is an element that has properties of both metals and non-metals. Metals appear on the left-hand side of the table, non-metals are on the right-hand side and metalloids are in a diagonal strip between them (see Figure C1.8).

Figure C1.8 ▼ Metals (white), non-metals (pink) and metalloids (blue) in the periodic table

																	H
																	He
Li	Be											B	C	N	O	F	Ne
Na	Mg											Al	Si	P	S	Cl	Ar
K	Ca	Sc	Ti	V	Cr	Mn	Fe	Co	Ni	Cu	Zn	Ga	Ge	As	Se	Br	Kr
Rb	Sr	Y	Zr	Nb	Mo	Tc	Ru	Rh	Pd	Ag	Cd	In	Sn	Sb	Te	I	Xe
Cs	Ba	La	Hf	Ta	W	Re	Os	Ir	Pt	Au	Hg	Tl	Pb	Bi	Po	At	Rn
Fr	Ra	Ac	Ru	Db													

Trends in observable properties of elements are evident in periods and groups.

QUESTION SET 1.3

Remembering

1 Describe how Mendeleev arranged the elements in the periodic table.

2 Identify where you would find the metals, non-metals and metalloids in the periodic table.

3 Explain the difference between a group and a period.

4 Identify the groups that have been given common names, such as the noble gases. Make a table showing the group number and the common name of these groups.

Understanding

5 Describe what all the elements in group 2 have in common.

6 Describe what all the elements in group 15 have in common.

Analysing

7 Suggest a reason why all the elements in the same group might undergo the same chemical reactions.

8 If beryllium reacts with fluorine gas to make beryllium fluoride, predict what would happen if you placed some magnesium metal in a gas jar containing chlorine gas. Explain your answer.

1.5 Elemental carbon

Carbon as an element exists in several different forms called **allotropes**. An allotrope is a different physical form of the same element. The atoms in each allotrope are arranged differently, giving it different properties.

The way carbon bonds together will be studied in detail in Chapter 3.

The best-known allotropes of carbon are diamond, graphite and charcoal (see Figure C1.9). Each of these is composed entirely of the element carbon but they are very different from one another. Diamond is the hardest known naturally occurring material, while graphite is an extremely soft substance. Graphite leaves layers of itself on paper, which is why it is used in pencils. Diamond is a transparent substance; graphite and charcoal are opaque and dark coloured. Diamond is a very poor electrical conductor, while graphite is an excellent conductor of heat and electricity. The properties of each substance are dependent on its bonding, and determine how it is used in society.

To learn more about nanotechnology, refer to Context 1, 'Matter in the universe', page 19.

Diamond

Graphite

Charcoal

Shutterstock.com/Africa Studio

Shutterstock.com/Slim Sepp

Shutterstock.com/Deyan Georgiev

▲ **Figure C1.9**
Allotropes of carbon: diamond, graphite and charcoal

Carbon is an element that has a lot of applications in **nanotechnology**, particularly the construction of carbon nanotubes, which have exceptionally good conductivity and are used in material construction, electronics and optics.

The way electrons become involved in bonding with other atoms will be covered in detail in Chapter 3.

1.6 Electron arrangement

The arrangement of electrons around the nucleus of an atom is used to predict chemical reactivity. This is because all chemical reactions involve the sharing or transfer of electrons between the atoms involved.

Movement of electrons and how this is used to analyse elements will be covered in section 1.8.

Energy levels

Electrons are found in specific **energy levels** (or **energy shells**) around the nucleus of the atom. Electrons in the same energy levels have the same amount of energy. The higher the energy level an electron occupies, the more energy the electron has. Every electron in an atom has to be in an energy level; electrons cannot exist between levels, although they can move between them when atoms absorb and release energy.

To see the relationship between electron arrangement and spectra, refer to Context 1, 'Matter in the universe', page 6.

Each energy level in an atom can only hold a certain number of electrons. The first energy level holds a maximum of two electrons; the second energy level can hold up to eight electrons. The general formula $2n^2$ can be used to determine the maximum number of electrons an energy level can hold, where n is the energy level number. For example, energy level 3 can hold $2 \times 3^2 = 18$ electrons. Table C1.2 shows the maximum number of electrons that can be found in each energy level.

Table C1.2 Maximum number of electrons in each energy level

Energy level (n)	Calculation of electron number ($2n^2$)	Number of electrons
1	2×1^2	2
2	2×2^2	8
3	2×3^2	18
4	2×4^2	32

Energy levels correspond to the horizontal periods in the periodic table. Period 1 contains the elements hydrogen and helium; so this period has two elements and two electrons in that shell. Period 2 contains eight elements and eight electrons in that shell. After this, the periodic table gets a little more complicated and the patterns are not as easy to identify. You will learn how to identify these more complicated patterns in your future chemistry studies.

Electron configuration

Electrons are arranged in specific ways that are unique to each atom. This is known as an atom's **electron configuration**. Electrons around a nucleus fill up in order and occupy the lowest energy levels first, so the single electron in hydrogen goes into the lowest energy level. The configuration is written as 1. The two electrons in helium also go into the lowest energy level. The electron configuration is written as 2.

With lithium, which has three electrons, the first two electrons go into the lowest energy level, which is now full. The remaining electron goes into the next highest energy level. The configuration of the lithium atom is written as 2,1.

Configurations are written to show how many electrons are in each level, and the order of the levels. The configurations for the first 10 elements are shown in Table C1.3.

Table C1.3 Electron configuration of the first 10 elements

Element	Number of electrons	Electron configuration
Hydrogen	1	1
Helium	2	2
Lithium	3	2,1
Beryllium	4	2,2
Boron	5	2,3
Carbon	6	2,4
Nitrogen	7	2,5
Oxygen	8	2,6
Fluorine	9	2,7
Neon	10	2,8

The first two energy levels are now full. The eight electrons of the next eight elements go into the third energy level. These configurations are shown in Table C1.4 on page 115.

Table C1.4 Electron configuration for the elements of period 3

Element	Number of electrons	Electron configuration
Sodium	11	2,8,1
Magnesium	12	2,8,2
Aluminium	13	2,8,3
Silicon	14	2,8,4
Phosphorus	15	2,8,5
Sulfur	16	2,8,6
Chlorine	17	2,8,7
Argon	18	2,8,8

Since the third energy level can hold up to 18 electrons, you would expect the electron configuration of potassium to be 2,8,9, but this is not a stable configuration. The reasons for this will be explained in your later chemistry studies. Instead, the 19th electron goes into the fourth energy level, giving a configuration of 2,8,8,1 as seen in Table C1.5. Likewise, the 20th electron of calcium goes into the fourth level.

Table C1.5 Electron configuration for the first two elements of period 4

Element	Number of electrons	Electron configuration
Potassium	19	2,8,8,1
Calcium	20	2,8,8,2

Elements in the first row of the transition metals have 21–30 electrons. Again, for reasons that will be explained in later chemistry studies, the last electrons of the transition metal elements go into the third not fourth energy level. This is the most stable arrangement of electrons. Some of the transition element configurations are shown in Table C1.6.

Table C1.6 Electron configuration for selected transition elements

Element	Number of electrons	Electron configuration
Scandium	21	2,8,9,2
Titanium	22	2,8,10,2
Zinc	30	2,8,18,2

Now that the third energy level is full, in the elements gallium and germanium, the electrons start to fill up the fourth level as seen in Table C1.7. Remember that there are already two electrons in the fourth energy level.

Table C1.7 Electron configuration for selected elements in period 4

Element	Number of electrons	Electron configuration
Gallium	31	2,8,18,3
Germanium	32	2,8,18,4

PERIODIC TABLE AND ELECTRON CONFIGURATIONS

Use this interactive periodic table to revise electron configurations and other information about the elements.

Subshells

The previous method of writing electronic configurations focuses only on what happens to electrons inside a whole energy level. However, each major energy level contains one or more **subshells**, designated s, p, d and f. Each subshell contains regions of physical space called **atomic orbitals**, inside which electrons move. Each atomic orbital cannot contain more than two electrons and if it does contain two electrons, they spin in opposite directions.

Table C1.8 shows how many orbitals are within each of the four subshells.

Table C1.8 Electrons, orbitals and subshells

Subshell designation	Number of orbitals	Number of electrons in subshell
s	1	2
p	3	6
d	5	10
f	7	14

The first energy level has only one subshell, the s subshell, and it is given the designation 1s. The second energy level has two subshells, s and p, given the designations 2s and 2p. The third energy level has 3s, 3p and 3d subshells. The fourth energy level has 4s, 4p, 4d and 4f subshells. The number in front represents the main energy level, and the letter represents a particular subshell.

These subshells have different energy levels. In a particular energy level, the s subshell always has the lowest energy, followed by the p subshell, then the d and f subshells. For example, the 2s subshell has lower energy than the 2p subshell. Figure C1.10 shows all the subshells in their energy order.

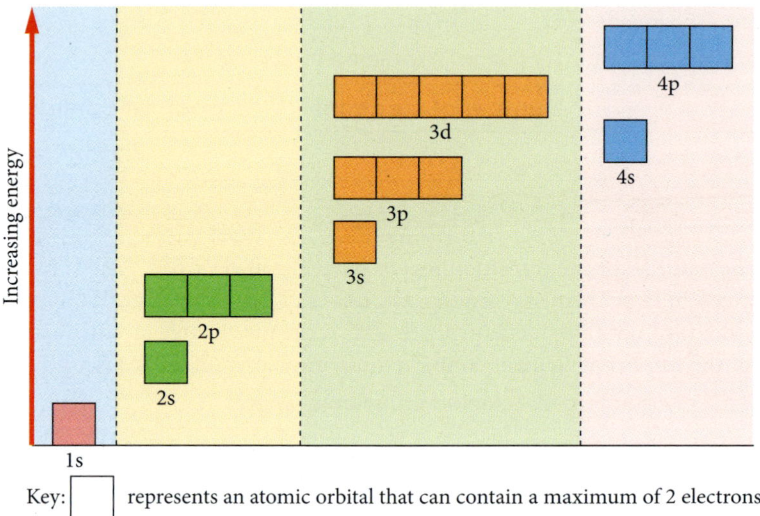

Figure C1.10 ▶ Energy order of the s, p, d and f subshells

The subshells can also be seen in the periodic table. Figure C1.11 shows how the blocks of the periodic table correspond to the subshells. Groups 1 and 2 are the s block, groups 13–18 are the p block, the transition metals are the d block and the **lanthanoids** and **actinoids** are the f block.

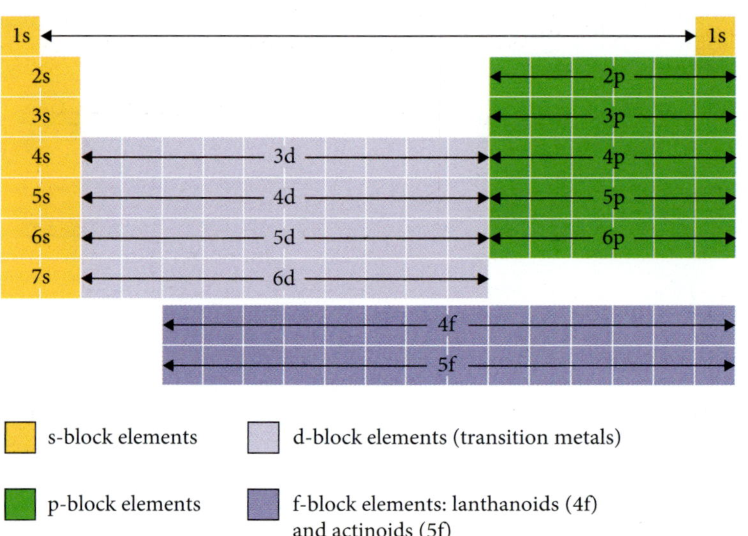

Figure C1.11 ▶ Subshells related to position in the periodic table

The periodic table can be used to predict which subshells have higher or lower energy than others. The subshell with the lowest energy is the 1s subshell. The order of the subshells in terms of their energy is 1s < 2s < 2p < 3s < 3p < 4s < 3d < 4p < 5s < 4d < 5p and so on. This is found by simply reading across the periodic table from left to right, following the rows down. You have probably noticed a quirk of the periodic table where the d subshell in the fourth row belongs in the third energy level. It is the 3d subshell. Each of the transition metal rows includes a subshell that belongs in the energy level above the row it is in. For example, the d subshell in the fifth row of the periodic table is the d subshell in the fourth energy level, subshell 4d.

If we use this subshell order, we fill up the subshells just as we filled up the energy levels in the previous method. Hydrogen has an electron configuration $1s^1$, with the superscript 1 representing the number of electrons in the subshell. Helium is $1s^2$. To write the configuration for lithium, the next electron needs to go into the next subshell, so lithium is $1s^2\ 2s^1$. Table C1.9 shows the electronic configuration for the first 20 elements, using this method.

Table C1.9 Electronic configuration of first 20 elements using subshell notation

Element	Atomic number	Electronic configuration
Hydrogen	1	$1s^1$
Helium	2	$1s^2$
Lithium	3	$1s^2\ 2s^1$
Beryllium	4	$1s^2\ 2s^2$
Boron	5	$1s^2\ 2s^2\ 2p^1$
Carbon	6	$1s^2\ 2s^2\ 2p^2$
Nitrogen	7	$1s^2\ 2s^2\ 2p^3$
Oxygen	8	$1s^2\ 2s^2\ 2p^4$
Fluorine	9	$1s^2\ 2s^2\ 2p^5$
Neon	10	$1s^2\ 2s^2\ 2p^6$
Sodium	11	$1s^2\ 2s^2\ 2p^6\ 3s^1$
Magnesium	12	$1s^2\ 2s^2\ 2p^6\ 3s^2$
Aluminium	13	$1s^2\ 2s^2\ 2p^6\ 3s^2\ 3p^1$
Silicon	14	$1s^2\ 2s^2\ 2p^6\ 3s^2\ 3p^2$
Phosphorus	15	$1s^2\ 2s^2\ 2p^6\ 3s^2\ 3p^3$
Sulfur	16	$1s^2\ 2s^2\ 2p^6\ 3s^2\ 3p^4$
Chlorine	17	$1s^2\ 2s^2\ 2p^6\ 3s^2\ 3p^5$
Argon	18	$1s^2\ 2s^2\ 2p^6\ 3s^2\ 3p^6$
Potassium	19	$1s^2\ 2s^2\ 2p^6\ 3s^2\ 3p^6\ 4s^1$
Calcium	20	$1s^2\ 2s^2\ 2p^6\ 3s^2\ 3p^6\ 4s^2$

Representing electrons

The arrangement of electrons around a nucleus can be represented visually in a number of different ways. Each of these highlights part of our understanding of how electrons are arranged around the nucleus of an atom.

Electron shell diagrams

A simple **electron shell diagram** shows the arrangement of electrons in their energy levels. This can be represented by showing each electron individually, or by using the numbers of electrons

Figure C1.12 ▼
Two versions of an electron shell diagram for calcium

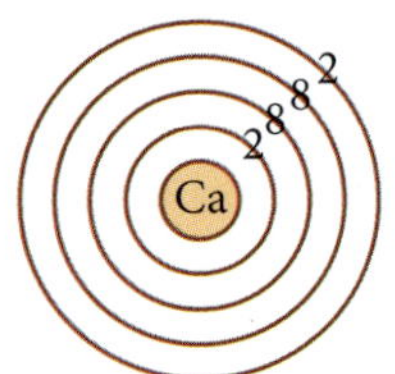

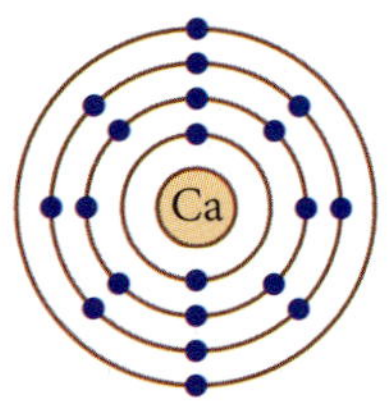

in each shell. The two different interpretations of an electron shell diagram for calcium can be seen in Figure C1.12. The lowest energy shell is closest to the nucleus.

Electron charge cloud diagrams

Electrons around a nucleus exist in regions of space. It is impossible to identify exactly where an electron is located in an atom, only the region of space in which it is likely to be. To represent this, **electron charge cloud diagrams** (see Figure C1.13) are used to show the probability of electrons being found in a particular place in the atom.

Unlike electron shell diagrams, this type of representation does not show how the electrons are arranged in energy levels, simply the region of space around the nucleus where an electron is likely to be located.

Atomic orbital diagrams

Each of the four orbitals, s, p, d and f, have a specific shape that can be mapped to show the physical space that electrons occupy. Figure C1.14 shows how the orbitals appear around a nucleus in an **atomic orbital diagram**. The electrons in a particular orbital are found somewhere within that region of space.

Figure C1.13 ▶
The electron cloud around the nucleus

To see how this information is used, refer to Context 1, 'Matter in the universe', page 6.

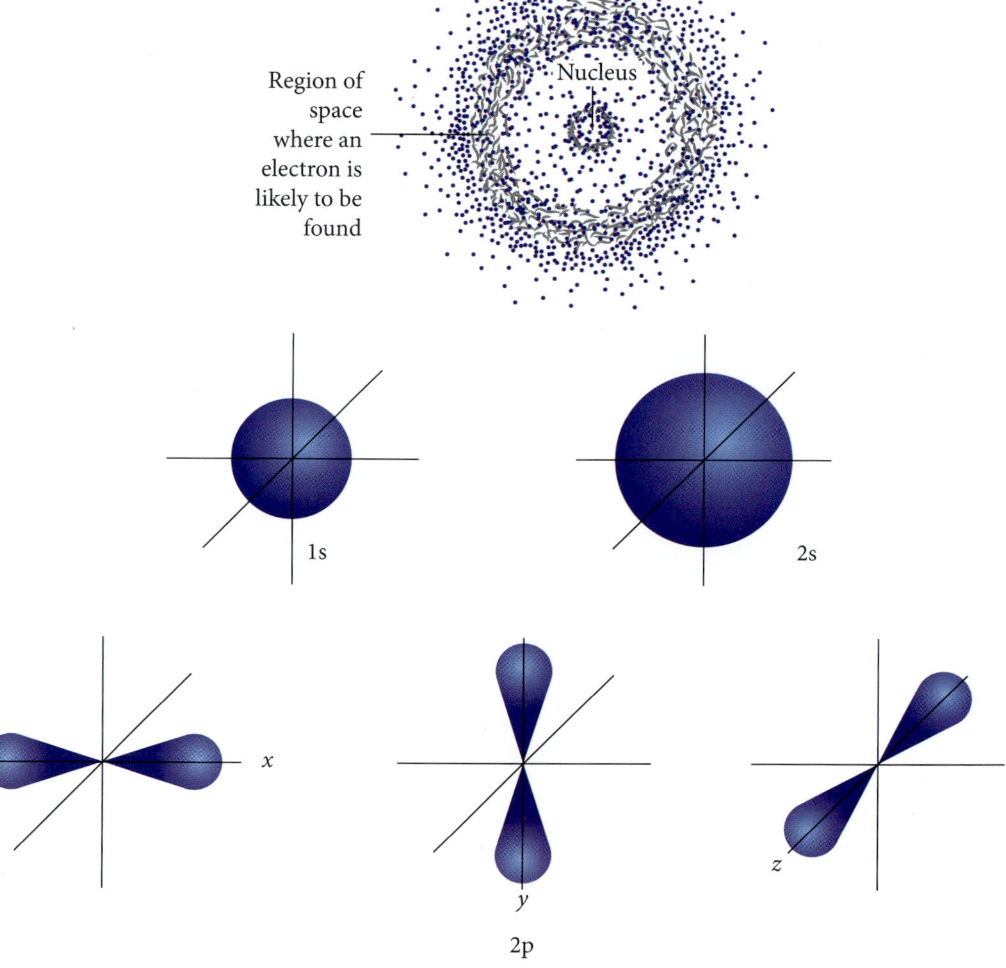

Figure C1.14 ▶
Atomic orbital diagrams

Electrons exist in distinct energy levels, which can be represented by electron shell diagrams or electron charge clouds.

QUESTION SET 1.4

Remembering

1. State how many electrons are found in the third energy level.
2. Describe how electrons fill up the energy levels in an atom.
3. Explain what is shown in an electron cloud diagram.

Understanding

4. Draw electron shell diagrams for the following elements.
   a Beryllium
   b Sulfur
   c Calcium
   d Helium
5. Write electron configurations for the following elements in either or both of the 2,8,1 and s, p, d, f form (check with your teacher).
   a Nitrogen
   b Phosphorus
   c Hydrogen
   d Scandium
   e Germanium
6. Identify the elements that have the following electron configurations.
   a 2,8,2 / $1s^2\ 2s^2\ 2p^6\ 3s^2$
   b 2 / $1s^2$
   c 2,8,14,2 / $1s^2\ 2s^2\ 2p^6\ 3s^2\ 3p^6\ 4s^2\ 3d^6$

Analysing

7. The following electron configurations are incorrect. Identify each element, explain why the configuration is incorrect and write the correct configuration for that element in either or both of the 2,8,1 and s, p, d, f form (check with your teacher).
   a 2,9
   b 2,8,10

1.7 Periodic table trends

The periodic table shows specific patterns for the elements. This is because the elements are arranged in order of atomic number, and into groups and periods.

The properties described in this section will help you when you study how atoms bond together in Chapter 3.

Valence shell electrons

The number of electrons in the **valence (or outer) shell** of the elements in groups 1, 2 and 13–18 can be determined from the periodic table. Table C1.10 lists the first four elements in group 1 and their electron configurations. It reveals a pattern: each of the elements in group 1 has one electron in its outer valence energy level. This is true for all elements in group 1.

Table C1.10 Group 1 electron configurations

Element	Electron configuration
Hydrogen	1
Lithium	2,1
Sodium	2,8,1
Potassium	2,8,8,1

To see examples of the use of different elements, refer to Context 1, 'Matter in the universe', page 9.

In groups 2 and 13–18, a similar pattern can be seen. The elements in group 2 have two electrons in their outer shell, the elements in group 13 have three electrons, the elements in group 14 have four electrons and so on. This method does not work for the transition metals in groups 3–12.

Atomic radius

Atomic radius is normally measured as the distance from the nucleus to the boundary of the cloud of electrons surrounding it.

- Atomic radius decreases from left to right across a period. The positive charge of the nucleus increases across the group. This is due to the extra protons in the nucleus. As the nucleus becomes more positive, the electrons in the outer energy level are more strongly attracted to the positive nucleus and they move closer together, decreasing the radius.
- Atomic radius increases down a group. The elements at the bottom of a group have more energy levels filled than those at the top of a group. With more energy levels filled, the distance between the nucleus and the outer energy level increases, so the atomic radius does as well.

These trends can be seen in Figure C1.15.

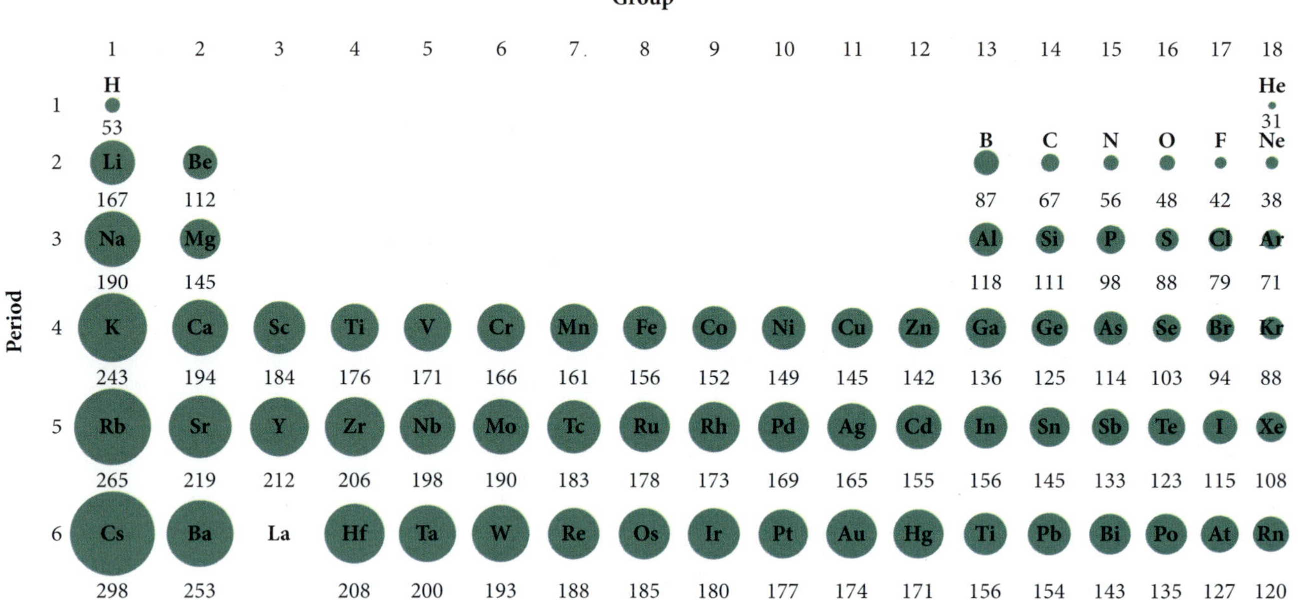

Figure C1.15 ▲
Trends in atomic radius (picometres) in the periodic table: atomic radius decreases across a period and increases down a group. Note that there are no data available for the calculated atomic radius of lanthanum.

Ionisation energy

Ionisation energy is the amount of energy needed to remove an electron from a neutral atom when it is a gas. An atom that has a low ionisation energy will become an **ion** (a charged atom) very easily.

- Ionisation energy increases from left to right across a period. Within a period, the electrons are in the same energy level so are all approximately the same distance from the positively charged nucleus. But as you go across a period, the number of protons increases, so the positive charge of the nucleus also increases. The attraction between the positive nucleus and the negative electrons becomes stronger. This makes it harder to remove an electron, so more energy is needed.
- Ionisation energy decreases down a group. Down a group, the electrons in their energy levels are getting further away from the nucleus and are less strongly bound to the positive nucleus. So it takes less energy to remove the electron from the atom because there is weaker attraction between the positive nucleus and the negative electrons.

Electronegativity

Electronegativity is the ability of an atom to attract electrons toward itself.

- Electronegativity increases from left to right across a period. This is because the nucleus is becoming more positive and so the electrons in the atom become more closely attracted to the nucleus and the atomic radius thus decreases. As the atom gets smaller, the atoms have a stronger attraction for electrons of nearby atoms.
- Electronegativity decreases down a group. As the number of energy levels increases, the electrons are further away from the nucleus so the attraction between the positive nucleus and the negative electrons is weaker, the atomic radius increases, and it is harder for an electron to be attracted to the atom.

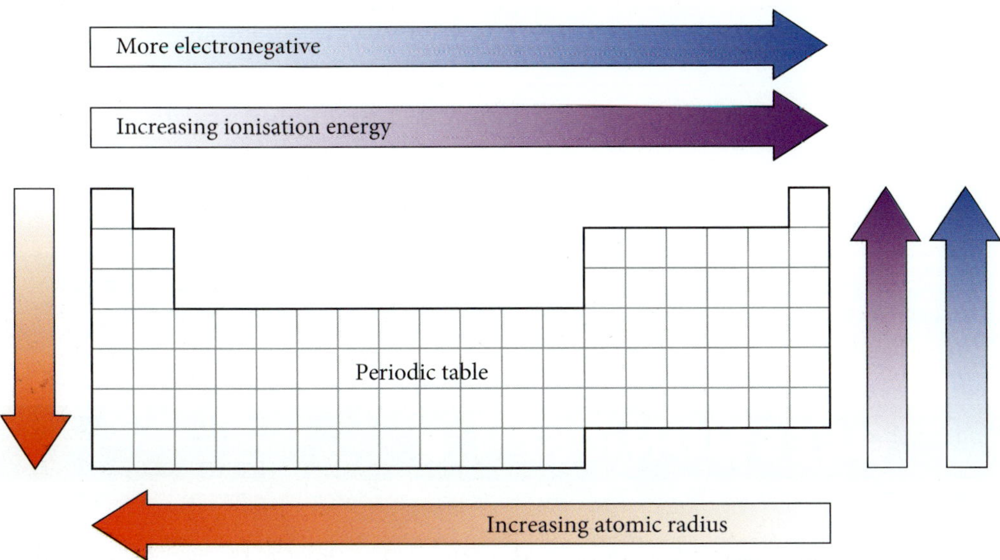

◀ Figure C1.16 Some trends in the periodic table

The periodic table can thus be used to determine which of two elements is more electronegative. For example, oxygen is more electronegative than sulfur because it is above it in group 16. Chlorine is also more electronegative than sulfur as it is further to the right of their period.

The most electronegative element is fluorine. The noble gases do not have an electronegativity as they have a full valence shell, are stable and do not need to attract electrons.

Electron affinity

Electron affinity is the ability of an atom in the gaseous state to accept an electron and form a negative ion. Electron affinity is different from electronegativity because it involves full addition of an electron, not just the ability to attract electrons.

- Electron affinity increases across a period from left to right. This is because the atomic radius decreases. The outer energy levels are more strongly attracted to the positive nucleus, so it is easier to add an electron to these atoms.
- Electron affinity decreases down a group. This means it is harder for atoms to add an electron. As there are more energy levels in these elements, the electrons are further away from the nucleus. The attraction between the positive nucleus and the negative electrons is weaker. It is harder to attract and add an electron to this atom.

Metallic character

Elements are classed as metals, non-metals or metalloids.

- Metallic character decreases across a period from left to right.
- Metallic character increases down a group.

Consistent with this, elements on the left of the periodic table are metals. Elements on the right of the table are non-metals. A strip of elements along the diagonal line shown in Figure C1.17 consists of elements classed as metalloids, which have characteristics of both metals and non-metals.

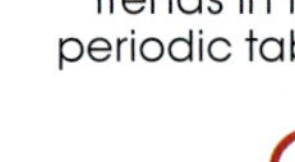

Figure C1.17 ▶ Electron affinity and metallic character trends in the periodic table

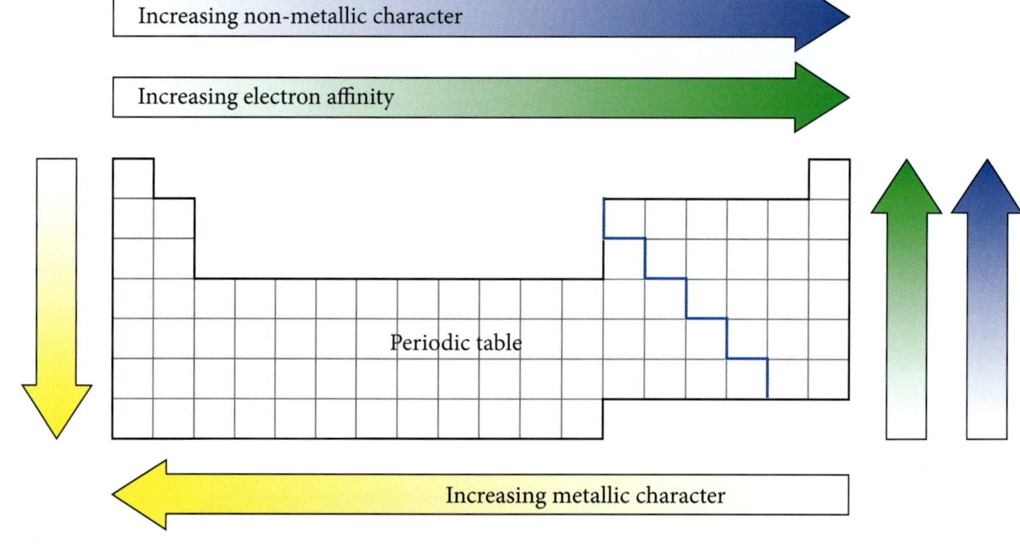

GRAPHING THE PERIODIC TABLE

Use this interactive activity to explore patterns in the periodic table, including atomic radius, electron affinity and electronegativity.

The structure of the periodic table is based on the electron configuration of atoms and shows trends.

QUESTION SET 1.5

Remembering

1 State how many electrons are in the outer shells of:
   a beryllium.
   b fluorine.
   c phosphorus.
   d sodium.
   e argon.

2 Define 'metalloid'.

3 Define 'ionisation energy'.

4 Describe the trend in metallic character in the periodic table.

Understanding

5 Explain, using electron configurations, how many electrons are found in elements in group 17.

Applying

6 State whether lithium or fluorine would have a bigger atomic radius.

7 In the following pairs of elements, state which element is more electronegative.
   a Nitrogen and oxygen
   b Magnesium and beryllium
   c Phosphorus and fluorine
   d Carbon and silicon

Analysing

8 Explain why atomic radius changes across a period and down a group.

1.8 Elemental spectra

In an atom, electrons are found in a specific arrangement around the nucleus. When all of the electrons are in in their lowest possible energy levels, the atom is said to be in the **ground state**.

Movement of electrons

When an atom absorbs energy, such as heat, the electrons in the energy levels around the nucleus gain this extra energy and can move up to a higher energy level. The energy levels for electrons in an atom occur at particular levels for a particular element. Electrons cannot exist between the energy levels. So the amount of energy they absorb can only be equal to the difference between one energy level and another. This movement of electrons can be seen in Figure C1.18. You can see that an electron may rise by one level or by more than one level.

For example, the energy levels in every magnesium atom are identical. When energy is added to a sample of magnesium, all the electrons will move between the same, set energy levels. So the amount of energy absorbed by any sample of the element will be the same.

As an atom can have multiple energy levels, it is possible for an electron to move up one, two or even more energy levels. An atom with electrons in upper energy levels is said to be in an **excited state**. Electrons in the excited, higher energy levels are unstable. After a very short time, less than one-millionth of a second, the electrons move down to their original energy levels. As they do, they release the energy that they previously absorbed. This is shown in Figure C1.19.

This energy is emitted as light. Energy is related to the **wavelength** of the light produced, so this light has a very specific wavelength. Because the electrons can move between a number of different levels, different wavelengths of light are emitted from a sample. As all magnesium atoms have the same energy levels, the light emitted by all magnesium atoms will have a consistent set of wavelengths.

Oxygen and other elements will have a different set of energy levels from magnesium. The energy levels in oxygen are still discrete and the electrons will still move between levels but they will absorb and emit light of different energies from the electrons of magnesium. The light emitted by oxygen will have different wavelengths from the light emitted by the magnesium.

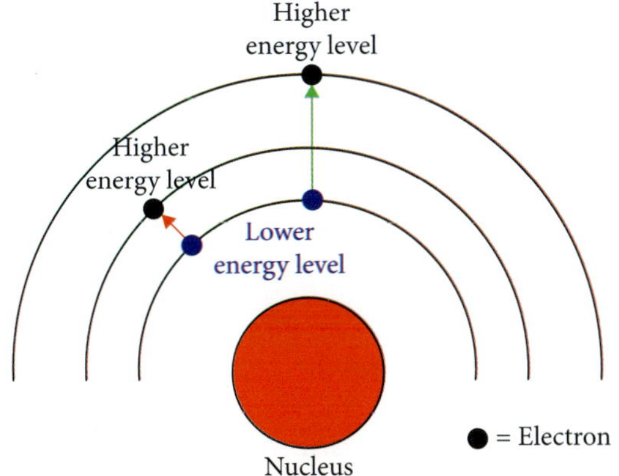

▲ **Figure C1.18** When electrons absorb energy, they can move between energy levels.

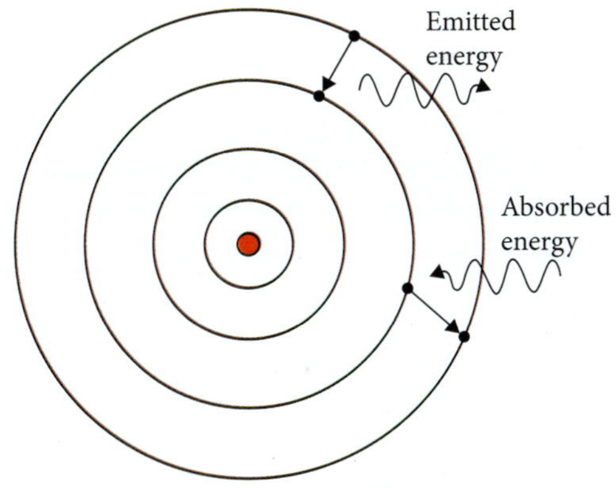

▲ **Figure C1.19** Absorption and emission of light by atoms due to electron movement

EXPERIMENT 1.1

FLAME TESTS

Some element ions give a characteristic colour when placed into the flame of a Bunsen burner. These different colours can be used to identify different cations (positive ions).

Aim

To observe and compare the colours of flames produced by different elements

Materials

- Solid samples of calcium nitrate, calcium chloride, sodium nitrate, sodium chloride, potassium nitrate, potassium chloride, strontium nitrate, strontium chloride
- 1.0 mol L^{-1} hydrochloric acid

▶

- Distilled water
- 2 × 100 mL beakers
- Platinum or nichrome wire loop
- Bunsen burner
- Matches

What are the risks in doing this experiment?	How can you manage these risks to stay safe?
The Bunsen burner and equipment will get hot and could cause burns.	Turn off the Bunsen burner or turn it to a yellow flame when not in use. Handle hot objects with care and do not place them directly onto bench tops; use a heatproof mat.
Use of chemicals	
Use of spectroscope	

In your write-up, add any more risks you can think of, as well as ways to manage them. In particular, expand on the three risks listed to identify specific risks involved with each of them. Ask your teacher to check your risk assessment before you proceed.

Procedure

1. Place approximately 20 mL of 1.0 mol L^{-1} hydrochloric acid into a 100 mL beaker.
2. Place approximately 20 mL of distilled water into a 100 mL beaker.
3. Dip the wire loop into the distilled water, then into the acid to clean it.
4. Hold the wire loop in the blue part of the Bunsen burner flame to remove any chemicals left on it. When the flame burns a normal blue colour, the wire loop is clean.
5. Dip the wire loop into the solid sample of calcium nitrate and hold it in the blue part of the Bunsen burner flame. Record the colour or colours observed.
6. Dip the loop into the distilled water and hydrochloric acid to clean it.
7. Repeat steps 5 and 6 for all of the solids, ensuring you record the colour of each flame.

Results

Create a table of your results showing the chemical name and the colour of the flame. You may want to compare your results with those of other groups as observation of colour is often different between different people.

Analysis of results

1. Describe any patterns you can see in your results.
2. Describe any problems you had with determining the colours. Compare your results and observations with those of other groups.
3. Research the colours you should have seen for these elements. Your teacher may provide you with the expected colours.

Discussion

1. Explain why the same elements will have the same colour flame when a flame test is conducted.
2. Explain why you may not have seen the expected colours in this experiment.

Conclusion

Write a conclusion linking elements and their flame colours.

Taking it further

A spectroscope can be used in a darkened room to view the flame. This will split the light emitted into its component wavelengths as described in the next section on emission spectroscopy. These emission spectra can be compared to known patterns to confirm the element identity.

Emission spectroscopy

The absorption and emission of light by an element can be used to identify it. When a sample of an element is heated as described in Experiment 1.1, it will absorb energy. Its electrons will move to higher energy levels and fall back down to the ground state, emitting light, which can be analysed. The process of analysing light is called **spectroscopy**. A **spectroscope** is a device used to take the light emitted from an element and separate it into its component wavelengths to produce a **line emission spectrum**. The light is composed of multiple wavelengths, which are dispersed through a prism and shone onto film to produce a spectrum as seen in Figure C1.20.

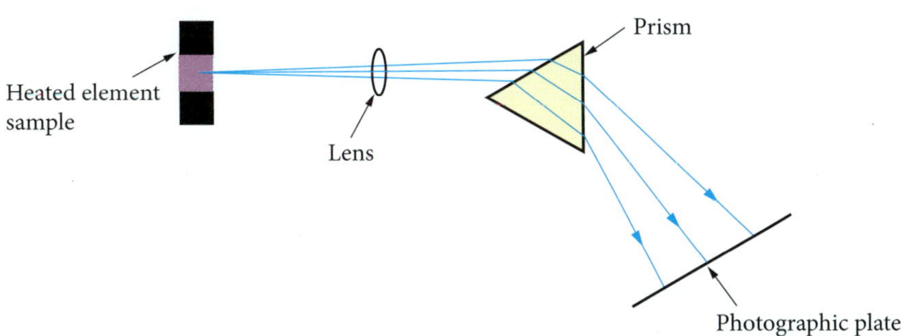

Figure C1.20 How a line emission spectrum is formed

SPECTROMETRY EXPLAINED

Use this interactive activity to find out more about how spectrometry works and how line emission spectra are produced.

Each element has a unique set of energy levels, so when the electrons move between them, it will involve absorption and emission of different amounts of energy. This means that every element will emit light of a different set of wavelengths from every other element. Some common element spectra are shown in Figure C1.21.

To see an example of applications, refer to Context 1, 'Matter in the universe', page 6.

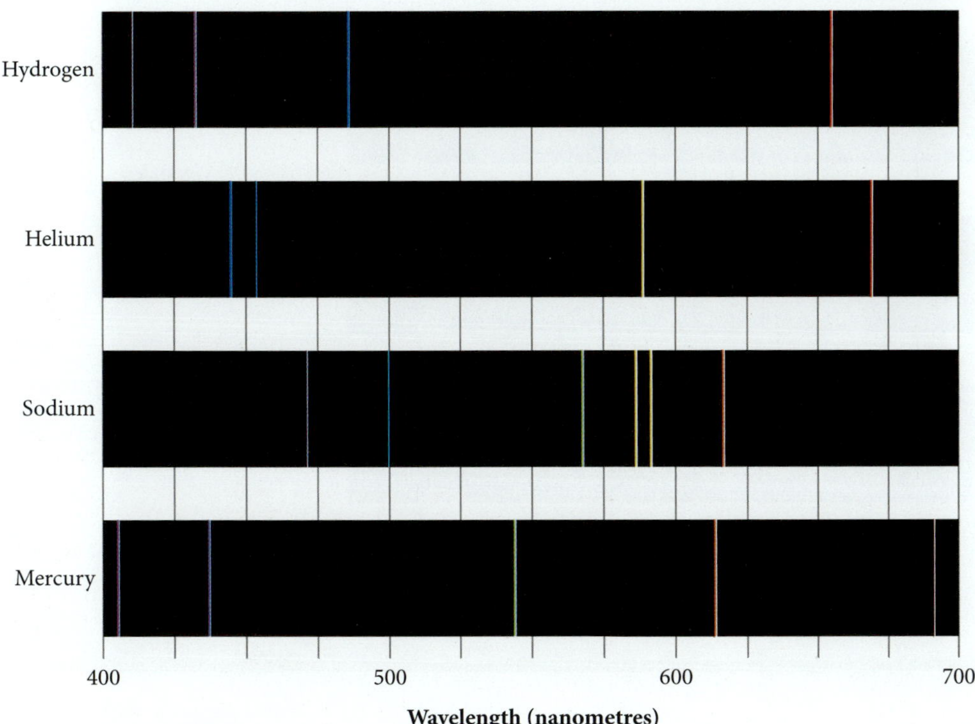

Figure C1.21 Emission spectra of some common elements

Because every atom of the same element has the same set of energy levels, the pattern produced for an element will always be the same. This is a way of identifying elements. An unknown element is heated and the light it emits is analysed and compared with known spectra.

QUESTION SET 1.6

Remembering

1 Explain 'ground state' and 'excited state' in reference to electrons in an atom. Use a diagram in your answer.

2 Describe what happens to electrons when an atom absorbs energy.

3 Describe how an atom can emit energy.

4 Describe a spectroscope and how it works. Use a simple diagram in your answer.

Understanding

5 Explain why all atoms of sodium will emit the same set of wavelengths of light when heated.

6 Explain why sodium and magnesium have different emission spectra.

Applying

7 Figure C1.22 shows the emission spectra for several known elements and one unknown element. Identify the unknown element, giving reasons for your answer.

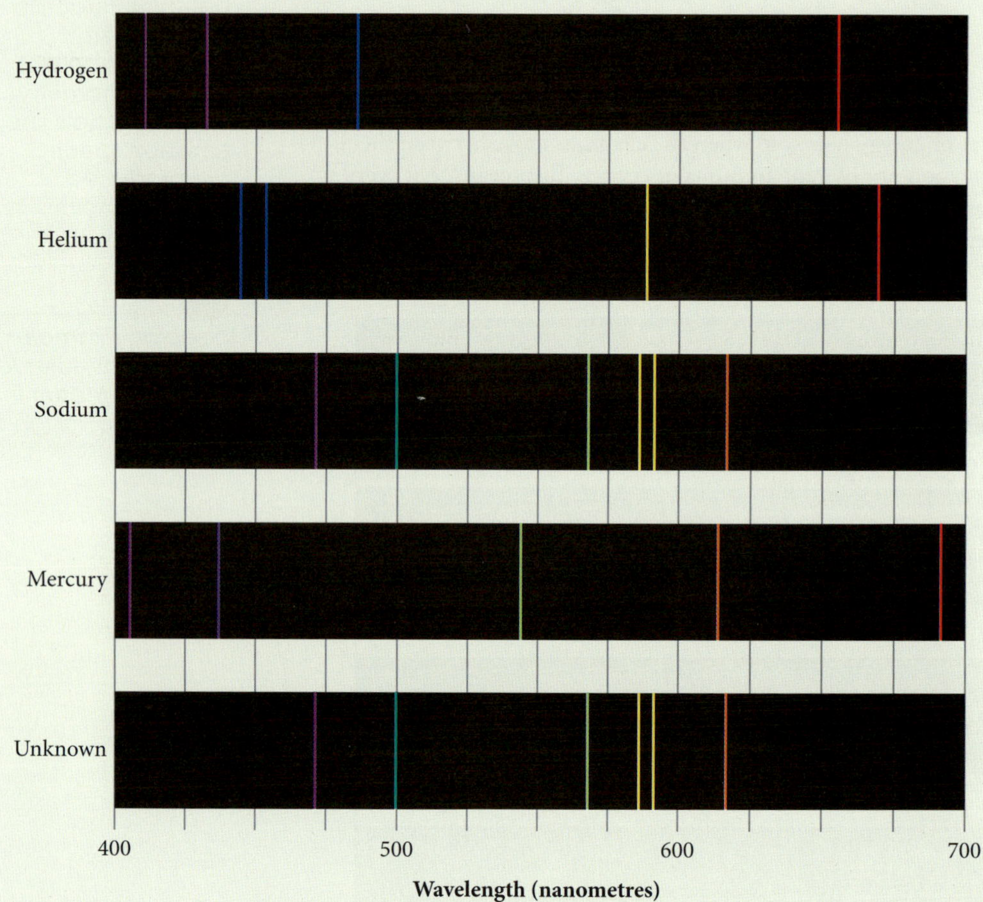

◀ **Figure C1.22** Emission spectra of some known elements and an unknown element

1.9 Atomic absorption spectroscopy

To see applications of atomic absorption spectroscopy, refer to Context 1, 'Matter in the universe', page 13, and Context 3, 'Water, the vital substance', page 74.

Atomic absorption spectroscopy is a related technique to emission spectroscopy, and it can also be used to perform **quantitative analysis**; that is, to find the amount of an element present. Atomic absorption spectroscopy uses the absorption of light by electrons in the atom to measure how much of an element is present in a sample of substance.

This process uses an atomic absorption spectrometer. The basic outline of the process is seen in Figure C1.23.

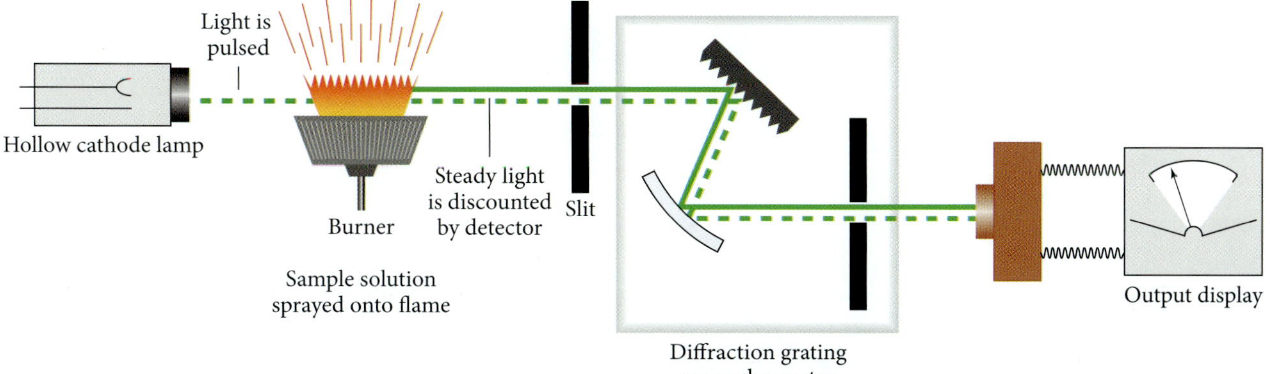

▲ **Figure C1.23**
The process of atomic absorption spectroscopy

First, the element being analysed is determined. This is important because the element in question is usually part of a sample of material such as food, paint or soil. As there are multiple elements present in these substances, shining normal light through them would be useless – all the elements would absorb the light. We need to focus on one element only.

The lamp for this process is made of the same element that is being tested. If zinc is being tested for, then the lamp is made of zinc. If mercury is being tested for, then the lamp is made of mercury. An electric current is passed through a gaseous sample of the element so it will emit light, as described previously. When the lamp is made of a single element, then the light emitted has only the unique set of wavelengths particular to that element.

The sample being tested is **vaporised**, changing the substances it contains into atoms. When the light from the lamp passes through the vaporised sample, only the element being tested for will absorb the light from the lamp. This is because it has the same energy levels as the atoms that emitted the light from the lamp. Other elements in the vaporised sample will not absorb this light because the energy levels of all other atoms are different and their electrons cannot absorb the energies of the light present.

The light passes through the sample and is focused through a slit before entering a **monochromator**. This selects just one wavelength of the light for analysis by the **detector**. The detector measures the intensity of the light, which is then displayed as a number. This number is not a concentration but rather a measure of the amount of light that passed through the sample without being absorbed. This is called an **absorbance** value.

Atomic absorption spectroscopy relies on electron transfer between atomic energy levels and can be used to identify elements.

A quantitative tool

To measure the amount of an element present, the absorbance of the sample is compared to that of known samples. This is done by constructing a **calibration curve**. First, a number of known concentrations of the element are prepared and their intensities are measured by atomic absorption spectroscopy. Then, a calibration curve of concentration against absorbance value is plotted. This allows the concentration of another sample to be compared and determined once its absorbance is measured.

WORKED EXAMPLE 1.3

To determine the concentration of mercury in a sample of fish, the absorbances of some mercury samples of known concentration were measured by atomic absorption spectroscopy. Table C1.11 shows the results obtained. The fish sample was then analysed. Its absorbance value was 0.57. Determine the concentration of the fish sample.

Table C1.11 Measurements of absorbance of known concentrations of mercury

Mercury concentration (parts per million, ppm)	Absorbance
2.0	0.15
4.0	0.30
6.0	0.46
8.0	0.61
10.0	0.74

Answer

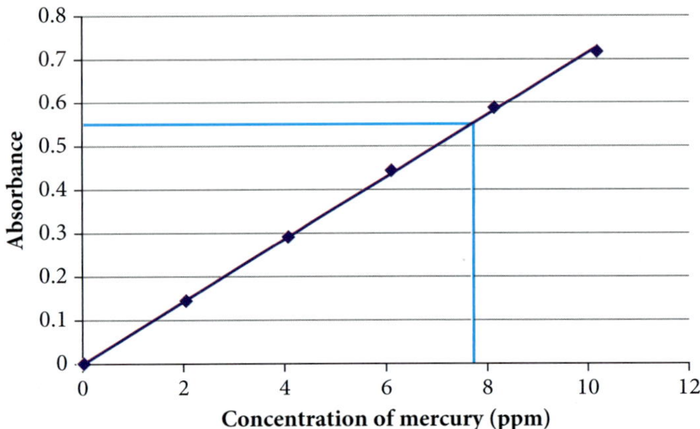

This gives an answer of 7.7 ppm.

Logic

Construct a calibration graph as described above.

By using interpolation on this graph, you can determine the concentration of the unknown sample.

Try these yourself

a Run-off water from a copper mine is suspected of having dangerous levels of copper. A calibration curve is shown in Figure C1.24 for known concentrations of copper.

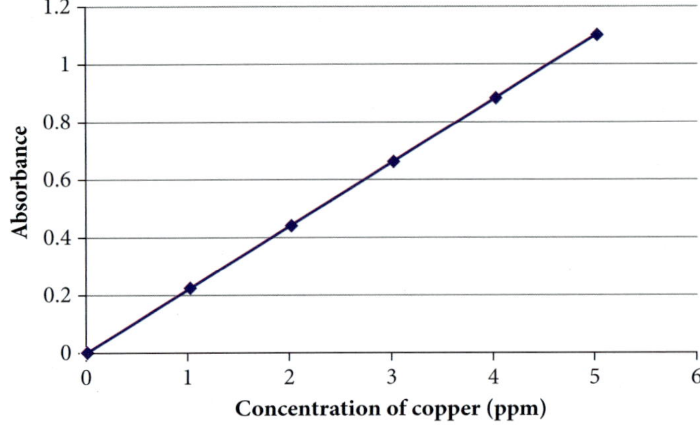

Figure C1.24 ▲
Calibration curve for determination of copper concentration

Samples of the water from the mine were analysed and the following absorbance values were obtained. For each absorbance reading, use the graph to determine the concentration.

i 0.75

ii 0.30

b A sample of soil was suspected of containing high levels of lead, which is dangerous for people working with that soil. To determine the concentration of lead in the soil, a lead lamp was used in the spectroscope to analyse the soil. Lead samples of known concentration were analysed and the data in Table C1.12 were obtained.

Table C1.12 Measurements of absorbance of known concentrations of lead

Lead concentration (ppm)	Absorbance
1.0	0.083
2.0	0.164
4.0	0.331
6.0	0.497

i Construct a calibration curve with the concentration of lead on the horizontal axis and the absorbance on the vertical axis.

ii The sample of soil gave an absorbance reading of 0.290. From the graph, determine the concentration of lead in the soil.

iii Safe levels of lead in the soil are less than 3.5 ppm. Explain whether this sample of soil would be safe to work with.

EXPERIMENT 1.2

CONSTRUCTING A CALIBRATION CURVE TO MEASURE CONCENTRATION OF COPPER(II) SULFATE

Copper sulfate solution is a blue colour; the intensity of the colour is directly related to the concentration of the solution. A number of methods can be used to determine the intensity of the colour. A simple light meter will determine the intensity of light passing through a solution. The more light that is absorbed, the more concentrated the solution is. A colorimeter shines light of a particular wavelength through the sample and measures an absorbance value in a similar way to an atomic absorption spectrometer.

In this experiment, you will measure the intensity of light passing through copper sulfate samples of different concentration and construct a calibration curve. You will then be provided with a sample of copper sulfate of unknown concentration and will use your calibration curve to determine the concentration of this solution.

Aim

To determine the concentration of a solution of copper sulfate through construction and use of a calibration curve

Materials

- Light source and light meter, colorimeter or colorimeter data probe with data logger or laptop
- 1.0 mol L^{-1} solution of copper sulfate – about 35 mL per group of students
- Distilled water
- 50 mL beaker
- 10 mL measuring cylinder

What are the risks in doing this experiment?	How can you manage these risks to stay safe?
Use of chemicals	
Use of light meter or data logger	

In your write-up, add any more risks you can think of, as well as ways to manage them. In particular, expand on the two risks listed to identify specific risks involved with each of them. Ask your teacher to check your risk assessment before you proceed.

Procedure

1 Collect approximately 35 mL of 1.0 mol L^{-1} copper sulfate from your teacher.

2 Regardless of the apparatus you are using, you will be using a piece of clear glassware to hold your sample of copper sulfate. Measure 10 mL into this piece of glassware, or fill the provided container to overflowing.

3 Take a reading of the absorbance of light by the sample. This will be provided as a read-out on your light meter or colorimeter.

4 Measure 8 mL of the 1.0 mol L^{-1} copper sulfate solution into a 10 mL measuring cylinder and make up to 10 mL with distilled water. This will provide you with a 0.8 mol L^{-1} copper sulfate solution.

5 Repeat steps 2 and 3 to determine an absorbance reading for this sample.

6 Measure 6 mL of the 1.0 mol L^{-1} copper sulfate solution into a 10 mL measuring cylinder and make up to 10 mL with distilled water. This will provide you with a 0.6 mol L^{-1} copper sulfate solution.

7 Repeat steps 2 and 3 to determine an absorbance reading for this sample.

8 Measure 4 mL of the 1.0 mol L^{-1} copper sulfate solution into a 10 mL measuring cylinder and make up to 10 mL with distilled water. This will provide you with a 0.4 mol L^{-1} copper sulfate solution.

9 Repeat steps 2 and 3 to determine an absorbance reading for this sample.

10 Measure 2 mL of the 1.0 mol L^{-1} copper sulfate solution into a 10 mL measuring cylinder and make up to 10 mL with distilled water. This will provide you with a 0.2 mol L^{-1} copper sulfate solution.

11 Repeat steps 2 and 3 to determine an absorbance reading for this sample.

12 Collect 10 mL of the unknown concentration sample from your teacher.

13 Repeat step 2 and 3 to determine an absorbance reading for this sample.

Results

1 Draw up a table of results showing your known concentrations and their absorbances. Include the result for the unknown concentration in this table.

2 Construct a fully labelled calibration curve for your known results.

3 Use the graph to determine the concentration of the unknown sample.

Analysis of results

1 Did your graph pass through the origin (0,0)? Why should you expect it to? Suggest reasons why your graph might not do this.

2 Compare your results to those of other groups. Did you all get the same answer? If possible, create a table showing a class set of results. Account for any differences.

Discussion

1 If your teacher can tell you the correct concentration of the unknown sample, discuss the accuracy of your results; that is, how close you got to the true value.

2 Identify one error that may have affected your results. Discuss its effect on the results. Identify the error as random or systematic.

Conclusion

Write a conclusion discussing the results of this experiment.

1.10 Mass spectrometry

To see an application of mass spectrometry, refer to Context 1, 'Matter in the universe', page 11.

Another method of analysing elements is **mass spectrometry**. Unlike emission and absorption spectroscopy, this method is not based on light or the promotion of electrons to a higher energy level. Mass spectrometry is based on the different masses of atoms in a sample. It can be used to determine what elements are present in a sample of material, or what isotopes are present in an element. The method is the same, but we will focus on how it is used to determine the **isotopic composition** of an element. An isotopic composition tells you which isotopes are present in an element and the percentage of each isotope present.

During mass spectrometry, the sample is bombarded with high-energy electrons or **ultraviolet light**. This knocks out or removes electrons, leaving the atom with an overall positive charge. All atoms can be turned into positively charged ions by this method, even elements that would not normally become positively charged.

A sample of carbon that contains two isotopes, carbon-12 and carbon-13, will show two positive ions, each with a different mass. Because they are isotopes, they will all have six protons, but will have either six neutrons (carbon-12) or seven neutrons (carbon-13).

The positive ions in the sample are accelerated through an electric field so they all move at high speeds. They then pass through a magnetic field where they undergo deflection according to their masses. Lighter ions are deflected more by the magnetic field; heavier ions are deflected less. This can be seen in Figure C1.25.

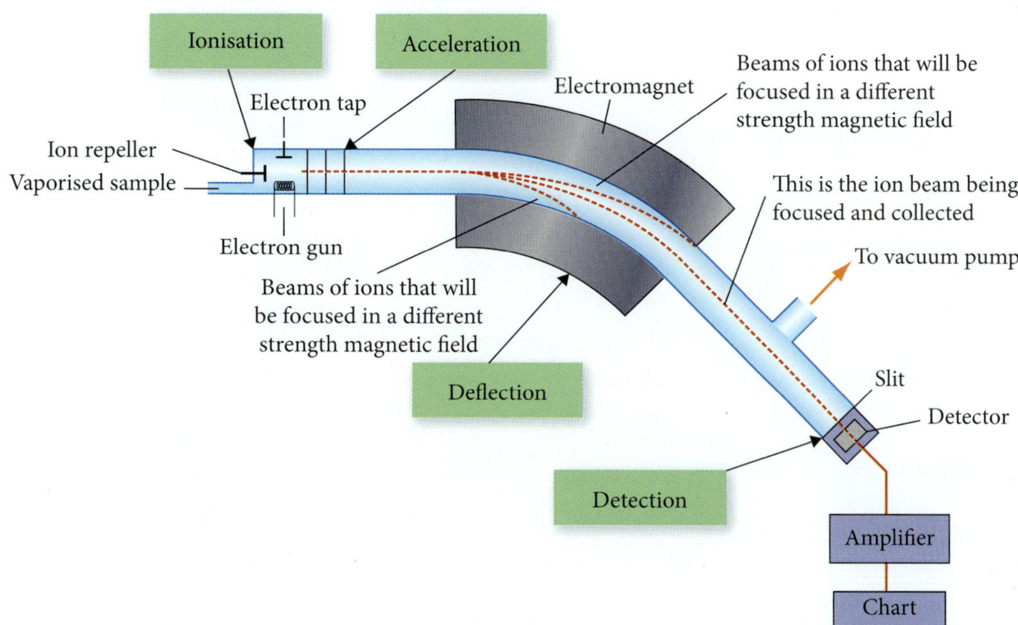

Figure C1.25
Deflection of different mass ions in a mass spectrometer

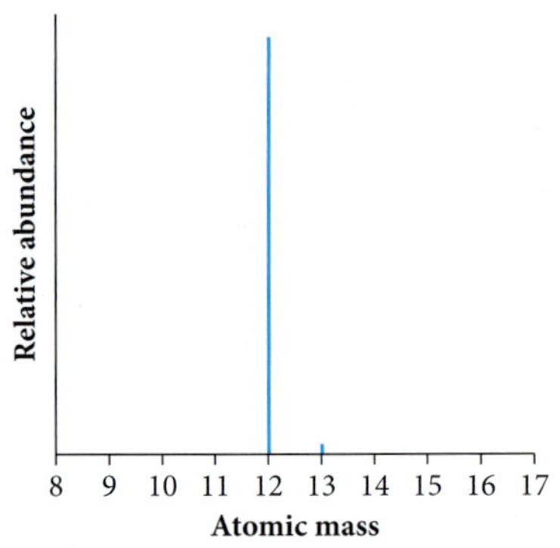

Figure C1.26 ▲
Mass spectrum of carbon isotopes

A quantitative tool

Detectors measure the amount of ions that strike them. This information is transformed into graphical form called a **mass spectrum** (Figure C1.26). This is a graph that shows the mass of the ions that are present and their relative abundance. It clearly shows that an ion with an atomic mass of 12 is present in the greatest concentration and there is a small amount of an ion with an atomic mass of 13. These are the two isotopes of carbon present in the sample.

WORKED EXAMPLE 1.4

Figure C1.27 shows the mass spectrum for molybdenum.

a Determine the number of isotopes and their relative atomic masses.

b Which isotope is present in the highest concentration?

c Which isotopes are present in equal amounts?

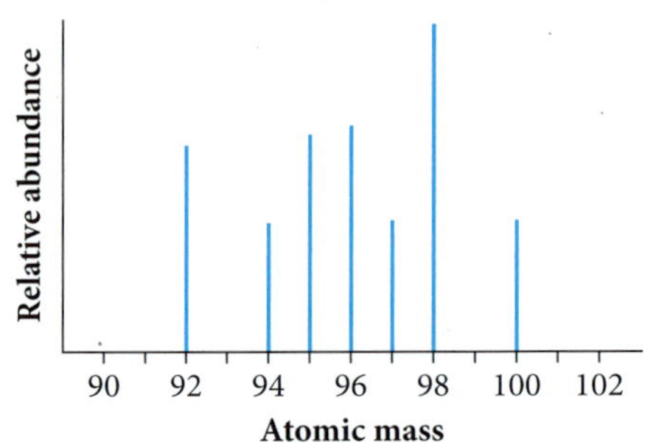

Figure C1.27 ▲
Mass spectrum of molybdenum

Answer	**Logic**
a There are seven isotopes of molybdenum. They have relative atomic masses of 92, 94, 95, 96, 97, 98 and 100.	There are seven peaks in the mass spectrum at 92, 94, 95, 96, 97, 98 and 100.
b The isotope with a mass of 98 has the highest concentration in the sample.	This isotope has the highest peak in the mass spectrum.
c The isotopes at 94, 95 and 100 are present in equal amounts.	Their peaks are the same height.

Try these yourself

Figure C1.28 shows the mass spectrum of an element.

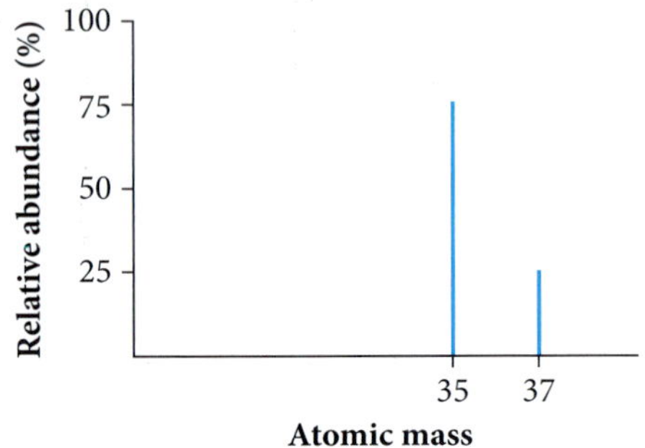

Figure C1.28 ▲
Mass spectrum of an element

a How many isotopes of this element are present?

b What are the relative atomic masses of this isotope?

c Which isotope is present in the greater amount?

d Use the graph to determine the relative abundance of each of the isotopes.

e Calculate the relative atomic mass of this element and state which element it is.

QUESTION SET 1.7

Remembering

1 Describe briefly the process of atomic absorption spectroscopy.

2 Explain how you would construct a calibration curve for the process of atomic absorption spectroscopy.

3 What is the name of the graph formed during the process of mass spectroscopy?

Understanding

4 Explain why the lamp in atomic absorption spectroscopy is made from the same element as the element being tested.

5 Explain why isotopes of different mass can be separated by a mass spectrometer.

Analysing

6 Cadmium is useful in small amounts, but dangerous in large amounts. A sample of paint was tested for its cadmium content. Samples of known concentrations of cadmium were analysed by atomic absorption spectroscopy and the results in Table C1.14 were obtained.

Table C1.14 Measurements of absorbance of known concentrations of cadmium

Cadmium concentration (mg L^{-1})	Absorbance
0.0	0.000
1.0	0.038
2.0	0.082
3.0	0.120
4.0	0.160
5.0	0.200
6.0	0.240
Unknown	0.110

a Construct a calibration curve for the known concentrations of cadmium.

b From the curve, determine the concentration of cadmium in the sample of paint.

7 Figure C1.29 shows the mass spectrum of an element.

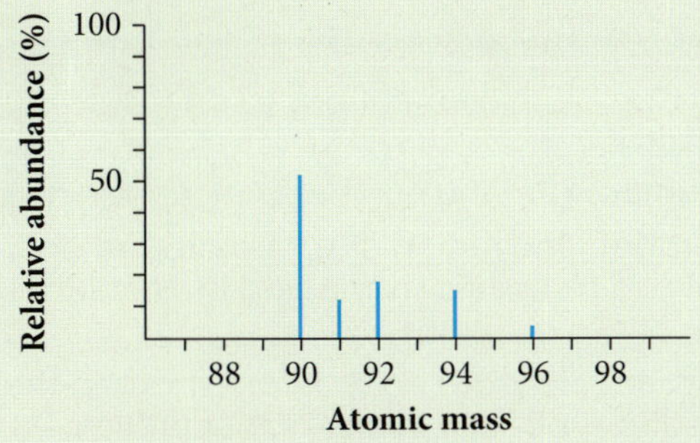

◀ **Figure C1.29** Mass spectrum of an element

a Identify the isotopes present in this element.

b Create a table showing the isotopes and their relative abundance.

c Calculate the relative atomic mass of this element and hence identify this element.

Reflecting

8 Identify some ways that atomic absorption spectroscopy and mass spectroscopy can help us learn more about what elements are in materials, or how we can identify them. Use examples in this section for ideas, or search the Internet for other uses.

CHAPTER CHECKLIST

You should know:

- an atom is made up of a nucleus containing positively charged protons, neutral neutrons, and negatively charged electrons in regions of space called energy levels surrounding the nucleus
- elements are substances composed of one type of atom and are represented by a symbol
- the atomic number of an atom indicates the number of protons and electrons; the mass number indicates the total number of protons and neutrons in the atom
- atoms are held together by forces of attraction between the positive nucleus and negative electrons surrounding the nucleus
- a stable nucleus is held together by the balance of repulsive forces between protons and the attractive strong nuclear force between all particles in the nucleus
- isotopes of an element have the same number of protons, but different numbers of neutrons. They may be stable or unstable, in which case they undergo radioactive decay
- relative atomic mass is used to compare the masses of atoms and elements. It is dependent on the relative abundance of an element's isotopes on Earth
- the periodic table is arranged in order of increasing atomic number and in groups and periods
- electrons are arranged in energy levels according to specific rules. The final arrangement is called an electron configuration. This arrangement can be represented in a variety of different visual ways
- movement of electrons between energy levels leads to analysis of elements through examination of the light they produce as spectra
- atomic absorption spectroscopy allows determination of the concentration of an element
- mass spectrometry can be used to determine the isotopic composition of an element.

You should be able to:

- use atomic representation to determine the number of protons, neutrons and electrons in an atom
- use information about the isotopes of an element to determine its relative atomic mass
- write electron configurations for elements using the periodic table
- identify trends on the periodic table and compare element properties such as electronegativity and metallic character
- use data and graphical results from atomic absorption spectroscopy and mass spectrometry to calculate concentrations of unknown substances and isotopic composition of elements.

CHAPTER GLOSSARY

$^{A}_{Z}X$ representation a format that describes the composition of the atom, including the element symbol (*X*), mass number (*A*) and atomic number (*Z*)

absorbance a measurement taken by a machine that compares the light passing into a substance with that exiting and gives a value

actinoids the period of the periodic table that, with the lanthanoids, make up the f block

alkali metals the common name of the elements found in group 1 of the periodic table

allotrope a different physical form of the same element

alpha particle a particle containing two protons and two neutrons, sometimes called a helium nucleus

atom the fundamental particle of matter; is composed of protons, neutrons and electrons

atomic absorption spectroscopy an analytical technique for determining the unknown concentration of an element based on the amount of light it absorbs

atomic number (*Z*) the number of protons in an atom

atomic orbital the region of space around an atom that has a specific shape and may contain a maximum of two electrons

atomic orbital diagram a diagram showing the space electrons occupy in one of four orbitals – s, p, d and f

atomic radius the distance from the nucleus to the boundary of the cloud of electrons surrounding it

calibration curve a graph constructed during atomic absorption spectroscopy that plots known concentrations against the absorbance values, used to determine the concentration of an unknown substance

chemical property a property of a substance relating to its ability to change to new substances during chemical reactions

detector a device used to measure light or particles, found in both atomic absorption spectroscopy and mass spectrometry processes

electron a negatively charged particle found in energy levels around the nucleus of an atom

electron affinity the ability of an atom in the gaseous state to accept an electron and form a negative ion

electron charge cloud diagram a visual representation of the region of space around a nucleus where an electron might be found

electron configuration the arrangement of electrons around an atom in their energy levels

electron shell diagram a visual representation of electrons in their energy levels around the nucleus

electronegativity the relative ability of an atom to attract electrons

electrostatic attraction a force that pulls particles together when they have an opposite charge

electrostatic repulsion a force that pushes particles apart when they have an identical charge

element a pure substance made up of atoms with the same atomic number

energy level a region of the atom in which electrons of the same energy can be found

energy shell see *energy level*

excited state when an electron is in a higher energy level than the ground state due to absorption of energy

ground state when all the electrons of an atom are in their lowest possible energy levels

group a vertical column in the periodic table that gives information on number of valence shell electrons and trends between atoms

ion a charged atom, either positive from losing electrons, or negative from gaining electrons

ionisation energy the amount of energy needed to remove an electron from a neutral atom when it is a gas

isotopes different forms of an element with the same number of protons but different numbers of neutrons

isotopic composition the number and amount of isotopes within a sample of an element

lanthanoids period of the periodic table that, along with the actinoids, make up the f block

line emission spectrum a pattern of lines showing the component wavelengths in light

mass number (*A*) the total number of protons and neutrons in an atom

mass spectrometry an analytical method that uses the different masses of particles to measure their relative abundance in a sample

mass spectrum a graph produced during mass spectroscopy that shows the mass and relative abundance of substances present

metalloid an element that has properties of both metals and non-metals

monochromator a device used in atomic absorbance spectroscopy to select light of a single wavelength

nanotechnology a branch of science dealing with particles in the range of 1–100 nm

neutron a neutral particle found in the nucleus of an atom

nucleus a region of the atom containing all the protons and neutrons; it occupies only a small part of the volume of the atom but contains most of the mass

period a horizontal row of elements in the periodic table that gives information on the number of energy levels occupied by electrons in an atom

periodic table a chart of the elements, arranged in increasing atomic number; it is organised into groups and periods to show trends in the elements

physical property an observable feature of a substance that can be measured without changing the identity of the substance, such as colour, density and hardness

proton a positively charged particle found in the nucleus of an atom

quantitative analysis analysis that measures values such as amount, concentration or volume rather than just identifying the substance

radioactive decay the spontaneous disintegration of an atom due to instability in the nucleus, during which particles or electromagnetic radiation are released

reactivity the likelihood of an element or substance undergoing a chemical reaction

relative atomic mass the mean mass of an element that takes into account the isotope masses and the relative abundance on Earth; it is measured against carbon-12

spectroscope a device used to separate light into its component wavelengths

spectroscopy the branch of chemistry involving absorption and emission of light from substances

strong nuclear force an attractive force that exists between particles in the nucleus; it is a short-range force acting only on adjacent particles

subshell a part of an energy level that contains orbitals of the same energy

synthetic element element that does not exist in nature, but has been made in a laboratory

transition elements elements found between groups 2 and 13 in the periodic table, also known as the d-block elements

ultraviolet light invisible, high-energy, high-frequency light of wavelengths 10 nm to 400 nm

valence shell the outermost shell of an atom that contains electrons

vaporised when a substance is heated so that it turns into its atomic form

wavelength a property of light related to the length of the wave, which can give properties of light such as colour

CHAPTER REVIEW QUESTIONS

Remembering

1 Define:
  a orbital.
  b allotrope.
  c isotope.
  d energy level.
  e calibration curve.
  f electronegativity.
  g excited state.

2 Compare the sizes and charges of the three particles found in an atom.

3 What information can be determined by the following in the periodic table?
  a The group an element is in
  b The period an element is in

4 How many valence electrons are in the following elements?
  a Oxygen
  b Chlorine
  c Magnesium
  d Selenium

5 Describe how trends in the atomic radius can be determined from the periodic table.

6 Copy and complete the following table, using your knowledge about trends in the periodic table.

Element pair	Highest ionisation energy	Highest electronegativity	More metallic character
Carbon and fluorine			
Sodium and lithium			
Silicon and nitrogen			
Beryllium and boron			

Understanding

7 An element has 14 electrons. Explain how you would fill up the energy levels to gain the final electron configuration.

8 Explain why electronegativity increases from left to right across the periodic table.

9 Explain why light is emitted from an atom when energy is applied to it in the form of heat in a flame test.

10 A sample of copper(II) nitrate is compared to a sample of barium nitrate during a flame test experiment. Explain why they emit light of different colours.

11 Draw electron shell diagrams to represent the arrangement of electrons in the following elements.
  a Helium
  b Carbon
  c Aluminium

12 Write electron configurations for the following elements.
  a Boron
  b Sulfur
  c Lithium
  d Vanadium

13 Identify the elements with the following electron configurations.
  a 2,6
  b 2,8,8
  c 2,8,15,2

Applying

14 a Copy and complete the following table. You may use a periodic table.

Element	Atomic number	Mass number	Number of protons	Number of neutrons
Carbon	6	14		
	17			18
Iron		56		
Copper		64	29	
	6			7

b Identify any isotopes from the table and explain why they are considered to be isotopes.

15 Use atomic representation to show the following information. You may need to extract information from the periodic table.

a Aluminium with atomic number 13 and mass number 27

b Tin with atomic number 50 and mass number 119

c Oxygen with 8 protons and 8 neutrons

d Iodine with 53 protons and 74 neutrons

e Rubidium with mass number 85

16 Silicon has three isotopes: silicon-28 with an abundance of 92.23%, silicon-29 with an abundance of 4.68% and silicon-30 with an abundance of 3.09%. Calculate the relative atomic mass of silicon.

Analysing

17 Figure C1.30 shows the line emission spectra of several common elements and one unknown element. State and explain the identity of the unknown sample.

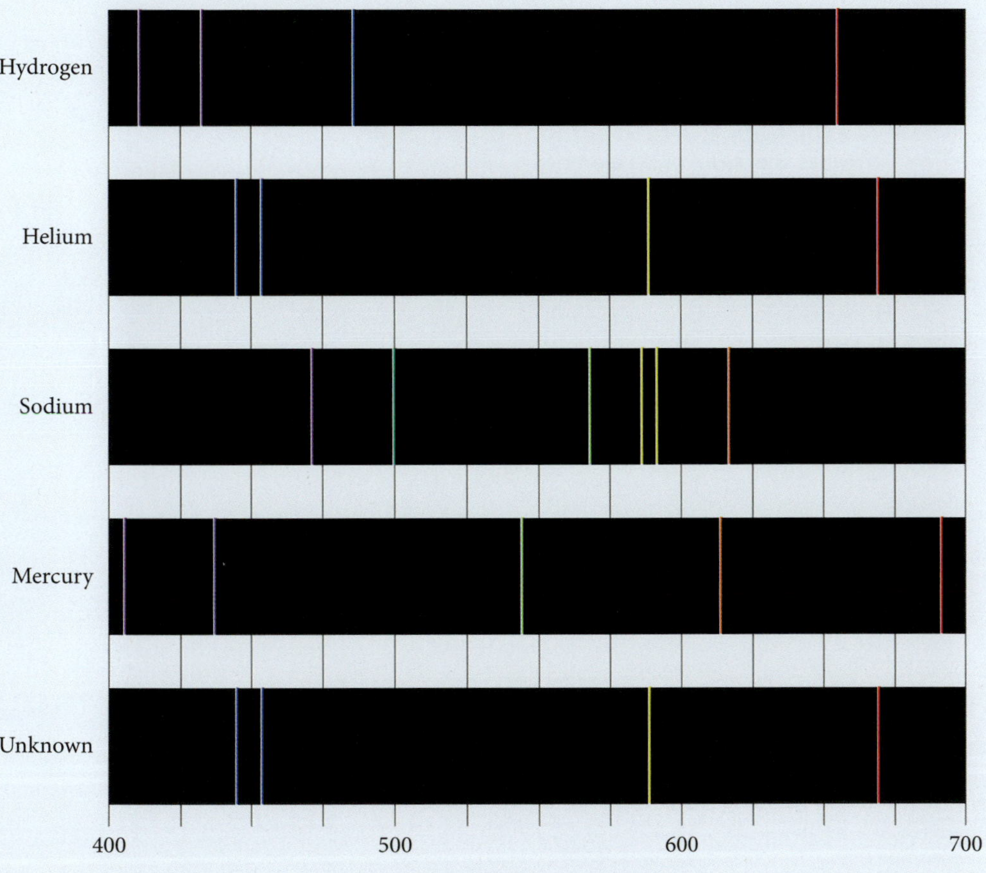

◀ **Figure C1.30** Emission spectra of known elements and an unknown element

18 The level of zinc in some food was labelled on the container. Students analysed samples of the food by atomic absorption spectroscopy to determine whether the level on the container was correct. Samples of known concentration of zinc were analysed and the following data were obtained.

Table C1.15 Measurements of absorbance of known concentrations of zinc

Zinc concentration (mg L^{-1})	Absorbance
1	0.03
2	0.06
5	0.16
10	0.32
15	0.48

a Explain what the lamp would have been made from for this analysis.

b Explain why the presence of a metal such as calcium in the food would not have interfered with this process.

c Construct a calibration curve with the concentration of zinc on the horizontal axis and the absorbance on the vertical axis.

d The sample of food gave an absorbance reading of 0.36. From the graph, determine the concentration of zinc in the food.

e The label on the container stated that the level of zinc in the food did not exceed 7.5 mol L^{-1}. Explain whether this statement was true.

19 Figure C1.31 shows the mass spectrum produced after the isotopic analysis of a particular element.

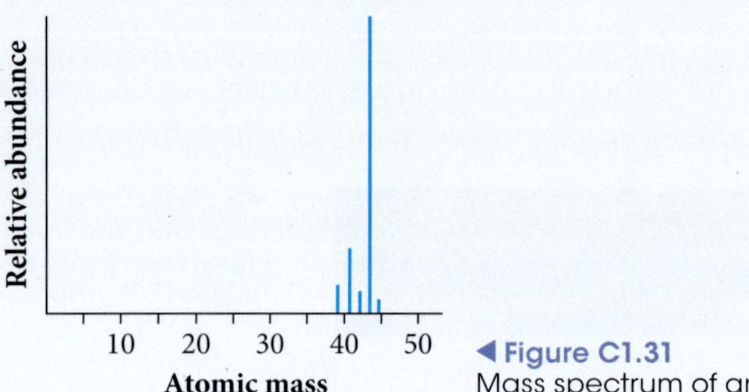

◀ Figure C1.31 Mass spectrum of an element

a Explain why the different isotopes can be separated by a mass spectrometer.

b How many isotopes were present in the element sample?

c Which isotope was present in the highest concentration?

Reflecting

20 Explain why different representations exist to show how electrons are arranged around a nucleus.

CHAPTER 2
CLASSIFYING AND SEPARATING SUBSTANCES

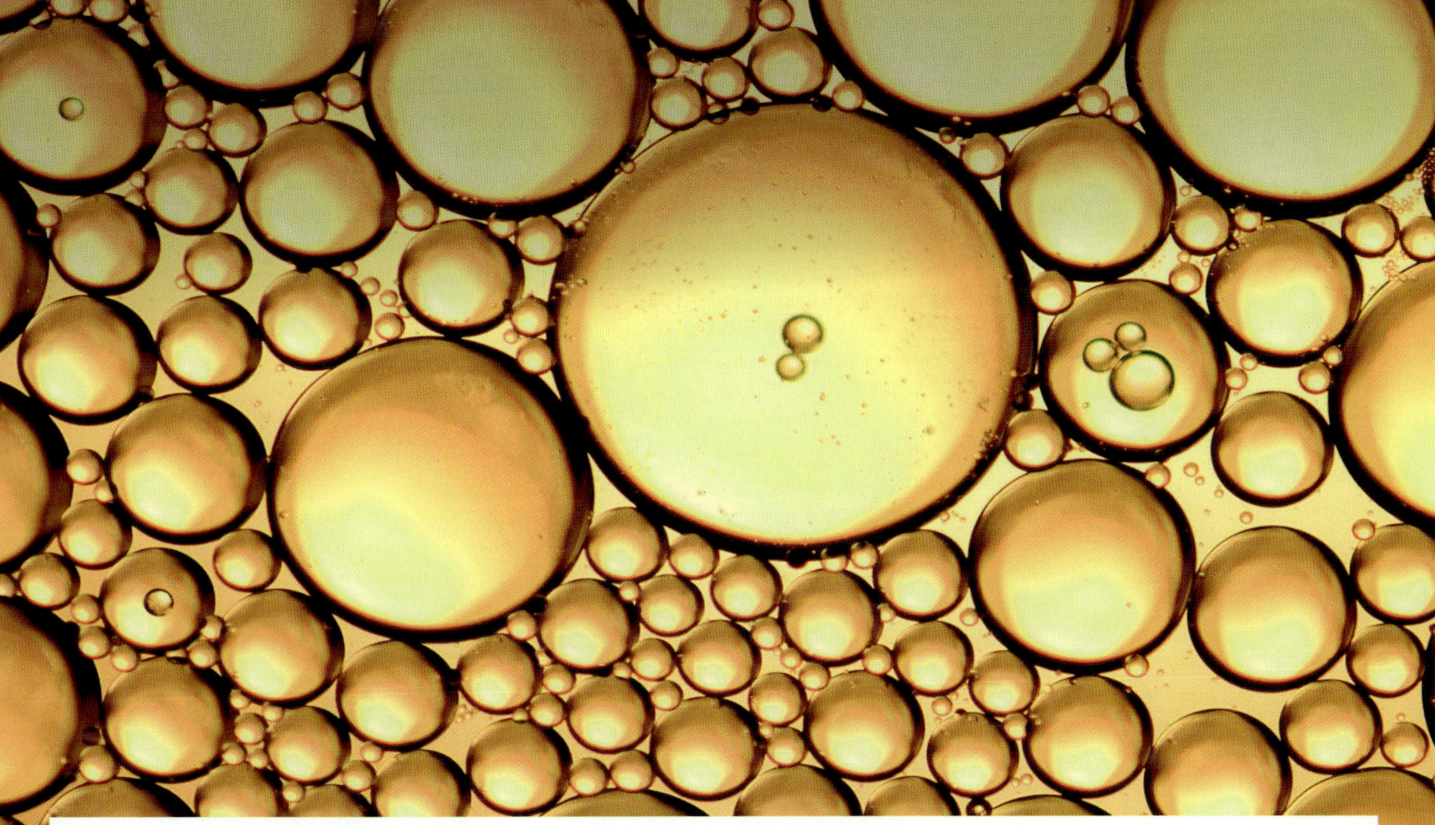

By the end of this chapter you will have covered the following material.

Science Understanding

- Materials are either pure substances with distinct measurable properties (for example, melting and boiling point, reactivity, strength, density) or mixtures with properties dependent on the identity and relative amounts of the substances that make up the mixture (ACSCH025)
- Differences in the properties of substances in a mixture, such as particle size, solubility, magnetism, density, electrostatic attraction, melting point and boiling point, can be used to separate them (ACSCH026)

AC

2.1 Classifying matter

All **matter** that exists naturally originally came from atoms that were formed in stars. These atoms combined in a variety of ways to form Earth and all its **materials**. All the materials we use as part of our daily life come from Earth. Some materials such as water are found naturally; others, such as plastics, have been synthesised.

As chemists seek to understand and investigate materials found on Earth, they also classify them. Materials can be classified according to how they are used, the properties they have or the state (solid, liquid, gas) they are in. Another way chemists look at materials is according to the type of **substances** that the materials are composed of and the way these substances are combined.

123RF/Adroat Komsawart

Figure C2.1 ▲ Earth and the Moon, as viewed from space.

Heterogeneous materials

Most of the familiar materials you see around you, such as wood, concrete, air and milk, are examples of **mixtures**. A mixture is matter that contains two or more different substances. Mixtures can also have different proportions of the same substances. You can see the differences in concrete when you examine it closely. Similarly, granite usually contains at least three different minerals. Different pieces of granite will probably have different amounts of each of these three minerals, but they are still granite.

Sometimes it is necessary to use a microscope to distinguish the different substances in a mixture. For example, it is only by looking at milk under a microscope that you can see the fat globules suspended in water.

Non-uniform mixtures such as granite and milk that contain physically separate materials are called **heterogeneous** mixtures.

Science Photo Library/Natural History Museum, London.

Figure C2.2 ▲ The composition of granite can vary.

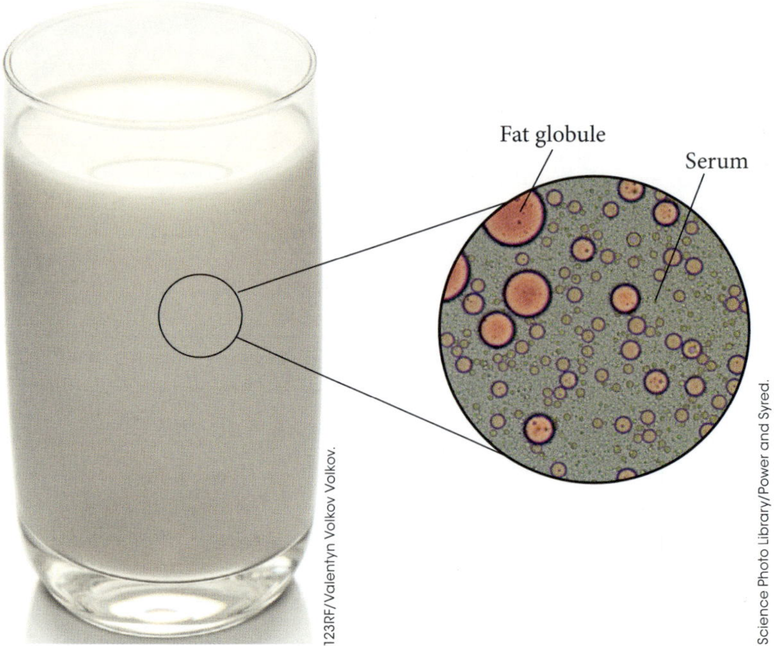

123RF/Valentyn Volkov Volkov.

Science Photo Library/Power and Syred.

Figure C2.3 ▶ It is only at the microscopic level that we can see that milk is a mixture.

Homogeneous materials

Materials that have uniform composition throughout are **homogeneous** mixtures. If you were to break a piece of homogeneous matter into smaller and smaller pieces and looked at the pieces under a microscope, then it would be impossible to distinguish one part of the material from another. Examples of homogeneous materials are raw sugar, salt water and window glass.

▲ Figure C2.4
Sugar, pure salt and saltwater, and glass are homogeneous materials.

While heterogeneous materials are always a mixture, only one type of homogeneous matter can be classified as a mixture; this is a **solution**. A solution consists of a **solute** (generally the dissolved material) in a **solvent** (dissolving material). Because the solute is distributed throughout the solvent as extremely small particles, a solution appears uniform throughout even under a powerful optical microscope. For this reason, solutions are classified as homogeneous.

The other type of homogeneous material is a pure substance. Pure substances are substances that are made up of only one type of particle. They can be divided into two groups based on the atoms they contain. **Elements** are pure substances that are composed of only one type of atom; for example, oxygen, carbon and gold. **Compounds** are pure substances that are composed of more than one type of atom chemically combined in fixed proportions; for example, water and carbon dioxide.

Materials are either pure substances (elements or compounds), which are made up of one type of particle and have constant composition, or mixtures, which are made up of two or more pure substances and have variable composition.

MIXTURES AND PURE SUBSTANCES

Visit this website to see a classification system for distinguishing between mixtures and pure substances.

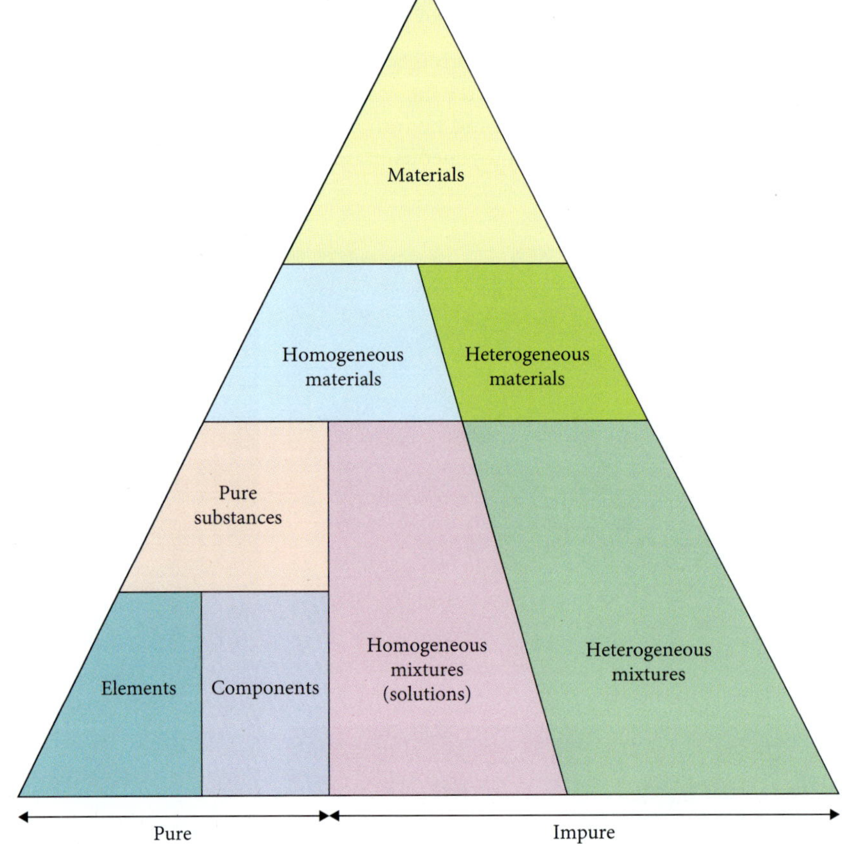

Figure C2.5 ▶ Classification of materials

To learn more about making use of available materials, refer to Context 2, 'Materials for a purpose', page 31.

WOW

A million dollar mixture

You might think that a pure substance is more valuable than an impure substance. However, impure diamonds can be worth more than pure diamonds if they are the right colour. Blue diamonds contain traces of boron while yellow diamonds contain traces of nitrogen. These impurities make the price of the diamonds far greater than ones without these impurities.

123RF.cornelius30.

Figure C2.6 Blue diamonds are very valuable.

QUESTION SET 2.1

Remembering

1 Copy and complete the following table. Distinguish between a mixture and a pure substance.

Mixture	Pure substance
	Homogeneous
Variable composition	
At least two different types of particles	
	Constant properties

Understanding

2 a Classify the following materials as heterogeneous or homogeneous.
   i Copper
   ii Soil
   iii Air
   iv Salt water
   v Carbon dioxide
   vi Salt
   vii Nitrogen
   viii Lemonade
   ix Blood
   x Mercury

  b Classify the homogeneous materials in part **a** as mixtures, elements or pure compounds.

3 Explain why every solution is a mixture but not every mixture is a solution.

Applying

4 Is whipped cream a heterogeneous or homogeneous mixture? Justify your decision.

5 Draw a dichotomous key to separate matter into four distinct groups.

Analysing

6 The following table gives information on four different materials.

Material	Transparent	Conducts heat	Conducts electricity	Malleable
A	Yes	No	No	No
B	No	Yes	Yes	Yes
C	Yes	Yes	No	No
D	No	No	No	Yes

  a Which material could be used to make a saucepan? Why?
  b Which material is best for an electrical cable?
  c Which material is best for making car windows in a cold area? Why?

2.2 Physical and chemical properties

All materials have particular characteristics – chemists call these properties. Properties can be divided into two groups – physical properties and chemical properties.

Materials are chosen and used for a specific purpose because they have properties that suit that purpose. A pure substance has a particular set of distinct, measurable properties that are unique to that substance. Chemists use this uniqueness to identify pure substances.

However, the properties of a mixture depend on the identity and relative amounts of substances that make up that mixture. Changing the relative amounts of substances in a mixture changes the properties of that mixture. For example, a solution of 10 g of salt in 100 mL of water has a higher boiling point that a solution of 1 g of salt in 100 mL of water.

A pure substance has distinct measurable properties that can used to identify it. The properties of a mixture depend on both the identity and relative amounts of substances that make up the mixture.

BOILING POINT ELEVATION AND FREEZING POINT DEPRESSION

Use this interactive to see how boiling points and freezing points of mixtures change as the relative amount of solute to solvent changes.

Physical properties

Physical properties are properties that can be determined without changing the chemical composition of a substance. Examples of physical properties are:

- melting point
- boiling point
- strength
- density
- malleability (ability to be beaten into sheets)
- ductility (ability to be drawn into wires)
- electrical conductivity
- thermal conductivity
- solubility
- state
- hardness.

While pure substances may have similar physical properties and so can be classified into broad groupings such as metals and non-metals, there are always differences that allow individual substances to be identified.

Consider the pure substances water and hydrogen peroxide. Their properties are listed in Table C2.1.

Table C2.1 Properties of water and hydrogen peroxide

Property	Water (H_2O)	Hydrogen peroxide (H_2O_2)
Colour	Colourless	Colourless
State (at 25°C)	Liquid	Liquid
Density (g mL^{-1})	1	1.4
Melting point (°C)	0	−0.4
Boiling point (°C)	100	150

While hydrogen peroxide could be easily mistaken for water because they are both colourless liquids, there are other properties that distinguish them. It would be disastrous to drink hydrogen peroxide as it harmful and corrosive.

Chemical properties

Chemical properties relate to the ability of a substance to react to form new substances. In determining chemical properties, the chemical composition of the original substance is changed. This is one way to distinguish physical and chemical properties. Common chemical properties include:

- decomposition by heat
- effect of light
- reactions with water, acids, bases, oxygen and other substances. For example, iron reacts with oxygen in air and water to form rust (Figure C2.7).

Although it is important to know whether a substance will react, for chemists it is just as important to find out if a substance does not react. For example, the fact that carbon dioxide is not flammable (does not react with oxygen) is why it is used in fire extinguishers.

The chemical properties of substances can be used to identify them and distinguish between different substances. Chemists use the properties of substances to identify unknown substances (for example, in forensic chemistry) or to identify if they come from the same source (see weblink 'Lava sampling').

123RF/Wouter Tolenaars.

Figure C2.7 ▲ Iron reacts with oxygen and water to form rust – a new compound.

Physical and chemical changes

Matter can undergo physical and chemical changes. **Physical changes** are changes in the physical properties, such as density, state and colour. In a physical change, there is no change in the chemical composition of the substance. Tearing paper, dissolving salt and freezing water are all examples of physical changes.

A change in which at least one new substance is formed is a **chemical change**. The chemical composition of the original substance has changed and the new substance formed has different chemical and physical properties. The types of chemical change a substance undergoes relate to its chemical properties.

Burning, digestion and fermenting are all examples of chemical changes. When copper is placed in colourless nitric acid, the liquid turns blue and a brown gas is given off. This indicates a chemical change has occurred.

Chemical changes are also referred to as chemical reactions. Common ways of determining whether a chemical change has occurred are if:

- a solid (called a precipitate) is formed; for example, when lead nitrate and sodium iodide solutions are mixed, a bright yellow solid is produced
- a gas is produced; for example, when magnesium metal is dropped into hydrochloric acid, a gas is produced
- there is a colour change; for example, when iron is exposed to air and water, it turns a reddish brown
- there is a significant change in temperature (energy is released or absorbed); for example, when sodium is dropped into water, heat is given off
- an insoluble solid disappears; for example, copper in nitric acid, as shown in Figure C2.8.

LAVA SAMPLING

Watch this video to see how the chemical properties of collected lava samples are used to determine whether two different volcanoes in Hawaii, USA, are connected.

To learn more about materials, refer to Context 2, 'Materials for a purpose', page 31.

Getty Images/Charles D. Winters.

Figure C2.8
The reaction of copper with nitric acid produces a blue solution and a brown gas.

Cycad seeds

The seeds of the cycad tree have a high nutrient content but are covered with a fleshy covering that contains toxic chemicals. The Djirbalngan people of North Queensland know this and use processes that change both chemical and physical properties to prepare the seeds for consumption. After the seeds are roasted to change the toxic chemicals, they are crushed. The crushed pieces are placed in a woven basket in running water for 12–24 hours to dissolve and wash away any remaining undesirable chemicals. The material is then baked into a bread-like substance.

Alamy/AfriPics.com.

Figure C2.9 ▶ A cycad seed contains toxins that can be chemically changed so that they are no longer harmful.

EXPERIMENT 2.1

OBSERVING CHEMICAL AND PHYSICAL PROPERTIES OF PURE SUBSTANCES

Physical properties can be observed and measured without changing the composition of a substance. They can be used to identify substances. Chemical properties can only be observed when the substance is undergoing a change in composition.

Aim

To observe the physical and chemical properties of different substances

Materials

- 5 g sulfur (S)
- 5 g iron filings (Fe)
- 5 g sodium hydrogen carbonate
- 5 g sodium chloride
- 5 g sucrose
- 2 × 1 cm strips magnesium ribbon (Mg)
- 30 mL of 2 mol L^{-1} hydrochloric acid (HCl)
- 5 g copper(II) chloride
- 3 × 5 cm squares of aluminium foil
- 50 mL beaker
- Glass stirring rod
- Watch glass
- Magnifying glass
- 12 test tubes and test-tube holder
- Magnet
- Spatula
- Bunsen burner
- 10 mL measuring cylinder
- 8 pieces of 10 cm × 10 cm paper
- Distilled water
- Forceps
- Test-tube tongs
- Metal tongs
- Matches

▶

What are the risks in doing this experiment?	How can you manage these risks to stay safe?
Hydrochloric acid is corrosive.	Wear safety glasses and protective clothing. Take care when pouring and clean up spills immediately. If spilt on skin, wash the affected area with plenty of water and notify your teacher.
Copper(II) chloride is toxic.	Wear appropriate safety gear. Dispose of it in the chemical waste jar provided.
Substances may spit out of heated test tube.	Wear appropriate gear. Do not point the test tube at anyone; use tongs.
Powdered sulfur is an irritant to eyes, nose and throat.	Stir the mixture carefully and do not breathe in the fumes.

In your write-up, add any more risks you can think of, as well as ways to manage them. Ask your teacher to check your risk assessment before proceeding.

Procedure

Part A: Observing physical properties

1 Obtain eight pieces of paper. Label each piece with: sulfur, iron filings, sodium hydrogen carbonate, sodium chloride, sucrose, magnesium, copper chloride or aluminium.

2 Using forceps, transfer one piece of magnesium and one piece of aluminium onto the appropriately labelled piece of paper. Using a clean spatula, transfer a small amount of each of the other substances onto the correspondingly labelled sheet.

3 Examine each substance carefully with the magnifying glass. Record your observations in the results table.

4 Test the effect of a magnet on each substance by moving the magnet under the sheet of paper. Record your observations.

5 Add a small amount of each substance to 5 mL of water in separate test tubes. Record your observations.

6 Save any solid samples remaining on the paper. Pour the copper chloride and water into the chemical waste jar. Dispose of solutions down the sink and add any solids to a waste container.

Part B: Observing chemical properties and changes

7 Combine the remaining iron and sulfur and mix them with a spatula. Test with a magnet. Record your observations in the results table.

8 Place a strip of magnesium in a clean, dry test tube and add 5 mL of HCl. Record observations. Feel the test tube and record any temperature difference.

9 Repeat step 8 with small amounts of sodium hydrogen carbonate, sodium chloride, sucrose, copper chloride and aluminium. Remember that no reaction is also a result.

10 Place a watch glass close to the Bunsen burner. Light the Bunsen burner. Grasp a piece of aluminium with the metal tongs and hold it in the flame. Record your observations. Place any remaining aluminium on the watch glass.

11 Place 2 g of sucrose into a clean, dry test tube. Grasp the test tube with a test-tube holder and heat it gently over the Bunsen flame. Record your observations. Make sure to check for odour.

12 If no change has been observed, then heat the sample more vigorously for 1 minute. Remove it from the flame and place it in test-tube rack to cool. Scrape some of the residue from the test tube with a spatula and examine it.

13 Test the solubility of the residue and record your observations.

14 Discard the cooled test tube with the remainder of the residue into a waste glass container.

15 One-third fill a 50 mL beaker with distilled water. Add a spatula full of copper chloride to the water and stir it with a glass rod.

16 Loosely crumple a piece of aluminium foil and place it in the beaker. Record your observations. Note any changes in temperature.

17 Dispose of beaker contents in a chemical waste jar. Do not pour them down the sink.

Results

1 Record your results for the physical changes in part A in a table like the one below.

Substance	Physical state	Colour	Odour	Solubility in water	Effect of magnet
Sulfur					
Iron filings					
Sodium hydrogen carbonate					
Sodium chloride					
Sucrose					
Magnesium					
Copper chloride					
Aluminium					

2 Record your results for the chemical properties and changes in part B in a table like the one below.

Substances	Observations
Fe and S – tested with magnet	
Mg and HCl	
Sucrose and HCl	
Sodium hydrogen carbonate and HCl	
Sodium chloride and HCl	
Copper(II) chloride and HCl	
Al and HCl	
Sucrose – heated	
Al – heated	
Copper(II) chloride and Al	

Analysis of results

1 Classify each result in part B as a physical change, a chemical change or no change.
2 For those changes identified as chemical describe the evidence to support the decision.
3 Compare the physical properties of:
   a heated aluminium with aluminium.
   b heated sucrose with sucrose.
4 Summarise the chemical properties of aluminium and sucrose.

Conclusion

In your own words, state the difference between a physical and chemical change.

QUESTION SET 2.2

Remembering

1 Copy and complete the following table to show the differences between physical and chemical changes.

Physical change	Chemical change
	New substance formed
Easy to reverse	
	Generally large energy change
Generally physical properties don't change	
Mass conserved	

2 Classify the following properties as physical or chemical.
- a Melting point
- b Ductility
- c Reactivity
- d Solubility
- e Flammability
- f Colour
- g Odour
- h Catches fire in air

3 List four indicators that a chemical reaction has taken place.

Understanding

4 Classify the following changes as physical or chemical.
- a Fading dye on cloth
- b Melting ice
- c Digesting food
- d Dissolving sugar in tea
- e Burning a candle
- f Making a cake
- g Evaporating water
- h Producing light in a glow stick

Applying

5 Read the wow box 'Cycad seeds' that describes the steps used by the Djirbalngan people to make the cycad seed edible.
- a List the steps used in the seed preparation.
- b Identify each step as a physical or chemical change.
- c Identify the physical or chemical properties involved in each change.

6 Helium does not react naturally with any substance and so does not form chemical compounds. Is it correct to say helium has no chemical properties? Explain.

Reflecting

7 Design an experiment to answer the question: Does water boil faster with or without salt added?

2.3 Separating mixtures

Most of the substances found naturally on Earth are mixtures. One of the most important substances, water, is a compound of hydrogen and oxygen. However, all natural sources of water are mixtures containing many dissolved salts, the most common of which is sodium chloride.

◀ **Figure C2.10**
Sea water is a mixture of water, sodium chloride and other compounds.

WOW

Salts from the Dead Sea

The Dead Sea contains such large amounts of dissolved salts that almost nothing can live in its waters. The Dead Sea's salts are valuable to both industry and agriculture. Countries adjacent to the Dead Sea extract the salts sodium chloride, potassium chloride, magnesium chloride and magnesium bromide from the water. The extraction process relies on one physical property – solubility. The salts are separated by differences in solubility at different temperatures.

Mixtures are combinations of two or more pure substances that have not been combined chemically, so they can be separated into their components by physical or mechanical means.

Chemists have developed many different separation techniques based on the properties of the materials they want to separate. Table C2.2 summarises mixture types, possible separation methods and properties the methods depend upon. The table does not list all possible separation techniques and you will learn about others during your Chemistry course.

The properties of the components in a mixture determine the technique used to separate them.

DISTINGUISH BETWEEN A MIXTURE AND A COMPOUND

Visit this website and try the animations, video and simulations of many of the separation techniques discussed.

Table C2.2 Separation method, typical mixture and properties

Separation method	Typical mixture separated by this method	Property used in separation
Electrostatic attraction	Mixture of electrostatic and non-electrostatic materials	Difference in electrical charge
Filtration	Mixture of insoluble solid and liquid	Difference in state and size of particles
Fractional distillation	Mixture of liquids	Significant but small difference in boiling points
Magnetic separation	Mixture of magnetic and non-magnetic material	Difference in attraction to a magnetic field
Separating funnel	Mixture of immiscible (undissolved) liquids	Difference in densities
Sieving	Mixture of solids or solids and liquids	Difference in particle size
Simple distillation	Mixture of liquids or liquids and solids	Big difference in boiling points
Vaporisation (evaporation or boiling)	Solution containing dissolved solids	Liquid has a much lower boiling point than dissolved solid

Many mixtures cannot be separated by only one technique and several techniques need to be applied in a particular sequence to obtain the desired products. For example, to separate a mixture of salt water, iron filings and sand, the techniques and order in which they were performed would determine the final products. You could use a magnet to remove the iron filings, and then filter the remaining combination of salt water and sand to separate out the sand, and then evaporate the water to obtain the salt. In this process, the water is not collected. If the water was also a required product, then simple distillation could be used instead of evaporation.

Separation by difference in particle size

There are two techniques that separate mixtures according to the size of the particles.

Sieving is one way of separating mixtures of solids, or solids and liquids with different-sized particles. The mixture is poured through a sieve (wire gauze in a frame) and particles that are smaller than the sieve pass through while the larger particles are trapped by the sieve.

Sieving is commonly used in cooking. For example, a mixture of rice and water is poured through a sieve to capture the rice and remove the water. Sieving is also used in the pharmaceutical, ceramics and mining industries. Sometimes, several sieves are stacked together with each one having smaller holes than the one above. In banks, mixtures of coins are

Science Source

◀ **Figure C2.11** Sieving is a common way of separating mixtures.

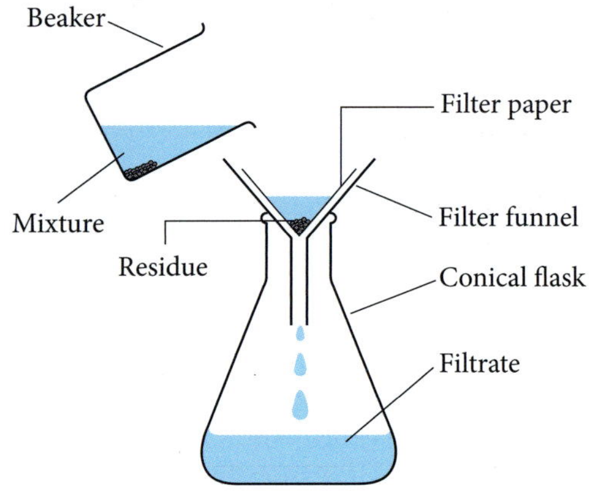

Figure C2.12 ▲
Filtration in the laboratory

separated by pouring them into a stack of sieves of different sizes. The stack is shaken and the smaller coins fall through until there is a tray of 5 cents on the bottom, a tray of $2 coins above that and so on.

Filtration is a separation technique that also depends on particle size. It is more commonly used for mixtures of solids and liquids, particularly when the solid particles are quite small. A typical filtration set up used in the laboratory is shown in Figure C2.12.

The liquid or solution is poured into the filter paper. The liquid passes through the filter paper and the solid, or **residue**, is trapped by the filter paper. The liquid or solution that passes through is called the **filtrate**.

Filtration is also used to separate larger particles suspended in air or to remove pollutants from exhaust gases produced by cars or power stations to prevent them being released into the atmosphere.

Separation by difference in boiling point

Three main separation techniques depend on differences in boiling point. **Vaporisation** is most commonly used to retrieve a solid that has been dissolved in a liquid (i.e., when there is a solution). The liquid component of the solution (solvent) is converted to vapour by either boiling the solution or allowing it to evaporate, leaving behind the dissolved solid (solute). Boiling is a much quicker process than evaporation. This technique is based on the large differences in boiling point of the solute and solvent.

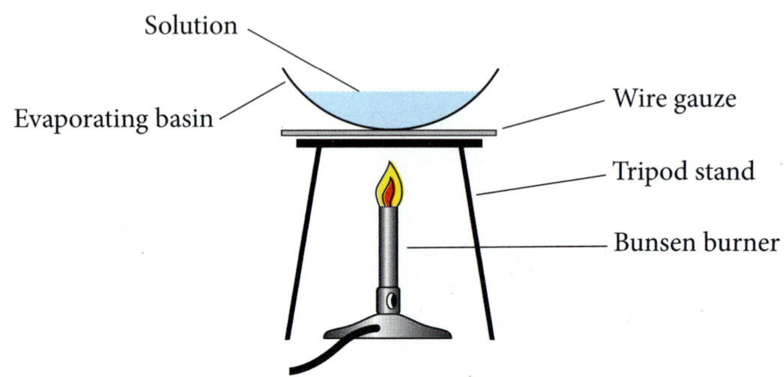

Figure C2.13 ▲
Vaporising a solution.

Figure C2.13 shows a laboratory set-up for vaporising a solution through boiling. Chemists frequently use boiling to remove all the solvent from the solution to obtain a dry solute. They call this evaporating to dryness.

Vaporising the liquid is effective when the dissolved solid is the desired product; for example, if you want to recover salt from sea water. However, this technique does not recover the liquid. If you want to recover the liquid, or the liquid and solid, then simple distillation should be used.

Distillation, or simple distillation, is a technique for separating two or more liquids or separating the liquid from the solids in a solution but also retrieving the liquid component. This technique relies on a difference of at least 50°C in boiling point between components to obtain an effective separation. Figure C2.14 shows a simple distillation apparatus.

Figure C2.14 ▶
Simple distillation apparatus

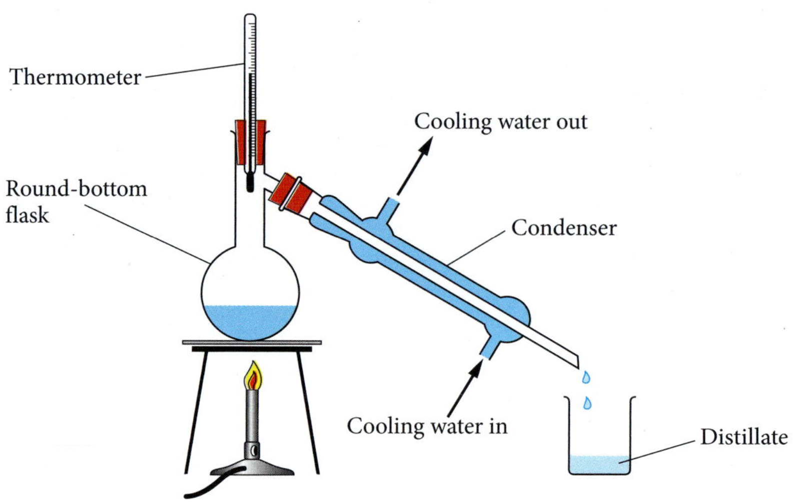

The mixture to be separated is placed in a round-bottom flask and heated to boiling. The vapour rises up the neck of the flask and then flows into the water-cooled condenser. The vapour condenses back to a liquid called the **distillate** and is collected in a beaker.

The liquid with the higher boiling point and any solids are left in the round-bottom flask. If the boiling points are not sufficiently different, then the distillate will still be a mixture.

Fractional distillation is the technique used to separate liquids that have boiling points that are close together. The mixture is heated and the components or fractions with different boiling points rise up the fractionating column to different heights. The component with the lowest boiling point is collected at the top of the column, while the component with the highest boiling point is at the bottom of the column. Figure C2.15 shows a fractional distillation column used to separate crude oil into its components.

Fractional distillation is widely used in industry for:

- separating crude oil into its components at oil refineries
- natural gas processing
- cryogenic air separation to obtain oxygen, nitrogen and argon from liquid air.

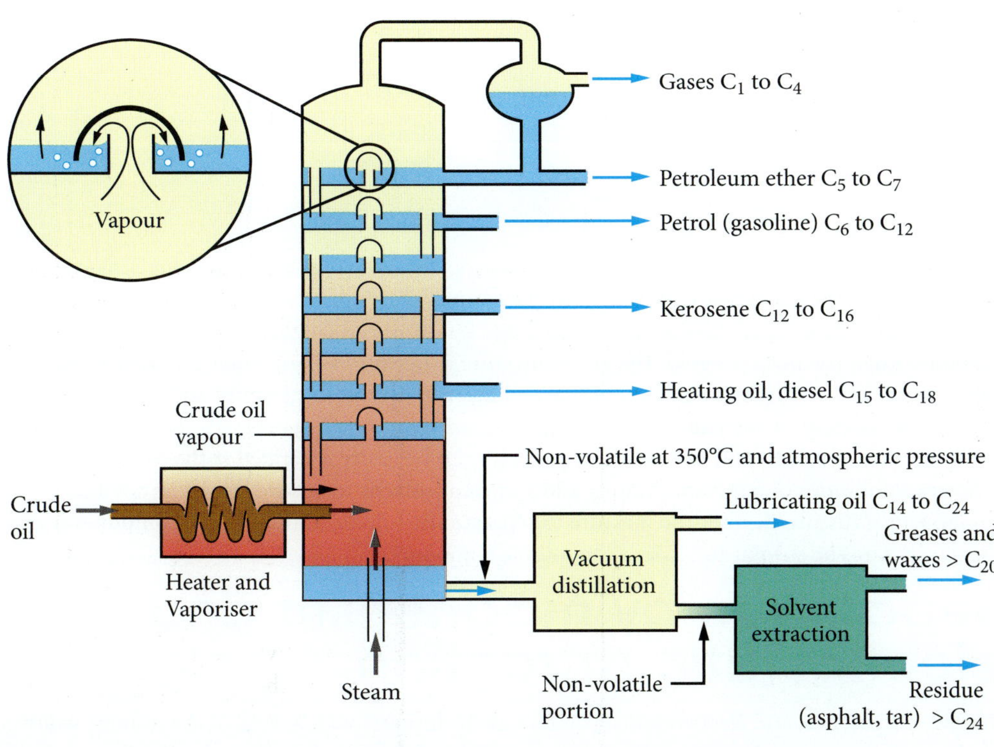

◀ **Figure C2.15** Fractional distillation of crude oil

WOW

Discovery of the noble gases

In 1894, Sir William Ramsay and Lord Rayleigh, using two different methods, separated all known gases from air and found a monatomic, chemically inert gaseous element remaining that they called argon. In 1898, Ramsay isolated the elements neon, krypton and xenon from air brought to a liquid state at high pressure and low temperature. The discovery of the noble gases changed the periodic table and led to new ideas about chemical bonding and atomic structure. Ramsay was awarded a Nobel Prize for his discovery.

Separation by density and solubility

A **separating funnel** as shown in Figure C2.16 can be used to separate two **immiscible** liquids, such as oil and water. Immiscible liquids separate into two layers, one (the less dense) on top of the other (the more dense), if left to stand. These liquids are not soluble in each other.

Figure C2.16 ▶ Two immiscible liquids in a separating funnel

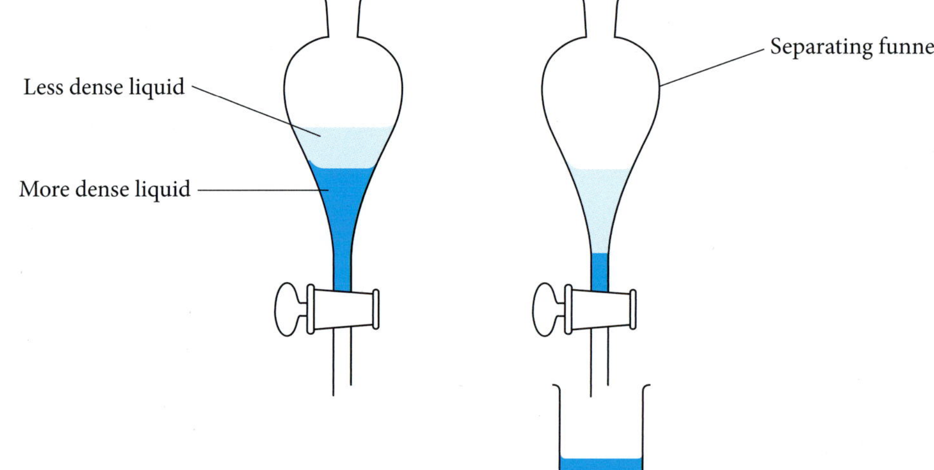

To learn more about making use of materials, refer to Context 2, 'Materials for a purpose', page 31.

When the stopcock at the base of the funnel is opened, the denser component runs into a beaker, leaving the less dense component in the funnel.

Solids of different densities are often separated by using running water and agitation. For example, when panning for gold, the gravel mixture is put into a pan, which is swirled in a running stream. The lighter particles are carried away by the running water and the heavier gold sinks to the bottom of the pan.

A mixture of solids with similar-sized particles can be easily separated if the solids have different solubilities in a solvent. Simply add sufficient solvent to ensure the entire soluble component is dissolved, filter the mixture to separate the soluble and insoluble components and then evaporate the solvent to retrieve the soluble component.

Separation by magnetism and electrostatic attraction

Magnetic separation and electrostatic separation are techniques widely used in the mining industry.

Magnetic separation uses the degree to which a substance is attracted to a magnetic field. Strongly magnetic materials such as iron, cobalt and nickel (and their ores) can be removed from low or non-magnetic materials by a low-intensity magnetic separator.

In Experiment 2.1, you used a magnet to determine which materials were magnetic, and you could have separated the iron and sulfur mixture by using magnetic separation. A simple example of magnetic separation can be seen in a junkyard where an electromagnet is used to separate out iron-based objects.

An **electrostatic separation** method separates particles on the basis of differences in electrical charge. As a mixture is brought into an electric field, differently charged particles will be attracted or repelled and follow different paths so they can be caught separately.

This separation technique is used in industrial plants that process mineral sands containing zircon, rutile and monazite.

SEPARATION OF MIXTURES BY DIFFERENT TECHNIQUES

Visit this website for theory, animations, videos and simulations of many separation techniques.

INVESTIGATION 2.1

SEPARATING SODIUM CHLORIDE AND COPPER(II) CHLORIDE

You have been given a mixture of sodium chloride and copper chloride. Both substances have similar solubilities in water but one is more soluble in ethanol than the other. Use this fact along with what you have learnt about different separation techniques to design and conduct an investigation to separate these substances as effectively as possible.

What is your aim?

Think about what you want to find out in this investigation.

What will you need?

Think about the equipment and chemicals will you need to complete the task. Be specific about quantities for all your materials; for example, chemicals and sizes of beakers. Remember to include basic materials such as matches, water and paper towel.

What are the risks?

Construct a table similar to the one below. Identify specific risks involved in the investigation and ways that you will manage the risks to avoid injuries or damage to equipment. Ask your teacher to check your risk assessment before you proceed.

What are the risks in doing this investigation?	How can you manage these risks to stay safe?

How will you carry out the investigation?

Hint: You will need to use the properties of solubility and boiling point. Write your method in a logical sequence of steps. Do not forget to be specific and include quantities.

What results will you collect?

Refer back to your aim. What data and observations do you need to record to determine whether you have achieved your aim?

What have you found?

Consider what the results show. Were they what you expected? Think about how you are going to determine how effective your separation was.

What do you conclude?

Reflect on what you found and consider whether or not you achieved your aim. What do the results show you about your investigation?

How could you improve the effectiveness of the separation?

Consider how successful your methodology was. What might you change or do differently to improve the separation?

QUESTION SET 2.3

Remembering

1 What is the relationship between hole size of a sieve and the effectiveness of the sieving process?

2 Describe the difference between simple distillation and fractional distillation.

Applying

3 Copy and complete the following table by giving the separation method and properties that enable separation.

	Separation method	Property enabling separation
Separating mud from water		
Separating salt and sand		
Separating oil and water		
Obtaining nitrogen and oxygen from liquid air		
Obtaining fresh water from salt water		
Separating gold from soil		

Analysing

4 The following lists the properties of beach sand and some minerals obtained from it.

Material	Magnetic	Electric charge	Density
Sand	No	No	Low
Rutile	No	Yes	High
Ilmenite	Yes	Yes	High
Zircon	No	No	High

a Using the information in the table, construct a flowchart to show how a mixture of sand, rutile, ilmenite and zircon could be separated.

b Research the techniques used by sand-mining companies and compare their techniques to the ones you used in part **a**.

5 A group of students is given a mixture of iron filings, oil, sugar, water and clear glass beads. They are asked to obtain pure samples of all components except the water.

The following flowchart shows the procedure they follow.

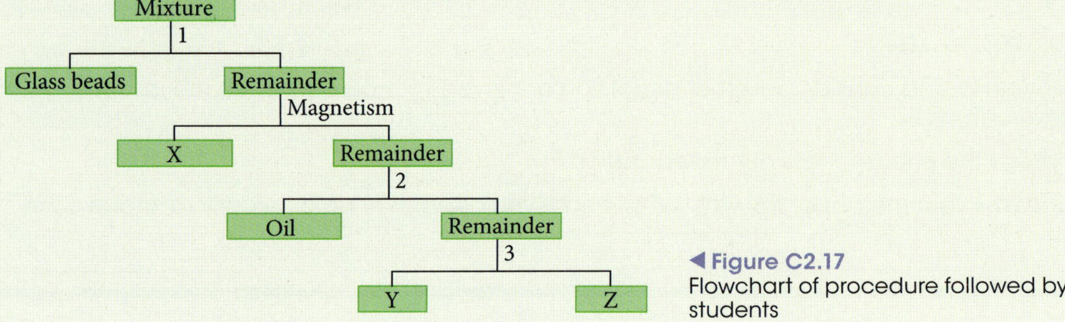

◀ **Figure C2.17** Flowchart of procedure followed by students

a Name the separation procedures 1, 2, 3 and the equipment they require.

b Identify substances X, Y and Z.

c The students are told that the oil sample is contaminated with paraffin wax. What advice would you give them on how they could separate this and still keep the oil?

CHAPTER CHECKLIST

You should know:

- how matter can be classified
- the difference between a mixture and a pure substance
- the difference between elements and compounds
- the difference between physical and chemical properties
- techniques for separating a mixture
- properties on which separation techniques are based.

You should be able to:

- classify different types of matter
- distinguish between, and identify, physical and chemical properties
- identify appropriate techniques for separating a given mixture
- plan and conduct a separation investigation
- identify the products of a separation process.

CHAPTER GLOSSARY

chemical change a change in the chemical composition of a substance to produce a new substance

chemical property a property of a substance relating to its ability to change to new substances during chemical reactions

compound a pure substance composed of more than one type of atom chemically combined in fixed proportions

distillate the vapour that condenses back to a liquid and which is collected after distillation

distillation a technique for separating a solution and retrieving the liquid component, based on differences in boiling point

electrostatic separation a technique for separating substances based on attraction to an electric field

element a pure substance made up of atoms with the same atomic number

filtrate the liquid or solution that passes through the filter paper during filtration

filtration a technique for separating a solid from a liquid or a solution, based on solubility

fractional distillation a technique similar to distillation, used to separate liquids that have similar boiling points

heterogeneous having non-uniform composition

homogeneous having uniform composition throughout

immiscible liquids that do not mix

magnetic separation a technique for separating substances based on attraction to a magnetic field

material a substance with particular qualities or that is used for specific purposes

matter a physical substance; anything that has mass and occupies space

mixture matter that contains two or more different materials or substances with varying composition

physical change a change in the physical properties of a substance but not the chemical composition of the substance

physical property a property that can be determined without changing the chemical composition of a substance

residue the solid remaining in the filter paper after filtration

separating funnel equipment used for separating immiscible liquids

sieving a technique for separating a mixture based on particle size

solute the substance that dissolves in a solution, usually present in smaller amount

solution a homogeneous mixture

solvent the substance in which the solute dissolves or the greater part of a solution

substance a homogeneous material made up of the same type of particle

vaporisation a technique for separating the liquid component of a solution, based on boiling point

CHAPTER REVIEW QUESTIONS

Remembering

1 List two differences between a mixture and a pure substance.

2 Describe how you could distinguish between a physical and chemical change.

3 Classify the following properties as physical or chemical.

   a Density
   b Boiling point
   c Reaction with oxygen
   d Solubility

4 Name the property or properties on which the following separation techniques rely.

   a Filtration
   b Fractional distillation
   c Electrostatic separation
   d Vaporisation

Understanding

5 Classify the following materials as heterogeneous mixture, solution, compound or element.

   a Paint
   b Tap water
   c Oxygen
   d Honey
   e Granite
   f Lead sheeting

6 Identify each of the following as a physical or chemical change.

   a Burning a piece of coal
   b Tearing a piece of paper
   c Tarnishing of silver
   d Fruit rotting
   e Dissolving instant coffee in hot water

7 Why is lack of reactivity considered to be a chemical property?

8 Explain the difference between a physical property and a physical change.

Applying

9 Four bottles contain sodium chloride, nitrogen, iron filings and water. What properties could be used to identify each of these substances?

10 For each of the mixtures listed below name the physical property and method that could be used to separate them.

   a Salt and iron filings
   b Iron filings and aluminium filings
   c Sugar and water
   d Oil and vinegar

Analysing

11 In an experiment, two clear liquids are combined and a white solid forms and settles to the bottom of the beaker.

   a Explain whether you could obtain the two original clear liquids.
   b Describe how you would separate the resultant mixture.

Reflecting

12 What are the properties of glass and plastic that make them interchangeable for some uses?

13 The element oxygen is a gas that makes up about 20% of the atmosphere. Oxygen is also the most abundant element in Earth's crust, which is solid. How can this be?

CHAPTER 3 METALLIC, IONIC AND COVALENT STRUCTURE AND BONDING

By the end of this chapter you will have covered the following material.

Science Understanding

- The properties of atoms, including their ability to form chemical bonds, are explained by the arrangement of electrons in the atom and in particular by the stability of the valence electron shell (ACSCH020)
- The type of bonding within substances explains their physical properties, including melting and boiling point, conductivity of both electricity and heat, strength and hardness (ACSCH027)
- Chemical bonds are caused by electrostatic attractions that arise because of the sharing or transfer of electrons between participating atoms; the valency is a measure of the number of bonds that an atom can form (ACSCH029)
- Ions are atoms or groups of atoms that are electrically charged due to an imbalance in the number of electrons and protons; ions are represented by formulae which include the number of constituent atoms and the charge of the ion (for example, O^{2-}, SO_4^{2-}) (ACSCH030)
- The properties of ionic compounds (for example, high melting point, brittleness, ability to conduct electricity when liquid or in solution) are explained by modelling ionic bonding as ions arranged in a crystalline lattice structure with forces of attraction between oppositely charged ions (ACSCH031)
- The characteristic properties of metals (for example, malleability, thermal conductivity, electrical conductivity) are explained by modelling metallic bonding as a regular arrangement of positive ions (cations) made stable by electrostatic forces of attraction between these ions and the electrons that are free to move within the structure (ACSCH032)
- Covalent substances are modelled as molecules or covalent networks that comprise atoms which share electrons, resulting in electrostatic forces of attraction between electrons and the nucleus of more than one atom (ACSCH033)
- Elemental carbon exists as a range of allotropes, including graphite, diamond and fullerenes, with significantly different structures and physical properties (ACSCH034)
- Carbon forms hydrocarbon compounds, including alkanes and alkenes, with different chemical properties that are influenced by the nature of the bonding within the molecules (ACSCH035)

AC

3.1 Bonding and the periodic table

The periodic table organises all of the known elements. Initially it was organised by its creator, Dmitri Mendeleev, on the basis of known properties of elements. As scientific understanding of elements and their atomic structure developed, the relationship between the atomic structure of elements, their properties and hence their position in the periodic table was better understood.

In Chemistry Chapter 1, you learnt about atomic structure and periodic trends in some of the properties of elements. In this chapter, you will use that information to examine the bonding that occurs between the same and different elements and relate this to the properties of the resultant substances.

Bonding and valency

The number of electrons in the valence shells of elements can be determined from the periodic table. Elements in the same group have the same number of electrons in the valence shell (**valence electrons**). The number of valence electrons increases from left to right across the groups of the periodic table. The noble gases (group 18) have full valence shells and so are chemically inert. Chemically, electrons are the most important part of the atom so elements that have the same number of electrons in their outer shell will have similar chemical properties.

Atoms tend to form **chemical bonds** to obtain a filled valence shell and become chemically stable. A chemical bond forms between atoms due to the electrostatic force of attraction between positive and negative charges. Figure C3.1 shows a chemical bond that is formed by electrons from different atoms being attracted to the two nuclei simultaneously. This attraction holds the two atoms together.

Chemical bonds are due to the electrostatic force of attraction between positive and negative charges in participating atoms.

Figure C3.1 ▶ A chemical bond forms when electrons from both atoms are attracted to both nuclei.

The electrons experience a force of attraction from both nuclei. The negative–positive–negative attraction holds the two particles together.

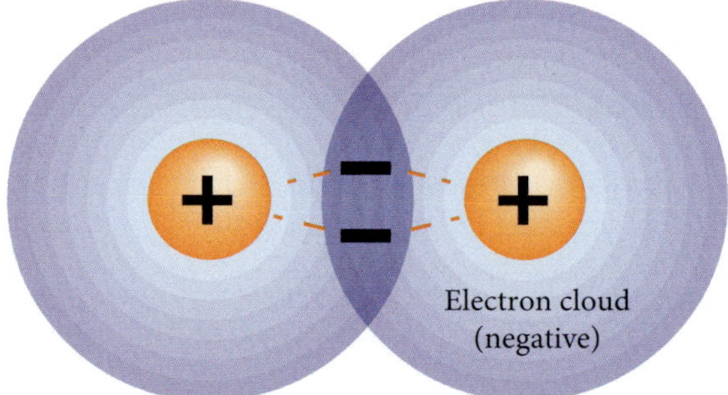

This attraction is called a chemical bond. One pair of electrons constitutes one bond.

The way an atom obtains a filled shell is by forming chemical bonds, which involves losing, gaining or sharing electrons. Which of these occurs depends on the position of the element in the periodic table and what element it combines with. This also determines the type of chemical bond formed. The properties of the resultant substance (element or **compound**) are directly related to the type of bonding within the substance.

The three types of chemical bonds that occur due to transfer or sharing electrons are:

- **metallic bonds**, which involve the electrostatic attraction between metal atoms that have released electrons and these electrons
- **ionic bonds**, which involve the electrostatic attraction between metal atoms that have lost electrons and non-metal atoms that have gained electrons
- **covalent bonds**, which involve electrostatic attraction between shared electrons and nuclei of non-metal atoms.

To learn more about making useful materials, refer to Context 1, 'Matter in the universe', page 15.

Figure C3.2 shows the relationship between metals and non-metals (including metalloids) and the type of bonding they participate in.

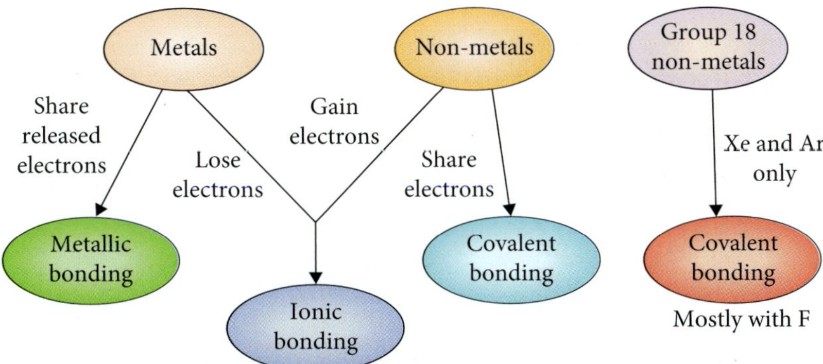

Figure C3.2 Types of bonding

Generally, except for pure metals, the number of valence electrons an atom has can be used to determine the number of bonds it forms. This is known as its **valency**.

BONDING

Watch this video, which shows the difference between each type of bonding.

Valency is the combining power of an atom and is equal to the number of hydrogen atoms that atom could combine with or displace from a compound. (Hydrogen has a valency of 1.)

Calculating valency of atoms

Some simple rules for calculating the valency of atoms are as follows.

1 Refer to the periodic table and determine the number of valence electrons an atom has.

2 If the number of valence electrons is four or fewer, then the valence is equal to the number of electrons. For example, magnesium has two valence electrons so has a valency of 2. Carbon has four valence electrons so has a valency of 4.

3 If the number of valence electrons is more than four, then the valency is calculated by subtracting the number of valence electrons from eight. For example, nitrogen has five valence electrons, so it has a valency of 3 (8 − 5). Chlorine has seven valence electrons so it has a valency of 1 (8 − 7).

Worked example 3.1 on page 162 does not consider the type of bonding involved or the actual formula, just the ratio of atoms based on the valency. There are other factors to consider when determining the actual formula. You will learn about these later on, but using valency provides a starting point for determining the formula of a compound.

Table C3.1 provides a summary of this information.

Table C3.1 Valency related to groups in the periodic table

Group number	1	2	3–12	13	14	15	16	17	18
Number of valence electrons	1	2	Variable	3	4	5	6	7	8
Valency	1	2	Variable	3	4	3	2	1	0

Once the valency is known, then the number of bonds an atom can form can be used to determine a ratio for combining different elements.

WORKED EXAMPLE 3.1

What would be a possible ratio of atoms for a compound formed by bonds between oxygen and nitrogen?

Answer

The ratio would be 2N:3O.

Logic

1 Determine the valency of each atom.

Oxygen has 6 electrons and needs 2 more to have a filled outer shell, so it has a valency of 2.

Nitrogen has 5 electrons and needs 3 more to have a filled outer shell, so it has a valency of 3.

2 Determine the ratio of the atoms.

Since nitrogen needs to form 3 bonds and oxygen needs to form 2, the lowest common multiple of 3 and 2 is 6.

So there will be 3 oxygen atoms (3 atoms × 2 bonds per atom = 6 bonds) and 2 nitrogen atoms (2 atoms × 3 bonds per atom = 6 bonds) and the ratio would be 2N: 3O.

Try these yourself

What would be the ratio of atoms formed by bonds between:

- **a** carbon and chlorine?
- **b** boron and hydrogen?
- **c** magnesium and oxygen?
- **d** aluminium and sulfur?

QUESTION SET 3.1

Remembering

1 Describe how chemical bonds hold atoms together.

2 Name the three different types of chemical bond that hold atoms together.

3 What factors determine the type of chemical bond formed between atoms?

Understanding

4 What is the relationship between an element's position in the periodic table, valence electrons and valency?

5 What information does valency provide about bonding?

6 What is the valency of the following elements?

- **a** Helium
- **b** Sodium
- **c** Phosphorus
- **d** Boron
- **e** Calcium

Applying

7 What would be the ratio of atoms formed by bonds between:

- **a** silicon and hydrogen?
- **b** sodium and nitrogen?
- **c** aluminium and fluorine?
- **d** phosphorus and oxygen?

3.2 Metals, bonding and properties

Elements are grouped on the periodic table on the basis of common properties. Metals make up 75% of the elements, while the remainder are non-metals or metalloids. Each of the three groupings has distinct physical properties by which it can be classified. The common properties of metals are listed in Table C3.2.

Table C3.2 Common properties of metals

Property	Description
Metallic lustre	Mirror-like shininess
Good conductor of heat	Allows heat to travel from one end to the other
Good conductor of electricity	Allows an electric current to easily pass through
Malleable	Able to be beaten into another shape or flattened into a thin sheet with a hammer without breaking
Ductile	Able to be drawn out into a wire

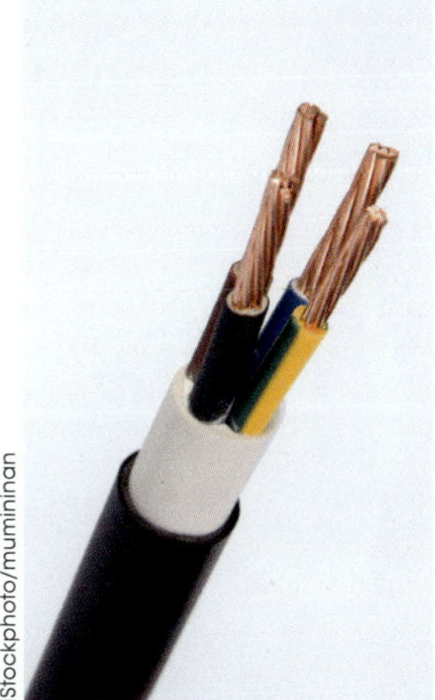
iStockphoto/mumininan

▲ **Figure C3.3** Copper wire is used in electrical wiring because it is a very good conductor.

WOW

Copper wire

Copper has been used in electrical wiring for communication since the telegraph services of the 1820s. Today, copper wire is still used for telephone, cable TV and ethernet connections. Advances in technology have led to changes in the way copper electrical wiring is structured and connected into circuits. Improvements have led to the development of copper wiring systems that can carry at least 1 gigabit (billion bits) of information per second, which equates to about 50 000 pages of text per second.

However, there are a number of other properties that vary from metal to metal. These are shown in Table C3.3.

Table C3.3 Some of the properties that vary from metal to metal

Property	Meaning	Comment
Density	The number of grams in each cubic centimetre of a material If a material is denser than water, then it will usually sink in water.	Density varies from metal to metal. For example, 1 cm^3 of gold weighs 19.3 g, but the same volume of aluminium weighs only 2.70 g. Gold is described as being a very dense metal, whereas aluminium is considered to be a 'light' (not very dense) metal.
Hardness	Resistance to being scratched The hardest substance cannot be scratched by any other material.	Most metals are fairly hard to scratch. Gold, lead and calcium are examples of soft metals; that is, they are easily scratched.
Melting point	The temperature at which the material melts; that is, turns from a solid to a liquid	Melting point (MP) varies considerably from metal to metal. Mercury is already a liquid at room temperature (its MP is −39°C) but a large number of metals require temperatures much higher than 1000°C to melt them.
Tensile strength	The ability of a material to withstand a stretching force For example, the materials in cables that are suspended between poles must have high tensile strength.	Tensile strength varies considerably from metal to metal. For example, steel (which is mostly iron) has greater tensile strength than aluminium.

Unlike metals, non-metals show a lot of variation in their properties. Non-metals tend to be classified by what they cannot do, such as conduct heat and electricity.

Metallic bonding

In order to understand why metals have the properties listed in Table C3.2, scientists have developed a model that can be used to explain the properties.

Metal atoms have few valence electrons, which are loosely held, as shown by their low ionisation energies, so to attain a noble gas electron configuration, they tend to lose electrons, which are negatively charged. When electrons are lost, the atom has a net positive charge and the resulting positively charged particle is called a **cation**.

In a metal, the atoms are surrounded on all sides by other metal atoms, which also tend to lose electrons. The valence electrons are attracted to the nuclei of the other atoms that are present and so are drawn into the space between the atoms by the combined action of all attractive forces. The electrons that are detached from their atoms are called **delocalised electrons**. When the electrons break away from their atom, they leave behind a cation. However, the overall piece of metal is uncharged because the total positive charge of the cations is equal to the total negative charge of the delocalised electrons.

The delocalised electrons are free to move randomly between atoms within the three-dimensional arrangement and are often referred to as a 'sea of electrons'. The electrostatic force of attraction between the positively charged metal ions and the negatively charged delocalised electrons holds the structure together by metallic bonding. Metallic bonds are **non-directional** because the electrons are free to move between cations.

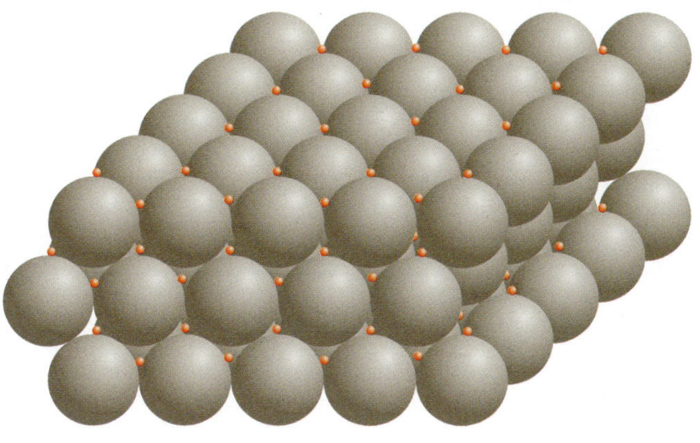

Key:

Figure C3.4 ▶ A model of a section of the metallic lattice of sodium

Metallic bonding is the electrostatic forces of attraction between cations and the electrons that are free to move within the regular lattice structure of a metal.

Explaining properties of metals

This model of the bonding within a metal can be used to explain the common properties listed in Table C3.2.

Metals are good conductors of electricity because of the highly mobile electrons within the lattice. When metals are connected into an electrical circuit, the mobile electrons have a net movement towards the positive terminal. Metals are also good conductors of heat because the mobile electrons acquire energy from the heat source and rapidly transfer it to cooler parts of the lattice.

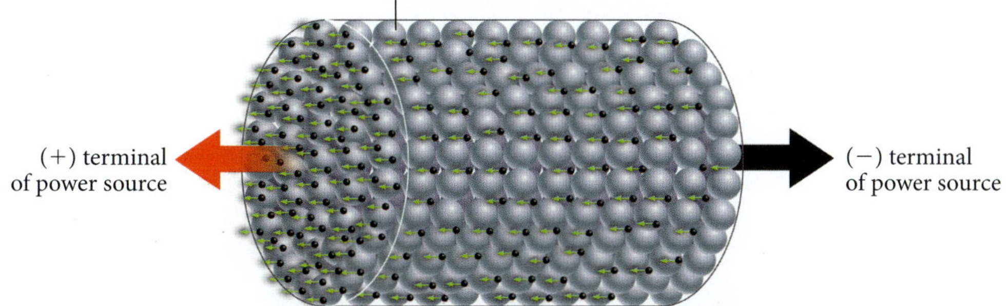

◀ **Figure C3.5**
Explaining the ability of a metal to conduct electricity

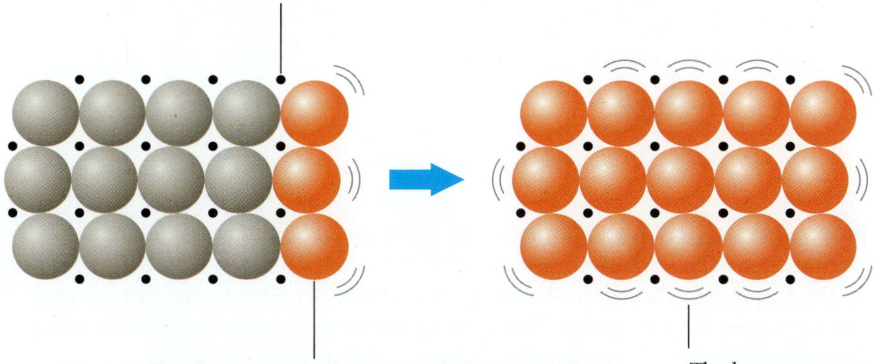

◀ **Figure C3.6**
Explaining the ability of a metal to conduct heat

The lustre of a metal is explained by light rays being reflected off the delocalised electrons and the close packing of the metal cations preventing light from passing through and so making the metal opaque. This is shown in Figure C3.8 on page 166.

Figure C3.7 ▶ Explaining the lustre of a metal

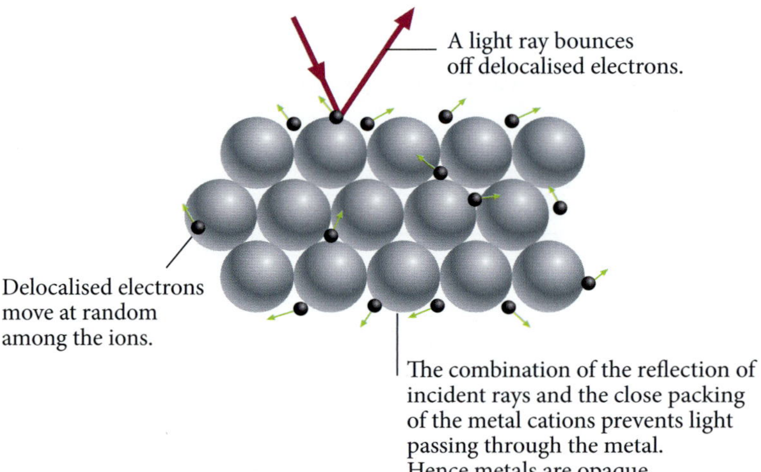

Metals are malleable and ductile because when the orderly array of cations is sheared, the mobile electrons adjust to the new arrangement, allowing one layer of cations to slide over another without disrupting the metallic bonding.

Figure C3.8 ▶ Explaining why metals are malleable and ductile

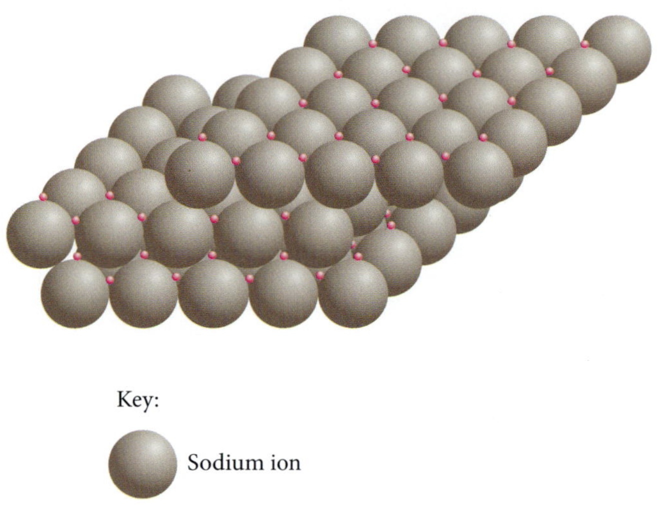

Some properties vary from metal to metal. The hardness of a solid is a measure of how difficult it is to scratch. The hardest known material in nature is diamond. When a substance is scratched, the particles in the surface layers are shifted out of position. For this to occur, the bonding forces between the particles in these layers must be disrupted. So the more strongly the particles are held in position, the harder it is to scratch. Therefore, scratch resistance is an indicator of the strength of bonds between particles. The fact that some metals are soft (e.g., gold) and so easily scratched, while others are hard and difficult to scratch, supports the inference that bonding forces in metals vary in strength.

Metals also generally have high melting points and boiling points. High melting and boiling points mean a lot of energy is required to overcome forces holding atoms together in their position. Melting and boiling points are a good indicator of the strength of the forces holding particles together in the solid and liquid form. The high melting and boiling points of many metals reflects the strong metallic bond, while the low melting points of sodium and mercury indicate that they have weaker metallic bonds. Generally, the bonding forces in metals vary in strength.

Density is a measure of mass per unit volume and is generally measured in grams per cubic centimetre (g cm^{-3}). Water has a density of 1 g cm^{-3} so objects that are less dense will float on water while those that are denser will sink.

METAL BASICS

Use this simulation to see the effect of force and heat on metal properties.

The density of a metal depends on:

- how closely the ions are packed
- the volume of each ion
- the mass of each ion.

The volume and mass of the metal ions are related to their position in the periodic table. Metals with higher atomic numbers and more filled electron shells will have greater masses and volumes. Generally, it would be expected that elements of lower atomic numbers have lower density while those with higher atomic numbers have higher density. While the packing of ions in metals is highly organised, the way they are packed varies and this makes a difference to the amount of space between ions. Packing also affects the strength of the bonding forces – the more closely packed, the stronger the forces are within that metal.

To learn about titanium, refer to Context 1, 'Matter in the universe', page 17.

QUESTION SET 3.2

Remembering

1 What are the major characteristics of the valency electrons in a metal?

2 Define 'metallic bonding.'

3 Explain why metals have delocalised electrons.

Understanding

4 Explain the following observations in terms of the metallic model.

  a You can draw out metals into a wire.

  b You should not stick a metal fork into a power socket.

  c You pick up a metal spoon that has been sitting in a cup of hot water and burn your hand.

REVISING METALS

Visit this website to revise the properties of metals, then take the test to check your understanding.

Applying

5 For each of the following purposes, suggest an important metallic property that makes it suitable for that purpose.

  a Copper is used in electrical wiring.

  b Iron is used in construction of bridges.

  c Aluminium is used in saucepans.

  d Jewellery is commonly silver and gold.

  e Electric light filaments are made of tungsten.

Analysing

6 Use the information in the table to answer the following questions.

Element	Melting point (°C)	Thermal conductivity at 25°C ($J\ s^{-1}\ m^{-1}\ K^{-1}$)	Electrical conductivity at 25°C ($MS\ m^{-1}$)
Sodium	98	141	21
Magnesium	650	156	22
Aluminium	660	237	37
Copper	1085	401	58.4

  a Compare the thermal and electrical conductivities of the metals. Is there a pattern in the data?

  b Explain any pattern identified in part **a** in terms of delocalised electrons.

  c Does there appear to be a correlation between position in the periodic table and melting point? Suggest why this may be so.

3.3 Properties of ionic compounds and ionic bonding

SODIUM REACTION

Watch the video to see what happens when sodium reacts with water and chlorine.

In section 3.2, you looked at what happens when only metal atoms bond together to form metals. Although metals are common, they are not usually found naturally in metallic form. This is because metal atoms generally tend to lose electrons to obtain a filled outer shell and readily combine with elements that tend to gain electrons to form compounds.

For example, sodium metal is very reactive and is never found naturally in its metallic state. When sodium is brought together with chlorine, it forms the **ionic compound** sodium chloride, which is common table salt.

Science Source/Charles D. Winters

▲ **Figure C3.9**
When sodium metal and chlorine gas meet

Science Source/ER Degginger

▲ **Figure C3.10**
Crystals of sodium chloride

The physical properties of sodium chloride are different from the properties of the sodium and chlorine from which it is composed.

Most ionic compounds:

- are hard and brittle, which means they are not easily scratched but will shatter on impact
- are non-conductors of electricity in the solid state but good conductors in a molten or an aqueous state
- have high melting and boiling points.

From these physical properties, we can infer that there are strong bonds between the particles in ionic compounds and they are made of charged particles, as shown by their ability to conduct electricity.

EXPERIMENT 3.1

COMPARING THE PROPERTIES OF A COMPOUND WITH THOSE OF ITS COMPONENT ELEMENTS

In this experiment, you will compare the properties of the elements with those of the resultant compound. The properties being compared experimentally are physical state, colour, odour, solubility in water, electrical conductivity and reaction with hydrochloric acid. You may also consult data tables to add melting point and density to the comparison table.

Aim

To compare the properties of the elements magnesium and oxygen with those of the compound magnesium oxide

Materials

- 2 strips of magnesium ribbon (20 cm and 1 cm)
- Steel wool
- Crucible tongs
- Pipe clay triangle
- Bunsen burner
- Tripod
- Test tubes, stoppers and test-tube rack
- 250 mL beaker
- Distilled water
- Crucible and lid
- Electrical conductivity apparatus
- Matches
- 1 mol L^{-1} hydrochloric acid (HCl)

What are the risks in doing this experiment?	How can you manage these risks to stay safe?
1 mol L^{-1} HCl is corrosive to skin and clothing.	Wear safety glasses and protective clothing. Take care when pouring and clean up spills immediately. If hydrochloric acid is spilt on the skin, wash the affected area with plenty of water and notify your teacher.
Magnesium burns with a very bright flame.	Do not look at the flame directly.
Magnesium oxide is an irritant.	Do not breath in the powder; work in a well-ventilated area.
A hot crucible retains heat, especially if contents have significant mass.	Use tongs to handle the crucible; place hot crucible on the heatproof mat. Do not touch the hot crucible. If you burn yourself, place the affected part under cold running water for 10 minutes and inform your teacher.
The Bunsen burner will get hot.	Do not use the Bunsen burner if the gas tube is damaged. Ensure long hair is tied back and the flame is away from flammable material. If you burn yourself, place the affected part under cold running water for 10 minutes and inform your teacher.

In your write-up, add any more risks you can think of, as well as ways to manage them.

Procedure

1 Thoroughly clean the surface of both strips of magnesium ribbon with steel wool. Record the appearance of the cleaned magnesium.
2 Coil the longer piece of magnesium ribbon so that it fits inside the crucible.
3 Place the crucible on a pipe clay triangle or a tripod over the Bunsen burner and carefully heat the crucible without the lid until the magnesium begins to glow.
   Warning: Do not look directly at the burning magnesium.
4 Place the lid on the crucible with tongs and heat strongly for about 10 minutes.
5 Remove the lid and heat for a further 5 minutes to ensure complete reaction.
6 Replace the lid and allow it to cool. This is the sample of magnesium oxide to be used for comparing properties.
7 Use the conductivity apparatus to test the electrical conductivity of magnesium metal, magnesium oxide and oxygen (air). Record the results.
8 Place 10 mL of distilled water in each of two test tubes. Add the 1 cm strip of magnesium to one and some of the magnesium oxide to the other. Stopper and shake. Record the results.
9 Consult data tables for the melting point and density of magnesium and magnesium oxide.
10 Add 10 mL of hydrochloric acid to each of three test tubes. To the first one add the strip of magnesium ribbon. To the second, add some magnesium oxide. Record your observations. Stopper the third test tube and shake it to aerate the water. Record your observations.

Results

Record your results in a table like the one below.

Properties	Magnesium	Oxygen	Magnesium oxide
Physical state			
Colour			
Electrical conductivity			
Solubility in water		0.0359 g O_2 per litre water at 25°C	
Reaction with HCl			
Melting point			
Density			

Analysis of results

What are the similarities and differences in the properties of the elements and compound?

Discussion

What other properties could be compared?

Conclusion

Summarise your findings about the comparison of properties of the elements and resultant compound.

Formation of ions and ionic bonding

When atoms lose or gain electrons to attain a full outer shell, they form a charged particle called an **ion**. The ions that form on loss of electrons are positive (+) and are called cations. Ions that form on gain of electrons are negative (−) and are called **anions**.

Sodium chloride is a combination of a metal and non-metal. The sodium atom with the electron configuration 2,8,1 has a tendency to lose one electron to become like neon (2,8), while the chlorine atom with the configuration 2,8,7 tends to gain one electron to become like argon (2,8,8). The sodium atom becomes a positively charged sodium ion and is represented as Na^+, while the chlorine atom becomes a negatively charged chloride ion and is represented as Cl^-. The resulting compound, sodium chloride (NaCl), is electrically neutral because the positive and negative charges balance each other. Figure C3.11 shows this diagrammatically.

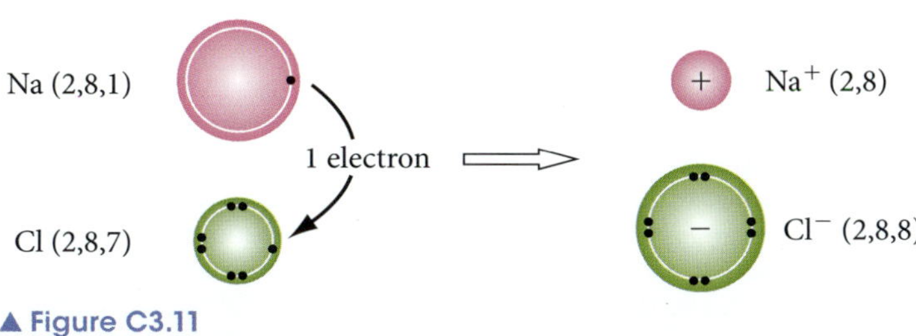

▲ **Figure C3.11**
Formation of sodium chloride

In Figure C3.11 only the valence electrons in the outer shell of the atoms have been drawn. This reaction can also be represented using **electron dot formula** (also called Lewis dot diagrams). These diagrams are often more convenient to use because the valence electrons can be represented as dots, as shown in Figure C3.12.

Key:
- • Outer-shell electron
- **Na** Nucleus and inner-shell electrons of a sodium atom
- **Cl** Nucleus and inner-shell electrons of a chlorine atom

▲ **Figure C3.12**
Electron dot formula

There is a strong electrostatic attraction between the positive cations and the negative anions and this attraction results in a chemical bond being formed. This type of chemical bond is called an ionic bond.

> Ions are atoms or groups of atoms that are electrically charged due to an imbalance in the number of electrons and protons. Ionic bonding is due to the electrostatic force of attraction between oppositely charged ions.

Ionic compounds have a crystalline structure in which the positive and negative ions are arranged in an orderly fashion with every positive ion being surrounded by negative ions and every negative ion being surrounded by positive ions, as shown in Figure C3.13. The electrostatic attraction between pairs of oppositely charged ions extends throughout the whole crystal. There are no separate units of NaCl, just a large array of cations and anions held together by ionic bonds.

IONIC COMPOUNDS

Watch this video describing the properties of ionic compounds.

When ionic compounds form, the number of electrons lost and the number of electrons gained must be equal. This means the total positive charge on the cation(s) must equal the total negative charges on the anion(s), so the formula of an ionic compound will always represent an electrically neutral substance

Figure C3.13 ▶ An ionic substance consists of an orderly array of positive and negative ions

a

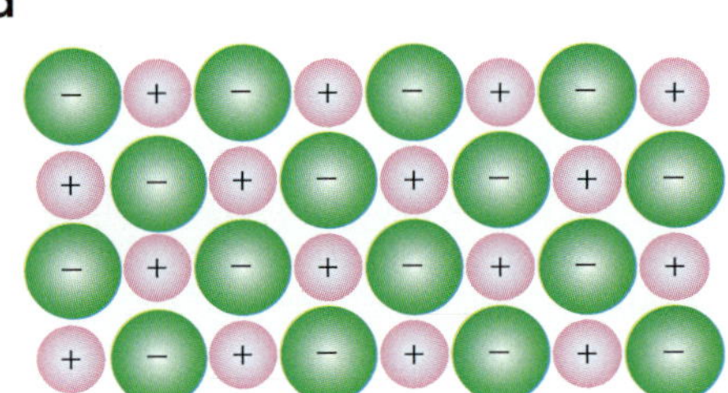

b

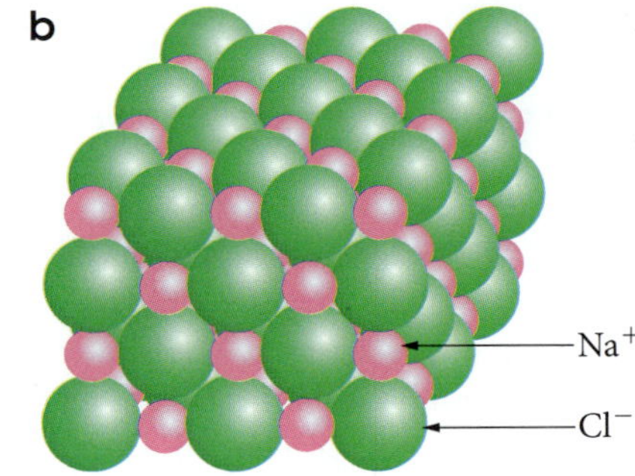

CRYSTALS

Visit this website and click on 'The caves' to see amazing photos.

Because there are no discrete molecules in an ionic compound, the formula gives the ratio of ions rather than the actual number of ions. A formula that gives the simplest ratio of atoms or ions is called an **empirical formula**. Formulas for ionic compounds are always empirical formulas.

WOW

Giant crystals

Under the right conditions, crystals can be gigantic. In April 2000, the largest natural crystals on Earth were discovered in a cave near Naica, in the state of Chihuahua, in Mexico. The selenite crystals are a crystalline form of the mineral gypsum (calcium sulfate). The larger crystals are up to 15 m in length and 1.2 m in diameter and are estimated to weigh 50 tonnes. The crystals have formed due to a unique combination of conditions. The temperature in the cave is 50°C with 100% humidity. To see pictures of these amazing crystals go to the weblink 'Crystals'.

Many Transition metal elements (in groups 3–12) can form multiple ions due to the small energetic difference between the 4s and 3d subshells. This means they can lose both the 's' electrons and a variable number of 'd' electrons.

Ions and the periodic table

For the main group elements, the periodic table can be used to deduce the charge on an ion because the number of electrons an element loses or gains depends on the number of its valence electrons. Figure C3.14 summarises this information.

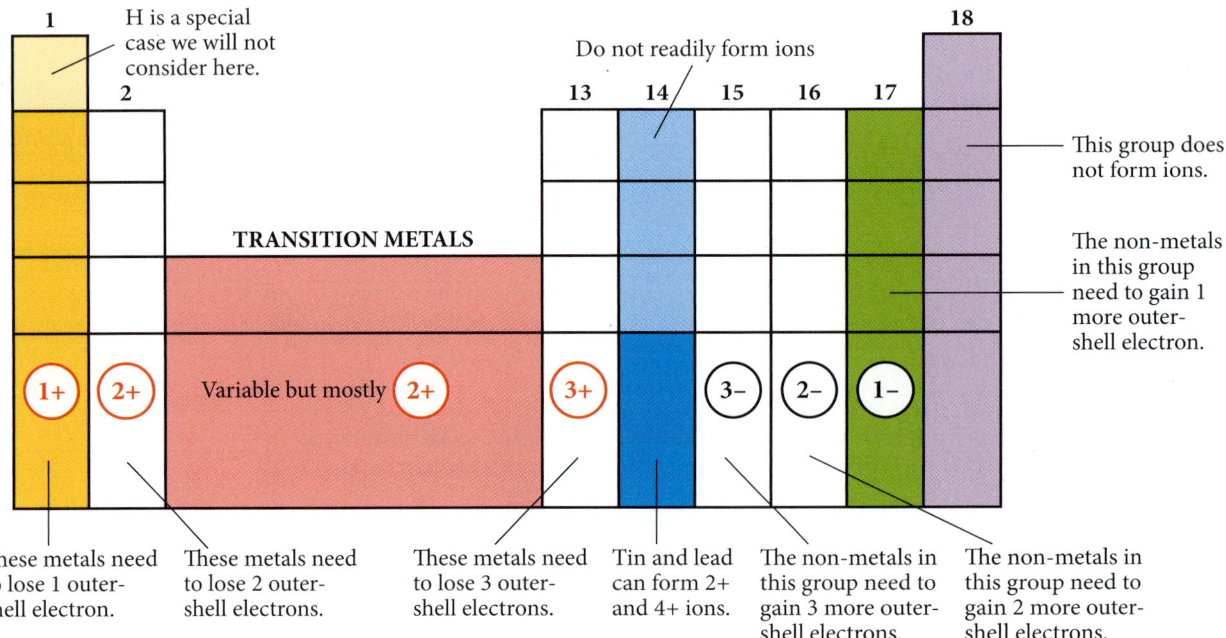

▲ Figure C3.14
The charges of ions formed from elements according to their group of the periodic table

Explaining properties of ionic compounds

In an ionic compound, each ion is strongly held in place by the attraction of adjacent oppositely charged ions. This bonding is described as non-directional because ions may be attracted to each other in any direction. The bonding is usually represented diagrammatically using lines.

The properties of ionic substances and an explanation of that property are given in Table C3.4.

Table C3.4 Properties of ionic compounds

Property	Explanation
High melting point	There is strong attraction between ions, so a lot of energy is needed to move the ions out of their fixed position in the lattice.
Conducts electricity if in the molten state and in solution	Charged particles are free to move in the molten and aqueous state but are held in a fixed position in the solid states.
Hard but brittle	Ions are held rigidly in place in the lattice but the application of stress brings ions of the same charge close together so the crystal can shatter.
Mostly soluble in water	Ions are attracted to the water molecules and move out of position.
Giant lattice of repeating ions	Electrostatic attraction between positive and negative charges causes ions of opposite charge to surround each other.

The non-directional bonding holds the ions in place so prevents ionic crystals from being malleable like metals, so ionic crystals will shatter when hit with a strong force. Ionic bonds are strong as shown by the hardness and high melting and boiling points of ionic substances.

The information in Table C3.4 is also shown schematically in Figure C3.15, which uses sodium chloride as the example.

IONIC BONDING

Watch this video, which shows how ionic bonding forms.

To learn why materials are mostly not found as elements, refer to Context 2, 'Materials for a purpose', page 36.

WOW

Checking out space

If an atom passes through light of sufficiently high energy, its outer electrons can absorb enough energy from the light to move right away from the nucleus. We say the atom has been ionised by the light.

Atoms of barium are relatively easy to ionise, because the electrostatic attraction between the outer electrons and the nucleus is not very strong. If some barium is ejected into space, it will instantly vaporise, forming a green cloud of barium atoms. Within a minute, the barium atoms will be ionised by the radiation from the Sun. This forms a bluish cloud of positively charged barium ions, which separates from the green cloud. This cloud can easily be seen at night. The bluish cloud then moves in response to the electrical and magnetic forces in space. It becomes elongated in the direction of the magnetic field lines of the magnetosphere.

Some barium releases are conducted from spacecraft far from Earth, and then tracked by telescopes. These barium ion clouds have enabled NASA scientists to study the magnetosphere and solar winds.

REVISING IONIC BONDING

Work through the tutorial to revise ionic bonding.

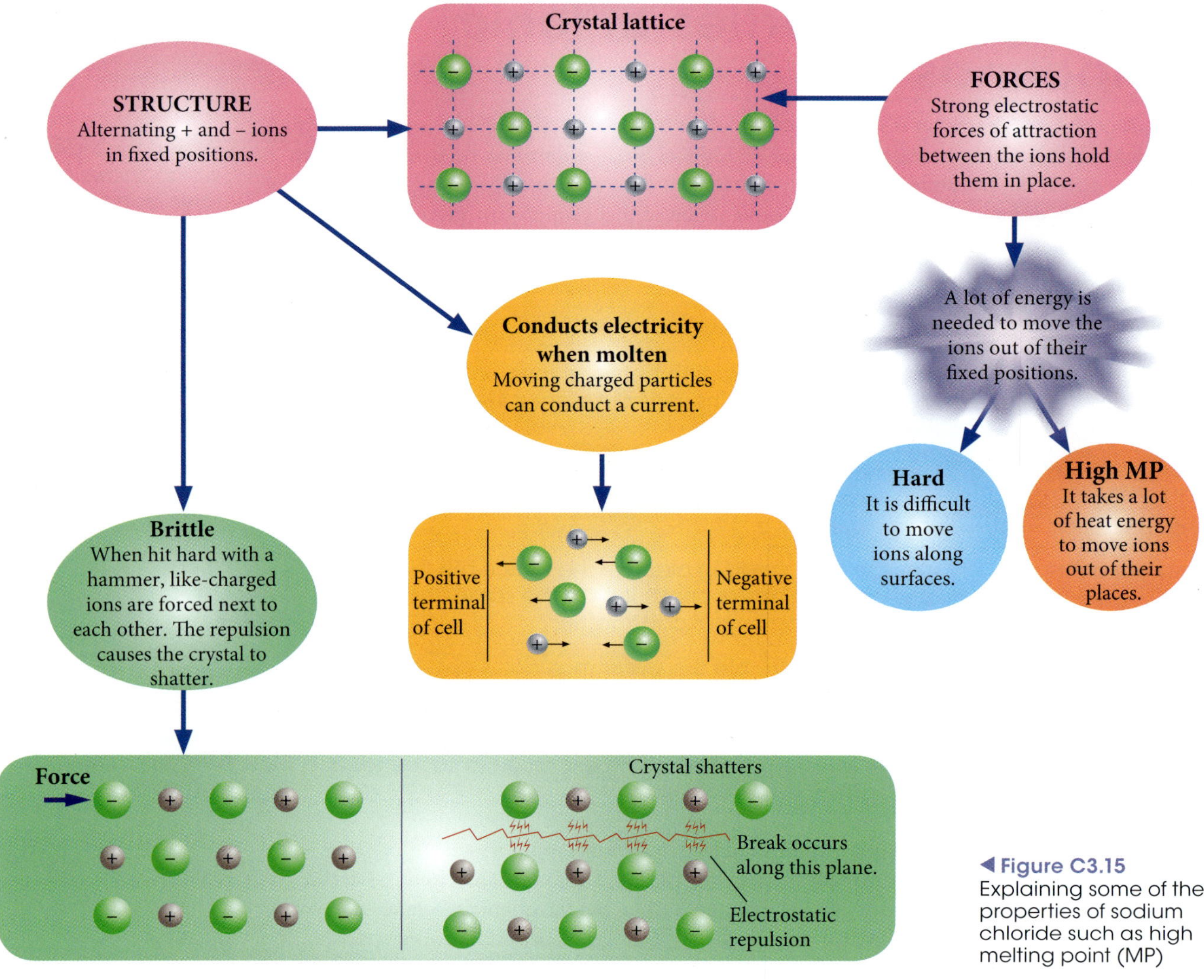

◀ **Figure C3.15** Explaining some of the properties of sodium chloride such as high melting point (MP)

QUESTION SET 3.3

Remembering

1 Describe what is meant by ionic bonding.

2 Explain why metals generally form cations and non-metals generally form anions when these two types of substances react together.

Understanding

3 List four properties of ionic solids and explain these properties using the ionic bonding model.

4 Explain why the formula of ionic solids is an empirical formula and does not give the actual number of ions present in the compound.

Applying

5 Give the electron configuration of the:

a potassium atom and ion.

b sulfur atom and ion.

c nitrogen atom and ion.

d calcium atom and ion.

6 Draw the electron dot formula of the nitrogen atom and ion, and calcium atom and ion.

7 State the charge that would be expected on an ion formed from an atom with an electron configuration of:

a 2,8,2; $1s^2\ 2s^2\ 2p^6\ 3s^2$.

b 2,5; $1s^2\ 2s^2\ 2p^3$.

8 Explain why the formula of:

a sodium bromide is NaBr, not Na_2Br or $NaBr_2$.

b aluminium fluoride is AlF_3, not AlF or Al_3F or Al_2F_6.

9 Identify the charged species that conduct electric current in:

a molten MgO.

b molten AgCl.

c a solution of LiI.

Analysing

10 Explain why hydrogen can form both a positive and a negative ion, and give the formula of the ion.

3.4 Names and formulas of ionic compounds

REVISE IONIC BONDING

Take the test to check your understanding of ionic bonding.

In the names and formulas of ionic compounds composed of only two elements, the positively charged ion is always listed first and the negatively charged ion is listed second. The name of the positive ion is the same as the name of the element from which it was formed. In the name of the negative ion, the end of the non-metal parent element is replaced with *-ide*. For example, chlorine becomes chlor*ide*, oxygen becomes ox*ide*, and bromine becomes brom*ide*; hence, the names sodium chloride, magnesium oxide and potassium bromide.

Transition metals have slightly different electron structures from other metals. As a result, many transition metals show more than one possible charge. A simple naming system is used to show which ion is present.

For example, the metal iron forms both Fe^{2+} and Fe^{3+} ions. Once, the ending of their name was used to indicate the charge. The one with a lower charge had *-ous* at the end of its name and the one with a higher charge had *-ic* at the end of its name. But now there is a much easier system in which the charge is shown as a Roman numeral. Two examples are shown in Table C3.5. Notice that there is no gap between the bracket and the symbol of the element. The former names are still sometimes used in the chemical industry.

Table C3.5 Naming transition metal ions

Ion	Fe^{2+}	Fe^{3+}	Cu^{+}	Cu^{2+}
Current naming convention	Iron(II)	Iron(III)	Copper(I)	Copper(II)
Former name	Ferrous	Ferric	Cuprous	Cupric

Therefore, there are two different ionic compounds formed between iron and chlorine – iron(II) chloride and iron(III) chloride. The particular compound present will be indicated by the formulas $FeCl_2$ and $FeCl_3$.

In many ionic compounds, the positive ion or the negative ion (or both) consists of two or more atoms that are strongly bonded together and act as a single entity. These ions are called **polyatomic ions**. In polyatomic ions, the outer-shell electrons have partially merged because the ions are sharing their outer-shell electrons. The charge on polyatomic ions is spread over the whole ion. As you can see in Table C3.6 on page 176, the names of the compounds involving polyatomic ions follows the convention discussed. A more comprehensive list is given in Appendix 4, page 422.

To learn why materials are mostly not formed as elements, refer to Context 2, 'Materials for a purpose', page 36.

Table C3.6 Some examples of common polyatomic ions and their charges

Name of ion	Formula	Valency	Example of a compound
Ammonium	NH_4^+	+1	Ammonium chloride
Hydroxide	OH^-	−1	Iron(III) hydroxide
Nitrate	NO_3^-	−1	Silver nitrate
Sulfate	SO_4^{2-}	−2	Copper(II) sulfate
Carbonate	CO_3^{2-}	−2	Calcium carbonate
Phosphate	PO_4^{3-}	−3	Sodium phosphate

Formulas for ionic compounds

When determining the formulas of ionic compounds, there is one simple rule: ionic compounds have no net charge. This means the total number of positive charges equals the total number of negative charges. Therefore, when determining the formula, you need to determine the ratio of ions that will achieve this.

WORKED EXAMPLE 3.3

a Write the formula for calcium sulfide.

b Write the formula for ammonium sulfate.

c Write the formula for iron(III) oxide.

Answers	**Logic**
a The formula is CaS.	**1** Determine the ions present. Calcium sulfide is made is of calcium ions and sulfide ions. Using Figure C3.14 on page 172, the charge on the ions can be determined. Ions present: Ca^{2+} and S^{2-} **2** Compare charges and determine the ratio that gives equal positive and negative charges. These ions have the same-sized charge. So, one sulfide ion balances one calcium ion, the total positive charge equals the total negative charge. **3** Write the formula. CaS The charges on the ions are not shown in the overall chemical formula.
b The formula is $(NH_4)_2SO_4$.	**1** Determine the ions present. Ions present: NH_4^+ and SO_4^{2-} **2** Compare charges and determine the ratio that gives equal positive and negative charges. If we have two NH_4^+ ions for each SO_4^{2-} ion, then the total positive charge will equal the total negative charge. (2 × 1+ = 2+, which matches 2−) **3** Write the formula. $(NH_4)_2SO_4$

Brackets are used when there is more than one polyatomic ion in a formula. The brackets show the total number of the entire ion.

c The formula is Fe_2O_3.

1 Determine the ions present.

Ions present: Fe^{3+} and O^{2-}

2 Compare charges and determine the ratio that gives equal positive and negative charges.

To make the two charges equal, determine the lowest common multiple of 2 and 3, which is 6. The charges need to add to 6+ and 6−. If we have two Fe^{3+} ions for every three O^{2-} ions, then the total positive charge will equal the total negative charge.

($2 \times 3+ = 6+$, which matches $3 \times 2- = 6-$)

3 Write the formula.

Fe_2O_3

Try these yourself

Write the formula for the following ionic compounds.

a Sodium fluoride

b Magnesium oxide

c Potassium hydroxide

d Copper(II) nitrate

e Aluminium sulfate

QUESTION SET 3.4

Remembering

1 a What is the overall charge on an ionic compound?

b How is this achieved?

Understanding

2 Write the name and give the formula of ions of the following elements.

a Bromide

b Sulfur

c Barium

d Potassium

e Nitrogen

3 Draw the electron dot formulas for the elements and ions in Questions **2a**, **c** and **e**.

Applying

4 Write the names of the following ionic compounds.

a KI

b $BaCl_2$

c CaH_2

d PbO_2

e Na_3PO_4

f $Zn(NO_3)_2$

5 Give the formulas of the following ionic compounds.

a Magnesium nitride

b Aluminium oxide

c Mercury(II) sulfide

d Ammonium hydroxide

e Copper(I) carbonate

f Zinc phosphate

Analysing

6 X represents different mystery elements. Deduce what charge X must have in each of the following compounds.

a XOH

b XI_2

c XNO_3

d X_2SO_4

e LiX

f ZnX_2

g Al_2X_3

PRACTICE

Practise naming and writing formulas for ionic compound with this worksheet.

3.5 Properties of covalent compounds and covalent bonding

In the previous sections you learnt that when metal atoms combine with metal atoms, the result is a metal held together by metallic bonding and that metals and non-metals combine to form ionic compounds held together by ionic bonding.

In this section, you will explore non-metals and their compounds. Non-metal elements such as oxygen and diamond, and compounds such as carbon dioxide and water, are commonly occurring substances. Although all these substances are held together by covalent bonding, the arrangement of the atoms and hence the properties of the resultant substances can be quite different. As a consequence, these substances (which may be elements or compounds) are divided into two distinct groups on the basis of their properties – **covalent molecular substances** and **covalent network substances**.

A covalent molecular substance is composed of discrete molecules in which the atoms are held together by covalent bonding. A covalent network substance is a three-dimensional network of covalently bonded atoms.

Non-metals show a lot of variation in their properties so they tend to be classified by what they cannot do, such as conduct heat and electricity.

Table C3.7 Common properties of each of these groups of substances

Covalent molecular substances	Covalent network substances
Low melting and boiling points (many are liquids and gases at room temperature)	Very high melting points (usually solids at room temperature)
Non-conductors of electricity in solid and liquid states	Non-conductors of electricity in solid and liquid states
Form solids that are generally soft	Form extremely hard and brittle solids
Variable solubility	Insoluble in water and most other solvents
Variable reactivity	Chemically unreactive
Examples are oxygen, water and propane	Examples are diamond and quartz (silicon dioxide)

To explain the properties of each of these groups of substances, we need to examine the bonding and arrangement of atoms.

Covalent bonding

Like all elements in the periodic table (except the noble gases), non-metals form bonds to attain a filled valence electron shell. They can achieve this by accepting an electron from a metal. However, non-metal atoms can also share electrons to attain a filled valence shell. The electrostatic force of attraction between the shared electrons and the positively charged nuclei holds the atoms together and is called covalent bonding. Covalent bonding is also referred to as **directional bonding** because there is a direct line between the bonded particles. An example of covalent bonding is shown for a hydrogen molecule in Figure C3.16 on page 179.

Covalent bonding is the electrostatic forces of attraction between electrons and the nucleus of more than one atom. Covalent substances may be molecules or networks.

Each hydrogen atom has one valence electron so needs one more to fill its valence shell. When two hydrogen atoms come close enough together the single electron of each atom will be attracted to both nuclei and drawn into the space between them. The electron shells of the individual atoms begin to merge and the two nuclei start to share the two electrons. Effectively, each hydrogen atom can now be considered to have two electrons and hence a filled valence shell. This shared pair is called a **bonding pair**.

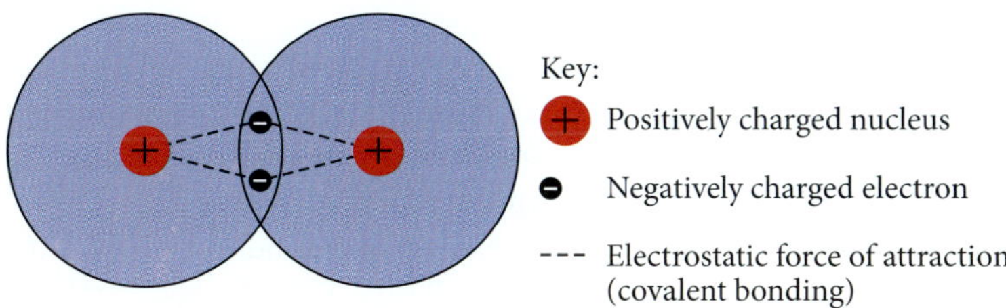

◀ **Figure C3.16**
Covalent bonding at work in a molecule of hydrogen (H_2)

Representing covalent bonds

A covalent bond can be represented in a number of different ways as shown in Figure C3.17. Fluorine forms F_2 by sharing one pair of electrons. An electron dot formula can be used to show shared and unshared electrons. The pairs of unshared valence electrons are called **non-bonding pairs** or **lone pairs**. Another representation used is to denote each pair of shared electrons by a dash. One dash means one shared (bonding pair) of electrons or one covalent bond.

COVALENT BONDING

Watch this video, which explains covalent bonding.

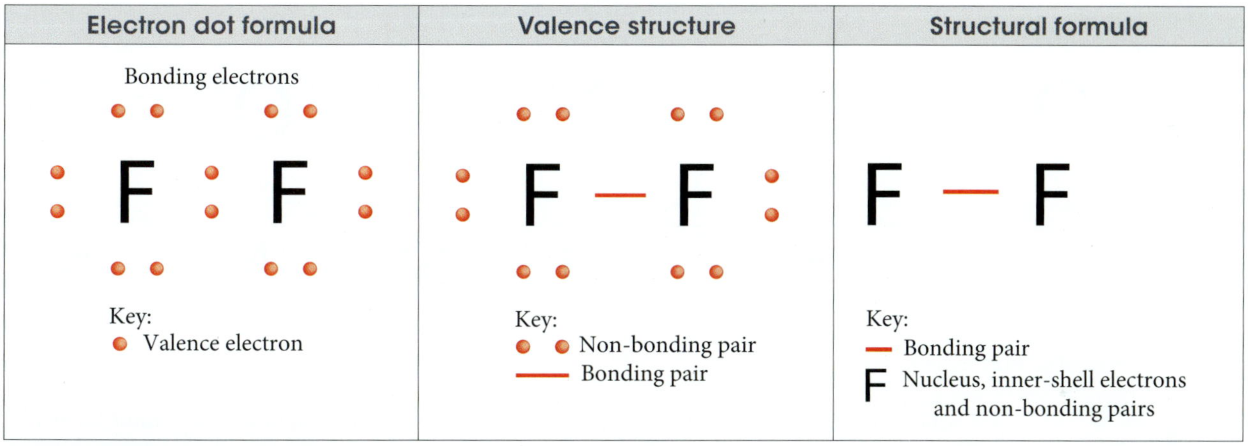

Electron dot formula	Valence structure	Structural formula
Bonding electrons :F:F: Key: ● Valence electron	:F—F: Key: ● ● Non-bonding pair — Bonding pair	F—F Key: — Bonding pair F Nucleus, inner-shell electrons and non-bonding pairs

▲ **Figure C3.17**
Representing covalent bonds

Covalent bonds and the periodic table

Covalent bonding occurs when both of the elements forming the substance need to gain electrons to attain noble gas configuration, such as H_2, H_2O and CO_2.

Hydrogen, metalloids and non-metals, except the noble gases, tend to form covalent bonds. The number of covalent bonds an atom forms is given by the valency (see Table C3.1) of the element, which is directly related to its position in the periodic table. This can be seen in the electron dot formulas of some simple molecules shown in Figure C3.18.

To learn about silicon, refer to Context 1, 'Matter in the universe', page 16.

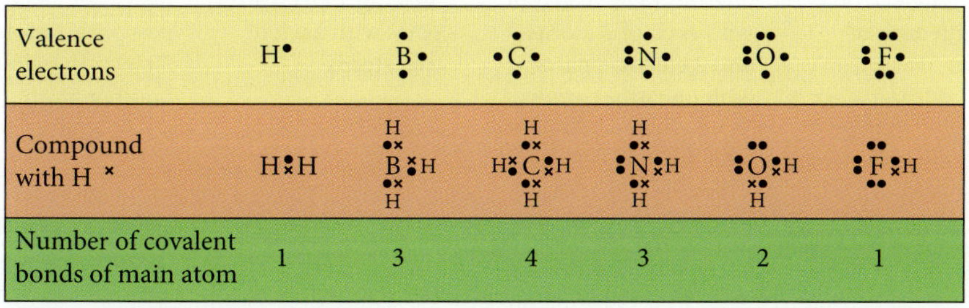

Valence electrons	H•	•B•	•C•	:N•	:O•	:F•
Compound with H ×	H:H	BH_3	CH_4	NH_3	H_2O	HF
Number of covalent bonds of main atom	1	3	4	3	2	1

Note: A cross is used for the H electron to show contribution to the bonding pair.

◀ **Figure C3.18**
Electron dot formulas of some elements and simple molecules

WOW

Sulfur

Sulfur is in the same group of the periodic table as oxygen so, like oxygen atoms, sulfur atoms need to share two pairs of electrons to gain the two electrons they need. But at ordinary temperatures, the sulfur molecule is a ring of eight atoms. Its formula is S_8.

Multiple covalent bonds

Sometimes when atoms come together, they share more than one pair of electrons. When this occurs, a **multiple covalent bond** will be formed.

Consider what happens when an oxygen atom bonds to another oxygen to form O_2 or when a nitrogen atom bonds to another nitrogen atom to form N_2.

Each oxygen atom has a valency of 2, so can form two covalent bonds. When two oxygen atoms come together, two pairs of electrons are drawn into the region between the two nuclei and their outer shells partly merge. This means there are four bonding electrons holding the nuclei closely together so this will be a stronger bond than that formed by one pair.

Since two pairs of electrons are shared, the bond is called a **double covalent bond**.

Nitrogen has a valency of 3, which means it can form three covalent bonds. When two nitrogen atoms come together, they share three pairs of electrons and form a **triple covalent bond**. This bond is even stronger than a double bond and is the main reason why nitrogen gas is relatively unreactive because a lot of energy is needed to break the triple bond in a chemical reaction.

Figure C3.19 shows how double and triple bonds can be represented.

Molecular formula	Electron dot formula	Valence structure	Structural formula
O_2	O :: O	O = O	O = O
N_2	: N ⋮⋮ N :	: N ≡ N :	N ≡ N

Figure C3.19 ▲
Ways of representing double and triple covalent bonds

Carbon, in group 14, has a valency of 4 so can form four covalent bonds. It cannot form a quadruple covalent bond because this type of bond cannot exist. A quadruple covalent bond would require four pairs of electrons to be located between the two nuclei. This cannot happen because the repulsive forces between the negative electrons becomes greater than the attractive forces between the electrons and the nuclei, preventing a fourth pair of electrons in the bond.

Carbon can achieve its four bonds in many ways, as shown in Figure C3.20.

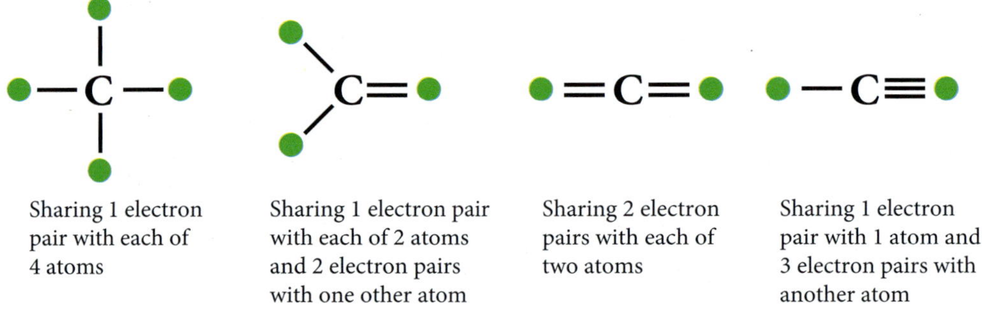

Figure C3.20 ▶
Different ways carbon can gain four electrons

ACTIVITY 3.1

MOLECULAR MODELS

Aim

To use a molecular model kit to construct models of molecular substances and represent these substances by electron dot formulas and structural diagrams

You will need

- Molecular model kit

What to do

1. Copy the following table into your notebook. Make sure you leave enough space to draw the diagrams.
2. Use the molecular model kit to construct a model of each of the substances listed.
3. Use your models to draw structural diagrams of the molecules constructed.
4. Draw an electron dot formula for each of the substances.

Name	Formula	Structural diagram	Electron dot formula
Water	H_2O		
Ammonia	NH_3		
Methane	CH_4		
Dihydrogen sulfide	H_2S		
Nitrogen	N_2		
Oxygen	O_2		
Carbon dioxide	CO_2		
Hydrogen cyanide	HCN		
Ethene	C_2H_6		

What did you discover?

1. Which of the substances had multiple covalent bonds?
2. What were the different shapes?
3. Compare the shapes of H_2O and H_2S. Suggest a reason for similarities or differences. (Hint: Consider their positions in the periodic table.)

Covalent network substances

Carbon and silicon both have a valency of four. This means they can form four covalent bonds. As a result, they can form three-dimensional networks of covalently bonded atoms. In these networks, the covalent bonding extends indefinitely throughout the whole crystal. The three-dimensional nature of the network means the atoms are held rigidly in position by very strong bonds.

Elemental carbon exists as a range of different forms or **allotropes**, including graphite, diamond and fullerenes, with significantly different structures and physical properties.

Carbon in the form of diamond is an example of a covalent network. Each carbon atom is covalently bonded to four other carbon atoms in an orderly pattern that continues throughout the whole structure to make a giant crystalline lattice as shown in Figure C3.21 on page 182. The bonding electrons are tightly bound and highly localised.

BUILD A MOLECULE

Use this interactive activity to make molecules.

The ball-and-stick model

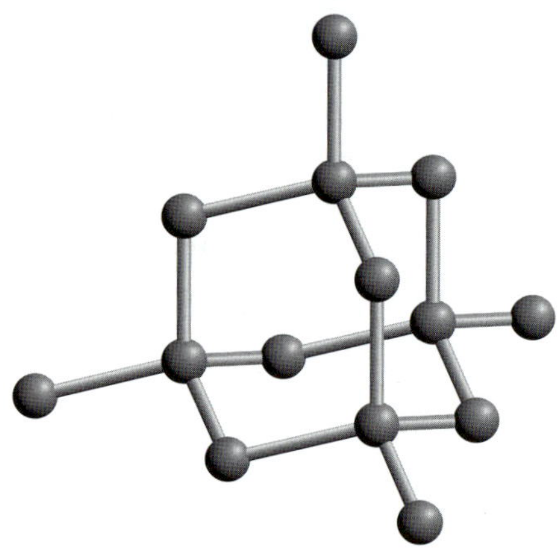

This model shows the geometry more clearly.

The 'solid ball' model

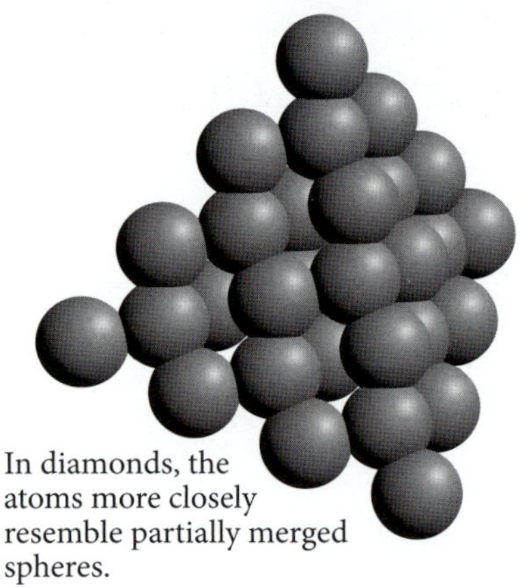

Sometimes this is called the 'space-filling' model.

Figure C3.21 ▶ Models of the structure of diamond

The properties of carbon in the form of diamond are consistent with those listed in Table C3.7.

Silicon has the same valency as carbon and exists in a structure similar to diamond.

There are also compounds that exist as covalent network substances. Silicon and carbon covalently bond to form the crystalline compound silicon carbide (SiC). This compound has a rigid diamond-like structure with alternating silicon and carbon atoms. Silicon carbide is commonly known as carborundum.

Another more common covalent network compound is silicon dioxide (SiO_2), or quartz. Each silicon atom is bonded to four oxygen atoms and each oxygen atom is bonded to two silicon atoms, as shown in Figure C3.22. This pattern extends throughout the entire crystalline lattice. Quartz is the main component of sand. The properties of SiO_2 and SiC are also consistent with those listed in Table C3.7.

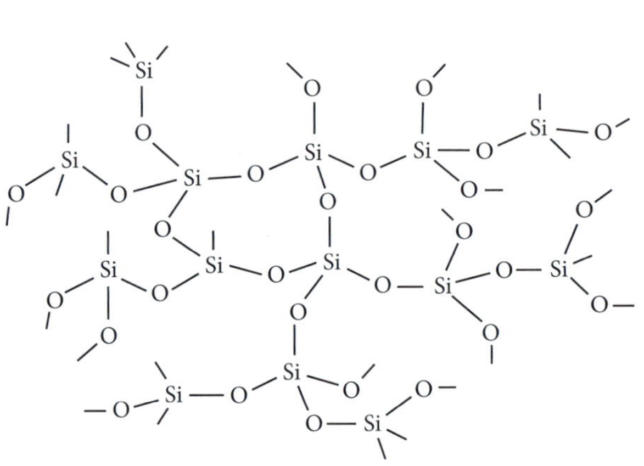

Science Photo Library/Roberto De Guglielmo

Figure C3.22 ▶ The covalent network structure of silicon dioxide and a photo of a quartz crystal

Graphite

The chemical formula of a covalent network compound represents the ratio of the atoms present in the compound.

Although graphite is a covalent compound and an allotrope of carbon, it has quite different properties from those of diamond. Graphite has a high melting and boiling point (like diamond), but it is soft and also a good conductor of electricity.

In graphite, the carbon atoms are arranged in flat parallel layers. Within each layer, each carbon atom is covalently bonded to three other carbon atoms, forming a hexagonal arrangement. Therefore, each layer is a two-dimensional network of carbon atoms. Weak bonding holds the layers together, as shown in Figure C3.23 on page 183.

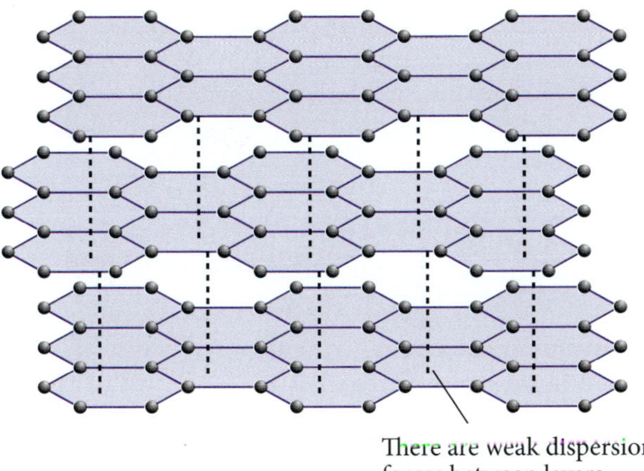

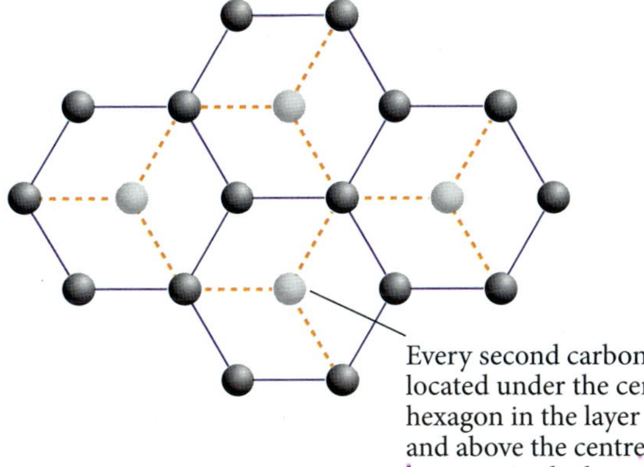

▲ **Figure C3.23**
Modelling the structure of graphite

In graphite, only three of carbon's four electrons are involved in covalent bonding. The fourth electron is delocalised, which means it is free to move along the layer occupied by its 'parent' atom. The attraction between the delocalised electrons and the layers above and below it hold the layers together.

The high melting and boiling point of graphite can be explained by the strong covalent bonds within the layers, while the softness and slippery feel can be explained by the weak bonds between layers that allow them to easily slide over each other. Graphite is an electrical conductor due to presence of the delocalised electrons that can move and therefore conduct electricity along the layers.

To learn about applications of fullerenes, refer to Context 1, 'Matter in the universe', page 19.

Fullerenes

Although fullerenes are not covalent network substances, they are an allotrope of carbon and are held together by covalent bonds so it would be appropriate to include them here.

Buckyballs

The first fullerene molecule was discovered in 1985 by a group of researchers at Rice University, Texas, USA. It was found to have the formula C_{60} and was named after Buckminster Fuller because it resembles the geodesic domes he designed. The molecules are more commonly referred to as buckyballs.

Fullerenes have been found in nature and in space. The discovery of fullerenes expanded the number of known allotropes of carbon. They have become the subject of intense research, especially in materials science, electronics and nanotechnology.

Fullerenes are molecules composed entirely of carbon. They are similar in structure and bonding to graphite. Each carbon is covalently bonded to three other carbons and the fourth electron is delocalised and free to move throughout the structure. They differ because they contain pentagonal and hexagonal rings that prevent them being planar. The most common and the first discovered is C_{60}. The structure resembles a soccer ball and is often referred to as an open cage structure, as shown Figure C3.24.

Science Photo Library/Russell Kightley

◀ **Figure C3.24**
A model of the C_{60} fullerene

ALLOTROPES OF CARBON

Visit this website to compare the properties and bonding of allotropes of carbon.

The nature of intermolecular bonds will be discussed in Chapter 6.

Like graphite, fullerenes are soft and slippery due to weak forces between molecules. They do not conduct electricity because there is no movement of electrons from one molecule to the next. Like diamond and graphite, they are insoluble in water because there is no attraction between the carbon atoms and water molecules. They typically have low melting points due to weak bonding between molecules.

Explaining the properties of covalent substances

Covalent molecular substances, whether elements or compounds, are composed of discrete molecules. Covalent bonding holds the atoms in the molecule together and is referred to as **intramolecular bonding** because it operates within the molecules.

The forces between molecules are quite weak compared to those within the molecule. These weak forces act to bond molecules together and are called **intermolecular bonding**.

Table C3.8 explains the properties of covalent molecular substances in terms of their bonding.

Table C3.8 Explaining properties of covalent molecular substances

Covalent molecular substances	Explanation
Low melting and boiling points (many are liquids and gases at room temperature)	The forces that hold the molecules together in the solid and liquid state are weak.
Non-conductors of electricity in solid and liquid states	There are no charged particles that can move through the substance
Form solids that are generally soft	The forces that hold the molecules together in the solid state are weak.
Tend to be malleable rather than shatter	The forces between molecules are weak so molecules are easily moved relative to each other.
Variable solubility	Solubility depends on the intermolecular forces, which vary between substances.

In a covalent network substance, the forces are strong and extend throughout the whole network. This explains why the properties of these substances are different from those of covalent molecular substances (Table C3.9).

Table C3.9 Explaining properties of covalent network substances

Covalent network substances	Explanation
Very high melting points (usually solids at room temperature)	The covalent bonds that hold the atoms together are strong.
Non-conductors of electricity in solid and liquid states	There are no charged particles that are free to move throughout the structure.
Extremely hard and brittle solids	The bonds are very strong between atoms so it is difficult to scratch, but an impact force disrupts the positions of the atoms and causes the network to shatter.
Insoluble in water and most other solvents	There is no attraction between the atoms in the network and water molecules.

COVALENT BONDING

Work through this tutorial, which provides an explanation of covalent bonding.

QUESTION SET 3.5

Remembering

1 a What is a covalent bond?
  b How many electrons share in the formation of a:
    i single bond?
    ii double bond?
    iii triple bond?

2 Which elements are most likely to form covalent network solids?

Understanding

3 a How many electrons are there in the valence energy level of each of carbon, hydrogen and chlorine atom in the molecule CH_3Cl?
  b Which noble gas electron configuration does each atom resemble?

4 Explain why covalent molecular substances and covalent network substances have different properties when they both have covalent bonding.

5 Explain why graphite can conduct electricity yet diamond cannot.

6 Give the molecular formulas for the covalent molecules formed between:
  a oxygen and fluorine.
  b boron and hydrogen.

7 Explain why the formula for methane is CH_4, not CH_3 or CH_5.

Applying

8 Draw electron dot formulas and structural diagrams for the following substances.
  a Br_2
  b HF
  c NF_3
  d CS_2
  e HCHO

9 Identify the bonding and non-bonding electron pairs in $SiCl_4$.

Analysing

10 Most substances that have an odour are covalent molecular substances. Suggest a reason for this.

3.6 Names and formulas of covalent compounds

Naming more complex covalent compounds will be considered in Units 3 & 4.

As you have already learnt, the valency of an element is an indication of the number of bonds it can form. Generally, this can be deduced from the periodic table as shown in Table C3.1. However, as with the transition metals, several non-metallic elements can display more than one valency in covalent compounds. Because of this, the names of covalent compounds need to provide an indication of the valency. Therefore, the convention for naming covalent compounds is different from the naming of ionic compounds.

Naming covalent compounds

The following rules apply to covalent compounds made up of only two elements.

1 Use the element name for the first element and change the end of the name of the second element to *ide*; for example, hydrogen fluoride.

2 The first named element is the one that is further to the left in the periodic table.

3 If both elements are in the same group, the one lower down the group is named first.
4 An exception to rules 2 and 3 is when oxygen is bonded to Cl, Br or I. In these compounds, oxygen is named last.
5 The number of atoms of each type is given by using the prefixes in front of each part of the name (though mono may be omitted from the first named element). For example, CO is carbon monoxide, CO_2 is carbon dioxide and N_2O_5 is dinitrogen pentoxide.

Prefix	mono	di	tri	tetra	penta	hexa	hepta	octa	nona	deca
Number of atoms	1	2	3	4	5	6	7	8	9	10

Sometimes, compounds are referred to by their common names rather than their systematic names. For example, H_2O, dihydrogen monoxide, is called water; NH_3, nitrogen trihydride, is called ammonia; CH_4, carbon tetrahydride, is called methane.

To learn about covalent compounds, refer to Context 2, 'Materials for a purpose', page 40.

Writing formulas for covalent compounds

Once the name is given, writing the formula is very straightforward. The order is given by the name, and the prefix of the element becomes the subscript for that element in the formula. For example, diphosphorus pentoxide is P_2O_5 and sulfur hexafluoride is SF_6.

QUESTION SET 3.6

Remembering

1 Explain how to determine which element is written first in the compound.

Applying

2 Write the formula of:
- a sulfur dioxide.
- b dinitrogen pentoxide.
- c carbon tetrachloride.
- d nitrogen trifluoride.
- e silicon tetrabromide.

3 Name the following compounds.
- a N_2O
- b NCl_3
- c SO_3
- d H_2S
- e N_2O_4

NAMING COVALENT COMPOUNDS

Practise naming covalent compounds with a worksheet on this website.

3.7 Comparing bonding and properties

Table C3.10 summarises information related to metallic, covalent and ionic bonds.

Table C3.10 Metallic, covalent and ionic bonds

Ionic bond	Covalent bond	Metallic bond
Formed by the attraction between one atom that has lost electrons and another atom that has gained electrons.	Formed by the mutual sharing of electrons between the same or different elements	Formed by the attraction between metal cations and a sea of mobile electrons
Strong because of electrostatic force of attraction	Fairly strong bond because the electron pair is strongly attracted by two nuclei	Strong because of the simultaneous attraction of the electrons by a large number of metal cations
Non-directional	Directional	Non-directional

EXPERIMENT 3.2

COMPARING DIFFERENT TYPES OF SUBSTANCES

Solids can be divided into four categories on the basis of their properties. Solids can be metallic, ionic, covalent molecular and covalent network. The property of electrical conductivity effectively distinguishes between metallic, ionic and covalent solids. To distinguish between covalent molecular and covalent network types, melting point must be considered.

In this experiment, you will classify a number of substances on the basis of their ability to conduct electricity in the solid, molten and/or aqueous states.

Note: Because of the amount of equipment and time involved, this experiment is best done by rotating through stations, each of which tests the properties of one substance.

Aim

To classify substances according to their physical properties

Materials

All stations

- Power pack, kit for measuring conductivity (Stations 1–5 need 2 kits)

Stations 1 and 2

- 2 crucibles with lids, Bunsen burner, tripod, pipe clay triangle
- Station 1 – sodium hydroxide pellets (enough to half-fill 2 crucibles)
- Station 2 – silver nitrate crystals (enough to half-fill 2 crucibles). Platinum electrodes work better at this station because of contamination of the electrodes in silver nitrate.

Stations 3–5

- 2 crucibles with lids, hot plate
- Station 3 – candle wax (enough to half-fill 2 crucibles)
- Station 4 – sulfur (enough to half-fill 2 crucibles). Place this station in a fume cupboard.
- Station 5 – camphor or naphthalene (enough to half-fill 2 crucibles). Place this station in a fume cupboard.

Stations 6–12

- 100 mL beaker
- Station 6 – 50 mL of distilled water
- Station 7 – 50 mL of kerosene
- Station 8 – 50 mL of ethanol
- Station 9 – 50 mL of 0.1 mol L^{-1} sucrose solution
- Station 10 – 50 mL of 0.1 mol L^{-1} sodium chloride solution
- Station 11 – 50 mL of 0.1 mol L^{-1} sodium hydroxide solution
- Station 12 – 50 mL of 0.1 mol L^{-1} hydrochloric acid solution

Station 13

- 1 piece each of copper sheet, tin foil, aluminium foil and quartz

What are the risks in doing this experiment?	How can you manage these risks to stay safe?
Hydrochloric acid and sodium hydroxide are corrosive.	Wear safety glasses, gloves and protective clothing. Take care when pouring. Clean up spills immediately; if spilt on skin, wash affected area with plenty of water and notify your teacher immediately.
Silver nitrate is toxic.	Wear appropriate safety gear. Dispose of in chemical waste jar provided.
Substances may spit out of heated crucible.	Wear appropriate safety gear; use tongs.
Sulfur and naphthalene are irritants to eyes, nose and throat.	Work in a fume cupboard. Do not breathe in the fumes.
Kerosene is flammable and can be an irritant.	Place kerosene well away from Bunsen burners and hotplates. Work in a well-ventilated area; do not breath the fumes.

In your write-up, add any more risks you can think of, as well as ways to manage them.

Procedure

1 At all stations, start the power pack at 2 V. If you get a reading, then record it. If not, turn the voltage up one step at a time until you get a reading or until you get to a maximum of 6 V. Record the result, either a current reading or no current.

2 At stations 1 and 2, place one of the crucibles on the tripod and heat it with a blue flame until it melts. Turn the Bunsen burner to a very low flame or off. Use one of the conductivity kits to test the molten substance. Use the other kit to test the solid in the other crucible. If the electrodes in the molten substance are covered in solid, then the solid will need to be melted before testing the conductivity.

3 At stations 3–5, there is a chance of the substance catching fire. If it does, place the lid on the crucible to put it out. Heat the crucible on the hot plate until the substance melts. Use one of the conductivity kits to test the molten substance. Use the other kit to test the solid in the other crucible.

4 At stations 6–12, dip the conductivity kit into the solution and record the result.

5 At station 13, place the electrodes against each sample.

Results

Record your results in a table similar to the following table.

Chemical	State	Conductivity	Classification

Analysing the results

What, pattern if any, is there to the conductivity results?

Discussion

Use theory to identify the types of bonding in each of the substances tested. Compare your results with the theoretical classifications and suggest any reasons for differences.

Conclusion

What generalisations can you make about particles in substances and conductivity?

The properties of a substance can be used to identify the type of bonding present. Table C3.11 summarises the main properties used to distinguish between metallic, ionic, covalent molecular and covalent network substances.

Table C3.11 Properties of substances

Property	Metallic	Ionic	Covalent molecular	Covalent network
Melting and boiling points	Variable but most commonly high	High	Low	High
Electrical conductivity	Good conductor in solid and liquid states	Non-conductor in solid state but good conductor in molten and aqueous state	Does not conduct	Does not conduct (except graphite)
Hardness and malleability	Hard and malleable	Hard and brittle	Mostly soft	Hard and brittle
Solubility	Generally insoluble	Generally soluble in water	Variable - depends on solvent	Insoluble

3.8 Hydrocarbons

SUMMARY OF THE DIFFERENT BONDING TYPES

Use this interactive activity to review the different bonding types.

Carbon is found in more places and in more forms than any other element. Carbon forms more compounds than any other element. The vast majority of compounds known to chemists are carbon compounds and there is a branch of chemistry, **organic chemistry**, devoted to carbon-based compounds. One of the main reasons for this is that carbon readily bonds to other carbon atoms (as seen in its allotropes) and each atom can form a total of four bonds from single to double to triple.

Chemists divide organic compounds into 'families' based on their structural features. This makes managing the huge number of carbon compounds easier. Families are organised according to which atoms or groups of atoms are bonded to carbon as well as whether multiple bonds are present.

The group of carbon compounds we are going to focus on is **hydrocarbons**. Hydrocarbons are compounds made up of only hydrogen and carbon. The simplest group of hydrocarbons are the **alkanes**.

Alkanes

The alkane family is very important because among its members are the fuels that supply us with energy every day. For most Australians, we use alkanes to run our cars, supply us with hot water, heat our homes and cook our foods. Alkanes also supply us with the raw materials to make many other substances, including plastics.

The simplest alkane is methane, CH_4, which you considered earlier. Next are ethane, C_2H_6, propane, C_3H_8, and butane, C_4H_{10}. Figure C3.25 shows models and structural formulas for propane and butane.

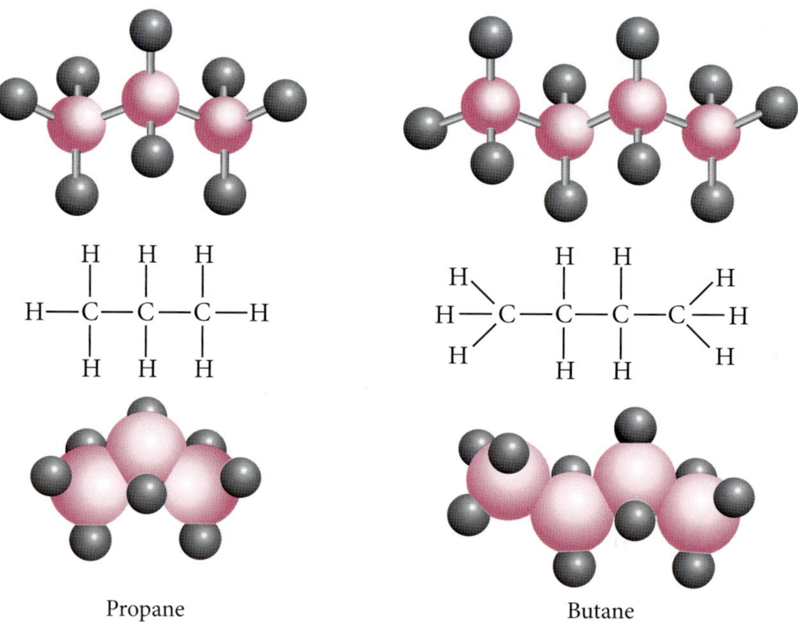

Figure C3.25 ▶ Models and structural formulas for propane and butane

Table C3.12 First ten members of the alkane family

Name	Formula
Methane	CH_4
Ethane	C_2H_6
Propane	C_3H_8
Butane	C_4H_{10}
Pentane	C_5H_{12}
Hexane	C_6H_{14}
Heptane	C_7H_{16}
Octane	C_8H_{18}
Nonane	C_9H_{20}
Decane	$C_{10}H_{22}$

The first ten alkanes are listed in Table C3.12. These are called straight-chain alkanes because the carbon atoms are joined in one continuous string. The chain is not really straight but more like a zig-zag because of the way carbon bonds are arranged tetrahedrally.

Alkanes are also classified as **saturated compounds**. This means they contain only carbon–carbon single bonds.

Each successive member of the series is formed by replacing one H with a C and then adding three more hydrogens. This pattern follows for all straight-chain alkanes, so the general formula for alkanes can be expressed as C_nH_{2n+2}, where n is a whole number.

Naming alkanes

As with ionic and covalent compounds, there is a systemic naming system for alkanes that allows us to work out the formula and structure of the compound.

The name of a carbon compound consists of a stem, which gives the length of the carbon chain, and an ending, which denotes the family of hydrocarbons the compound belongs to. All compounds that are in the alkane family have names ending in -ane. The stems of the first ten members of the series are: meth- C, eth- C_2, prop- C_3, but- C_4, pent- C_5, hex- C_6, hept- C_7, oct- C_8, non- C_9, dec- C_{10}.

The combination of the stem and ending -ane leads to the names given in Table C3.12.

The alkanes considered so far have been straight-chain alkanes. These are compounds in which all the carbon atoms are in one string so no carbon atom is attached to more than two other carbon atoms.

However, many carbon compounds contain branched chains. In branched-chain compounds, at least one carbon atom is attached to more than two other carbon atoms.

The rules for naming branched alkanes are:

1 The end of the name indicates to which hydrocarbon family the compound belongs. Alkanes always end in -ane.

2 Determine the longest continuous carbon chain and use the name of the corresponding alkane. (This chain may not necessarily be drawn linear.) This chain is the main chain and represents the parent structure. For example, if the longest chain had six carbon atoms the parent structure would be hexane.

3 Other atoms or groups of atoms attached to the main (parent) chain are **substituents** and form 'branches'. If the substituent is a carbon group, then it is called an alkyl group. The alkyl group is named according to the number of carbon atoms and given the ending -yl. For example, a substituent group CH_3CH_2– would be called ethyl.

No. C atoms	1	2	3	4	5	6	7	8
Stem	meth-	eth-	prop-	but-	pent-	hex-	hept-	oct-

Other common substituents are F fluoro, Cl chloro, Br bromo, I iodo and NO_2 nitro.

```
 6      5        4        3       2        1
H3C—CH2—CH2—CH—CH2—CH3
                          |
                         CH2
                          |
                         CH3
```

Number the carbon atoms in the main (parent) chain so the branch or branches have the lowest possible number/s.

The position at which the group is attached to the main (parent) chain is specified by the number of the carbon to which it is attached. (The number is separated by a hyphen.)

In the example above, the branch is 3-ethyl and the compound is 3-ethylhexane. (Note: The alkane name is written as one word.)

4 List the names of the substituents in alphabetical order (ignore di-, tri- etc.).

In the example below, there are two branches: 3-ethyl and 2-methyl. The 3-ethyl group is listed before the 2-methyl. Words and numbers are separated by a hyphen.

```
 6      5        4        3      2       1
H3C—CH2—CH2—CH—CH—CH3
                          |       |
                         CH2    CH3
                          |
                         CH3
```

The name of this compound is 3-ethyl-2-methylhexane.

Where there is more than one substituent of the same type, use the prefixes di- (two), tri- (three), tetra- (four), penta- (five) etc. (Separate numbers by a comma.)

$$
\begin{array}{cccccc}
6\ \ 5 & 4 & 3 & 2 & 1 \\
H_3C-CH_2- & CH- & CH- & CH- & CH_3 \\
 & | & | & | & \\
 & CH_3 & CH_2 & CH_3 & \\
 & & | & & \\
 & & CH_3 & &
\end{array}
$$

The above compound contains two methyl groups so 'dimethyl' will be part of its name, which is 3-ethyl-2,4-dimethylhexane.

Drawing structural formulas of branched alkanes

Drawing structures given the systematic name is quite simple.

1 Use the main name to draw the longest carbon chain and number the carbon chain, starting from 1 at an end of the chain (which end doesn't matter).

For example, in 2,2-dimethylbutane, the main chain is butane and the carbon atoms are numbered from 1 to 4.

$$
\begin{array}{ccccccc}
4 & & 3 & & 2 & & 1 \\
C & - & C & - & C & - & C
\end{array}
$$

2 Identify the substituents that form the branches and to which carbon atom they are attached.

For example, in 2,2-dimethylbutane, there are two methyl groups, both attached to carbon number 2.

3 Add the substituent group(s) to the parent chain.

$$
\begin{array}{ccccccc}
 & & & & C & & \\
 & & & & | & & \\
C & - & C & - & C & - & C \\
 & & & & | & & \\
 & & & & C & &
\end{array}
$$

4 Add hydrogen atoms to complete the structure. (Check that each carbon atom has four bonds.)

$$
\begin{array}{ccccccc}
 & & & & CH_3 & & \\
 & & & & | & & \\
CH_3 & - & CH_2 & - & C & - & CH_3 \\
 & & & & | & & \\
 & & & & CH_3 & &
\end{array}
$$

Properties of alkanes

Alkanes are covalent molecular substances, so their properties will be similar to those previously identified for covalent molecular substances. Some physical properties of alkanes are summarised in Table C3.13. These properties can be explained by the covalent molecular nature of the molecules.

Table C3.13 Some physical properties of alkanes

Property	Alkanes
Appearance	Colourless and odourless
Melting point and boiling point	Low melting point and boiling point, which increases with molecular mass, the first four being gases at room temperature
Electrical conductivity	Non-conductors
Hardness	Solid alkanes are soft
Malleability	Solid alkanes are malleable
Solubility	Insoluble in water but soluble in some other solvents

Chemical reactivity

Alkanes do not react with most substances. They only have two important reactions: combustion and substitution.

1 Combustion: Combustion is the most significant reaction that alkanes undergo. When ignited in a plentiful supply of oxygen, alkanes react readily to produce water, carbon dioxide and a large quantity of heat. This makes alkanes important as fuel. For example:

$$C_3H_8(g) + 5O_2(g) \rightarrow 3CO_2(g) + 4H_2O(g)$$

$$2C_8H_{18}(g) + 25O_2(g) \rightarrow 16CO_2(g) + 18H_2O(g)$$

2 Substitution: In a substitution reaction, a hydrogen atom is replaced with an atom of another element. This reaction usually only occurs with chlorine or bromine and it will not occur unless sufficient energy is supplied. Usually, the reaction will occur if the mixture is subjected to UV light. For example, in the presence of UV light, the following reactions will occur.

$$CH_4(g) + Cl_2(g) \rightarrow \underset{\text{chloromethane}}{CH_3Cl(g)} + HCl(g)$$

$$C_6H_{14}(g) + Br_2(soln) \rightarrow \underset{\text{bromohexane}}{C_6H_{13}Br(soln)} + HBr(soln)$$

Substitution reactions can continue until all the hydrogen atoms in the compound have been replaced by halogen, so in the case of methane the final product would be CCl_4, which is tetrachloromethane. The name of the product gives the number of carbon atoms (methane) and the number of halogen atoms (tetrachloro).

Alkenes

The **alkene** family contains hydrocarbons in which one pair of carbon atoms is joined by a double bond and all the other carbon atoms are joined by single bonds. All compounds that are in the alkene family have names ending in *-ene*. The simplest alkene is ethene, C_2H_4, and the next member is propene, C_3H_6. Examples of the structures of alkenes are shown in Figure C3.26.

Compounds that contain multiple bonds between carbon atoms are called **unsaturated compounds**. Alkenes are therefore unsaturated hydrocarbons.

The names and formula of the first nine alkenes is given in Table C3.14.

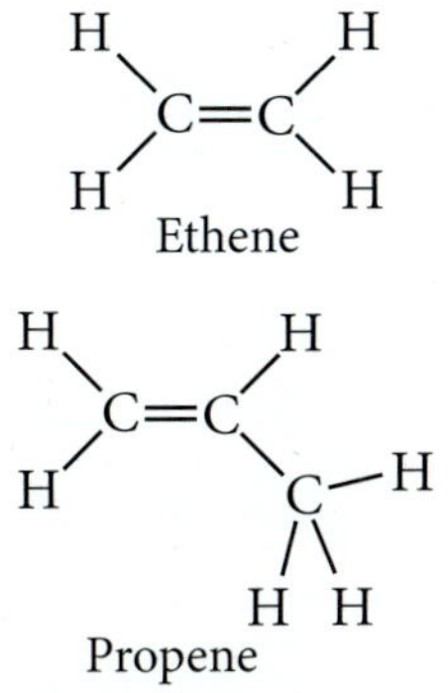

H H
C=C
H H
Ethene

H H
C=C
H C—H
H H
Propene

▲ **Figure C3.26** Ethene and propene

Table C3.14 The first nine members of the alkene family

Name	Formula
Ethene	C_2H_4
Propene	C_3H_6
Butene	C_4H_8
Pentene	C_5H_{10}
Hexene	C_6H_{12}
Heptene	C_7H_{14}
Octene	C_8H_{16}
Nonene	C_9H_{18}
Decene	$C_{10}H_{20}$

From Table C3.14, it can be seen that alkenes have a general formula of C_nH_{2n}. As with alkanes, each successive member of the series is formed by adding CH_2 to the formula of the previous member.

WOW

Ethene

Ethene gas is a ripening hormone that stimulates the production of certain enzymes that cause fruit to ripen. The sweet smell produced by ripe fruit is actually ethene gas. In an effort to make fruit last longer, new plastics have been developed that absorb ethene, preventing the fruit from ripening.

Naming alkenes

Naming alkenes is not quite as straightforward as naming alkanes. This is because the double bond can occupy a number of different positions. While this does not change the molecular formula of the compound, it does change the structural formula, which shows the position of the double bond. This can be seen in the example shown in Figure C3.27.

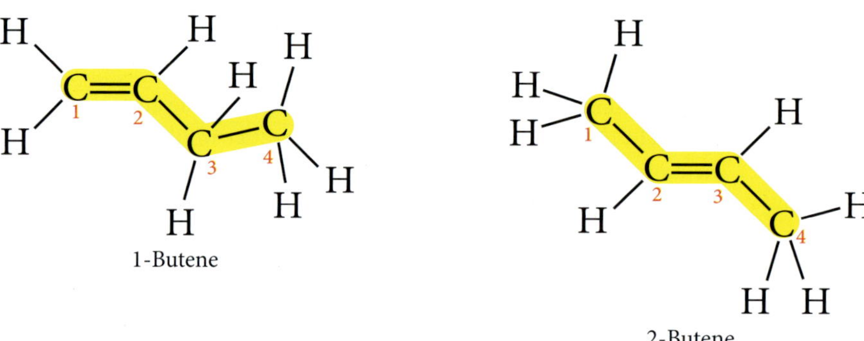

Figure C3.27 ▶ The structural isomers of butene

Compounds that have the same molecular formula but different structural formula are called structural **isomers**.

When naming alkenes, the location of the double bond needs to be included. Alkenes are named according to the following rules.

1 Take the usual stem name (such as *eth-* and *but-*), which indicates the number of carbon atoms in the chain, and add the ending *-ene*.

2 Determine the location of the double bond by numbering the carbon atoms from the end of the chain closest to the double bond (Figure C3.27).

3 Show the location of the double bond by putting in front of the name the number of the carbon at which the double bond starts followed by a hyphen (-).

To name branched alkenes, add the following rules to those for naming branched alkanes.

1 When naming alkenes the longest continuous carbon chain containing the carbon–carbon double bond becomes the main (parent) chain. The name corresponding to the alkane stem with the ending -ene forms the main name.

2 Branched alkenes are named in a similar way to branched alkanes, except the lowest number is assigned to the double bond and *not* the substituent group.

$$\begin{array}{c} \quad\quad\quad\quad\quad CH_3 \\ \quad\quad\quad\quad\quad | \\ CH_3-CH-C=CH_2 \\ | \quad\quad\quad\quad \\ CH_3 \quad\quad\quad\quad \end{array}$$

2,3-Dimethyl-1-butene

Properties of alkenes

As might be expected, the physical properties of alkenes are similar to those of alkanes. As a family, alkenes show similar trends in melting point, boiling point and state under normal conditions. At room temperature, the first three (C_2-C_4) are gases, C_5-C_{15} are liquids and the remainder are solids. All alkenes are insoluble in water but will dissolve in other solvents similar to alkanes.

Chemical reactivity

The presence of the double bond makes alkenes very reactive. They have two important reactions: combustion and addition.

Like alkanes, alkenes undergo combustion in a plentiful supply of oxygen to produce carbon dioxide and water. For example:

$$C_4H_8(g) + 6O_2(g) \rightarrow 4CO_2(g) + 4H_2O(g)$$

Alkenes undergo a reaction that adds more atoms to the compound by breaking the double bond. This type of reaction is called an **addition reaction**. Alkenes undergo a large number of addition reactions. Some of the more important are shown here.

Reaction with hydrogen

The alkene is converted to an alkane with the aid of a catalyst. For example, ethene is converted to ethane:

$$CH_2{=}CH_2(g) + H_2(g) \rightarrow CH_3CH_3(g)$$

PROPERTIES OF ALKANES AND ALKENES

View this slide show to find out more about the physical and chemical properties of alkanes and alkenes.

Reaction with a halogen

When a halogen such as chlorine or bromine reacts with an alkene, halogen atoms add across the double bond. For example:

$$CH_3{-}CH{=}CH_2(g) + Cl_2(g) \rightarrow CH_3CHCl{-}CH_2Cl(g)$$

Reaction with water

In this reaction, water, in the presence of a sulfuric acid catalyst, adds an H and OH across the double bond to produce a different group of organic compound – an alcohol. For example, ethene is converted to ethanol:

$$CH_2{=}CH_2(g) + H_2O(g) \rightarrow CH_3{-}CH_2OH(l)$$

Polymerisation

In the presence of a suitable catalyst and reaction conditions, ethene molecules will react with themselves to form long chains. This type of reaction is called addition polymerisation.

Polymerisation will be studied in Units 3 & 4.

The following points should be noted about addition reactions.

- When a symmetrical reagent, for example H_2 or Cl_2, is added, the product of the reaction has the same group added to each carbon.
- When an asymmetrical reagent, for example H_2O or HBr, is added to a symmetrical alkene, there is only one possible product.
- When an asymmetrical reagent, for example H_2O or HBr, is added to an asymmetrical alkene, there are two possible products. In reality, one product predominates because of Markovnikoff's rule, which states that the hydrogen atom will add across the double bond to the carbon that already has the greater number of hydrogen atoms.

To learn about fuels, refer to Context 2, 'Materials for a purpose', page 43.

Benzene

Benzene is the simplest member of the family of **aromatic hydrocarbons**. This group originally got their name because many of them have a strong aroma (smell).

Benzene is a colourless liquid with molecular formula C_6H_6, so it must be an unsaturated molecule. The structure is a six-sided flat ring with all bond angles of 120°. The carbon atoms are joined in a regular hexagon with one hydrogen atom attached to each carbon, as shown in Figure C3.28 on page 196.

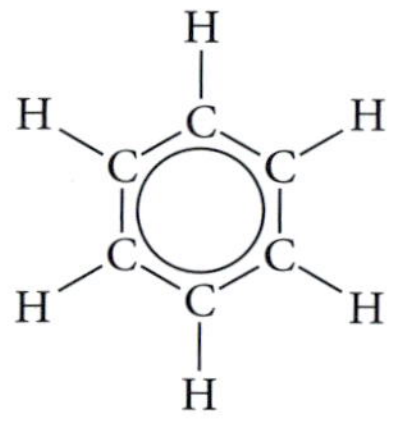

Figure C3.28 ▶ Structure of a benzene ring

On the basis of this representation, it was thought that benzene contained alternating single and double bonds so should undergo addition reactions similar to the alkenes and cycloalkenes. However, this rarely happens. The reactions of benzene are also completely different from those of other double-bonded hydrocarbons.

Further investigation of the benzene ring found that all the carbon–carbon bonds had the same bond length, which is somewhere between a single bond and a double bond. The structure of benzene can be explained through two types of bonding. Each carbon atom is joined to an adjacent carbon atom by a single bond with the electron pair located between the nuclei of the adjacent carbon atoms. Each carbon is also joined to a hydrogen atom by a single bond with the electron pair located between the carbon and hydrogen atoms. These three bonds utilise three of the valance electrons of each carbon atom.

As carbon has four valence electrons, the remaining six valence electrons (one from each carbon) make a cloud of **delocalised electrons** above and below the ring as shown in Figure C3.29. This is similar to the delocalisation of electrons found in the structure of metals.

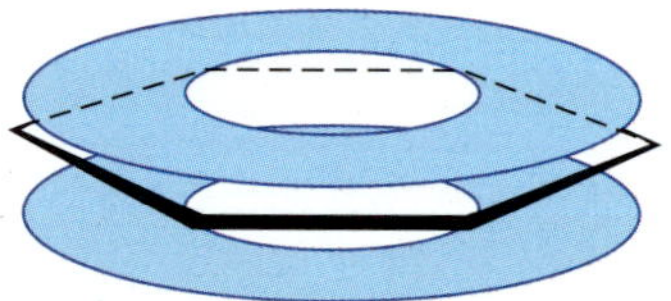

Figure C3.29 ▶ Bonding in benzene

WOW

Kekulé's dream

August Kekulé is credited with solving the puzzle of the bonding in benzene. It is said he had a dream in which a dancing snake formed a ring and bit its own tail. When he woke up, he worked out the structure of benzene, thus solving one of the biggest puzzles in 19th century chemistry.

In chemical structures, the benzene ring is usually represented in either of the two forms shown in Figure C3.30.

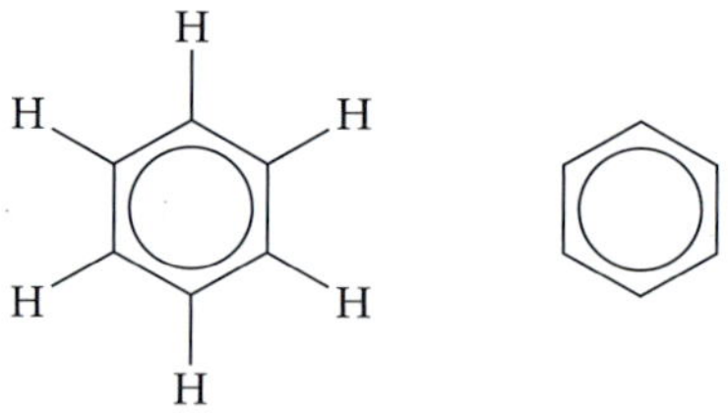

Figure C3.30 ▶ Different ways of drawing benzene

Properties of benzene

As with alkanes and alkenes, benzene has a low melting point, boiling point and density. It is also insoluble in water.

Chemical reactivity

The cyclic delocalisation of electrons in the benzene molecule makes it extremely stable. Benzene tends to undergo reactions in which the stable ring is preserved because reactions that disrupt the bonding due to the delocalised electrons are less favourable and will only occur at higher temperatures and under more vigorous conditions.

Most reactions of benzene are substitution reactions (like alkanes) where one or more of the hydrogens are replaced by a different atom or group of atoms.

Some examples of substitution reactions that benzene undergoes are:

- alkylation – substitution with an alkyl group

$$C_6H_6 + CH_3Cl \xrightarrow{AlCl_3} C_6H_5CH_3 + HCl$$

Benzene Chloromethane Methylbenzene Hydrogen chloride

- halogenation – substitution with a halogen

$$C_6H_6 + Br_2 \xrightarrow{FeBr_3} C_6H_5Br + HBr$$

Benzene Bromine Bromobenzene Hydrogen bromide

- nitration – substitution with a nitro group.

$$C_6H_6 + HNO_3 \xrightarrow{H_2SO_4} C_6H_5NO_2 + H_2O$$

Benzene Nitric acid Nitrobenzene Water

QUESTION SET 3.7

Remembering

1 Give the general formula for:
  a alkanes.
  b alkenes.
  c benzene.
2 Explain the difference between saturated and unsaturated compounds.
3 Explain why benzene is extremely stable.

Understanding

4 Which of the following is an alkane and which is an alkene? Explain your choice.
  C_6H_6 C_6H_{10} C_6H_{12} C_6H_{14}
5 Draw a structural formula for each of the following.
  a Butane
  b Octane
  c 1-Propene
  d 3-Hexene
  e 2-Methylpentane
  f 2,3-Dimethylbutane
  g 2-Methyl-2-pentene
  h 3,3-Dimethyl-1-butene
6 a Explain why the physical properties of alkanes and alkenes are similar but their chemical properties are different.
  b Would you expect benzene to be soluble or insoluble in water? Justify your response.
7 Name the following compounds.
  a C_4H_{10}
  b C_9H_{20}
  c $H_3C-CH_2-CH=CH-CH_3$
  d $H_2C=CH-CH_2-CH_3$
  e 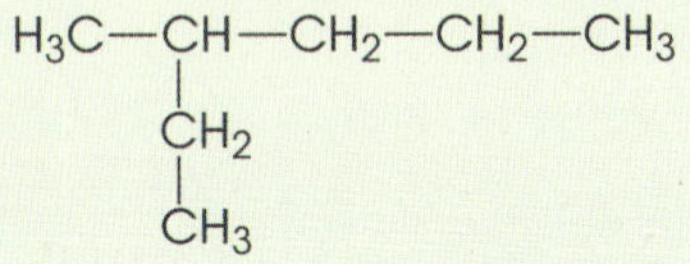
  f 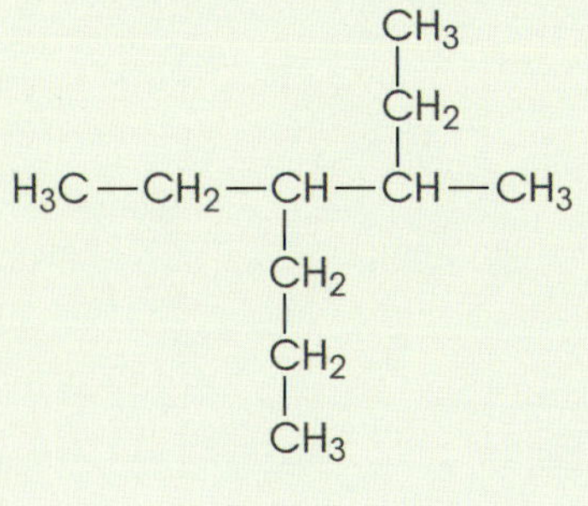
  g 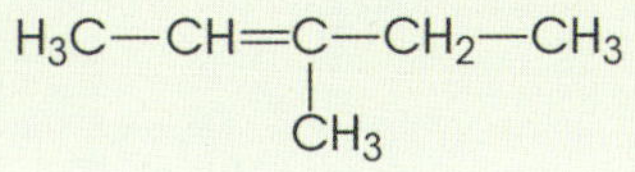

h

$$\begin{array}{l} \qquad\qquad\quad CH_3 \\ \qquad\qquad\quad | \\ H_3C—CH—C{=}CH_2 \\ \qquad\ \ | \\ \qquad CH_2 \\ \qquad\ \ | \\ \qquad CH_3 \end{array}$$

8 Explain the differences between substitution and addition reactions.

Applying

9 Draw the structural formula and name the straight chain isomers of C_8H_{16}.

10 Write balanced equations for the:

a combustion of pentane.

b reaction of butane with chlorine gas (under UV light).

c reaction of propene with water.

d reaction of ethene with bromine.

e reaction of benzene with chlorine.

f alkylation of benzene with $CH_3CH_2CH_2Cl$.

Analysing

11 A certain gaseous hydrocarbon is bubbled through bromine. The bromine is decolorised. What can you conclude about the unknown hydrocarbon?

CHAPTER CHECKLIST

You should know:

- the different ways in which chemical bonds form
- the different types of chemical bonds – ionic, metallic and covalent
- the relationship between valence electrons, valency and an element's position in the periodic table
- why metals have delocalised electrons
- the relationship between properties and bonding for metals, ionic compounds, covalent molecular and covalent network substances
- the different allotropes of carbon and reasons for differences in their properties
- the relationship between anions, cations and an element's position in the periodic table
- the similarities and differences between covalent molecular and covalent network substances
- the general structure and formula of alkanes, alkenes and benzene
- reasons for differences in reactions of alkanes, alkenes and benzene.

You should be able to:

- use the periodic table to determine valency of an element
- identify different types of chemical bonds from chemical formulas
- predict the type of bond that will form from given elements
- explain the properties of a metal in terms of metallic bonding
- predict the formation of anions, cations and the formula of the resultant compound
- write the name and formula of ionic compounds
- explain the properties of ionic compounds in terms of ionic bonding
- write the names and formulas of molecular compounds
- explain the properties of covalent molecular and network substances in terms of their bonding
- draw electron dot formulas and structural formulas for different types of compounds
- use properties to identify type of bonding
- draw structural formulas and name straight-chain and branched alkanes and alkenes
- write balanced equations for reactions of alkanes, alkenes and benzene.

CHAPTER GLOSSARY

addition reaction a reaction that involves the breaking of a multiple bond and the addition of new atoms to the compound

alkane a compound of carbon and hydrogen containing only single bonds

alkene a compound of carbon and hydrogen that contains carbon–carbon double bonds

allotrope different physical form of the same element

anion a negatively charged atom or groups of atoms; formed by the addition of electron(s)

aromatic hydrocarbon hydrocarbon containing one or more benzene rings

bonding pair a pair of electrons shared by two atoms to form a covalent bond

cation a positively charged atom or groups of atoms; formed by the loss of electron(s)

chemical bond the electrostatic force of attraction between the protons and electrons of participating atoms

compound a pure substance composed of more than one type of atom chemically combined in fixed proportions

covalent bond the electrostatic force of attraction between shared negatively charged electrons and positively charged nuclei; results in a molecule or covalent lattice

covalent molecular substance a discrete molecule in which the atoms are held together by covalent bonding

covalent network substance a three-dimensional network of covalently bonded atoms

delocalised electron an electron that is detached from its atom and is free to move about within the structure

directional bonding bonding that is in a direct line between adjacent particles; typical of covalent bonding

double covalent bond a covalent bond in which two pairs of electrons are shared

electron dot formula representation of an atom or ion using dots or crosses to show only valence electrons

empirical formula the simplest whole number ratio of the elements in a chemical compound

hydrocarbon a chemical compound composed of only hydrogen and carbon

intermolecular bond the electrostatic forces of attraction between molecules in close proximity

intramolecular bond the electrostatic force of attraction that holds atoms together within a molecule

ion a charged atom, either positive from losing electrons, or negative from gaining electrons

ionic bond the electrostatic force of attraction between positively charged cations and negatively charged anions; results in an ionic compound with a three-dimensional lattice structure

ionic compound a compound that is made up of positive ions (cations) and negative ions (anions)

isomers (structural) compounds that have the same molecular formula but different structural formulas

metallic bond the electrostatic force of attraction that holds metal atoms together; acts between negatively charged delocalised valence electrons and positively charged metal cations

multiple covalent bond a covalent bond in which more than one pair of electrons are shared

non-bonding or lone pair a valence pair of electrons present in a covalent substance that is not involved in the covalent bond

non-directional bonding bonding that occurs in most directions; typical of metallic and ionic bonding

organic chemistry the branch of chemistry that studies carbon compounds

polyatomic ion an ion that consists of two or more atoms that are strongly bonded by covalent bonding

saturated compound a compound that contains only carbon–carbon single bonds

substituent an atom or group of atoms that has replaced a hydrogen atom on a hydrocarbon

transition metals the metals in groups 3–12 of the periodic table

triple covalent bond a covalent bond in which three pairs of electrons are shared

unsaturated compounds compounds in which there is at least one multiple bond (double or triple) between carbon atoms

valence electron an electron in the valence (outermost) shell of an atom in its ground state

valency the combining power of an atom equal to the number of hydrogen atoms that atom could combine with or displace from a compound (hydrogen has a valency of 1)

CHAPTER REVIEW QUESTIONS

Remembering

1 a List the three main types of chemical bond.
  b Describe each type of bond and represent it with a diagram.

2 Draw up a table comparing metallic, ionic, covalent molecular and covalent network substances, using the properties of electrical conductivity, melting and boiling point, and hardness.

3 Explain the differences in bonding and properties of diamond, graphite and fullerenes.

4 Describe the similarities and differences between alkanes, alkenes and benzene.

5 Explain what is meant by an isomer and give an example.

Understanding

6 Phosphorus trichloride is a liquid with a boiling point of 74°C; it does not conduct electricity. Calcium chloride is a solid with a melting point of 772°C; when molten it conducts electricity. Explain, in terms of bonding, why these compounds have such different properties.

7 Classify each of the following solids as ionic, covalent molecular, metallic or covalent network: magnesium, tetrabromomethane, barium chloride, phosphorus triiodide, silicon dioxide, lithium sulfide, iodine and diamond.

Applying

8 Name the following compounds.
  a Ag_2O
  b $AlCl_3$
  c K_2SO_4
  d NH_4Br

e $Mg(OH)_2$

f $FeCO_3$

9 Write the correct formula for each of the following compounds.

a Magnesium chloride

b Calcium carbonate

c Copper(II) nitrate

d Ammonium chloride

e Potassium sulfide

f Lead(IV) oxide

10 a What is the valence of element X in each of the following compounds?

i X_2S_3

ii $X(NO_3)_2$

b What is the valence of Y in each of the following compounds?

i MgY_2

ii $(NH_4)_3Y$

11 Name the following compounds.

a PCl_3

b SF_4

c N_2O_3

d NO

e Cl_2O_7

12 Write the formula for each of the following compounds.

a Boron trichloride

b Phosphorus pentaiodide

c Hydrogen bromide

d Dinitrogen tetroxide

e Silicon dioxide

13 Draw electron dot formulas for the following molecules.

a CaF_2

b HCl

c Na_2O

d NH_3

14 Name the following hydrocarbons.

a $CH_3CH_2CH_2CH_3$

b $CH_3—CH_2—CH_2—CH_2—CH_2—CH_3$

c $CH_2{=}CH—CH_2—CH_2—CH_3$

d $CH_3—CH{=}CH—CH_3$

e $CH_3—CH_2\text{-}CH{=}CH—CH_3$

f
```
              CH3
              |
H3C—CH—CH—CH3
       |
       CH3
```

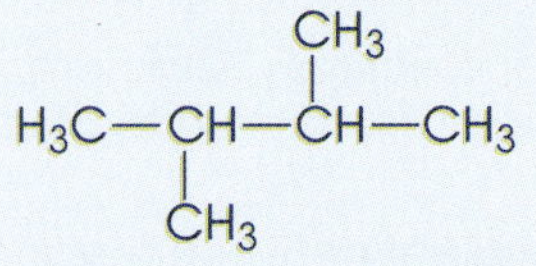

g
```
H3C—CH2—CH2—C=CH—CH3
             |
             CH2
             |
             CH3
```

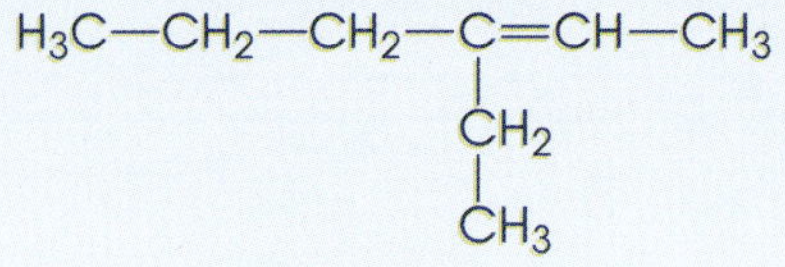

h

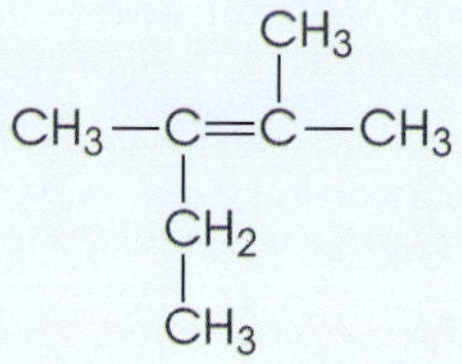

$$CH_3-C(CH_2CH_3)=C(CH_3)-CH_3$$

15 Draw a structural formula for each of the following.

a Ethane

b Hexane

c 3-Octene

d 1-Pentene

e 3,4-Dimethyl-2-pentene

f 3-methyl-1-pentene

g 2-Ethyl-3-methyl-1-hexene

16 What is wrong with the following names? Give the correct one.

a 5-Hexene

b 4-Heptene

17 a Write equations for the combustion of:

i pentane.

ii 2-heptene.

b Write equations for the reactions of the following alkanes in the presence of UV light.

i Propane with $Br_2(aq)$

ii Octane with $Cl_2(g)$

18 Write balanced equations for the following reactions and draw the structural formulas of the products.

a 1-Pentene with Cl_2

b Propene with H_2O (with H_2SO_4 catalyst)

c 3-Hexene with bromine water

d Benzene with iodine

e Benzene with nitric acid

Analysing

19 Five solids A–E have the properties listed below. The relevant properties of sodium chloride and copper are also given.

Solid	Melting point (°C)	Relative conductivity of		Soluble in		'Hammer' test
		Solid	Liquid	Water?	Hexane?	
A	327	5	2	No	No	Flattens
B	2030	0	0	No	No	Shatters
C	91	0	0	No	Yes	Forms powder
D	734	0	0.2	Yes	No	Forms powder
E	2870	0	0	No	No	Shatters
NaCl	801	0	0.2	Yes	No	Forms powder
Cu	1083	60	4	No	No	Flattens

Note: The hammer test describes what occurs when the material is continually hit with a hammer.

a Classify each of the solids A–E as ionic, covalent molecular, covalent network or metallic.

b Explain why sodium chloride and copper have the conductivity properties listed in the table.

c For either covalent molecular compounds or covalent network compounds, explain why they have the melting points, conductivities and solubilities shown in the table.

20 Four bottles containing clear liquids were on the laboratory shelf but the labels had fallen off. The missing labels were found on the floor and had the names: hexene (C_6H_{12}), pentane (C_5H_{12}), water (H_2O) and methylated spirits.

A group of chemistry students was asked to devise a series of tests to identify the contents of each of the four bottles. The students labelled the bottles A–D and conducted their tests. Their results are shown in the following table.

Bottle	Soluble in water?	Flammable	Decolorises bromine in the absence of UV light
A	Yes	No	No
B	No	Yes	Yes
C	No	Yes	No
D	Yes	Yes	No

Using the information from the tests, identify which bottle contained which chemical, giving reasons for your decision.

CHAPTER 4
CHEMICAL REACTIONS

By the end of this chapter you will have covered the following material.

Science Understanding

- All chemical reactions involve the creation of new substances and associated energy transformations, commonly observable as changes in the temperature of the surroundings and/or the emission of light (ACSCH036)
- Endothermic and exothermic reactions can be explained in terms of the Law of Conservation of Energy and the breaking and reforming of bonds; heat energy released or absorbed can be represented in thermochemical equations (ACSCH037)
- Fuels, including fossil fuels and biofuels, can be compared in terms of energy output, suitability for purpose, and the nature of products of combustion (ACSCH038)

AC

To revise chemical and physical changes, refer to Chemistry section 2.2 on page 143.

4.1 Chemical reactions

Chemical reactions are important in our lives, in our homes and in industry. Chemical reactions occur every time we light a gas stove or a barbecue. Rusty garden furniture is an example of a chemical reaction. Cooking food causes the ingredients to chemically change and form new substances. In industry, new materials are continuously being developed – these require chemical reactions to change the starting materials into new materials, such as WikiCell, an edible form of packaging that will significantly decrease our need for plastics.

Some materials require very little processing, but most materials are produced by chemical reactions; for example, paints, fuels, finishes for cupboards and benchtops, household appliances and cleaning agents.

Photosynthesis, respiration, rusting and combustion are all chemical reactions. Photosynthesis occurs in plants when in the light and respiration occurs in both plants and animals all the time. Iron rusts when it is left out in the weather. Petrol burns in the combustion chamber of cars. In each case, chemical changes have occurred.

123RF/achiartistul

▲ **Figure C4.1**
A car engine – the site of fuel combustion, a chemical reaction that releases energy

123RF/firina

▲ **Figure C4.2**
Green leaves and stems – the site of photosynthesis, a reaction that builds molecules that store energy

Chemical equations

A chemical reaction can be represented by a chemical equation, which can use words or formulas. In a chemical equation, the reactants are written on the left-hand side of the arrow and the products are written on the right-hand side of the arrow. The single arrow always points in the direction of the reaction; that is, from the reactants to the products.

For example, the word equation for photosynthesis is:

$$\text{carbon dioxide} + \text{water} \rightarrow \text{glucose} + \text{oxygen}$$

Carbon dioxide and water are the reactants. Glucose and oxygen are the products. The arrow points towards the products. The '+' is used when there is more than one reactant or more than one product.

WORKED EXAMPLE 4.1

Respiration is the combustion reaction of glucose and oxygen in body cells to form carbon dioxide and water. The word equation is:

glucose + oxygen → carbon dioxide + water

a Identify the reactants.

b Identify the products.

Answers	**Logic**
a Glucose and oxygen are the reactants.	They are on the left-hand side of the arrow.
b Carbon dioxide and water are the products.	They are on the right-hand side of the arrow.

Try these yourself

Write word equations and identify the reactants and products for the following reactions.

a Rust (iron(III) oxide) forms when iron reacts with oxygen in the air.

b Petrol (octane) reacts with oxygen in the combustion chamber of cars to form carbon dioxide and water.

Balancing chemical equations

Chemists also represent chemical reactions by using chemical symbols and formulas for the reactants and products. Numbers are written in front of chemical formulas to balance the equation so that there is the same number of atoms of each element on each side. The number '1' is not written because it is assumed when no other number appears before the formula.

Generally, equations are written as the lowest whole number ratio between reactants and products. This ensures that the law of conservation of mass is maintained.

The states for each of the reactants and products are included: (s) for solid, (l) for liquid, (g) for gas and (aq) for aqueous solution. In some sources, (aq) is not included; if a reactant or product does not have a state included, it is assumed that it is an aqueous solution (dissolved in water).

Chemists use balanced chemical equations to summarise information about chemical reactions. These equations concisely communicate information about reactants, products and numbers of atoms of each element present.

For example, as shown in Figure C4.3, magnesium reacts with hydrochloric acid to form magnesium chloride and hydrogen according to the equation:

$$Mg + HCl \rightarrow MgCl_2 + H_2$$

The equation is balanced so that there is the same number of atoms of each element on each side of the equation.

$$Mg + 2HCl \rightarrow MgCl_2 + H_2$$

- There is one 'atom' of magnesium at the beginning of the reaction and one 'atom' of magnesium at the end.
- There are two 'atoms' of hydrogen at the beginning of the reaction and two 'atoms' of hydrogen at the end.

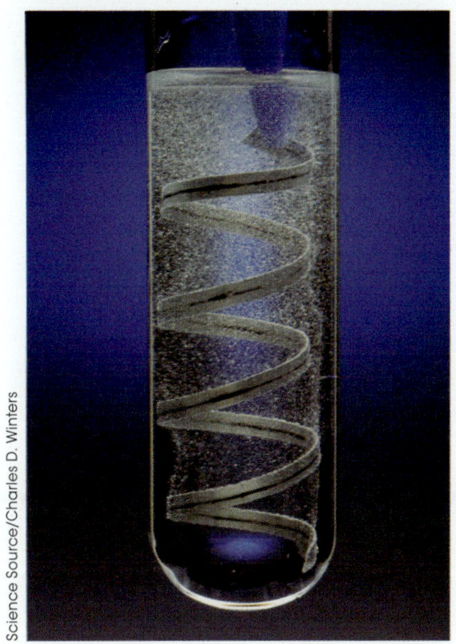
Science Source/Charles D. Winters

◀ **Figure C4.3** Magnesium reacting with hydrochloric acid

To learn about the Bronze Age, refer to Context 2, 'Materials for purpose', page 33.

- There are two 'atoms' of chlorine at the beginning of the reaction and two 'atoms' of chlorine at the end.

The states for each of the reactants and products are included.

$$Mg(s) + 2HCl(aq) \rightarrow MgCl_2(aq) + H_2(g)$$

Magnesium is a solid (s); hydrogen is a gas (g). Both hydrochloric acid and magnesium chloride are dissolved in water.

The method for balancing is still the same for complex reactions.

WORKED EXAMPLE 4.2

Write a balanced chemical equation, including states, for the reaction of octane (C_8H_{18}) with oxygen in a car's combustion chamber to produce carbon dioxide and water.

Answer

$2C_8H_{18}(l) + 25O_2\ (g) \rightarrow 16CO_2(g) + 18H_2O(l)$

Logic

1 Write reactants and products using their chemical formulas.

$C_8H_{18} + O_2 \rightarrow CO_2 + H_2O$

2 Balance the equation.

Start with the elements that only appear in one substance on each side of the equation.

$C_8H_{18} + O_2 \rightarrow 8CO_2 + 9H_2O$

There are $(8 \times 2) + 9 = 25$ atoms of oxygen on the right-hand side and 2 atoms of oxygen on the left-hand side. When balancing equations, whole numbers are used, so, all quantities in the equation need to be doubled.

$2C_8H_{18} + 25O_2 \rightarrow 16CO_2 + 18H_2O$

There is now the same number of 'atoms' of each element on each side of the equation.

3 Add the states for each of the reactants and products.

$2C_8H_{18}(l) + 25O_2\ (g) \rightarrow 16CO_2(g) + 18H_2O(l)$

Octane and water are liquids; oxygen and carbon dioxide are gases.

Try these yourself

Write balanced chemical equations for:

a rusting of iron when it reacts with oxygen to form iron(III) oxide.

b respiration of glucose ($C_6H_{12}O_6$) with oxygen to produce carbon dioxide and water.

CREATIVE CHEMISTRY – BALANCING SYMBOL EQUATIONS

Use this interactive activity for practice at balancing chemical equations.

QUESTION SET 4.1

Remembering

1 What is the purpose of a chemical equation?

2 Identify the steps required to write balanced chemical equations.

Understanding

3 Balance the following chemical equations.

a $Fe(s) + HCl(aq) \rightarrow FeCl_2(aq) + H_2(g)$

b $NaHCO_3(s) + H_2SO_4(aq) \rightarrow Na_2SO_4(aq) + CO_2(g) + H_2O(l)$

c $CuFeS_2(s) + SiO_2(s) + O_2(g) \rightarrow Cu_2S\ (s) + FeSiO_3(s) + SO_2(g)$

d $PbS(s) + O_2(g) \rightarrow PbO(s) + SO_2(g)$

e $C_3H_8(g) + O_2(g) \rightarrow CO_2(g) + H_2O(l)$

Analysing

4 Identify whether the following equations are balanced. Correctly balance the equations that are not balanced and justify your answer.

a $H_2(g) + O_2(g) \rightarrow H_2O(l)$

b $H_2SO_4(aq) + Ba(OH)_2(aq) \rightarrow BaSO_4(s) + 2H_2O(l)$

c $C_6H_{12}O_6(aq) \rightarrow 2C_2H_5OH(l) + CO_2(g)$

d $2C_2H_5OH(l) + 6O_2(g) \rightarrow 4CO_2(g) + 6H_2O(l)$

e $TiO_2(s) + 2Cl_2(g) + 2C(s) \rightarrow TiCl_4(aq) + 2CO(g)$

5 Write balanced chemical equations to represent each of these chemical reactions.

a The Haber process is an important chemical reaction where gaseous ammonia (NH_3) is produced by reacting hydrogen gas with nitrogen gas.

b Magnesium hydroxide is an ingredient in common antacids. It reacts with the dilute hydrochloric acid in the stomach to produce a solution of magnesium chloride and water.

c Zinc metal reacts with dilute sulfuric acid to form dilute zinc sulfate and hydrogen gas.

To learn why materials are mostly not found as elements, refer to Context 2, 'Materials for purpose', page 36.

4.2 Reactivity

Chemicals react to achieve stability; that is, the products are more stable than the reactants. During chemical reactions, bonds are broken and new bonds are formed. Energy is needed to break bonds, to chemically break down the forces of attraction that occur between:

- metal ions and the delocalised electrons in metallic lattices
- cations and anions in ionic lattices
- non-metallic elements in covalent molecules and covalent lattices.

Energy is released when bonds are formed.

To see examples of chemical reactions used throughout history and in industry, refer to Context 2, 'Materials for a purpose', pages 34, 43, 45 and 46.

Spontaneous chemical reactions

Spontaneous reactions are reactions that occur at room temperature. They do not require any extra energy to be added for the reaction to occur. More reactive substances are more likely to undergo reactions at room temperature. Some spontaneous reactions occur quickly and others occur more slowly. However, all of them occur at room temperature, as shown in Experiment 4.2 on page 211.

Spontaneous chemical reactions are reactions that occur at room temperature without the input of energy such as heat or electricity.

To see examples of how reactivity affects how materials are formed, refer to Context 2, 'Materials for a purpose', page 37.

Science Photo Library

Figure C4.4 ▲ Batteries spontaneously change chemical energy to electrical energy.

To see more examples of corrosion, refer to Context 2, 'Materials for a purpose', page 36.

Metals can react spontaneously. All metals have some common properties; for example, they all have relatively high melting points. However, they do not have identical melting points. Similarly, they may have similar chemical properties, but they are not identical.

Some metals react spontaneously with water, acid and oxygen. The generalised equations for these reactions are:

$$\text{metal} + \text{water} \rightarrow \text{a salt} + \text{hydrogen}$$
$$\text{metal} + \text{acid} \rightarrow \text{a salt} + \text{hydrogen}$$
$$\text{metal} + \text{oxygen} \rightarrow \text{metal oxide}$$

123RF/Jose Angel Astor

Figure C4.5 ▶ Metals in cars react with oxygen in the air to corrode.

EXPERIMENT 4.1

REACTION OF METALS WITH OXYGEN

Corrosion is the spontaneous reaction of a metal with oxygen in a moist environment.

Aim

To compare the reactivity of metals by comparing their reactions in a moist environment

Hypothesis

Write a hypothesis for this experiment.

Materials

- 5 cm × 0.5 cm piece of copper
- 5 cm × 0.5 cm piece of zinc
- 5 cm × 0.5 cm piece of aluminium foil
- 5 cm piece of magnesium
- 5 cm iron nail (ungalvanised)
- 5 medium test tubes
- Test-tube rack
- 5 cm × 5 cm piece of sand paper

What are the risks in doing this experiment?	How can you manage these risks to stay safe?
Metal residue may remain on your hands.	Wash your hands after handling metals.

Procedure

1 Clean each of the metals with the sand paper.
2 Place a piece of each metal into separate test tubes.
3 Place the test tubes into the test-tube rack.
4 Add water to each of the test tubes until the piece of metal is covered.
5 Record observations daily for 1 week.

Results

Record your observations in an appropriate format. For example, you could take a series of digital photos of each test tube.

Analysis of results

Write a balanced chemical equation for any reaction that has occurred.

Discussion

1 Which metal was the most reactive? Justify your answer.
2 Which metal was the least reactive? Justify your answer.
3 Do your results support your hypothesis? Justify your answer.
4 Rank the metals from most reactive to least reactive. Justify your answer.

EXPERIMENT 4.2

REACTIVITY OF METALS

Aim

To compare the reactivity of metals by observing their reactions with water and acids

Hypothesis

Write a hypothesis for this experiment.

Materials

- 3 × 5 cm × 0.5 cm pieces of copper
- 3 approx. 5 cm long ungalvanised iron nails
- 3 × 5 cm pieces of magnesium
- 3 × 5 cm × 0.5 cm pieces of zinc
- Bottle of calcium metal
- 50 mL of 1 mol L^{-1} hydrochloric acid
- 50 mL of 4 mol L^{-1} hydrochloric acid
- 5 test tubes
- Spatula

What are the risks in doing this experiment?	How can you manage these risks to stay safe?
Acid could splash or spill onto your hands.	Wash your hands immediately if they come in contact with acid and after using acids.
Vigorous bubbling may splash into your eyes.	Wear safety glasses.

Procedure

Part A: Reacting metals with water

1 Place one-quarter of a spatula of calcium into a test tube.
2 Place a piece of each of the other metals into separate test tubes.
3 Add about 3 cm depth of water to each test tube.
4 Record your observations.

Part B: Reacting metals with 1 mol L^{-1} hydrochloric acid

1 For any metals that did *not* violently react with water, place a small quantity or piece of each into separate test tubes.
2 Add about 3 cm depth of 1 mol L^{-1} hydrochloric acid to each test tube.
3 Record your observations.

►

Part C: Reacting metals with 4 mol L^{-1} hydrochloric acid

1 For any metals that did *not* violently react with 1 mol L^{-1} hydrochloric acid, place a piece of each into separate test tubes.
2 Add about 3 cm depth of 4 mol L^{-1} hydrochloric acid to each test tube.
3 Record your observations.

Results

Draw a table to show your results.

Analysis of results

Write a balanced equation for any chemical reaction that occurred in parts A, B and C.

Discussion

1 Which metal was the most reactive? Justify your answer.
2 Which metal was the least reactive? Justify your answer.
3 Do your results support your hypothesis? Justify your answer.
4 Rank the metals from most reactive to least reactive. Justify your answer.
5 Why were the metals that reacted with water not tested with the hydrochloric acid?

Representing chemical reactions

There are a variety of ways to represent chemical reactions, depending on the detail that is required. So far, balanced chemical equations have been used. These give general information about the reactants and the products. However, **ionic equations** give information about each **species** present in the reaction. Each of the aqueous species are broken up into their ions. The species that change are the ones that are taking part in the reaction. The species that are the same on both sides of the ionic equation are called **spectator ions**. They are present but do not take part in the chemical reaction. **Net ionic equations** are more specific again; they only include information about the species that take part in the reaction. Net ionic equations do not show the spectator ions and must be balanced for charge.

Net ionic equations only show the species that are taking part in the chemical reaction; spectator ions are not included.

WORKED EXAMPLE 4.3

Tin reacts with hydrochloric acid to form tin(II) chloride and hydrogen gas:

tin + hydrochloric acid → tin(II) chloride + hydrogen

Write the net ionic equation for the reaction.

Answer

$Sn(s) + 2H^+(aq) \rightarrow Sn^{2+}(aq) + H_2(g)$

Logic

The balanced equation is:

$Sn(s) + 2HCl(aq) \rightarrow SnCl_2(aq) + H_2(g)$

The ionic equation is:

$Sn(s) + 2H^+(aq) + 2Cl^-(aq) \rightarrow Sn^{2+}(aq) + 2Cl^-(aq) + H_2(g)$

Both the tin and hydrogen undergo chemical changes during the reaction. The chloride ion remains the same on both sides of the equation. It does not take part in the chemical reaction. The chloride ion is a spectator ion, so it will not appear in the net ionic equation.

Try these yourself

Write net ionic equations for each of the following reactions.

a $2Al(s) + 3H_2SO_4(aq) \rightarrow Al_2(SO_4)_3(aq) + 3H_2(g)$

b Copper metal reacts with nitric acid to form copper(II) nitrate, nitrogen dioxide gas and water.

QUESTION SET 4.2

Remembering

1 Define:

  a spontaneous chemical reaction.

  b spectator ion.

2 Identify two different ways that chemists represent chemical reactions.

Understanding

3 Justify whether reactivity is a chemical or physical property.

Applying

4 Write net ionic equations for each of the following reactions:

  a $Fe(s) + CuSO_4(aq) \rightarrow Cu(s) + FeSO_4(aq)$

  b $Na_2CO_3(aq) + MgCl_2(aq) \rightarrow MgCO_3(s) + 2NaCl(aq)$

5 Write net ionic equations for the reaction of:

  a calcium metal and water.

  b magnesium and hydrochloric acid.

  c zinc and sulfuric acid.

  d lead nitrate solution and potassium iodide solution to form potassium nitrate solution and solid lead iodide.

4.3 Energy and chemical reactions

For more information about activation energy, refer to Chemistry section 9.2 on page 374.

For an example of an exothermic reaction, combustion, refer to Context 2, 'Materials for a purpose', page 43.

In a chemical reaction bonds are broken in the reactants, and bonds are formed to make the products. These changes involve energy. Metallic lattices consist of metallic ions with delocalised electrons. It requires energy to overcome the forces of attraction between the metal ions and the delocalised electrons. Ionic compounds are lattice structures consisting of positive ions (cations) and negative ions (anions). Energy is required to overcome the electrostatic forces of attraction between the cations and anions. Covalent molecules are made up of non-metallic atoms held together by covalent bonds. Energy is required to break these covalent bonds. Therefore, energy needs to be put in to break the bonds in the reactants.

Energy is stored in the chemical bonds of the reactants and the products. The law of conservation of energy states that energy cannot be created or destroyed; it can only be changed from one form to another. So if more energy is stored in the bonds of the reactants than the products, then when the products are formed, the extra energy is released as heat, sound or light. An example of this is the burning of magnesium. It produces a very bright white light. A small amount of energy is needed to get this reaction going, the reason for which will be discussed in Chemistry Chapter 9. However, once the reaction is going, no further energy input is required. The burning of magnesium is an **exothermic** reaction because energy is released overall (see Figure C4.6 on page 214):

$$2Mg(s) + O_2(g) \rightarrow 2MgO(s) + \text{light}$$

Another exothermic reaction is the reaction between magnesium and hydrochloric acid. This differs from the burning of magnesium in two ways.

- Heat is produced instead of light.
- The reaction is spontaneous at room temperature; no energy is needed to start the reaction:

$$Mg(s) + 2HCl(aq) \rightarrow MgCl_2(aq) + H_2(g) + \text{heat}$$

These reactions are written as thermochemical equations because they include both the chemicals and the energy term in the equation.

To see examples of endothermic reactions used in the extraction of metals and in catalytic cracking, refer to Context 2, 'Materials for a purpose', pages 33 and 43.

Physical processes, such as changes of state, can also involve the release of energy. For example, water vapour (steam) cools down when it forms liquid water – it releases energy as heat. This phase change is an exothermic process. Here, the gaseous water molecules are losing kinetic energy and are slowing down. Hence, forces between the water molecules are being formed so that the water molecules are packed together more closely:

$$H_2O(g) \rightarrow H_2O(l) + \text{heat}$$

Energy is released during an exothermic reaction because the energy of the reactants is greater than the energy of the products.

▲ **Figure C4.6**
Burning magnesium releases energy in the form of light

Other chemical reactions are **endothermic** – the products contain more chemical energy than the reactants. Hence, energy must be continuously added to allow the chemical reaction to proceed. The decomposition of copper carbonate is an example of this type of reaction. This reaction only occurs while the copper carbonate is being heated. The products contain more chemical energy than the reactants so extra energy is put into the reaction. Heat energy is converted to chemical energy.

$$CuCO_3(s) + \text{heat} \rightarrow CuO(s) + CO_2(g)$$

Energy is absorbed during an endothermic reaction because the energy of the reactants is less than the energy of the products.

Physical processes such as the boiling of water can also involve the input of energy. Hence, the boiling of water is an endothermic process. The water molecules are gaining kinetic energy and the forces between the water molecules are being broken as molecules move faster and further apart. The energy required to change a liquid to a gas is called the latent heat of vaporisation. The latent heat of vaporisation for water is 2.26 kJ g^{-1}:

$$H_2O(l) + \text{heat} \rightarrow H_2O(g)$$

The energy required to change a solid to a liquid is called the latent heat of fusion.

WOW

Respiration – more than you think!

The general equation used for the respiration reaction that occurs in our body cells summarises more than 20 reactions into:

$$C_6H_{12}O_6 + 6O_2 \rightarrow 6CO_2 + 6H_2O + \text{energy}$$

A lot of the energy released in this process is stored in small molecules called adenosine triphosphate (ATP). Some is released as heat and helps to keep us warm.

9780170246644

EXPERIMENT 4.3

ENDOTHERMIC AND EXOTHERMIC REACTIONS

Aim

To observe a variety of chemical and physical processes and classify them as endothermic or exothermic

Materials

- 4 g solid sodium hydroxide
- 10 g solid potassium nitrate
- Granulated zinc
- 50 mL of 4 mol L^{-1} hydrochloric acid
- 5 g sodium carbonate
- 50 mL of 2 mol L^{-1} ethanoic (acetic) acid
- 4 styrofoam cups
- Thermometer or temperature probe
- Stirring rod

What are the risks in doing this experiment?	How can you manage these risks to stay safe?
Acid could splash or spill onto your hands.	Wash your hands immediately if they come in contact with acid and after using acids.
Vigorous bubbling may cause splashes in your eyes.	Wear safety glasses.

Procedure

Part A

1 Pour 100 mL of water into a styrofoam cup and use the thermometer or temperature probe to measure its temperature.

2 Add 4 g of solid sodium hydroxide to the styrofoam cup, stir it and record the final temperature.

Part B

3 Pour 100 mL of water into a styrofoam cup and measure its temperature.

4 Add 10 g of solid potassium nitrate to the styrofoam cup, stir it and record the final temperature.

Part C

5 Pour 50 mL of hydrochloric acid into a styrofoam cup and measure its temperature.

6 Add 4 or 5 pieces of granulated zinc to the styrofoam cup, stir it and record the final temperature.

Part D

7 Pour 50 mL of ethanoic (acetic) acid into a styrofoam cup and measure its temperature.

8 Add 5 g of sodium carbonate to the styrofoam cup, stir it and record the final temperature.

Results

Draw a table to clearly show your results.

Analysis of results

Draw an appropriate graph to visually display the results.

Discussion

1 Justify which of the processes were endothermic and which were exothermic.

2 Write equations to communicate these processes.

3 Why were styrofoam cups used for these experiments?

Enthalpy is the heat absorbed in a chemical reaction at constant pressure. The change in enthalpy, more commonly referred to as the heat of reaction, is given the symbol ΔH. It is measured in kJ. Since enthalpy is defined as the heat absorbed during a chemical reaction, by definition ΔH is positive. Therefore, a negative ΔH indicates that energy has been released during a chemical reaction.

Exothermic reactions have negative ΔH values. Endothermic reactions have positive ΔH values. This information can also be represented diagrammatically.

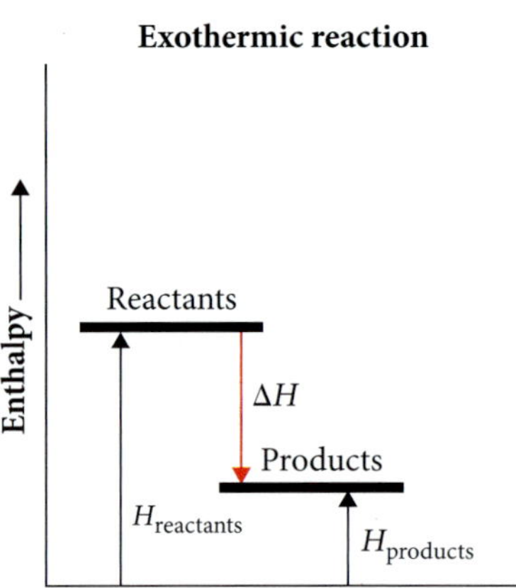

▲ Figure C4.7
In an exothermic reaction, energy is released. The energy of the reactants is greater than the energy of the products

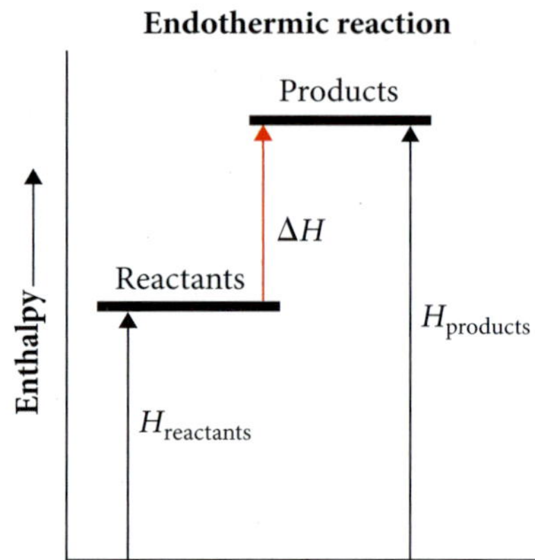

▲ Figure C4.8
In an endothermic reaction, energy is absorbed. The energy of the reactants is less than the energy of the products

WORKED EXAMPLE 4.4

Write thermochemical equations for the following reactions.

a Magnesium reacts with oxygen to form solid magnesium oxide. This is an exothermic reaction during which 1202 kJ is released.

b Solid copper carbonate decomposes to form solid copper oxide and carbon dioxide. This is an endothermic reaction during which 44 kJ is absorbed.

Answers	**Logic**
a $2Mg(s) + O_2(g) \rightarrow 2MgO(s)$ $\Delta H = -1202$ kJ	This is an exothermic reaction, so the ΔH value is negative.
b $CuCO_3(s) \rightarrow CuO(s) + CO_2(g)$ $\Delta H = +44$ kJ	This is an endothermic reaction, so the ΔH value is positive.

Try these yourself

Write thermochemical equations for the following reactions.

a Liquid water is formed from hydrogen gas and oxygen gas in a reaction that releases 242 kJ heat.

b Ammonia gas (NH_3) absorbs 92 kJ of heat to form nitrogen and hydrogen gases.

c Calcium oxide solid is formed when 180 kJ of heat is absorbed by solid calcium carbonate. Carbon dioxide is the other product.

d Ammonia gas reacts with oxygen to form water vapour and nitrogen oxide (NO) gas. The reaction releases 905 kJ of energy.

e Liquid water freezes to form ice. The latent heat of fusion is 6.02 kJ mol^{-1}.

Temperature

Temperature is a common indicator of the change in energy of a substance. For example, when water is heated, it absorbs heat energy and its temperature increases. Conversely, when water is cooled, it releases heat energy and its temperature decreases.

Temperature scales

Temperature can be measured on various scales. In Australia, temperature is commonly reported in degrees Celsius (°C). The Celsius temperature scale can have both positive and negative values. On a hot day, the temperatures are higher than 35°C. In some parts of Australia, temperatures can be as low as −10°C. Water freezes at 0°C and boils at 100°C at standard pressure, which is 100 kPa or 1 bar.

Scientists often use the **absolute temperature scale**, which is measured in kelvin (K). This scale only has positive values. There is a linear relationship between the Celsius temperature scale and the absolute temperature scale:

$$\text{temperature (°C)} = \text{temperature (K)} - 273$$

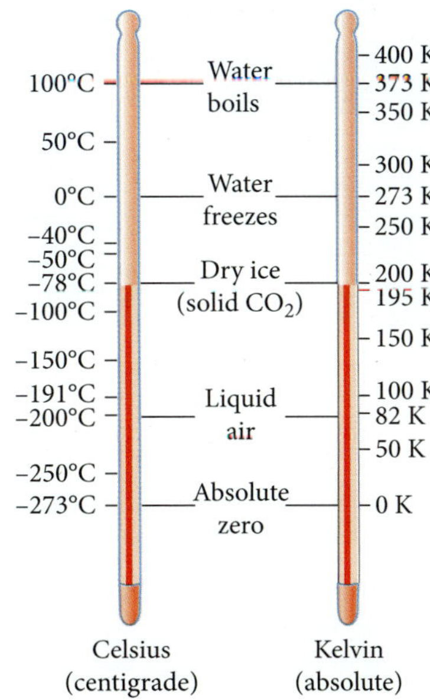

◀ Figure C4.9 Celsius and absolute temperature scales

To learn more about the absolute temperature scale, refer to Chemistry section 8.2 on page 346.

WORKED EXAMPLE 4.5

What is the:

a freezing point of pure water in kelvin?

b boiling point of pure water in kelvin?

Answers	**Logic**
a The freezing point of water is 273 K.	The freezing point of water is 0°C. temperature (°C) = temperature (K) − 273 Rearrange the equation to make temperature (K) the subject: temperature (K) = temperature (°C) + 273 temperature (K) = 0 + 273 = 273 K
b The boiling point of water is 373 K.	The boiling point of water is 100°C. temperature (°C) = temperature (K) − 273 Rearrange the equation to make temperature (K) the subject: temperature (K) = temperature (°C) + 273 temperature (K) = 100 + 273 = 373 K

Try these on your own

1 Convert the following temperatures from °C to K.

- **a** 25°C
- **b** −4°C

2 Convert the following temperatures from K to °C.

- **a** 635 K
- **b** 138 K

Figure 4.10 ▶ Sand and grass have different capacities to retain heat

To see how the specific heat capacity of water affects our bodies, refer to Context 3, 'Water, the vital substance', page 65.

Specific heat capacity

Each substance has a different ability to hold heat. On a hot day, sand feels hot when we walk on it but grass feels cool. Even though the Sun is shining on each of them and they are receiving the same amount of heat energy, they feel very different. They have different capacities to hold heat.

When chemists compare the capacity of different substances to hold heat in chemistry, they use a standard unit of measure called the **specific heat capacity**. The specific heat capacity is the heat needed to increase the temperature of 1 g of the substance by 1 K. Hence, values can be compared because the mass of the substances and the increase in temperature are the same. The only variable is the material's capacity to hold heat.

The specific heat capacity is the heat needed to increase the temperature of 1 g of a substance by 1 K.

EXPERIMENT 4.4

COMPARING HEAT CAPACITIES

A common way to measure energy changes in the laboratory is to measure the change in temperature of water or a solution.

Aim

To compare the specific heat capacities of vegetable oil and water

Materials

- 150 g water
- 50 g vegetable oil
- 3 × 150 mL beakers
- Hot plate
- Thermometer or temperature probe
- Waste bottle for disposal of vegetable oil

What are the risks in doing this experiment?	How can you manage these risks to stay safe?
Vegetable oil is flammable.	Use a hot plate to heat the oil, not a Bunsen burner.
Splashing could occur as one liquid is added to another.	Wear safety glasses and pour carefully. Wash off any spills onto your skin immediately with water.

Procedure

1 Pour 50 g of water into beaker A and heat it until its temperature reaches 50°C.
2 Pour 50 g of water into beaker B and measure its temperature.
3 Pour the water from beaker A into beaker B, stir it and measure the final temperature of the water.
4 Empty beaker B.
5 Pour 50 g of vegetable oil into beaker A and heat it until its temperature reaches 50°C.

▶

6 Pour 50 g of water into beaker B and measure its temperature.
7 Pour the vegetable oil from beaker A into beaker B, stir it and measure the final temperature of the mixture.
8 Empty beaker B into a waste bottle.

Results

1 Record the initial temperature of the water in beaker B for each experiment.
2 Record the final temperature of the water and water–oil mixture in beaker B for each experiment.

Analysis of results

In which experiment did beaker B have the greater temperature change?

Discussion

Justify whether water or vegetable oil had the greater specific heat capacity.

The specific heat capacity of water is 4.18 J K^{-1} g^{-1}. This means that 4.18 joules of energy is needed to increase the temperature of 1 g of water by 1 K.

The total amount of energy needed to heat a substance is dependent on the:

- mass of the substance
- substance (i.e., its specific heat capacity)
- increase in temperature required.

This can be written as a mathematical equation:

$$q = mC\Delta T$$

where:

q is the quantity of heat involved, measured in joules (J)
m is the mass of the substance, measured in grams (g)
C is the specific heat capacity of the substance, measured in J K^{-1} g^{-1}
ΔT is the change in temperature (final temperature − initial temperature), measured in kelvin (K).

If the temperature change is positive, then the temperature has increased and the substance has gained heat. If the temperature change is negative, then the temperature has decreased and the substance has released heat.

For example, if the temperature of 100 g of water increased from 18°C to 60°C, then we can calculate how much heat was added.

$$m = 100 \text{ g}, C = 4.18 \text{ J K}^{-1} \text{ g}^{-1}, \Delta T = 60 - 18 = 42°\text{C} = 42 \text{ K}$$

Next, substitute data into the equation:

$$q = mC\Delta T = 100 \times 4.18 \times 42 = 17\,556 \text{ J} = 18 \text{ kJ}$$

The mathematical equation for the total amount of energy needed to heat a substance is $q = mC\Delta T$.

WORKED EXAMPLE 4.6

In an experiment, 26.8 kJ of heat was used to increase the temperature of 180 g of water. The initial temperature of the water was 22°C. What was the final temperature of the water?

Answer

The final temperature of the water is 57.6°C.

Logic

First, extract the data from the question and your textbook.

$m = 180$ g

$C = 4.18$ J K^{-1} g^{-1}

$q = 26.8$ kJ $= 26\,800$ J

Next, rearrange the equation and substitute data.

$$\Delta T = \frac{q}{m \times C} = \frac{26800}{180 \times 4.18} = 35.6 \text{ K}$$

$\Delta T = 35.6 \text{ K} = 35.6°\text{C}$.

$\Delta T = T\text{(final)} - T\text{(initial)}$

Rearrange the equation and substitute data.

$T\text{(final)} = \Delta T + T\text{(initial)} = 35.6 + 22 = 58°\text{C}$

Try these yourself

Use the following data to answer the questions.

Substance	Specific heat capacity (J K^{-1} g^{-1})
Ethanol	2.46
Vegetable oil	2.00

a How much heat is needed to increase the temperature of 15 g of ethanol by 40°C?

b What mass of vegetable oil was heated by 2346 J, if the temperature of the oil increased from 30°C to 60°C?

QUESTION SET 4.3

Remembering

1 Define:

a endothermic.

b specific heat capacity.

2 Write the equation to show the relationship between temperature in degrees Celsius and temperature in kelvin.

Applying

3 Write a balanced chemical equation, including enthalpy of reaction, for the reaction between carbon monoxide and oxygen to form carbon dioxide. 566 kJ of heat is released during this reaction.

4 The mean temperature range for Bundaberg, Queensland, over the period 1959 to 2012 was 10.2°C. What was the mean temperature range over this period for Bundaberg in kelvin?

Table C4.1 Specific heat capacities of some materials

Substance	Specific heat capacity (J K^{-1} g^{-1})
Aluminium	0.897
Copper	0.385
Pyrex glass	0.750
Tin	0.228
Vegetable oil	2.00

5 A student placed 10 g pieces of tin and aluminium in the sun and left them there for 2 hours. Which would feel hotter? Use the information in Table C4.1 to justify your choice.

6 Use the information in Table C4.1 to calculate the amount of energy needed to increase the temperature of 500 g of vegetable oil by 70°C.

Analysing

7 A student wanted to heat some water in the science laboratory using as little heat as possible. Use the information in Table C4.1 to justify whether the student should put the water into a Pyrex glass beaker or a copper beaker during the heating process.

4.4 Combustion reactions

Combustion reactions are reactions where substances are burnt in the presence of oxygen and release large amounts of energy. Generally, combustion refers to the burning of fuels to release the energy needed for our daily activities. The general equation used for respiration is a combustion reaction that occurs in our body cells. Glucose and oxygen react to produce carbon dioxide and water:

$$C_6H_{12}O_6 + O_2(g) \rightarrow CO_2(g) + H_2O(l)$$

During this process, energy is released to enable us to breathe, talk, walk and carry out all of the other activities that we do each day.

Similarly, fossil fuels are burnt to release energy that is converted to electricity and to run our cars. **Fossil fuels** are produced over millions of years from decayed animal and plant matter. Coal, crude oil and natural gas are all fossil fuels. They are the predominant sources of energy used in Australia.

iStockphoto/gpflman

▲ **Figure C4.11**
Burning coals is an example of the combustion of a fossil fuel.

Many fuels contain compounds made from carbon and hydrogen. These compounds are called hydrocarbons. Complete combustion occurs when these fuels react with oxygen at standard atmospheric pressure to produce carbon dioxide and water; that is, the oxides of carbon and hydrogen are produced. However, sometimes incomplete combustion occurs and carbon monoxide or carbon and water are the products. Incomplete combustion occurs when there is not enough oxygen for complete combustion.

For example, methane (CH_4), a major component of natural gas, undergoes complete or incomplete combustion depending on the amount of oxygen available.

Complete combustion produces carbon dioxide and water:

$$CH_4(g) + 2O_2(g) \rightarrow CO_2(g) + 2H_2O(l)$$

Incomplete combustion produces carbon monoxide or carbon and water:

$$CH_4(g) + \frac{3}{2}O_2(g) \rightarrow CO(g) + 2H_2O(l)$$

$$CH_4(g) + O_2(g) \rightarrow C(s) + 2H_2O(l)$$

The formation of carbon or carbon monoxide requires less oxygen per mole of methane than the formation of carbon dioxide.

To see the importance of fuel combustion in our lives, refer to Context 2, 'Materials for a purpose', page 43.

Heat of combustion

The **heat of combustion** is the heat released when a fuel undergoes complete combustion at standard atmospheric pressure. Hence, combustion reactions are always exothermic reactions. It is difficult to directly measure the heat released when a fuel undergoes combustion, so an indirect method is used by applying the law of conservation of energy. A known quantity of water will be heated by the combustion of a fuel. These experiments are referred to as calorimetric experiments since they measure changes in heat. Hence:

$$\text{heat absorbed by water} = \text{heat released by combustion of fuel}$$

Heat of combustion is the heat released when a fuel undergoes complete combustion at standard atmospheric pressure.

WORKED EXAMPLE 4.7

A student clamped a conical flask with 150 g of water above a spirit burner that contained hexane. The temperature of the water was measured before and after the experiment and was found to increase by 20°C. The spirit burner containing hexane was weighed before and after the experiment. The mass of hexane that underwent combustion was 0.26 g. How much heat is released by the complete combustion of 1 g of hexane?

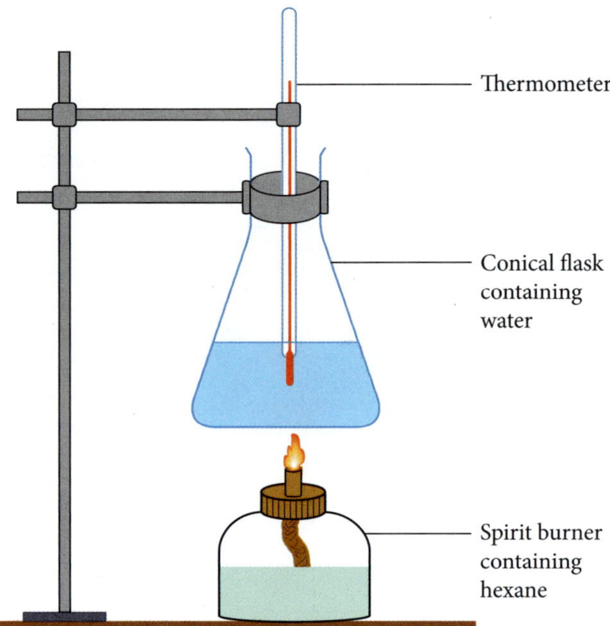

◀ **Figure C4.12**
Equipment set-up to determine the heat of combustion for hexane

Answer

1 g of hexane releases 48 kJ.

Logic

1. Write down all of the data for water and determine the heat absorbed by the water.

   $m = 150$ g

   $C = 4.18$ J K^{-1} g^{-1}

   $\Delta T = 20$°C

   $q = mC\Delta T = 150 \times 4.18 \times 20 = 12\,540$ J

2. Law of conservation of energy:

   heat absorbed by the water = heat released by combustion of hexane

3. Write down all of the data for hexane and use ratios to determine the heat released by 1 g of hexane.

   Heat released by combustion of hexane = 12 540 J

   Mass of hexane that was combusted = 0.26 g

   0.26 g of hexane releases 12 540 J (i.e. 12.54 kJ)

   1 g of hexane releases 48 230 J

Try these yourself

A student clamped a flask with 100 mL of water above a spirit burner that contained 1-propanol (C_3H_7OH). The temperature of the water increased by 30°C. A mass of 0.372 g of 1-propanol was used during this experiment. How much heat is released by 1 g of 1-propanol?

INVESTIGATION 4.1

HEAT OF COMBUSTION OF ETHANOL

You are to use a method similar to that outlined in Worked example 4.7 to determine the heat of combustion for 1 g of ethanol. Then, compare your value to the accepted value in the scientific literature, which is 29.6 kJ g^{-1}.

What is your aim?

Write an aim for your investigation.

What will you need?

Identify the pieces of equipment that you will need to conduct this investigation.

What are the risks?

Construct a table similar to the one below. Identify specific risks involved in the investigation and ways that you will manage the risks to avoid injuries or damage to equipment. Ask your teacher to check your risk assessment before you proceed.

What are the risks in doing this investigation?	How can you manage these risks to stay safe?
Ethanol is a flammable liquid.	
Glassware may get hot during this investigation.	

How will you carry out your investigation?

1. Research a method for conducting this investigation.
2. Set up the equipment for this investigation.
3. Draw a diagram or take a photo of your equipment.
4. Annotate your photo or diagram to indicate any safety precautions needed to ensure the investigation is safe.
5. Show your teacher your annotated photo or diagram and get their approval before you commence your investigation.

What results will you collect?

1. Identify the measurements that you will need to collect during this investigation.
2. Present the results in an appropriate way so that they will be easy to analyse.

How will you analyse your results?

1. How much heat was absorbed by the water during the investigation?
2. How much heat was released by the ethanol during the investigation?
3. How much heat is released by 1 g of ethanol?

What have you found?

Theoretically, ethanol releases 29.6 kJ g^{-1}. Compare this with the value that you determined from this investigation.

a Is your value greater than, equal to or less than the theoretical value?

b Account for your answer to part **a**.

Ideas for improvement or further investigation

Research ways to modify your investigation to minimise sources of error.

INVESTIGATION 4.2

ASSESSING FOOD LABELLING INFORMATION

Food labels in Australia contain a panel with nutrition information, which includes the energy content per serve and the size of a typical serve. You are to choose two foods such as cereals or nuts to conduct an investigation to assess the accuracy of the information on the label.

What is your aim?

1 Identify two foods that you will investigate. Read the information on the labels and identify a statement with regard to energy content of each food.

2 Write an aim for your investigation that will enable you to assess the accuracy of the statement regarding energy content.

What will you need?

Think about the equipment you will need to enable you to perform an investigation into the energy content of foods. For example, if your food is a solid, you will need to modify the previous methods you used for flammable liquids.

What are the risks?

Construct a table similar to the one below. Identify specific risks involved in the investigation and ways that you will manage the risks to avoid injuries or damage to equipment. Ask your teacher to check your risk assessment before you proceed.

What are the risks in doing this investigation?	How can you manage these risks to stay safe?
Some students may be allergic to nuts or other ingredients in foods.	
Equipment may become hot.	

How will you carry out your investigation?

1 Research a method for conducting this investigation.

2 Set up the equipment for this investigation.

3 Draw a diagram or take a photo of your equipment.

4 Annotate your photo or diagram to indicate any safety precautions needed to ensure the investigation is safe.

5 Show your teacher your annotated photo or diagram and obtain their consent before you commence your investigation.

What results will you collect?

1 Identify the measurements that you will need to collect during this investigation.

2 Present the results in an appropriate way so that they will be easy to analyse.

How will you analyse your results?

1 How much heat was absorbed by the water during the investigation?

2 How much heat was released by the food during the investigation?

3 How much heat is released by a typical serving size of the food?

What have you found?

Compare your answer to the information on the label.

What do you conclude?

Could you use your data to make a valid assessment of the manufacturer's claim? Justify your answer.

Ideas for improvement or further investigation

Are there any modifications that you could make to your method to minimise sources of error?

Importance of chemical calculations

On 23 July 1983, Air Canada Flight 143 was flying from Montreal to Edmonton when it ran out of fuel. The Boeing 767 plane had become the world's biggest glider! Based on its height, it would land 15 km short of the nearest airport at Winnipeg. The plane was directed to land at the disused Gimli Air Force base. The pilots had very limited control of the plane but were able to successfully land it.

The plane ran out of fuel because of incorrect chemical calculations. Canada was changing over from the old imperial measurement system to the new metric system. Incorrect calculations had been used to determine the amount of fuel needed for the journey.

To see properties and uses of fuels, refer to Context 2, 'Materials for a purpose', page 38.

Biofuels

Biofuels are fuels that are produced from biodegradeable materials such as crops, rather than from fossil fuels. Biofuels are classified as renewable sources of fuels since they can be continually produced from crops or algae. Biogasoline or green gasoline has a similar chemical composition to unleaded petrol, while biodiesel has a similar composition to diesel. The major component of both biogas and natural gas is methane.

Biodiesel has a slightly different chemical composition from diesel. Therefore, these two fuels have slightly different chemical properties. The heat of combustion for diesel is about 42.6 kJ g^{-1} while biodiesel's is about 37.2 kJ g^{-1}. Biodiesel's heat of combustion will differ slightly as different crops are used for its production. Canola oil, soybean oil and peanut oil all have different heats of combustion. Biodiesel has lower CO_2 and CO emissions than does diesel. However, it has slightly higher emissions of the oxides of nitrogen, commonly written as NO_x, since there are several oxides of nitrogen.

The word equation for the production of biodiesel is:

$$\text{vegetable oil} + \text{methanol} \rightarrow \text{biodiesel} + \text{glycerol}$$

Methanol can be sourced from fossil fuels or by processing crops such as wheat.

The heat of combustion of some common fossil fuels are:

- Natural gas 53.6 kJ g^{-1}
- Coal 9.8–27.9 kJ g^{-1} depending on the type of coal (brown or black)
- Petroleum 48 kJ g^{-1}
- Crude oil 44.9–46.3 kJ g^{-1} depending on source of crude oil
- Diesel 42.6 kJ g^{-1}

The heat of combustion of some biofuels are:

- Biogas ~33 kJ g^{-1}
- Biodiesel 37.2 kJ g^{-1}
- Bioethanol 29.6 kJ g^{-1}

Heat of combustion is only one property of these fuels. Another is the comparison of the emission of CO_2 if it is assumed that these fuels undergo complete combustion. Since each of these fuels is a mixture of different chemicals, these properties can only be given as approximate values because they will depend on the precise composition of the fuel. This is dependent on where the fossil fuel is sourced from or the biodegradeable material from which the biofuel is sourced. For example, biogas sourced from sewage would have a slightly different composition from biogas sourced from landfill sites composed of household garbage.

For more information on the comparison of biofuels with fossil fuels, refer to Context 2, 'Materials for a purpose', page 54.

HOW IT'S MADE – BIODIESEL PRODUCTION

Watch this video about the processes involved in the production of biodiesel.

QUESTION SET 4.4

Remembering

1 Define the following terms.
  a Fossil fuels
  b Law of conservation of energy

Understanding

2 Compare the products of complete and incomplete combustion of hydrocarbons.

3 Explain why incomplete combustion reactions have a lower heat of combustion than complete combustion reactions for the same fuel.

Applying

4 Write chemical equations for the:
  a complete combustion of 1-hexene.
  b incomplete combustion of ethane.

Analysing

5 Justify whether the following reaction is a complete or incomplete combustion reaction.

$C_8H_{18}(l) + 11O_2(g) \rightarrow 5CO_2(g) + 3CO(g) + 9H_2O(l)$

CHAPTER CHECKLIST

You should know:

- the way to represent chemical reactions by chemical equations
- chemical equations can be either complete balanced chemical equations or net ionic chemical equations
- during chemical reactions, bonds are broken and new bonds are formed
- energy is absorbed to break bonds and released when bonds are formed
- spontaneous chemical reactions occur at room temperature at different rates
- more reactive metals are more likely to react spontaneously
- exothermic reactions release energy, the energy of the products being less than the energy of the reactants. They have a negative ΔH value
- endothermic reactions absorb energy, the energy of the products is greater than the energy of the reactants. They have a positive ΔH value
- the definitions of specific heat capacity, heat of combustion and biofuel
- how to compare biofuels and fossil fuels
- the absolute temperature scale only has positive values and is measured in kelvin.

You should be able to:

- write balanced chemical equations
- write net ionic chemical equations
- write thermochemical equations in which the energy term is included in the equation or written separately as a ΔH value
- convert temperatures from degrees Celsius to kelvin and vice versa
- use specific heat data to determine the change in energy when the mass or temperature of a material is changed
- write equations for both complete and incomplete combustion reactions of hydrocarbons
- perform calorimetric calculations
- use knowledge of calorimetric calculations to assess information about the energy content of foods
- identify independent and dependent variables
- plan valid investigations.

CHAPTER GLOSSARY

absolute temperature scale a temperature scale that only has positive values since 0 K is defined as the point where there is no thermal motion for any known species. There is a linear relationship between this scale and the Celsius scale:

$$\text{temperature (°C)} = \text{temperature (K)} - 273$$

biofuel energy source produced from renewable sources such as crops

combustion a reaction with oxygen to form the oxides of each of the elements present; with adequate oxygen, combustion of a hydrocarbon will produce carbon dioxide and water

endothermic a chemical or physical change where the chemical energy of the products is greater than the chemical energy of the reactants; hence, heat energy is absorbed

enthalpy the total energy possessed by a chemical substance at a constant pressure; it is usually expressed as the change in enthalpy

exothermic a chemical or physical change where the chemical energy of the products is less than the chemical energy of the reactants; hence, heat energy is released

fossil fuel an energy source, such as crude oil and coal, produced from the decayed remains of animals and plants

heat of combustion heat released when a fuel undergoes complete combustion at standard atmospheric pressure

ionic equation equation that gives information about all species in the reaction; aqueous solutions are written as separate ions; for example, NaCl(aq) is written as Na^+ and Cl^-

net ionic equation an equation that only contains the ions that take part in the chemical reaction; that is, spectator ions are excluded

species a general term that refers to atoms, ions and molecules

spectator ion an ion that is present in the same form on each side of the equation

specific heat capacity heat energy required to raise the temperature of 1 gram of a substance by 1 kelvin

CHAPTER REVIEW QUESTIONS

Remembering

1 Define the following terms.
  a Net ionic equation
  b Exothermic
  c Enthalpy
  d Renewable fuel

2 Identify two fuels.

Understanding

3 Identify whether the following processes are endothermic or exothermic and justify your answers.
  a Water freezing to become ice
  b Sublimation of dry ice (solid CO_2) to gaseous carbon dioxide
  c $HCl(aq) + NaOH(aq) \rightarrow NaCl(aq) + H_2O(l)\ \Delta H = -55\text{ kJ}$
  d Photosynthesis

4 Explain why biodiesel and diesel have different heats of combustion.

Applying

5 Copy and complete the following table.

	Temperature	
	(°C)	**(K)**
Melting point of mercury		234
Freezing point of helium	−272	
Boiling point of ethanol		351
Melting point of sodium chloride		1074

6 a Arsenic poisoning of water supplies in Bangladesh has been caused by the following reaction.

$FeAsS(s) + O_2(g) + H_2O(l) \rightarrow FeSO_4(aq) + H_3AsO_3(aq)$

Balance this equation.

b Write a thermochemical equation for the complete combustion of propane. The enthalpy for this reaction is -2220 kJ mol^{-1}.

7 Balance the following equations.

a $Fe_2O_3(s) + H_2(g) \rightarrow Fe(s) + H_2O(l)$

b $Al(s) + Cl_2(g) \rightarrow AlCl_3(s)$

c $Cu(s) + AgNO_3(aq) \rightarrow Ag(s) + Cu(NO_3)_2(aq)$

d $V_2O_5(s) + Ca(s) \rightarrow CaO(s) + V(s)$

e $H_2O_2(l) \rightarrow H_2O(l) + O_2(g)$

Analysing

8 Identify whether each of unknowns A–E is a solid, liquid or gas at 25°C. Justify your answers.

Unknown	Melting point (K)	Boiling point (K)
A	2041	4098
B	171	239
C	330	398
D	265	332
E	302	944

Table C4.2 Specific heat capacities of some substances

Substance	Specific heat capacity (J K^{-1} g^{-1})
Copper	0.385
Pyrex glass	0.750
Styrofoam	1.3

9 Refer to Table C4.2 to justify:

a why styrofoam cups were used instead of Pyrex glass beakers in Experiment 4.3.

b whether copper, Pyrex glass or styrofoam should be used to hold water when it is placed over a lit spirit burner when conducting a heat of combustion investigation.

10 A student conducted an investigation to determine the heat of combustion (kJ g^{-1}) of 1-butanol. The specific heat capacity of water is 4.18 J g^{-1} K^{-1}. The student collected the following data:

- mass of water heated 150 g
- initial temperature of water 16.3°C
- final temperature of water 39.6°C
- mass of 1-butanol used 0.48 g

a Use the student's data to calculate the heat of combustion (kJ g^{-1}) for 1-butanol.

b The theoretical value for the heat of combustion for 1-butanol is 36.1 kJ g^{-1}. Justify why the student's value is different from the theoretical value.

Evaluating

11 Evaluate this statement: 'The kelvin temperature scale is a useful scale for chemists.'

12 Compare the properties of diesel and biodiesel and assess their use and potential use as fuels.

Reflecting

13 It is much easier to write information about a chemical reaction as a word equation. Assess this statement from the point of view of a:

a member of the public. b chemist.

14 Exothermic reactions are more important than endothermic reactions. Assess this statement with reference to specific examples.

CHAPTER 5 CALCULATING CHEMICAL QUANTITIES

By the end of this chapter you will have covered the following material.

Science Understanding

- The relative atomic mass of an element is the ratio of the weighted average mass per atom of the naturally occurring form of the element to 1/12 the mass of an atom of carbon-12; relative atomic masses reflect the isotopic composition of the element (ACSCH024)
- A mole is a precisely defined quantity of matter equal to Avogadro's number of particles; the mole concept and the Law of Conservation of Mass can be used to calculate the mass of reactants and products in a chemical reaction (ACSCH039)

AC

5.1 Relative mass of atoms and substances

To revise particles in the atom and relative atomic mass, refer to Chemistry sections 1.1 and 1.3 on pages 104 and 109.

In Chemistry Chapter 1, you learnt that the mass of an atom is its relative atomic mass (represented by the symbol A_r). The relative atomic mass of an atom is the mass relative to that of a carbon-12 atom. This means the mass of a carbon atom is considered to be exactly 12 and all others are compared to this.

Relative mass of substances

Chemical symbols and formulas, such as He, Na, Cl_2 and H_2O, are short-hand representations for elements and compounds. Symbols and formulas may also represent a group of atoms or formula units. Just as relative atomic mass is used to describe the mass of an atom, **relative molecular mass** (symbol M_r) is used to describe the mass of one molecule of a molecular substance on a scale in which the mass of an atom of the carbon-12 isotope is exactly 12.

> **The relative molecular mass (M_r) of a substance is the mass of one molecule of the substance on a scale in which the mass of an atom of the carbon-12 isotope is exactly 12. It is the sum of the relative atomic masses as given by the molecular formula.**

Figure C5.1 ▶ Examples of molecules

Hydrogen (H_2)

Each molecule contains 2 H atoms only.

Oxygen (O_2)

Each molecule contains 2 O atoms only.

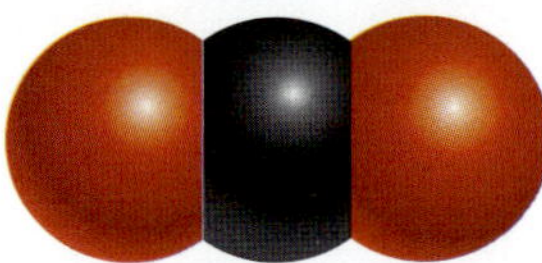

Carbon dioxide (CO_2)

Each molecule contains 1 atom of C and 2 atoms of O.

Water (H_2O)

Each molecule contains 2 atoms of H and 1 atom of O.

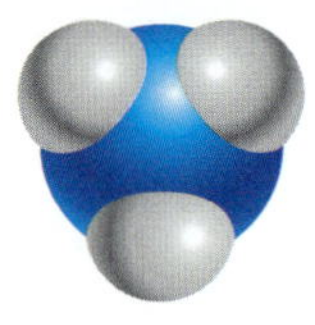

Ammonia (NH_3)

Each molecule contains 1 atom of N and 3 atoms of H.

Pentane (C_5H_{12})

Each molecule contains 5 atoms of C and 12 atoms of H.

The relative molecular mass of an element or compound is calculated by adding the relative atomic masses of all the component atoms of the molecule.

WORKED EXAMPLE 5.1

Calculate the relative molecular mass of ethane, C_2H_6.

Answer

The relative molecular mass is 30.0.

Logic

1 Identify the number of each type of atom.

Number C atoms = 2

Number of H atoms = 6

2 Multiply the number of each type of atom by their relative atomic mass (from the periodic table)

Mass C = $2 \times A_r(C) = 2 \times 12.0 = 24.0$

Mass H = $6 \times A_r(H) = 6 \times 1.0 = 6.0$

3 Add the masses of each element to obtain the relative mass of the compound.

$M_r(C_2H_6) = 24.0 + 6.0 = 30.0$

This can be simplified to

$M_r(C_2H_6) = 2 \times A_r(C) + 6 \times A_r(H) = 2 \times 12.0 + 6 \times 1.0 = 30.0$

Try these yourself

Calculate the relative molecular mass of:

a phosphorus (P_4).

b carbon dioxide (CO_2).

c nitric acid (HNO_3).

d glucose ($C_6H_{12}O_6$).

Not all pure substances exist as molecules. Ionic compounds consist of a crystalline lattice of cations and anions combined in a definite fixed ratio, as indicated by the formula of the compound, such as Al_2O_3. Therefore, the term 'relative molecular mass' would be incorrect when applied to ionic substances and 'relative formula mass' is used instead. This is also given the symbol M_r. The **relative formula mass** is the mass of one formula unit of an ionic compound on a scale in which the mass of an atom of the carbon-12 isotope is exactly 12.

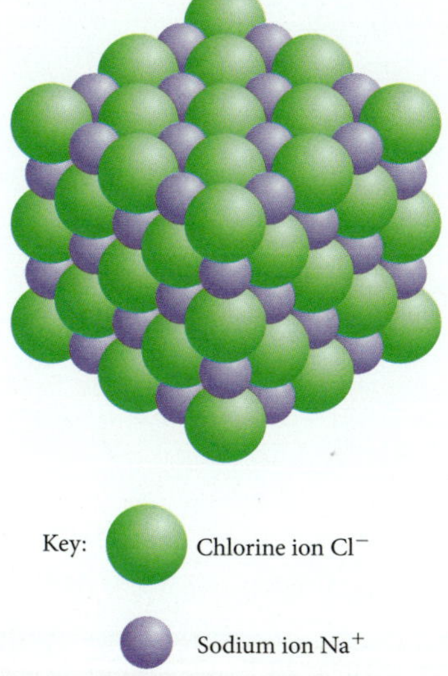

Figure C5.2 Example of an ionic compound

The relative formula mass of a compound (M_r) is the sum of the relative atomic masses of the atomic species as given by the formula of the compound.

WORKED EXAMPLE 5.2

Calculate the relative formula mass of calcium hydroxide, $Ca(OH)_2$.

Answer

The relative formula mass is 74.1.

Logic

1 Identify the number of each type of atom.

Number Ca atoms = 1

Number of H atoms = 2

Number of O atoms = 2

2 Multiply the number of each type of atom by their relative atomic mass (from the periodic table)

Mass Ca = $1 \times A_r(Ca) = 1 \times 40.1 = 40.1$

Mass H = $2 \times A_r(H) = 2 \times 1.0 = 2.0$

Mass O = $2 \times A_r(O) = 2 \times 16.0 = 32.0$

3 Add the masses of each element to obtain the mass of the compound.

$M_r(Ca(OH)_2) = 40.1 + 2.0 + 32.0 = 74.1$

This can be simplified to

$M_r(Ca(OH)_2) = 1 \times A_r(Ca) + 2 \times A_r(H) + 2 \times A_r(O) = 1 \times 40.1 + 2 \times 1.0 + 2 \times 16.0 = 74.1$

Try these yourself

Calculate the relative formula mass of:

a aluminium nitride (AlN).

b barium nitrate ($Ba(NO_3)_2$).

c sodium carbonate (Na_2CO_3).

d copper sulfate ($CuSO_4$).

QUESTION SET 5.1

Understanding

1 Calculate the relative molecular mass of the following.

a Nitrogen (N_2)

b Hydrogen sulfide (H_2S)

c Diphosphorus pentoxide (P_2O_5)

d Sucrose ($C_{12}H_{22}O_{11}$)

e Sulfuric acid (H_2SO_4)

2 Calculate the relative formula mass of the following.

a Potassium chloride (KCl)

b Magnesium fluoride (MgF_2)

c Calcium phosphate ($Ca_3(PO_4)_2$)

d Copper(II) nitrate ($Cu(NO_3)_2$)

e Ammonium sulfate ($(NH_4)_2SO_4$)

Applying

3 A scientist found contamination in a sample of gas being prepared. The contaminant gas was sampled and found to have a relative molecular mass of 28. Which of the following gases could it be: H_2O, N_2, CO, CO_2 or NO_2?

5.2 Percentage composition

The chemical formula of a compound gives information about the elements present in a compound and the ratio in which the atoms of those elements are present. For example, in phosphoric acid (H_3PO_4), H, P and O are present in the ratio 3:1:4.

But it is important to remember that the ratio of the atoms is *not* the same as the ratio of the masses of the atoms because atoms have different relative atom masses. Sometimes, scientists need to know the ratio by mass or the percentage composition.

The **percentage composition** of a compound is the percentage by mass of each of the different elements in the compound.

Alamy/Eric Taylor

iStockphoto/Ngataringa

iStockphoto/pioneeer111

◀Figure C5.3 Bauxite, aluminium oxide (alumina) and aluminium have different percentage compositions of aluminium

Knowledge of percentage composition provides useful information in many different instances. For example, a chemist can determine which mineral ores contain the higher percentage of a particular metal and therefore provide advice to mining and metal extraction industries. A chemist often compares the percentage composition of an unknown compound with the percentage composition calculated from a known formula. If the percentages match, then the identity of the unknown can be confirmed.

Calculating percentage composition

The percentage composition can be calculated from the chemical formula and the relative atomic masses of the elements in the compound. This composition is independent of how much of the compound there is.

WOW

Martin Kalproth

German chemist Martin Kalproth (1743–1817) achieved prominence through his development of new procedures for analysing compounds. Although early chemists, including Antoine Lavoisier, recalculated analysis results so they totalled 100%, Kalproth believed that when using pure reagents and attention to detail, a less than 100% result suggested the presence of another element. He was the first to discover uranium and zirconium and characterised them as distinct elements. He also elucidated the composition of numerous compounds containing tellurium, strontium, cerium and chromium.

WORKED EXAMPLE 5.3

Calculate the percentage composition of aluminium oxide (Al_2O_3).

Answer

The percentage composition is 52.9% Al and 47.1% O.

Logic

1 Calculate the relative mass of each element in the compound.

Mass of Al = 2 × 27.0 = 54.0

Mass of O = 3 × 16.0 = 48.0

2 Calculate the relative formula (or molecular) mass.

$$\text{Formula mass of } Al_2O_3 = \text{mass of Al} + \text{mass of O} = 54.0 + 48.0 = 102.0$$

3 Calculate the % of each element.

$$\%\ Al = \frac{\text{mass of Al in compound}}{\text{formula mass of } Al_2O_3} \times 100 = \frac{54.0}{102.0} \times 100 = 52.9\%$$

$$\%\ O = \frac{\text{mass of O in compound}}{\text{formula mass of } Al_2O_3} \times 100 = \frac{48.0}{102.0} \times 100 = 47.1\%$$

Try these yourself

Calculate the percentage composition of each element in:

a methane (CH_4).

b sodium chloride (NaCl).

c calcium cyanide ($Ca(CN)_2$).

d ammonium oxalate ($(NH_4)_2C_2O_4$).

QUESTION SET 5.2

Remembering

1 Explain percentage composition.

2 Give an example where a chemist would want to know percentage composition.

Understanding

3 Calculate the percentage composition of each element in:

a $COCl_2$.

b K_2SO_4.

4 The herbicide 2,4,5-T (2,4,5-trichlorophenoxyacetic acid) has the formula $C_8H_5O_3Cl_3$. What is the percentage of chlorine by mass in this compound?

Applying

5 Ore minerals of copper are chalcopyrite ($CuFeS_2$), bornite (Cu_5FeS_4), covellite (CuS) and chalcocite (Cu_2S)

  a Which mineral has the highest percentage by mass of copper?

  b Which has the lowest percentage by mass of copper?

  c If a mining company wanted to get the best yield of both copper and iron, which mineral would be the best to mine?

To learn more about the use of copper in the Bronze Age, refer to Context 2, 'Materials for a purpose', page 33.

5.3 The Avogadro constant and the mole

So far you have considered the relative masses of elements and compounds, but how do chemists know how many atoms or molecules are in a sample? The individual atoms and molecules are too small to even see with anything but the most specialised equipment, so chemists devised a counting unit that was practical for counting and handling a standard number of particles.

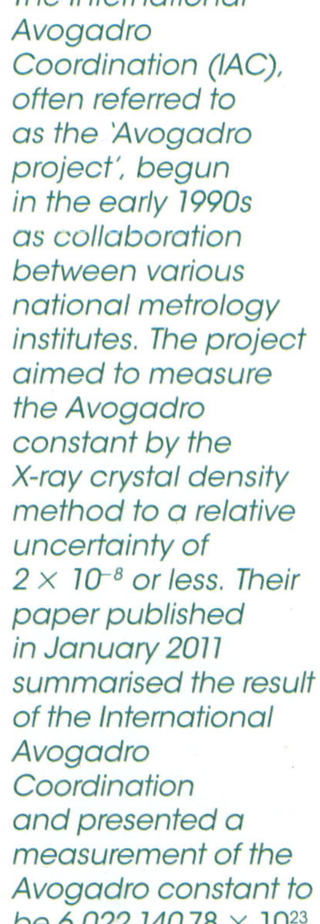

The International Avogadro Coordination (IAC), often referred to as the 'Avogadro project', begun in the early 1990s as collaboration between various national metrology institutes. The project aimed to measure the Avogadro constant by the X-ray crystal density method to a relative uncertainty of 2×10^{-8} or less. Their paper published in January 2011 summarised the result of the International Avogadro Coordination and presented a measurement of the Avogadro constant to be $6.022\,140\,78 \times 10^{23}$.

The Avogadro constant

Chemists deal with amounts of substances in terms of their masses, so they use a relationship between mass and number of particles.

The relative masses of the atoms of an element are devised by comparing them to the mass of carbon-12. One atomic mass unit (**amu**) is equal to one-twelfth the mass of one atom of carbon-12. Chemists were able to work out that one carbon-12 atom, with a relative atomic mass of 12 amu, has an actual mass of 1.99×10^{-23} grams. However, no laboratory scale can measure the mass of an individual atom. For everyday use, chemists needed to be able to measure quantities of substances in a larger unit, such as a gram, but they still need to know how many particles there are.

Science Photo Library/Andrew Lambert Photography

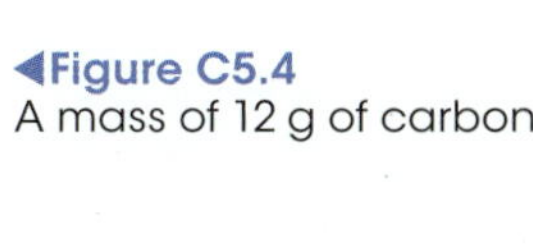

◀Figure C5.4
A mass of 12 g of carbon

AVOGADRO PROJECT

Visit this website to learn more about the Avogadro project.

Chemists chose a number of particles that would have an equivalent mass in grams to the mass of one atom in atomic mass units. The same number fits all elements because equal numbers of different atoms always have the same mass ratio. The number was determined to be about 6.02×10^{23} and is called the **Avogadro constant** (represented by the symbol N_A) in honour of Amadeo Avogadro, a 19th-century Italian scientist.

The Avogadro constant (N_A) is the number of atoms (6.02×10^{23}) in exactly 12 grams of the carbon-12 isotope.

The Avogadro constant is a scaling factor between macroscopic and microscopic (atomic scale) observations of nature.

WOW

The Avogadro constant

The Avogadro constant was initially determined by French physicist Jean Perrin, who proposed it be named after Amadeo Avogadro to honour his contributions to chemistry. One of Avogadro's most important contributions was his resolution of the confusion surrounding atoms and molecules (although he did not use the term 'atom'). Avogadro believed that particles could be composed of molecules and that molecules could be composed of still simpler units, atoms.

The Avogadro constant is a number so large it is difficult to comprehend. If you had this number of oranges, then they would form a sphere the size of the Earth. Alternatively, this number of grains of rice would cover Australia to a depth of approximately 1 km.

AVOGADRO NUMBER

Watch this video on Avogadro number and the mole.

The Avogadro constant and the mole

Just as it is convenient to group sheets of paper into reams (500 sheets) or eggs into cartons of a dozen, chemists measure the amount of substances in a batch size called the **mole** (mol) and represent it by the symbol n.

Chemists have chosen 1 mole as a standard unit for large numbers of atoms, ions or molecules. So, the mole is the SI base unit representing the chemical quantity of a substance.

A mole is the amount of substance containing 6.02×10^{23} particles of that substance.

Converting between moles and number of particles

The relationship between the number of moles (n) of a substance and the number of particles (atoms, ions or molecules) is given by:

$$\text{Number of moles } (n) = \frac{\text{number of particles}}{\text{number of particles per mole } (N_A)}$$

$$n = \frac{\text{number of particles}}{6.03 \times 10^{23}}$$

Alternatively, this relationship can be shown using the diagram shown in Figure C5.5.

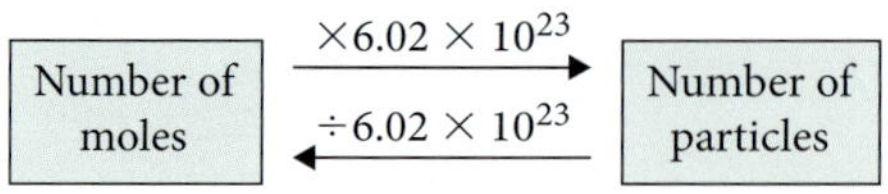

Figure C5.5 ▶ Converting between moles and number of particles

Number of moles	$\times 6.02 \times 10^{23}$ → ← $\div 6.02 \times 10^{23}$	Number of particles

Worked example 5.4 shows how to convert from number of particles to number of moles. Worked example 5.5 shows how to convert from number of moles to number of particles.

WORKED EXAMPLE 5.4

How many moles of magnesium are there in 1.45×10^{23} atoms of magnesium?

Answer

The number of moles of magnesium is 0.24 mol.

Logic

To convert from particles to moles, divide the number of particles given by N_A.

$$n(\text{Mg}) = \frac{1.45 \times 10^{23}}{6.03 \times 10^{23}}$$

$$= 0.24 \text{ mol}$$

Try these yourself

Calculate the number of moles of each of the following substances, given the number of particles.

- **a** 3.3×10^{25} atoms of carbon
- **b** 1.2×10^{22} molecules of water
- **c** 6.6×10^{21} ions of sodium
- **d** 3.01×10^{24} formula units of calcium carbonate
- **e** 9.05×10^{23} molecules of hydrogen gas

WORKED EXAMPLE 5.5

How many atoms are there in 0.25 mol of neon atoms?

Answer

There are 1.5×10^{23} neon atoms.

Logic

To convert from moles to particles, multiply the number of moles given by N_A.

Number of atoms of Ne $= 0.25 \times 6.02 \times 10^{23}$

$= 1.505 \times 10^{23}$ atoms

Try these yourself

Calculate the number of particles in each of the following substances, given the number of moles.

- **a** 0.1 mol of methane molecules
- **b** 6.0 mol of hydrogen atoms
- **c** 0.5 mol of aluminium ions
- **d** 0.082 mol of argon gas
- **e** 2.1 mol of ammonia molecules

Moles and chemical formulas

In Chapter 3, you learnt that in the molecular formulas of elements and compounds the subscripts represent the numbers of atoms of each element present in a molecule of the substance. For example, in one molecule of water (H_2O), there are two atoms of hydrogen and one atom of oxygen. In two molecules of water, there are double the number of hydrogen (4) and oxygen (2) atoms, and so on.

The subscripts also represent the number of moles of each atom present in one mole of molecules of the substance. So, if there were one mole of water molecules, there would be two moles of hydrogen atoms and one mole of oxygen atoms. Similarly, if there were 2 moles of water molecules, then there would be 4 moles of hydrogen atoms and 2 moles of oxygen atoms.

To learn more about fuels, refer to Context 2, 'Materials for a purpose', page 43.

This also applies to ionic compounds. One mole of sodium sulfate (Na_2SO_4) formula units contains 2 moles of sodium ions (Na^+) and 1 mole of sulfate ions (SO_4^{2-}).

Using the relationship between moles and the Avogadro constant, we can calculate the number of individual atoms.

WORKED EXAMPLE 5.6

a Calculate the number of moles of carbon atoms and oxygen atoms in 5 mol of carbon dioxide (CO_2).

b How many moles of atoms are present altogether?

c How many individual atoms are present?

Answers	**Logic**
a There are 5 mol of C and 10 mol of O.	From the formula, it can be seen that 1 mol of CO_2 contains 1 mol C and 2 mol O. For 5 mol CO_2 $n(\text{C}) = 5 \times 1 \text{ mol C} = 5 \text{ mol of C}$ $n(\text{O}) = 5 \times 2 \text{ mol O} = 10 \text{ mol of O}$
b There are 15 mol of atoms.	The total number of moles of atoms is the sum of number of moles of individual atoms $n(CO_2) = 5 + 10 = 15 \text{ mol of atoms}$
c There are 9.03×10^{24} individual atoms present.	Using $n = \dfrac{\text{number of particles}}{6.02 \times 10^{23}}$ Total number of individual atoms $= 15 \times 6.02 \times 10^{23}$ $= 9.03 \times 10^{24}$

Try these yourself

How many:

a moles of K^+ ions and Cl^- ions are there in 3 mol of KCl?

b moles of H atoms are there in 6 mol of $C_6H_{12}O_6$?

c moles of atoms are there in 0.5 mol of ammonia (NH_3)?

d individual oxygen atoms are there in 2.5 mol of calcium carbonate ($CaCO_3$)?

e individual atoms are there in 1.25 mol of ethanol (C_2H_5OH)?

QUESTION SET 5.3

Remembering

1 Define the Avogadro constant.

2 Define the mole.

Understanding

3 Explain the relationship between the Avogadro constant and the mole.

Applying

4 Calculate the number of moles of each of the following substances, given the number of particles.

- **a** 4×10^{24} atoms of helium
- **b** 1.03×10^{22} molecules of chlorine gas
- **c** 1.204×10^{23} ions of copper
- **d** 8.22×10^{21} formula units of aluminium oxide

5 Calculate the number of particles of each of the following substances, given the number of moles.
   a 2.5 mol of oxygen molecules
   b 0.33 mol of gold atoms
   c 1.6 mol of iron ions
   d 0.035 mol of silver nitrate
6 Copper(II) carbonate has the formula $CuCO_3$. In 3 mol of $CuCO_3$ how many:
   a moles of Cu^{2+} ions are there?
   b moles of CO_3^{2-} ions are there?
   c moles of oxygen atoms are there?
   d individual atoms of oxygen are there?
7 Octane (C_8H_{18}) is an important constituent of petrol. If we have 3×10^{26} molecules of octane, how many moles of each of the following are present?
   a C_8H_{18} molecules
   b C atoms
   c H atoms

Analysing

8 How many moles of oxygen atoms are in each of the following?
   a 2.5 mol of H_2SO_4 molecules
   b 0.2 mol of CH_3COOH molecules
   c 3.02×10^{24} formula units of $Al_2(CO_3)_3$

5.4 Moles and mass

Chemists often need to calculate the number of moles present in a given mass of a substance. The mass of 1 mole of a particular element or compound is numerically equal to its relative atomic, molecular or formula mass in grams. The mass of 1 mole of a substance is called the **molar mass**, symbol *M*. The units of molar mass are grams per mole (g/mol or g mol^{-1}).

For example, the molar mass of magnesium (Mg) is 24.31 g mol^{-1}, the molar mass of oxygen gas (O_2) is (2×16.0) 32 g mol^{-1} and the molar mass of CO_2 is ($1 \times 12 + 2 \times 16$) 44 g mol^{-1}.

One mole of any substance has a mass equal to the relative atomic, molecular or formula mass of that substance expressed in grams.

Roland Smith

◀**Figure C5.6** One mole of different substances

WORKED EXAMPLE 5.7

Calculate the mass of 1 mole (molar mass) of ethane, C_2H_6.

Answer

The molar mass of ethane is 30.0 g mol^{-1}.

Logic

In Worked example 5.1, the relative molecular mass of ethane was calculated to be 30.0.

Therefore, $M(C_2H_6)$ = 30.0 g mol^{-1}

Try these yourself

Calculate the molar mass of:

a bromine gas (Br_2).

b sulfuric acid (H_2SO_4).

c calcium hydroxide ($Ca(OH)_2$).

d ammonium nitrate (NH_4NO_3).

e ethanol (CH_3CH_2OH).

It is important to remember that the molar mass (*M*) is an actual mass that can be measured in grams.

Converting between moles and mass

If you know the amount of a given substance in grams, then you can calculate the number of moles and vice versa. For example, if you have 18 g of water, then you must have 1 mol of water because the molar mass is 18.0 g mol^{-1}. If you want to measure out 2 mol of water, then you would measure (2 × 18) 36 g of water.

The relationship between the number of moles (*n*), mass (*m*) and the molar mass (*M*) of a substance is given by:

$$\text{Number of moles} = \frac{\text{mass (g)}}{\text{molar mass (g mol}^{-1}\text{)}}$$

$$n = \frac{m}{M}$$

This formula can be rearranged if you want to calculate the mass of a given number of moles:

$$m = n \times M$$

Alternatively, this can be represented schematically as shown in Figure C5.7.

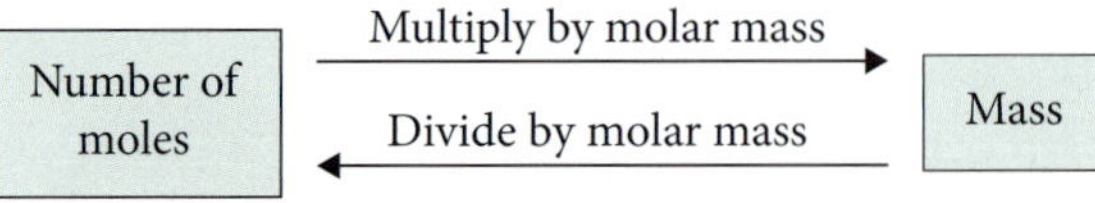

Figure C5.7 ▶ Converting between the number of moles and mass

WORKED EXAMPLE 5.8

How many moles of copper(II) sulfate ($CuSO_4$) are there in 12.2 g of copper(II) sulfate?

Answer

There are 0.078 mol in 12.2 g of copper(II) sulfate.

Logic

1 Write the relationship.

$$n = \frac{m}{M}$$

n = ?, m = 12.2 g, M can be calculated

2 Calculate the molar mass of copper(II) sulfate.

$M(CuSO_4) = M(Cu) + M(S) + 4 \times M(O) = 63.6 + 32 + 4 \times 16$

$M(CuSO_4) = 156.6\ g\ mol^{-1}$

3 Substitute.

$$n(CuSO_4) = \frac{12.2}{156.6}$$

$$= 0.0779\ mol$$

Try these yourself

Calculate the number of moles of:

a 13 g zinc.

b 7.5 g chlorine gas.

c 3.35 g lead(II) iodide.

d 3.5 kg copper.

WORKED EXAMPLE 5.9

What is the mass of 0.75 mol of sodium hydroxide (NaOH)?

Answer

There are 30 g of sodium hydroxide in 0.75 mol.

Logic

1 Write the relationship.

$$n = \frac{m}{M}$$

$n(NaOH)$ = 0.75 mol, m = ?, M can be calculated

2 Calculate the molar mass of sodium hydroxide.

$M(NaOH) = 23.0 + 16 + 1 = 40\ g\ mol^{-1}$

3 Rearrange to find m.

$m = n \times M$

Substitute.

$m(NaOH) = 0.75 \times 40$

$= 30\ g$

Try these yourself

Calculate the mass of:

a 4.2 mol of iron.

b 0.045 mol of carbon monoxide (CO).

c 0.25 mol of sodium carbonate (Na_2CO_3).

d 1.34 mol of ammonia (NH_3).

EXPERIMENT 5.1

MEASURING MOLAR QUANTITIES

In the chemistry laboratory, the amount of substance is usually measured in mass and sometimes in volume. The same number of moles of different substances have different masses because the substances are made up of different types and numbers of atoms.

Aim

To determine the number of moles in a measured quantity of substance

Materials

- Sulfur powder
- Water
- Aluminium foil
- Copper wire
- Sodium chloride
- Copper(II) sulfate ($CuSO_4$)
- Sucrose
- Carbon powder
- Electronic scales
- Measuring cylinder
- Eyedropper
- Spatula
- 2 × 50 mL beakers

What are the risks in doing this experiment?	How can you manage these risks to stay safe?
Copper(II) sulfate is toxic.	Wear appropriate safety gear, clean up spills and dispose of chemicals as directed.
Powdered sulfur is an irritant to eyes, nose and throat.	Do not agitate the powder when measuring it.

In your write-up, add any more risks you can think of, as well as ways to manage them.

Procedure

Part A

1 Calculate the molar mass of each of the substances provided.

2 Measure out 1 mole of carbon powder, water and sodium chloride.

3 Compare the apparent amounts of each of these substances.

Part B

Use your understanding of the mole to solve each of the following problems. List the steps you followed to solve each problem and show relevant calculations.

1 How many moles of carbon are in 1 level teaspoon of carbon?

2 How many atoms of copper are in a 1 cm piece of copper wire?

3 How many atoms of aluminium are in a 2 cm^2 piece of aluminium foil?

4 What contains the greater number of moles – a level teaspoon of sugar or a level teaspoon of copper(II) sulfate?

5 How many molecules are there in 1 drop of water?

Part C

Collate the class results for each problem in part B.

Discussion

1 Compare your results with those of other groups in the class and suggest possible reasons for any variation.

2 How accurate do you consider the results to be? Support your answer with experimental data.

3 What assumptions were made about the samples provided?

4 Suggest how the accuracy of the experiment could be improved.

QUESTION SET 5.4

Remembering

1 What is the difference between relative molecular mass and molar mass?

Applying

2 Calculate the molar mass of:

a hydrogen peroxide (H_2O_2).

b magnesium bromide ($MgBr_2$).

c calcium carbonate ($CaCO_3$).

d pentane (C_5H_{12}).

3 What is the mass of each of the following?

a 7.1×10^{-3} mol of neon gas

b 5.43 mol of lead

c 3.14×10^{-4} mol of phosphorus (P_4)

d 1.236 mol of magnesium sulfate ($MgSO_4$)

e 0.452 mol of potassium phosphate (K_3PO_4)

4 How many moles are there in 10.00 g of:

a water.

b acetic acid (CH_3COOH).

c ethane (C_2H_6).

d sodium bromide.

e barium chloride.

5 a How many moles of carbon dioxide (CO_2) are in 0.028 g of carbon dioxide?

b What mass of CO_2 contains exactly 3.0 mol of CO_2.

Analysing

6 The artificial sweetener NutraSweet® is the compound aspartamine, which has the molecular formula $C_{14}H_{18}N_2O_5$.

a What is the mass of 1.00 mol of aspartamine?

b How many moles of aspartamine are in 6.22 g of the substance?

c What is the mass of 0.245 mol of aspartamine?

d How many molecules are there in 4.28 mg of aspartamine?

7 a How many molecules of oxygen gas are there in 4.00 g of O_2?

b How many atoms of oxygen would be obtained if the molecules in part **a** were split in half?

8 a How many moles of water molecules would be in the body of an 80 kg person? Assume the body is 80% water.

b How many water molecules would this be?

9 A single molecule of a certain compound has a mass of 2.20×10^{-22} g. What is the molar mass of this substance?

10 A mixture of hydrogen and helium contains twice as many helium atoms than hydrogen molecules. What is the mass of hydrogen in 6 g of the mixture?

To see an application of calculating quantities in uranium mining, refer to Context 2, 'Materials for a purpose', page 46.

5.5 Empirical and molecular formulas

The **empirical formula** of any compound is the simplest whole number ratio in which the atoms of the elements are present. For an ionic compound, this is the formula that is given because these have an infinite lattice structure. In contrast, a molecular substance consists of individual molecules, so its molecular formula states the actual number of each type of atom present in the molecule.

For example, the compound hydrogen peroxide has the molecular formula H_2O_2. It is composed of two atoms of hydrogen and two atoms of oxygen bonded together. The empirical formula of hydrogen peroxide is HO because this is the simplest whole number ratio. However, for water (H_2O), the empirical and molecular formulas are the same because the ratio of atoms cannot be simplified any further.

Other examples of empirical and molecular formulas are shown in Table C5.1.

Table C5.1 Empirical and molecular formula of different compounds

Substance	Molecular formula	Empirical formula
Methane	CH_4	CH_4
Octane	C_8H_{18}	C_4H_9
Glucose	$C_6H_{12}O_6$	CH_2O
Calcium nitride	Not applicable	Ca_3N_2
Silicon dioxide	Not applicable	SiO_2

It is possible for different compounds to have the same empirical formula.

Determining empirical and molecular formulas

The empirical formula of a substance can be determined experimentally. The term 'empirical' means obtained from experiment. This can be done by experimentally determining the percentage composition and then applying the mole concept to determine the ratio of particles. (Remember, the ratio of particles and moles of particles is the same.)

WORKED EXAMPLE 5.10

An unidentified compound was analysed in the laboratory and found to have the following percentage composition by mass: 26.09% C, 4.35% H and 69.56% O.

a What is its empirical formula?

b If the relative molecular mass of the compound is 92, what is its molecular formula?

Answers

a The empirical formula is CH_2O_2.

Logic

1 Assuming 100 g of the compound, determine the mass of each element.

Element	C	H	O
% (given)	26.09	4.35	69.56
Mass (in 100 g)	26.09	4.35	69.56

Sometimes dividing by the smallest number does not give a ratio close to a whole number. For example, if the ratio was 2.33 to 1, multiplying by 3 gives a ratio of 7 to 3. Other common ratios are 1.33 to 1, or 4 to 3, and 1.67 to 1, or 5 to 3. When these ratios occur, another step is needed in the calculation.

2 Find the number of moles of each element by dividing the mass by the molar mass $\left(n = \frac{m}{M}\right)$.

Element	C	H	O
Mass (in 100 g)	26.09	4.35	69.56
Moles (n)	26.09/12.01 = 2.17	4.35/1.01 = 4.31	69.56/16.00 = 4.35

3 Find the simplest whole number ratio by dividing the number of moles by the smallest number then rounding to a whole number

Element	C	H	O
Mass (in 100 g)	26.09	4.35	69.56
Moles (n)	2.17	4.31	4.35
Simplest ratio	2.17/2.17 = 1 = 1	4.31/2.17 = 1.99 = 2	4.35/2.17 = 2.00 = 2

Empirical formula = CH_2O_2

b The molecular formula is $C_2H_4O_4$.

$$\begin{aligned}\text{Relative empirical formula mass} &= 1 \times 12.01 + 2 \times 1.01 + 2 \times 16.00 \\ &= 12.01 + 2.02 + 32.00 \\ &= 46.03\end{aligned}$$

To calculate the molecular formula, determine how many empirical units there are in the molecule.

$$\frac{\text{relative molecular mass}}{\text{relative empirical mass}} = \frac{92}{46.03} = 2$$

Therefore, the molecular formula contains 2 empirical formulas.

$$\begin{aligned}\text{Molecular formula} &= 2 \times CH_2O_2 \\ &= C_2H_4O_4\end{aligned}$$

Try these yourself

a Find the empirical formula of the following compounds, given their percentage composition by mass:

i 7.00% C, 93.00% Br

ii 64.9% C, 13.5% H, 21.6% O

iii 49.5% C, 5.2% H, 16.6% O, 28.8% N

b If the relative molecular mass of the compound in Question **a(ii)** is 371, what is its molecular formula?

EXPERIMENT 5.2

DETERMINING THE EMPIRICAL FORMULA OF MAGNESIUM OXIDE

The empirical formula of a substance can be calculated from experimental data.

In this experiment, you will carry out the reaction:

$$\text{magnesium} + \text{oxygen} \rightarrow \text{magnesium oxide}$$

By determining the masses of each element and using the relationship:

$$\text{number of moles} = \frac{\text{mass}}{\text{molar mass}}$$

the number of moles of each element can be determined and from this the empirical formula.

Aim

Write an aim for this experiment.

Materials

- 1 strip of magnesium ribbon (about 20 cm)
- Steel wool
- Crucible tongs
- Pipe clay triangle
- Bunsen burner
- Tripod
- Balance
- Crucible and lid
- Matches

What are the risks in doing this experiment?	How can you manage these risks to stay safe?
Burning magnesium emits a bright light.	Do not look directly at burning magnesium.
Magnesium oxide is an irritant to eyes, nose and throat.	Lift only the edge of the crucible lid. Work in a well-ventilated area.

In your write-up, add any more risks you can think of, as well as ways to manage them.

Procedure

1 Heat the crucible and lid strongly for 5 minutes over a hot Bunsen flame. Allow it to cool.
2 Weigh the crucible and lid.
3 Thoroughly clean the surface of the magnesium ribbon with steel wool.
4 Record the appearance of the cleaned magnesium.
5 Coil the piece of magnesium ribbon so that it fits inside the crucible. Weigh the crucible, lid and magnesium ribbon.
6 Place the crucible on a pipe clay triangle on a tripod over the Bunsen burner and carefully heat the crucible without the lid until the magnesium begins to glow. Use tongs to then put the lid on before the magnesium bursts into flames. (Warning: Do not look directly at the burning magnesium.)
7 Once the lid is in place, heat the crucible strongly for about 10 minutes, lifting the lid occasionally to admit oxygen. Try to prevent any magnesium oxide smoke escaping.
8 Remove the lid and heat the crucible for a further 5 minutes to ensure complete reaction.
9 Replace the lid and allow the reaction mixture to cool. Reweigh the crucible, lid and magnesium oxide.

Results

Record your results in a results table like the one below.

Mass of crucible and lid	
Mass of crucible, lid and magnesium	
Mass of magnesium	
Mass of crucible, lid and magnesium oxide	
Mass of magnesium oxide	
Mass of oxygen	

Analysing the results

1. Calculate the number of moles of magnesium and oxygen.
2. What is the ratio of moles of magnesium to moles of oxygen?
3. Using the mole ratio, determine the empirical formula of magnesium oxide.

Discussion

1. Compare your results with those of other groups in the class and suggest possible reasons for any variations.
2. The actual formula is MgO, which is a ratio of 1:1. How does this compare with your results?
3. What are the likely sources of error in this experiment?
4. Suggest how the accuracy of this experiment could be improved.

Conclusion

Write a conclusion for this experiment.

THE REDUCTION OF COPPER OXIDE

Watch this video, which demonstrates the reduction of copper(II) oxide to copper.

QUESTION SET 5.5

Remembering

1. Describe the difference between an empirical formula and a molecular formula.

Understanding

2. Give the empirical formula of each of the following compounds.
   a. $COCl_2$
   b. C_4H_{10}
   c. N_2O_4
   d. $Na_4P_2O_6$

Applying

3. An oxide of chlorine contains 18.4% oxygen by mass. What is the empirical formula of the compound?
4. Nicotine was analysed to determine its composition. It was found to contain the following percentages by mass: 74.0% C, 8.7% H and 17.3% N. Its relative molar mass was found to be 162. What is the:
   a. empirical formula?
   b. molecular formula?
5. Calculate the formula of the compound formed when 20.7 g of lead combines with oxygen to form 23.9 g of lead oxide.

Analysing

6 A compound of mass 7.24 g containing only carbon, hydrogen and oxygen was burnt completely in air producing 10.38 g CO_2 and 5.66 g H_2O.

a Calculate the number of moles of CO_2 and H_2O produced.

b Calculate the number of moles of C and H.

c Calculate the masses of C and H.

d Calculate the mass of O in the original compound.

e Determine the empirical formula of the original compound.

7 Magnesium bromide crystallises to form a hydrated salt with the formula $MgBr_2.xH_2O$. When 7.30 g of the hydrated salt was heated to remove the water of crystallisation, the anhydrous salt remaining weighed 4.60 g. Find the value of *x* and give the formula for the hydrated salt.

8 A group of students was trying to determine the formula of an oxide of copper. They weighed a dried crucible, placed some of the metal oxide in it and reweighed it. They then placed the crucible in a reduction tube and heated it while a stream of hydrogen was passed over it. The oxide was reduced to the pure metal while the oxygen reacted with the hydrogen and was removed.

The crucible was allowed to cool with hydrogen still passing over it and then reweighed.

The following results were obtained.

Mass of dry crucible = 6.10 g

Mass of crucible + metal oxide = 14.95 g

Mass of crucible + metal = 13.95 g

a Calculate the mass of metal in the sample.

b Calculate the mass of oxygen in the sample.

c Find the empirical formula of the compound.

d Why was hydrogen passed over the sample while it was still cooling?

EMPIRICAL AND MOLECULAR FORMULAS

For more practice calculating empirical and molecular formulas, read through the worked examples and try the problems.

5.6 Chemical equations, moles and mass

When a chemical reaction occurs, new substances are formed. As you have already learnt, chemical reactions can be represented using chemical equations. Balanced chemical equations represent the relationship between the number of particles that react and the number of particles that are produced.

Remember that chemical equations must be balanced because atoms are conserved.

Moles and chemical reactions

The reaction between magnesium and oxygen can be represented by the balanced equation:

$$2Mg(s) + O_2(g) \rightarrow 2MgO(s)$$

The coefficient of each species in the equation indicates the ratio in which the species react. In this case, two atoms of magnesium react with one molecule of oxygen gas to produce two formula units of magnesium oxide.

The chemical equation can also be read in terms of moles, because the number of particles is directly proportional to the number of moles through the Avogadro constant. This means the coefficients of the species in the equation also indicate the molar ratio of reaction; that is, 2 moles of magnesium react with 1 mole of oxygen gas to produce 2 moles of magnesium oxide.

The study of the amounts of reactants and products in a chemical reaction is called **stoichiometry**. This word comes from the Greek words *stoicheion*, meaning 'element', and *metron*, meaning 'measurement'.

Consider the reaction for the complete combustion of methane gas:

$$CH_4(g) + 2O_2(g) \rightarrow CO_2(g) + 2H_2O(g)$$

The relationships between the number of moles of reactants and products is represented in Table C5.2. Although the numbers are different, the ratio remains the same.

Table C5.2 Ratio of particles involved when methane reacts with oxygen gas

CH_4	$+ 2O_2(g)$	$\rightarrow CO_2(g)$	$+ 2H_2O(g)$
1 mol	2 mol	1 mol	2 mol
2 mol	4 mol	2 mol	4 mol
0.5 mol	1 mol	0.5 mol	1 mol
5 mol	10 mol	5 mol	10 mol

If we know the number of moles of any species in the reaction, then we can use the balanced equation to work out the number of moles of all the other species in that reaction.

WORKED EXAMPLE 5.11

Propane gas (C_3H_8) burns in oxygen to produce carbon dioxide and water.

If 2.5 mol of propane is burnt, calculate the number of moles of:

a oxygen gas needed for complete combustion.

b carbon dioxide produced.

c water produced.

Answers

a 12.5 mol of O_2 are needed.

b 7.5 mol of CO_2 are produced.

c 10 mol of H_2O are produced.

Logic

1 Write a balanced equation and determine the reaction ratio from the coefficients.

$C_3H_8(g) + 5O_2(g) \rightarrow 3CO_2(g) + 4H_2O(g)$

1 mol + 5 mol → 3 mol + 4 mol

2 Use the ratio to calculate the number of moles from the reacted amount given.

a 1 mol of C_3H_8 reacts with 5 mol of O_2.

$$n(O_2) = \frac{5}{1} \times n(C_3H_8)$$

$$n(O_2) = \frac{5}{1} \times 2.5 = 12.5 \text{ mol of } O_2$$

b 1 mol of C_3H_8 produces 3 mol of CO_2.

$$n(CO_2) = \frac{3}{1} \times n(C_3H_8)$$

$$n(CO_2) = \frac{3}{1} \times 2.5 = 7.5 \text{ mol of } CO_2$$

c 1 mol of C_3H_8 produces 4 mol of H_2O.

$$n(H_2O) = \frac{4}{1} \times n(C_3H_8)$$

$$n(H_2O) = \frac{4}{1} \times 2.5 = 10 \text{ mol of } H_2O$$

Try these yourself

a If 4.5 mol of magnesium were burnt in oxygen gas, calculate the number of moles of:

i oxygen required for complete combustion of the magnesium.

ii magnesium oxide produced.

b Iron reacts with hydrochloric acid (HCl) to produce iron(II) chloride and hydrogen gas:

$$Fe(s) + 2HCl(aq) \rightarrow FeCl_2(aq) + H_2(g)$$

- i How many moles of Fe are needed to completely react with 1.5 mol of HCl?
- ii How many moles of H_2 are produced when 3 mol of HCl react?
- iii If 4 mol of $FeCl_2$ are produced, how many moles of Fe react?

c Butane (C_4H_{10}) undergoes combustion in the presence of oxygen gas to produce carbon dioxide and water. If 5.5 mol of butane react, calculate the number of moles of:

- i oxygen consumed
- ii carbon dioxide and water produced.

ANTOINE LAVOISIER – CONSERVATION OF MASS

Watch this video, which shows 'Antoine Lavoisier' explaining his Law of conservation of mass.

Notice that although atoms and mass are conserved, moles are not. This is because moles count the number of molecules and formula units, not necessarily individual atoms. So 1 mole of a substance may contain many moles of individual atoms. (Refer to 'Moles and chemical formulas' in section 5.3.)

Mass and chemical equations

Since chemical equations indicate a relationship between numbers of moles of reactants and products, and as masses are related to moles, chemical equations can also be used to determine relationships between masses of reactants and products.

Remember chemical reactions obey the **Law of conservation of mass** developed by Antoine Lavoisier. This states that the mass of the products equals the mass of reactants.

The Law of conservation of mass states that in a chemical reaction, mass is neither created nor destroyed. The total mass of the products equals the total mass of the reactants.

For example, in the reaction between magnesium and oxygen, the mass–mass relationships between reactants and products are:

$$\begin{array}{lclcl} 2Mg(s) & + & O_2(g) & \rightarrow & 2MgO(s) \\ 2\text{ mol} & + & 1\text{ mol} & \rightarrow & 2\text{ mol} \\ 48.6\text{ g} & + & 32\text{ g} & \rightarrow & 80.6\text{ g} \end{array}$$

As Lavoisier demonstrated, mass is conserved.

Figure C5.8 ▶ Marie and Antoine Lavoisier

◀**Figure C5.9**
A chemical industry

WOW

Green chemistry

Minimising waste is one of the fundamental principles of green chemistry. By developing alternative synthesis processes, chemists aim to maximise the amount of desired product and reduce the amount of waste, which leads to a more profitable industry and less impact on the environment.

Application of the relationships between masses of products and reactants is very important in chemistry. It is a vital in calculating how much of a desired product is made and how much by-product (waste) is also made. This is important in all chemical manufacturing, including the plastic, metal, pharmaceutical, fertiliser and petroleum industries.

To see an application of relating reactants and products in uranium mining, refer to Context 2, 'Materials for a purpose', page 46.

Mass–mass calculations

Mass–mass problems use the mass of a known substance and a balanced chemical equation to calculate the mass of all other species (reactants and products) involved in the reaction.

When solving problems involving mass–mass stoichiometry, it is important to remember that the mass relationships between reactants and products are derived from the mole ratios in a balanced chemical equation.

Figure C5.10 shows the steps involved in mass–mass calculations.

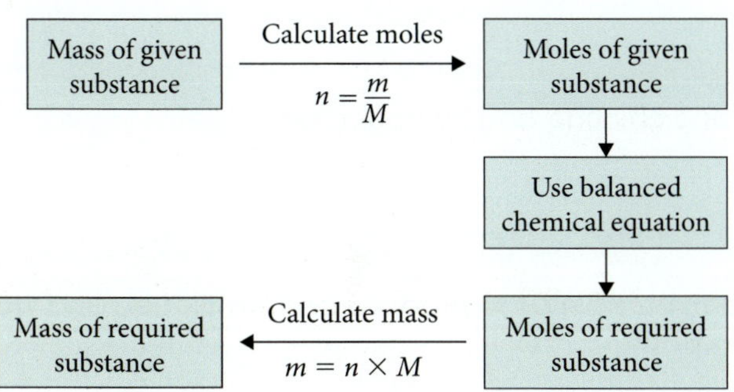

◀**Figure C5.10**
Mass-mass calculations

WORKED EXAMPLE 5.12

Sulfuric acid is formed when sulfur dioxide and water react with oxygen in the presence of a catalyst. If 3.20 g of SO_2 react, calculate the mass of:

a O_2 that reacts.

b sulfuric acid that is produced.

Answers

a The mass of oxygen reacted is 0.8 g.

b The mass of sulfuric acid produced is 4.9 g.

Logic

1 Write a balanced equation.

$2SO_2(g) + O_2(g) + 2H_2O(l) \rightarrow 2H_2SO_4(l)$

2 Calculate the number of moles of SO_2 using the relationship $n = \frac{m}{M}$.

$m = 3.20$ g, $n = ?$, M can be calculated.

$M = 32.1 + 2 \times 16 = 64.1$ g mol^{-1}

$n = \frac{3.20}{64.1} = 0.0499$ mol

3 Use the balanced chemical equation to calculate the number of moles of O_2 and H_2SO_4

a From the equation 2 mol of SO_2 react with 1 mol of O_2.

$n(O_2) = \frac{1}{2} \times n(SO_2)$

$n(O_2) = \frac{1}{2} \times 0.0499 = 0.025$ mol of O_2

b 2 mol of SO_2 produce 2 mol H_2SO_4.

$n(H_2SO_4) = \frac{2}{2} \times n(SO_2)$

$n(H_2SO_4) = \frac{2}{2} \times 0.0499 = 0.0499$ mol of H_2SO_4

4 Calculate the mass of the required substance using $m = n \times M$.

a Mass of O_2

$n = 0.025$ mol, $m = ?$, M can be calculated

$M = 2 \times 16 = 32$ g mol^{-1}

$m = 0.025 \times 32 = 0.8$ g

b Mass of H_2SO_4

$n = 0.0499$ mol, $m = ?$, M can be calculated

$M = 2 \times 1.02 + 32.1 + 4 \times 16 = 98.14$ g mol^{-1}

$m = 0.0499 \times 98.14 = 4.9$ g

Try these yourself

a Calcium carbonate decomposes when heated strongly. The decomposition is represented by the equation:

$$CaCO_3(s) \rightarrow CaO(s) + CO_2(g)$$

If 25 g of calcium carbonate is decomposed, calculate the mass of:

i CaO produced.

ii CO_2 produced.

b Zinc reacts with hydrochloric acid to form zinc chloride and hydrogen gas.

If 2.6 g zinc react, calculate the mass of:

i zinc chloride produced.

ii hydrogen gas produced.

c Octane (C_8H_{18}, a major component of petrol) burns in air to produce carbon dioxide and water. If 45.6 g of octane is burnt, calculate the mass of:

i oxygen that reacts.

ii carbon dioxide produced.

d In the wine industry, glucose ($C_6H_{12}O_6$) from grapes is fermented by yeast to produce ethanol (C_2H_5OH) and carbon dioxide.

If 45 g of glucose is fermented, calculate the mass of ethanol produced.

QUESTION SET 5.6

Remembering

1 What is stoichiometry?

2 State the law of conservation of mass.

3 Why do reactants and products always need to be related using a balanced chemical equation?

Understanding

4 Copy and complete the table to determine the missing numbers of moles.

$Fe_2O_3(s) + 3CO(g) \rightarrow 2Fe(s) + 3CO_2(g)$

Fe_2O_3	CO	Fe	CO_2
1 mol			
		3 mol	
	0.6 mol		
			0.09 mol

5 For the reaction $2H_2O_2(l) \rightarrow 2H_2O(l) + O_2(g)$, what is the relationship between total mass of products and total mass of reactants?

Applying

6 Nitrogen and hydrogen react to form ammonia (NH_3).

- **a** How many moles of hydrogen are needed to form 20 mol of ammonia?
- **b** How many grams of ammonia will be produced when 3.41 g of hydrogen react with excess nitrogen?

7 If 80.5 g of O_2 was formed when potassium chlorate ($KClO_3$) decomposed, then how many grams of potassium chloride, the other product, would be formed?

8 Calcium phosphate ($Ca_3(PO_4)_2$) and water are produced when phosphoric acid (H_3PO_4) reacts with calcium hydroxide. If 30.0 g of H_3PO_4 react, then calculate the mass of:

- **a** calcium hydroxide that reacts.
- **b** calcium phosphate produced.

9 TNT (trinitrotoluene ($C_7H_5N_3O_6$)) is an explosive. It is produced by a reaction between toluene (C_7H_8) and nitric acid. Water is a by-product of the reaction. In order to produce 1000 kg of TNT, calculate the mass of:

- **a** toluene.
- **b** nitric acid.

REVISION

View the slideshow to revise concepts dealt with in this chapter. The slideshow also covers limiting reagents, which could be covered as extension work.

Analysing

10 Washing soda ($Na_2CO_3.10H_2O$) is commonly used as a water-softening agent and for cleaning up grease. If it is not kept in an airtight container, then it loses some of its water of crystallisation to form $Na_2CO_3.H_2O$.

A retailer buys a 10 kg bag of washing soda for $30. While it is standing in the storeroom, the bag punctures and the crystals turn into powder as water of crystallisation is lost. The retailer sells 1 kg bags of the powder for $5.00 per bag. Does the retailer make a profit?

CHAPTER CHECKLIST

You should know:

- the relative atomic mass (A_r) is used to compare the mass of atoms and elements. It is dependent on the relative abundance of its isotopes on Earth.
- the relative molecular or formula mass (M_r) of a substance is the mass of one molecule or formula unit of the substance on a scale in which the mass of an atom of the carbon-12 isotope is exactly 12
- the percentage composition of a compound is the percentage by mass of each of the different elements in the compound
- the Avogadro constant (N_A) is the number of atoms (6.02×10^{23}) in exactly 12 grams of the carbon-12 isotope
- a mole is the amount of substance containing about 6.02×10^{23} particles of that substance
- one mole of any substance has a mass equal to the relative atomic, molecular or formula mass of that substance expressed in grams
- the empirical formula of any compound is the simplest whole number ratio in which the atoms of the elements are present
- a molecular substance consists of individual molecules so its molecular formula states the actual number of each type of atom present in the molecule
- balanced chemical equations show relationships between the number of moles of reactants and products
- the law of conservation of mass.

You should be able to:

- calculate relative molecular and formula masses of substances
- calculate percentage composition of a compound
- convert between moles and the number of particles (atoms, molecules, formula units) of a given substance and vice versa
- convert between moles, mass and molar mass
- calculate empirical and molecular formulas
- work out the number of moles of reacting species from a balanced equation
- calculate mass of reactants and products in a chemical reaction given the amount of one substance in the reaction.

CHAPTER GLOSSARY

amu atomic mass unit; 1/12 the mass of one atom of carbon-12

Avogadro constant (N_A) the number of atoms in exactly 12 grams of the carbon-12 isotope

empirical formula the simplest whole number ratio of the elements in a chemical compound

law of conservation of mass the law that states that in a chemical reaction matter is neither created nor destroyed

mole (n) the amount of substance containing about 6.02×10^{23} particles of that substance

molar mass (M) the mass of one mole of a substance

percentage composition the percentage by mass of each different element in a compound

relative formula mass (M_r) the mass of one formula unit of an ionic compound on a scale in which the mass of an atom of the carbon-12 isotope is exactly 12

relative molecular mass (M_r) the mass of one molecule of a substance relative to the mass of an atom of the carbon-12 isotope taken as exactly 12

stoichiometry the study of the amounts of substances in a chemical reaction

CHAPTER REVIEW QUESTIONS

Remembering

1 What is the relationship between relative atomic mass and relative molecular mass?

2 What is the difference between relative molecular mass and relative formula mass?

3 Describe how percentage composition provides useful information to chemists.

4 Explain the relationship between the Avogadro constant and moles.

5 What is the difference between relative molecular mass and molar mass?

6 a Explain the difference between empirical formula and molecular formula.

b Is it correct to refer to the formula of an ionic compound as the empirical formula? Why or why not?

7 In a chemical reaction, mass is neither created nor destroyed. What law is this?

Understanding

8 Calculate the relative molecular or formula mass of:

a sulfur dioxide (SO_2).

b potassium nitrate (KNO_3).

c zinc hydroxide ($Zn(OH)_2$).

d acetic acid (CH_3COOH).

9 Calculate the percentage composition of:

a calcium in calcium hydroxide ($Ca(OH)_2$).

b oxygen in aluminium carbonate ($Al_2(CO_3)_3$).

10 Calculate the percentage composition of each element in:

a nickel(II) sulfate ($NiSO_4$).

b phosphoric acid (H_3PO_4).

11 Calculate the number of moles in:

a 3.01×10^{24} carbon atoms.

b 1.05×10^{22} carbon monoxide molecules.

c 1.204×10^{28} ethanoic acid molecules (CH_3COOH).

12 Calculate the number of:

a atoms in 0.30 mol of nickel metal.

b molecules in 4.0 mol of ethane (C_2H_6).

c formula units in 7.5 mol of potassium bromide (KBr).

13 Why is it incorrect to apply the term 'molecular mass' to ionic substances?

14 How many moles are contained in:

a 1.01 g of neon gas (Ne).

b 2.538 g of iodine molecules (I_2).

c 5.10 g aluminium oxide (Al_2O_3).

15 Calculate the mass of:

a 2.05 mol of rubidium atoms.

b 1.5 mol of sulfuric acid molecules.

c 1.0 mol of glucose ($C_6H_{12}O_6$).

d 0.1 mol of sodium chloride (NaCl).

16 A chemist finds that a certain compound has an empirical formula of CH_3O. What information about the compound does this formula give the chemist?

17 The equation for the combustion of ethane is:

$$2C_2H_6 + 7O_2 \rightarrow 4CO_2 + 6H_2O$$

Calculate the number of moles of:

a CO_2 produced in the combustion of 1 mol of C_2H_6.

b H_2O produced in the combustion of 3 mol of C_2H_6.

c O_2 consumed in the combustion of 0.6 mol of C_2H_6.

18 How many moles of oxygen gas would be obtained if 0.24 mol of mercury(II) oxide is decomposed by heating into mercury and oxygen gas?

Applying

19 Urea has the formula CH_4ON_2. In 2 mol of urea, calculate the number of:

a moles of H atoms.

b moles of O atoms.

c individual atoms of N.

20 In an experiment to find the formula of a compound, the following results were obtained:

Mass of crucible = 35.030 g

Mass of crucible + copper = 38.205 g

Mass of crucible + copper(II) oxide = 39.005 g

Find the empirical formula of the compound.

21 In an experiment, 0.98 g of magnesium combined with chlorine gas to form 3.8 g of magnesium chloride. Show that these results support an empirical formula of $MgCl_2$ for this compound.

22 Dilute hydrochloric acid was added to zinc until no further reaction occurred.

a If 0.012 mol of zinc chloride was formed, what mass of zinc reacted?

b What mass of zinc chloride would be produced if 0.040 mol of hydrochloric acid reacted with excess zinc?

Analysing

23 The mass of the bones in an average adult is about 11 kg. Calcium phosphate ($Ca_3(PO_4)_2$) makes up 50% of this mass. What is the mass of phosphorus in the bones of an average adult?

24 What mass of zinc has the same number of atoms as 1 g of hydrogen gas?

25 Elemental analysis of a compound of iron gave the following percentages by mass: Fe 20.09%, S 11.55%, H 5.04%, O 63.32%.

a Calculate the empirical formula for the compound.

b If all the hydrogen comes from water of crystallisation, how many molecules of water are there in a formula unit of this compound? Hence, deduce the formula of the compound in the usual hydrate form (for example, $Na_2CO_3.10H_2O$).

26 A silver coin was analysed and found to contain only silver and copper. A mass of 1.580 g of this coin was dissolved in concentrated nitric acid and the resultant solution was diluted. Reaction of the solution with excess hydrochloric acid yielded 1.050 g of silver chloride. Calculate the percentage of silver in the sample.

27 A sample of iron wire reacts with oxygen to form iron(III) oxide. If 3.58 g of the wire yields 5.00 g of the iron(III) oxide, calculate the percentage purity of the iron.

CHAPTER 6

WATER AND THE INTERMOLECULAR FORCES

By the end of this chapter you will have covered the following material.

Science Understanding

- Observable properties, including vapour pressure, melting point, boiling point and solubility, can be explained by considering the nature and strength of intermolecular forces within a substance (ACSCH055)
- The shapes of molecules can be explained and predicted using three-dimensional representations of electrons as charge clouds and using valence shell electron pair repulsion (VSEPR) theory (ACSCH056)
- The polarity of molecules can be explained and predicted using knowledge of molecular shape, understanding of symmetry, and comparison of the electronegativity of elements (ACSCH057)
- The shape and polarity of molecules can be used to explain and predict the nature and strength of intermolecular forces, including dispersion forces, dipole-dipole forces and hydrogen bonding (ACSCH058)
- Data from chromatography techniques (for example, thin layer, gas and high-performance liquid chromatography) can be used to determine the composition and purity of substances; the separation of the components is caused by the variation of strength of the interactions between atoms, molecules or ions in the mobile and stationary phases (ACSCH059)
- Water is a key substance in a range of chemical systems because of its unique properties, including its boiling point, density in solid and liquid phases, surface tension, and ability to act as a solvent (ACSCH061)
- The unique properties of water can be explained by its molecular shape and hydrogen bonding between molecules (ACSCH062)
- The solubility of substances in water, including ionic and molecular substances, can be explained by the intermolecular forces between species in the substances and water molecules, and is affected by changes in temperature (ACSCH065)

AC

6.1 Water and intermolecular forces

Water has unique physical and chemical properties. For a small molecule, it has a big influence on life on Earth. Water's unique properties arise partly from the strength of the intermolecular forces between the water molecules. These forces are affected by the shape of the water molecules and are important for some practical applications.

WOW

Small but powerful

Intermolecular forces are extremely small. Yet they are used by geckos to perform amazing feats. A gecko can hang from a polished surface, supporting its entire body weight, with only one toe. Aristotle wrote about these feats 2300 years ago. Now scientists think they have determined how they can happen. There are 14 000 tiny 'hairs' called setae on every square millimetre of a gecko's footpad. Each individual seta has a diameter of 5 micrometres, thinner than human hair. These tiny setae have even smaller curved pads that create billions of interactions between the gecko's foot and the surface. This explains why a 5 cm long gecko can support the equivalent weight of a 9-year-old child.

The study of how geckos adhere to a surface has influenced the design and fabrication of bio-inspired dry and reversible adhesive surfaces.

Science Photo Library/Paul D. Stewart

Figure C6.1 ▲ The feet and toes of gecko lizards use powerful intermolecular forces to support their weight

GECKO ADHESIVE FIT FOR SPIDERMAN

Watch this video to learn about making new adhesive materials that don't feel sticky.

Forces between molecules

You need to heat water to over 3000°C to break its covalent bonds and isolate hydrogen and oxygen. When ice melts at 0°C and water vaporises at 100°C, water is merely changing state. These relatively low temperatures are evidence that covalent bonds are not breaking. The bonds between the water molecules, the **intermolecular forces**, are broken. The low melting point shows that the intermolecular forces are much weaker than the covalent bonds within the molecule.

Intermolecular forces between molecules are much weaker than covalent bonds within molecules. The structure and composition of molecules determines which intermolecular forces are present. These forces determine the properties of the material.

6.2 Shapes of molecules

To explain the properties of materials, chemists need to understand their underlying structure. Molecules exist in a three-dimensional space, but we generally represent molecules in two dimensions. The structure and bonding in molecules is described by using either electron dot formulas or **valence structures**.

Electron dot formulas

Atoms tend to be more stable when there are eight valence electrons around the central atom. This is known as the octet rule. In electron dot formulas, each of the outer valence electrons is shown as a dot (or '×') around the chemical symbol, as shown in Figure C6.2 on page 259. Eight valence electrons would be shown by four pairs of dots circling the central nucleus. They are not shown as pairs until four positions are occupied. So, for the element nitrogen, there would be one pair and three separate electrons.

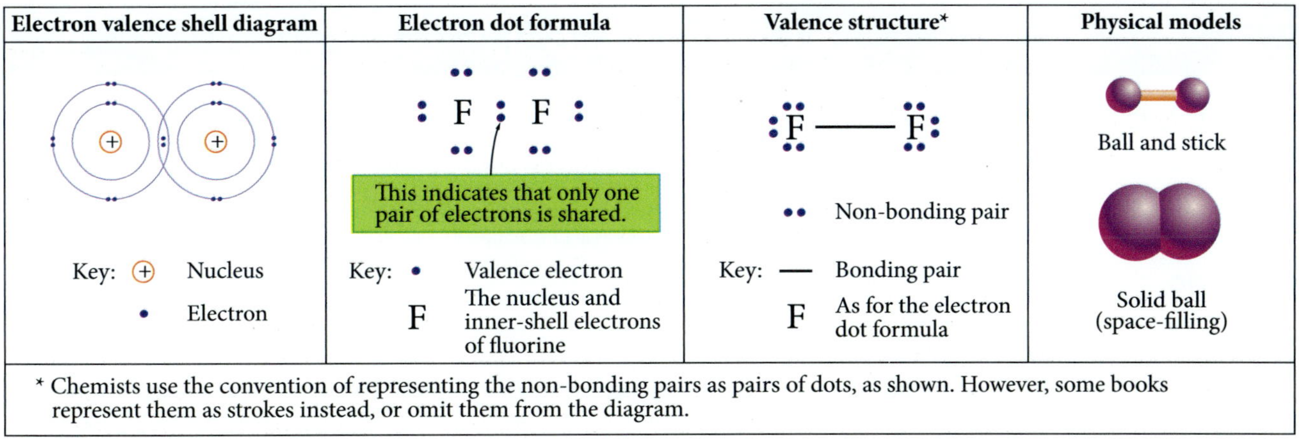

Electron valence shell diagram	Electron dot formula	Valence structure*	Physical models
Key: ⊕ Nucleus; • Electron	This indicates that only one pair of electrons is shared. Key: • Valence electron; F The nucleus and inner-shell electrons of fluorine	•• Non-bonding pair. Key: — Bonding pair; F As for the electron dot formula	Ball and stick; Solid ball (space-filling)

* Chemists use the convention of representing the non-bonding pairs as pairs of dots, as shown. However, some books represent them as strokes instead, or omit them from the diagram.

▲ Figure C6.2
Representing the shape of molecules

In covalent bonds, the electrons shared between two atoms are called the **bonding pair** of electrons, so the dots are placed between the symbols for the two elements. A single bond is formed when one pair of electrons is shared. A double bond occurs when two pairs of electrons (two sets) are shared between the two atoms. The pairs of valence electrons not involved with bonding are called the **lone pair** or non-bonding pair of electrons. The lone pairs are shown around the element as a pair of dots or crosses.

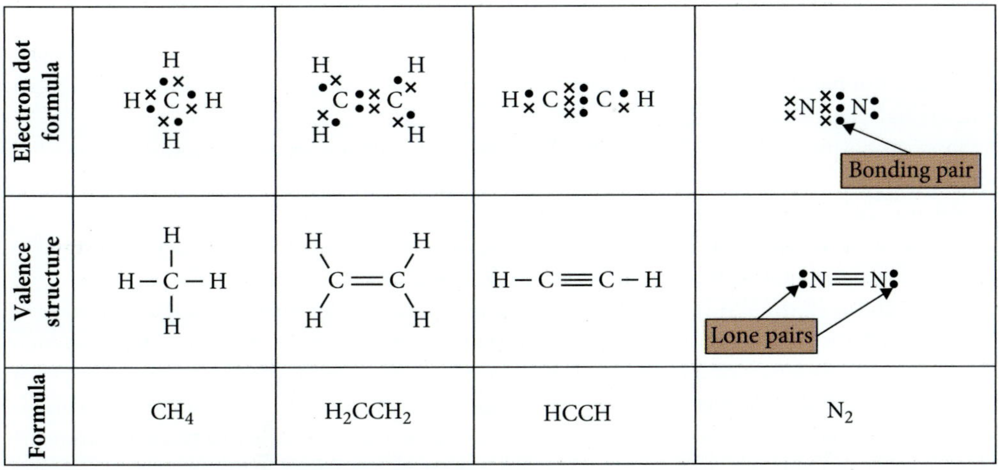

Electron dot formula				Bonding pair
Valence structure	H–C–H (with H above and below)	H₂C=CH₂	H–C≡C–H	:N≡N: Lone pairs
Formula	CH_4	H_2CCH_2	HCCH	N_2

◀ Figure C6.3
Electron dot formulas and valence structure for the molecules CH_4, H_2CCH_2, HCCH and N_2. Note that dots or 'x's are used to distinguish which electrons are from the different atoms.

Valence structures

Valence structures can be used as an alternative to drawing electron dot formulas. A single bond, shown by a line joining the two atoms, has one pair of electrons. A double bond, shown by two short lines, has two pairs of electrons. A triple bond has three short lines or three pairs of electrons. Lone pairs of electrons are drawn around the element as either two dots, a short line or another line that points away from the element. Valence structures have the advantage of showing the structure of the molecules and are often called the **structural formulas**.

To revise covalent bonding and multiple bonds, refer to Chemistry section 3.5 on page 178.

Electron dot formulas use pairs of dots to represent the bonding between atoms. Valence structures use lines to show the bonding, and non-bonding or lone pairs of electrons are shown as a pair of dots, crosses or a line and can be used to represent the shape of molecules.

ACTIVITY 6.1

REPRESENTING MOLECULES

Aim

To draw electron dot formulas and the valence structures for a range of molecules

What to do

1 Determine the maximum number of covalent bonds an atom of each of the following elements can form: H, C, N, O, F, Ne, P, Cl

2 Draw electron dot formulas for fluorine (F_2), hydrogen fluoride (HF), water (H_2O), carbon tetrachloride (CCl_4), phosphine (PH_3), carbon dioxide (CO_2), hydrochloric acid (HCl), nitrogen dioxide (NO_2) and neon (Ne).

What did you discover?

Can you use the formula to predict the arrangement of atoms around the central atom? Which atoms are more likely to be central to the molecules?

a

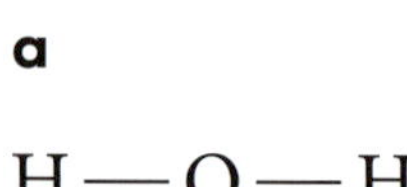

b

H H
\ /
O

c

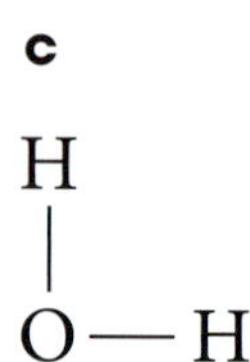

Figure C6.4 ▲ The possible shape of the water molecule: a) two-dimensional shape, b) V-shaped molecule and c) two hydrogens at right angles. Which one is correct?

Predicting the shape of molecules

It is important to know the shape of molecules because the shape will determine the intermolecular bonding and properties of the substance. The valence structure of water shows a two-dimensional shape, in which oxygen is central to two hydrogen atoms. The three-dimensional shape could be the three atoms in a row, a V-shape molecule or two hydrogen atoms at right angles to the central oxygen.

Valence shell electron pair repulsion (VSEPR) theory has been used for over 50 years to determine the shape and hence the function of many molecules. It gives an understanding of how proteins may bond to an enzyme, essential for many medical applications.

Although the 'R' in VSEPR stands for the word 'repulsion', the theory is based upon the **Pauli exclusion principle**, not electron repulsion. Pauli's exclusion principle states that each orbital can only have 0, 1 or 2 electrons. If two electrons are present, then they spin in opposite directions around the nucleus. The discovery of this electron spin is fundamental to the Pauli exclusion principle. Two electrons of the same spin have a zero probability of being found in the same location and will be found in locations as far apart as possible.

VSEPR (valence shell electron pair repulsion) theory is based upon the number of electron pairs surrounding the central atom. The electron pairs arrange themselves *as if* they repel each other.

The three-dimensional structure of simple molecules depends upon the number of electron pairs around the central atom. The electron dot formulas are used to indicate electron pairs. These may be bonding and/or lone pairs. The electron pairs will be at a maximum distance from each other. It helps to think of the electron pairs as points on a sphere with the central atom in the middle.

Lone pairs are closer to the single nucleus and so occupy more space. Bonding pairs are shared by two atoms and are attracted by the two nuclei. Hence, they occupy less space and cause less repulsion than lone pairs.

The common shapes of molecules are summarised in Table C6.1 on page 262.

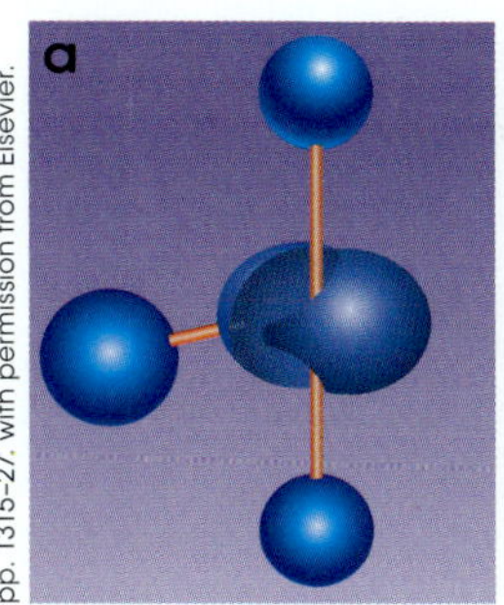

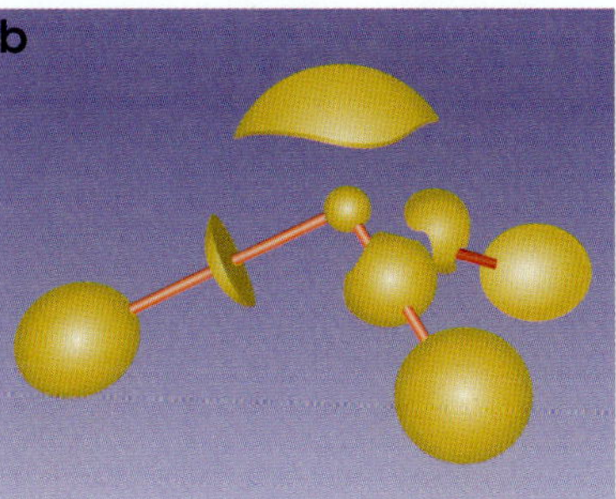

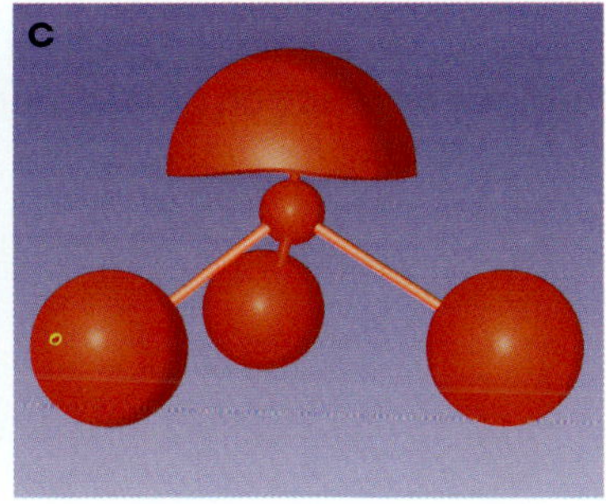

▲ **Figure C6.5**
The electron density surrounding molecules calculated by the VSEPR theory. a) ClF_3; b) NH_3; c) PF_3. Note that lone pairs take up more space than bonding pairs of electrons.

To work out the shape of a molecule:

- draw the electron dot formula
- count the lone pairs and bonding pairs. Remember that lone pairs are electron pairs that are not shared, while bonding pairs form the covalent bonds
- treat the electrons in a covalent bond, whether a single, double or triple bond, as one set of electrons
- use Table C6.1 on page 262 to predict the shape based upon the number of bonding and lone pairs.

The simplest example to show the application of the VSEPR theory is BeF_2 (Figure C6.6). Here the central Be atom has only two electron pairs, one to each F atom. The three atoms are in a line, which means that the bonding pairs are at 180° from each other. The shape is linear.

The carbon atom in methane (CH_4) has four bonding pairs to the hydrogen atoms. The four hydrogen atoms and their bonding pairs of electrons are furthest apart in a tetrahedral shape (the angle between the bonds is 109.5°).

The nitrogen atom in ammonia (NH_3) has four electron pairs, which includes three bonding and one lone pair. The hydrogen atoms are at three corners of the tetrahedral and the lone pair occupies the final spot. This results in a pyramid shape for ammonia (the angle between the bonds is 107.3°) (see Figures C6.7 and C6.8).

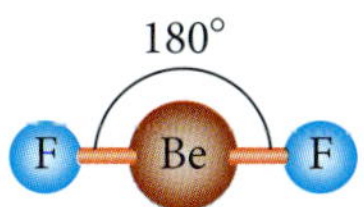

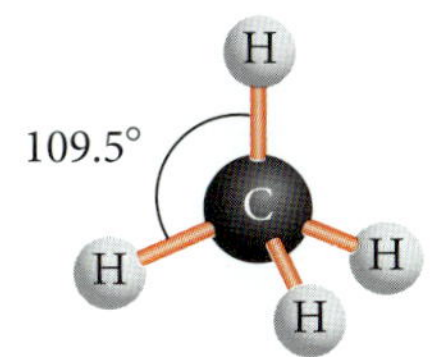

▲ **Figure C6.6**
The molecule BeF_2 is linear and CH_4 is tetrahedral.

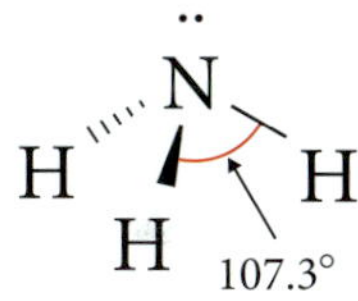

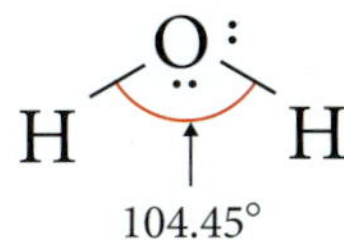

▲ **Figure C6.7**
Ammonia is pyramid-shaped and water is bent or V-shaped.

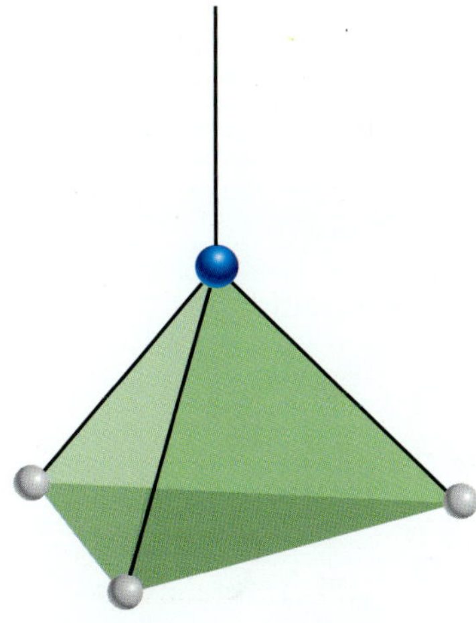

▲ **Figure C6.8**
Explaining the pyramid shape of ammonia

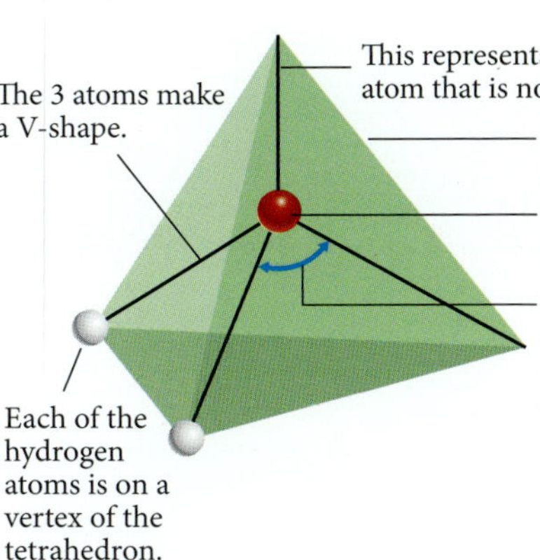

▲ **Figure C6.9**
Explaining the V-shape of water

In Chemistry Chapter 3, the structure of pure carbon in diamond form was described as tetrahedral, with C atoms at each vertex of the tetrahedron. The electronic configuration of a carbon atom is $1s^2\ 2s^2\ 2p^2$.

Oxygen is the central atom in water. Oxygen has two lone pairs and two bonding pairs of electrons. These four pairs are directed to the corners of the tetrahedron. Two of these are taken up by oxygen's lone pairs. This results in a V-shape for water molecules. Water is often called a bent molecule (see Figure C6.9).

Table C6.1 Shapes of molecules

Generic formula (M, X = atoms; E = lone pair of electrons)	Number of electron pairs	Lone pairs on M	Molecular shape
MX	1	0	Linear 180° HCl
MX_2	2	0	Linear 180° CO_2
MX_3	3	0	Trigonal planar 120° SO_3
MX_2E	3	1	V-shaped 115° NO_2
MX_4	4	0	Tetrahedral 109.5° CH_4
MX_3E	4	1	Trigonal pyramidal 107.3° NH_3
MX_2E_2	4	2	V-shaped 104.45° H_2O
MX_6	6	0	Octahedral $Cu(H_2O)_6^{3+}$

VSEPR AND THE 3D STRUCTURES

Visit this website to see the VSEPR structures of some common molecular geometries.

ACTIVITY 6.2

MOLECULAR MODELS

Three-dimensional molecules are best viewed with the aid of models.

Aim

To produce a series of models that demonstrates the shape of molecules

You will need

- Molecular model kits or straws and beads
- Digital camera (optional)

What to do

1 Create a table with column headings HF, CO_2, H_2O, CH_4, C_2H_6, C_2H_4, NH_3, NO_2.

2 a For each compound, draw the electron dot formulas.

  b Note the number of bonding and lone pairs.

3 a Draw the valence structure.

  b Note the number of lone pairs and bonding pairs.

4 Use the table to predict the shape of the molecules.

5 Make the model of the molecule.

What did you discover?

Compare your model with your prediction. How did it compare? Was it easier to see in three dimensions?

Extension

Use the Internet, computer programs or tablet apps to find 'ball and stick' and space-filling representations of the molecules. Why would chemists have so many ways to represent the molecules?

QUESTION SET 6.1

Remembering

1 What is the maximum number of covalent bonds the following atoms can form?

  a Cl

  b O

  c N

  d C

  e H

  f S

2 Why does oxygen always have two lone pairs of electrons?

3 Draw the electron dot formulas for HCl, NH_3 and CH_4 and answer the following.

  a Identify how many bonding pairs and non-bonding pairs there are in each molecule.

  b Describe the shape of each of these molecules.

  c Explain why they have this shape.

Understanding

4 Explain why H_2S has the same shape as H_2O.

5 Give explanations for why a water molecule is V-shaped but carbon dioxide is linear.

Applying

6 Draw the electron dot and valence structures and then describe the shape you would expect for each of the following molecules.

a SiH_4
b CH_3Cl
c H_2O_2
d CS_2
e H_2Se
f HCN
g $COCl_2$
h CO

7 Phosphine (PH_3) is a toxic gas that consists of only P and H.

a Predict the shape of phosphine by:
 i writing the electronic configuration of P.
 ii drawing the electron dot formula of P and hence deduce the formula for phosphine.
 iii determining the number of bonding and non-bonding pairs.
 iv predicting the shape of the phosphine.

b Why would this molecule have a different shape from PF_5?

6.3 Uneven electron sharing

To revise electronegativity, refer to Chemistry section 1.4 on page 111.

In Chemistry 3, you considered molecules such as N_2 or Cl_2 made from the same non-metal element. In these molecules, the electrons in the covalent bonds are evenly shared between the two atoms. This is not true with certain **heteroatomic** molecules that have more than one type of element present. With molecules such as HCl, NH_3 and H_2O, the electrons are not evenly shared. That is because the three elements Cl, N and O tend to attract the electrons more, due to their **electronegativity** (electron-attracting power).

Bonding electrons in HCl molecules are strongly attracted to the chlorine nucleus because chlorine is more electronegative than hydrogen. This means the electrons will spend more time at the chlorine end of the molecule and this end of the molecule will be slightly more negative. This is represented in diagrams with a delta negative sign (δ−). The hydrogen end will be slightly more positive and is represented by a delta positive sign (δ+). This is shown in Figure C6.10.

The hydrogen−chlorine bond is a **polar covalent bond**. The molecule is considered to be a **dipole**: it has two poles of opposite charges at either end, rather like a magnet with a north and a south pole. Diatomic molecules with a net dipole are considered to be **polar**. Non-polar molecules such as hydrogen have a uniform charge over the whole molecule.

Common polar bonds are those between O—H, N—H, C—O and C—Cl. The most electronegative elements in order of decreasing strength are:

$$F > O > N = Cl > Br > C = S = I > P = H > Si$$

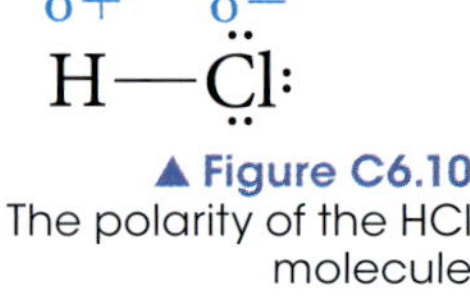

▲ Figure C6.10 The polarity of the HCl molecule

Uneven sharing of bonding electrons can occur when H or C forms covalent bonds with strong electronegative elements such as O, N, Cl and F. When this happens, the covalent bond is a polar covalent bond.

Polar and non-polar polyatomic molecules

For molecules with more than two different atoms, the presence of a polar covalent bond does not necessarily mean the molecule is polar.

$\ddot{O}=C=\ddot{O}$

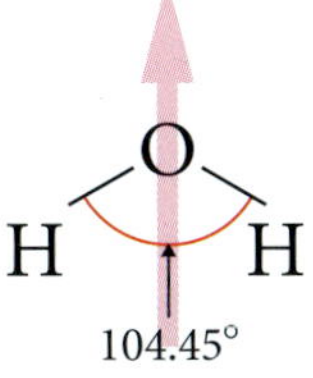

◀ **Figure C6.11** The carbon dioxide molecule is non-polar, despite the presence of polar covalent bonds. The water molecule is polar. The dipoles add together in water but cancel out in carbon dioxide. The arrow points from the positive end to the negative end of the dipole.

Carbon dioxide (CO_2) is a linear molecule with a central carbon atom and oxygen on each side. The molecule is symmetrical. The C—O bond is polar, with the dipole pointing towards the oxygen because it is more electronegative, but there is no net separation of charge in the overall molecule. Each oxygen will attract the electrons, but as they are on either end of the molecule, the charge is distributed over the whole molecule. It is like having an equally strong person on each end of a rope in tug of war; the rope doesn't move as the pull from each end is equal. The poles cancel each other out. Although the molecule has polar covalent bonds, there is no net separation of the charges so no net dipole. This molecule is non-polar.

Contrast this with water as shown in Figure C6.11. In water, the O—H bond is polar. Oxygen is central and hydrogen is at the bottom of the Λ-shape. This means that the bottom end of the molecule is slightly positive. Water is asymmetrical; there is a net dipole present so water is polar. The presence of polar covalent OH bonds and a V-shape means that the water molecule is polar.

To determine symmetry, it may help to consider the symmetry of your body. Vertically you are symmetrical, with an arm and leg on either side. But horizontally above and below your waist you are not symmetrical (asymmetrical).

The ammonia (NH_3) and tetrachloromethane (CCl_4) molecules have different shapes and polar bonds. NH_3, a pyramidal-shaped molecule, is polar. The N atom at the top of the pyramid is slightly negative due to its greater electronegativity. The hydrogen end of the molecule is slightly positive. Due to the asymmetry, it is a polar molecule. CCl_4 has polar bonds due to differences in electronegativity between C and Cl. It is tetrahedral in shape with a central carbon. The dipole in each of the C–Cl bonds points to the Cl atom. The negative regions on chlorine in CCl_4 cancel themselves out due to the symmetry of the molecule. CCl_4 molecules are non-polar.

N H H H 107.3°

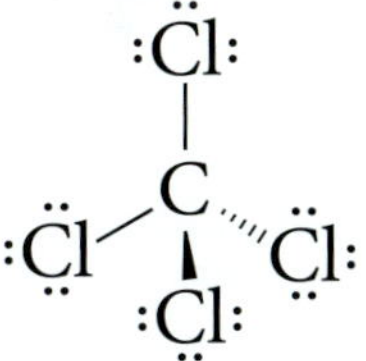

▲ **Figure C6.12** The molecular shape of ammonia (NH_3) and tetrachloromethane (CCl_4)

Molecules can be non-polar despite the presence of polar bonds. For the molecule to be polar, there must be a polar bond present and there must be asymmetry in the molecular shape.

QUESTION SET 6.2

Remembering

1 List the four elements that form polar covalent bonds with carbon or hydrogen.

2 Define the following terms.
  - a Dipole
  - b Polar covalent bond
  - c Polar

Understanding

3 The most electronegative elements in order of decreasing strength are F > O > N = Cl > Br > C = S = I > P = H > Si. Consider the following pairs of elements: O–H, Cl–H, C–H, P–H, N–O, N–Cl, N–H, S–H, C–S, F–O, Cl–O, C–O, P–O, C–Cl.
  - a In covalent bonds between these elements, indicate which element would be slightly more negative and which would be more positive by labelling with δ− and δ+ where appropriate.
  - b Classify the pairs of elements given above as highly polar, having some polarity or non-polar.

Applying

4 Consider the following molecules.

a SiH_4

b CH_3Cl

c CS_2

d PH_3

i Draw electron dot formulas and valence structures.

ii Determine the shape of the molecule.

iii Predict if the molecule has polar bonds.

iv Show the direction of the dipole, if present, by drawing an arrow towards the negative end of each polar bond.

v State whether the molecule has a net dipole, making it polar.

5 Determine whether the following are polar.

a Ammonia

b Iodine chloride

c Dichloromethane (CH_2Cl_2)

d Nitrogen trichloride

e Sulfur difluoride

Analysing

6 Hydrogen peroxide is polar. Predict the shape of H_2O_2. Explain why at first this appears unusual.

6.4 Three types of intermolecular forces

All bonding is due to attraction of opposites (**electromagnetic attraction**). There are three types of intermolecular forces – **dispersion forces**, **dipole–dipole forces** and **hydrogen bonds**. Intermolecular forces occur between the molecules. They can occur between molecules of the same substance and between molecules of different substances.

Intermolecular forces are much weaker than the intramolecular forces that exist within molecules. Intramolecular forces are generally considered to be covalent. Covalent, ionic and metallic bonding are much stronger than the intermolecular forces.

The strength of intermolecular forces increases with increasing polarisation of molecules. The weakest is the dispersion force, then the dipole–dipole force and then a special case of the dipole–dipole force, the hydrogen bond.

Dipole–dipole forces

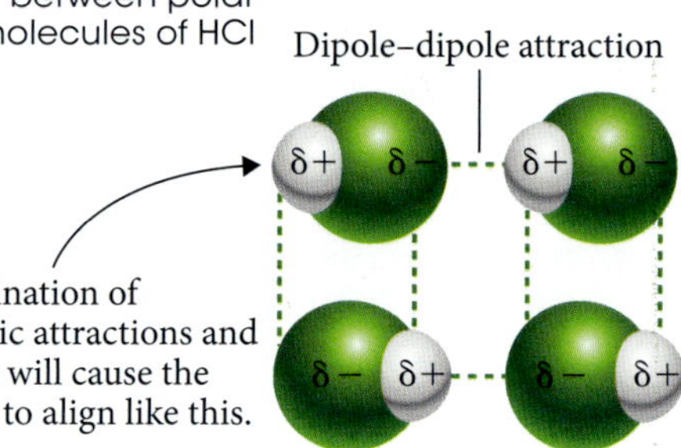

Figure C6.13 ▼ Dipole-dipole attraction between polar molecules of HCl

In polar molecules, there is a net dipole. One end of the molecule is slightly negative while the other end is slightly positive. Hydrochloric acid (HCl) is a polar molecule. The presence of dipoles creates an attraction between neighbouring molecules, a dipole–dipole force, which is an attraction between two dipoles in two individual molecules. Molecules line up so that the slightly positive (δ+) end of one molecule attracts the slightly negative end (δ−) of another molecule. The dipole–dipole force (dipole-dipole attraction) is relatively strong because the dipole is permanent.

Hydrogen bonds

Boiling and melting points give an estimation of the strength of the intermolecular forces between molecules. Figure C6.14 shows a general increase in boiling points as you go down the periodic table groups, if you could exclude period 2 hydrides.

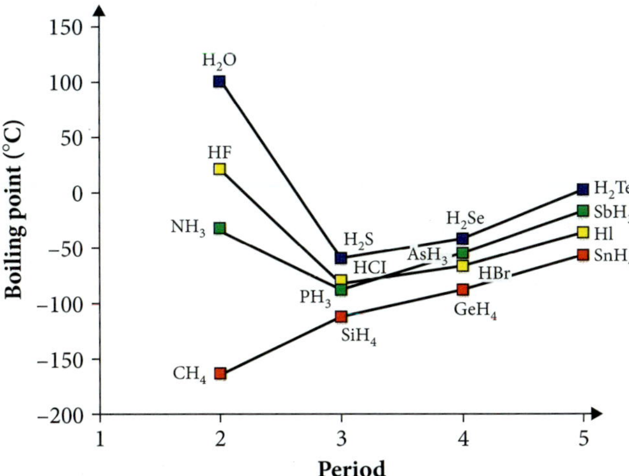

◀ **Figure C6.14**
Boiling point of hydrides of groups 14–17

When nitrogen, oxygen or fluorine is attached to a hydrogen atom, the resulting molecule has an unusually high boiling point. When these very electronegative elements bond to hydrogen, a special type of dipole–dipole force occurs called a hydrogen bond. This is the strongest of the weak intermolecular forces being about one-tenth the strength of a covalent bond. In diagrams, it is shown as a dotted line between the molecules.

Hydrogen bonds are due to the hydrogen nucleus (a single proton) being extremely small. When hydrogen is bonded to the most electronegative elements (fluorine, oxygen and nitrogen), the charge over the polar covalent bond is at a maximum. Water displays hydrogen bonding. Oxygen's strong ability to attract electrons means that the electrons in the O–H covalent bond are more attracted to oxygen. Oxygen becomes slightly negatively charged and hydrogen is left with a slightly positive charge. The hydrogen atom is so small that it also means that two molecules can get very close to each other. Similarly, in ammonia the nitrogen attracts the electrons of the covalent bonds, leaving the hydrogen atoms more positive. These can attract adjacent nitrogens. Hydrogen bonding then occurs between the hydrogen in –OH, –NH or –FH and the lone pairs on nitrogen, oxygen or fluorine on the adjacent molecule.

The dipole–dipole force occurs when there is a permanent dipole present. The hydrogen bond is a special case of the dipole–dipole force.
A hydrogen bond occurs when an H atom that is bonded to an O, N or F atom in one molecule becomes attracted to the lone pair of electrons of N, O or F of an adjacent molecule.

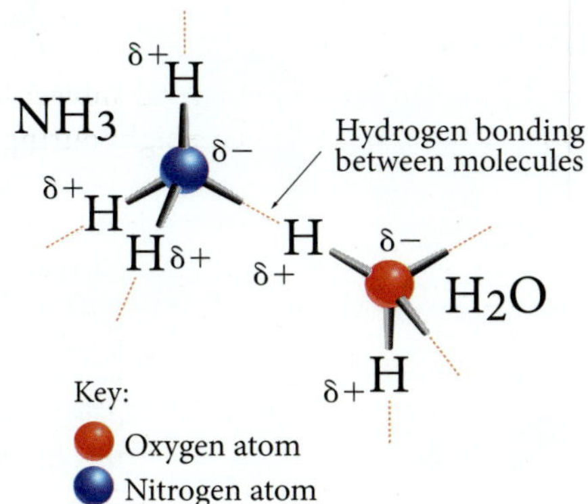

◀ **Figure C6.15**
Hydrogen bonding occurs between an H on one molecule and the lone pair of electrons on N (or O or F) in another molecule.

Shocking news: Hydrogen bond about to be redefined?

The definition of a hydrogen bond may be about to change. It is not as clear cut or simple as it appears. Scientists have found that hydrogen bonding is occurring in more subtle situations. Solid hydrogen sulfide appears to have a structure resembling ice. Some researchers have discovered evidence that hydrogen sulfide can form hydrogen bonds with ethylene. Even more unusual is that non-polar substances such as methane and krypton have been found to bond with hydrogen. But maybe the most controversial is the evidence of shared electron density between hydrogen-bonded molecules. This means that there is some covalent nature to this bond.

To see how different types of intermolecular forces relate to the properties of substances, refer to Context 3, 'Water: the vital substance', page 59.

Dispersion forces

Polar molecules can form dipole–dipole or hydrogen bonds between the molecules. Non-polar molecules cannot, so there must be some other force that enables non-polar molecules, such as dry ice, liquid nitrogen and oils and fats, to form liquids and solids. This force is a dispersion force.

The simplest molecule, hydrogen (H_2), can be used to explain this force. The force is extremely weak because hydrogen has an extremely low melting point. The hydrogen molecules have a cloud of electron density surrounding them. When the electrons are on one side of the molecule there is, for an instant, a temporary dipole. One end of the molecule will be very slightly more positive and the other end will be very slightly more negative. The temporary positive end attracts a neighbouring molecule's electrons, inducing them to come towards the positive area. This forms the weak electrostatic attractions that are dispersion forces. The fluctuating changes in charge result in a very weak force between molecules.

Dispersion forces must exist in every molecule, as every molecule has the rapidly fluctuating electrons orbiting the molecule. The force is due to the rapid movement of electrons in the cloud of electron density surrounding the molecule.

Dispersion forces are much weaker than other intermolecular forces. When dipole–dipole or hydrogen bonding is present, the effect of dispersion forces is discounted.

Dispersion forces are also present in the noble gases (see Figure C6.17). Electrons in two adjacent atoms occupy positions so that the atoms form temporary dipoles. With more electrons, there are more chances for dispersion forces to occur. Molecules are electrically neutral, so when there are more electrons, there are more protons. So, increased molecular weight means an increased chance of dispersion forces and a higher melting point. The electrons in larger atoms are also further from the nucleus and more readily form temporary dipoles. This explains why the larger noble gas argon has a boiling point of −186°C, much higher than that of helium −272°C. A similar trend of increasing dispersion forces with increasing molecular weight is seen with the halogens.

Figure C6.17 ▼ Noble gases can explain how dispersion forces occur through instantaneous dipoles induced due to the movement of electrons.

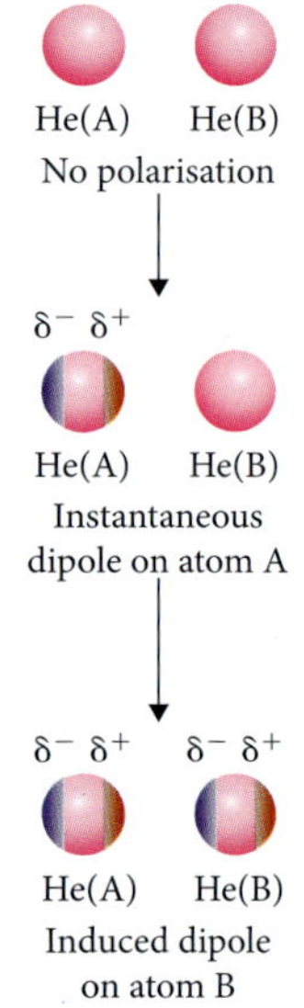

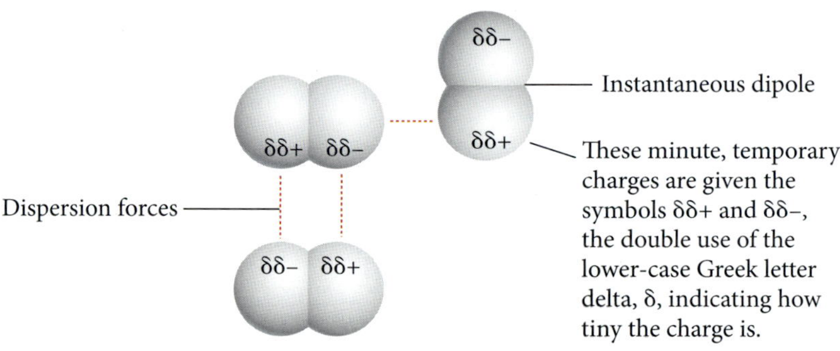

Figure C6.16 ▶ The dispersion forces between hydrogen molecules in solid hydrogen, showing instantaneous dipole–dipole interactions

Dispersion forces are due to instantaneous temporary dipoles that occur with the movement of electrons. The more electrons present (the larger the molecular weight), the bigger the dispersion force will be between the molecules.

A gecko's amazing ability of their toes to stick to almost anything relies on dispersion forces between each seta and the surface. With a large surface area for dispersion forces to occur, the overall force can be strong.

Table C6.2 Boiling points of noble gases and halogens

Noble gas	Boiling point (°C)	Halogen	Boiling point (°C)
Helium	−269	Fluorine	−188
Neon	−246	Chlorine	−34.6
Argon	−186	Bromine	58.8
Krypton	−152	Iodine	183
Xenon	−108	Astatine	337
Radon	−62		

INTERMOLECULAR FORCES AND BOILING POINTS

Visit this website to learn how intermolecular forces affect boiling points.

6.5 The importance of bonding

Bonding is responsible for many of the properties of materials and biomolecules. The nucleic acids in the DNA double helix are held together by hydrogen bonds. This allows them to unzip easily, enabling replication and transcription. Hydrogen bonding helps maintain the structures of proteins, enzymes, ribosomes and cell membranes.

Analytical techniques such as chromatography work because of surface interactions and the intermolecular bonds of different substances to different surfaces or solvents. Utilising this knowledge allows chemists to determine the organic substances in mixtures.

To learn more about in the interaction between molecules, refer to Chemistry section 8.2 on page 346.

Water and hydrogen bonding

There are three types of bonding in water. Strong covalent bonding exists between the H and O within the molecule. Hydrogen bonding and dispersion forces occur between the water molecules. Hydrogen bonding is responsible for the unique properties of water.

The unique properties of water include unusual melting and boiling points, density in solid and liquid phases, **surface tension** and an ability to act as a very good solvent. Hydrogen bonding explains all the properties that relate to temperature. You need a relatively large amount of energy to break the bonds in water, which is why water has a relatively high boiling point, 100°C. For a certain amount of energy, fewer molecules escape into the air. The fewer vaporised water molecules means lower **vapour pressure**, the downwards pressure exerted by the gas molecules on the surface of the liquid.To change state from liquid to gas (high latent heat) or to increase the average temperature of the water (high specific heat) energy must be absorbed. The change is endothermic.

There is a negative correlation between intermolecular force strength and vapour pressure. Compounds that have strong intermolecular forces will have low vapour pressure.

The surface tension of water is high as hydrogen bonding occurs in all directions and is relatively strong. It holds the molecules together.

The cohesive nature of water helps plants take up nutrients from the soil. Root hairs draw up water containing dissolved nutrients such as nitrogen, phosphorus and potassium from the soil. **Capillary action** helps the water to move up the xylem of the plant to the leaf surface. Capillary action relies on the adhesion of water molecules to the sides of the xylem vessel and the cohesive properties of water. The cohesion of water molecules means dissolved nutrients are pulled all the way up tall trees, defying gravity.

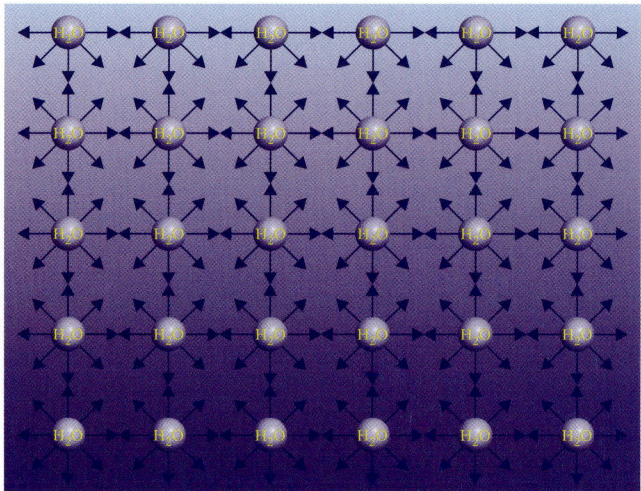

Figure C6.18 In the bulk of liquid water, each molecule has forces operating in all directions. However, on the surface, there is an imbalance – the liquid molecules are attracted to each other and exert a net force that pulls them together.

Capillary action is due to three main forces. These are the:

- **cohesive force** – the intermolecular force between molecules in a substance that helps to maintain a certain shape of the liquid.
- surface tension – this is due to the cohesive forces at the surface of material and results in the surface of the fluid being under tension
- **adhesive force** – these are the forces of attraction between unlike molecules.

Capillary action can be seen when part of a towel is dipped in water – the whole towel eventually gets soaked. Capillary action can also be seen in narrow glass vessels, where adhesive forces between water molecules and the surface of the glass and cohesive forces in water cause the surface of the water to curve to form a **meniscus**. The water continues to be drawn up, until gravity overcomes the adhesive force.

To see how hydrogen bonding affects water, refer to Context 3, 'Water: the vital substance', page 67.

INVESTIGATION 6.1

CAPILLARY ACTION

Aim

To determine the relationship between the diameter of a tube and the height of the water column

What do you need

- Beaker with coloured water
- Glass tubes with diameters in the range 1–4 mm

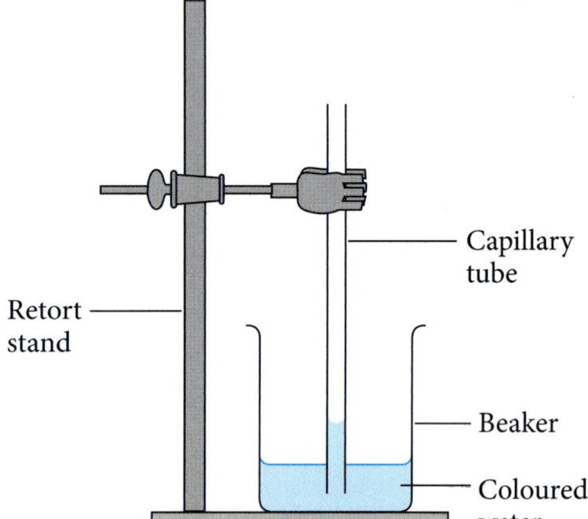

Figure C6.19 Experimental set-up for testing capillary action

What to do

1 Complete the following risk assessment to identify any risks that may occur during the investigation and ways to reduce them. Ask your teacher to check your risk assessment before you proceed.

What are the risks in doing this investigation?	How can you manage these risks to stay safe?

2 Place the glass tubes upright in the beaker of coloured water. The water should be drawn up the surface of the tubes.
3 Measure the height of the water column.

What did you discover?

1 Was there a correlation between the size of the capillary tube and the height of the water column?
2 Did you always notice the water curving up to form a meniscus? Was the surface tension always strong enough to support a column of water defying gravity?

Extension

Use the Internet and other resources to investigate why mercury has a concave meniscus. How do solutes in water affect capillary action?

As with all substances, as water cools, its particles slow down and pack closer together so the density increases. Therefore, cooler water has a higher density than warmer water. When the temperature is less than 4°C, the molecules do not have as much kinetic energy and no longer move between the spaces, but form more hydrogen bonds, as shown in Figure C6.20. The shape of the water molecule means a regular hexagonal shape is formed. Water has the unique property of expanding as it freezes. The density decreases as it cools from 3°C to 0°C. Ice is about 8% less dense than water.

Another property of water is its ability to act as a solvent. Pure water is liquid from 0°C to 100°C. This means that it can act as a solvent over this temperature range. Water dissolves and transports a range of material across the whole planet and also through each cell of a living organism. Water does this by forming hydrogen bonds or **ion–dipole** bonds with a wide range of substances from salts to organic molecules such as proteins.

To revise how the density of water and ice influences the physical and biological processes on Earth, refer to Context 3, 'Water, the vital substance', page 63.

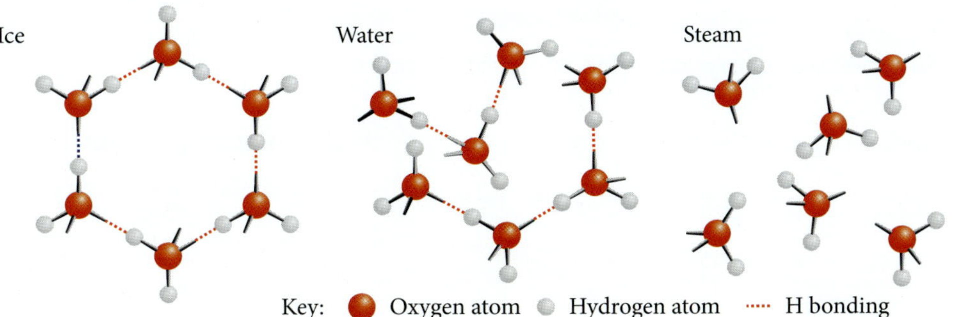

Figure C6.20 The structure of ice compared to that of liquid water and steam showing hydrogen bonding

An understanding of hydrogen bonding, dipole–dipole and dispersion forces is important because it allows chemists to predict properties such as melting and boiling points and whether one substance will dissolve in another substance.

Water of crystallisation

Water of crystallisation occurs when water molecules are attracted to the ions of a salt. Many ionic compounds that crystallise out of an aqueous solution include water molecules in a regular array through the crystal lattice. The water molecules are in set ratio to the ionic substance. These compounds are called hydrated. For example, $CuSO_4.5H_2O$ has five water molecules of crystallisation per unit of copper(II) sulfate. The bonds between the ions and water are called ion–dipole bonds; the ion and the dipole in the water molecules form a weak bond. Water or other molecules that form dipole bonds to a metal atoms are referred to as **ligands**. The water ligands take the positions around the central ion and will follow the VSEPR theory. $CuSO_4.5H_2O$ is an octahedral shape. The central ion is a cation, so the negative oxygen of water points towards the Cu^{2+} as shown in Figure C6.21.

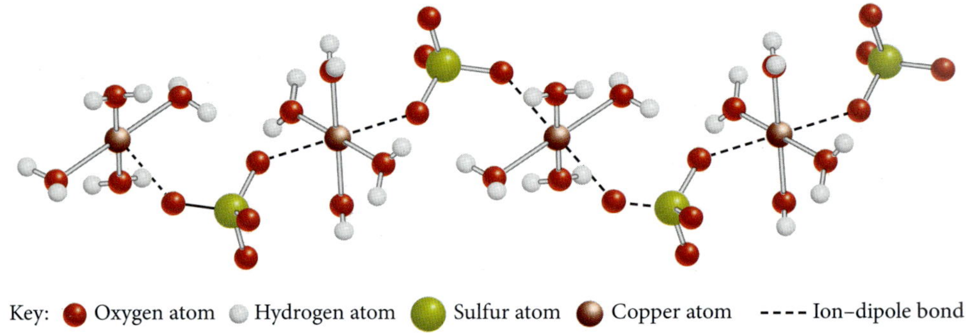

Figure C6.21 ▶ The crystals of copper(II) sulfate consist of an octahedral arrangement of water molecules and sulfate ions around the central copper ion.

The water of crystallisation can be evaporated by heating the hydrated compound. When all the molecules have been removed, the compound is said to be **anhydrous**. To do this, a known mass of the hydrated compound is heated so that the mass of water removed can be determined. From this, the empirical formula of the hydrate can be calculated.

EXPERIMENT 6.1

WATER OF CRYSTALLISATION

Aim

To determine the number of water molecules of crystallisation of hydrated sample

Materials

- Hydrated samples (2–3 g) of sodium carbonate, barium chloride, calcium chloride, copper(II) sulfate, magnesium sulfate
- Crucible and lid
- Desiccator
- Matches
- Balance
- Bunsen burner
- Pipe clay triangle
- Tripod
- Crucible tongs
- Heatproof mat

▶

What are the risks in doing this experiment?	How can you manage these risks to stay safe?
A hot crucible retains heat, especially if its contents have significant mass.	Avoid inhaling fumes when heating. Use tongs to handle the crucible; place the hot crucible on the heatproof mat. Do not touch the hot crucible. If you burn yourself, place the affected part under cold running water for 10 minutes and inform your teacher.
The Bunsen burner will get hot.	Do not use the Bunsen burner if the gas tube is damaged. Ensure long hair is tied back and the flame is away from flammable material. If you burn yourself, place the affected part under cold running water for 10 minutes and inform your teacher.
Barium salts are moderately toxic.	Avoid breathing fine dust. Wear safety glasses and wash your hands after the experiment. Dispose of all chemicals according to your teacher's directions. Do not pour them down the sink.
Copper(II) sulfate is slightly toxic	Wear safety glasses and wash your hands after the experiment. Dispose of all chemicals according to your teacher's directions. Do not pour them down the sink.
Ammonium chloride is slightly toxic.	Use in a well-ventilated area. Wear safety glasses and wash your hands after experiment. Dispose of all chemicals according to your teacher's directions. Do not pour them down the sink.
Sodium carbonate may be a slight skin irritant.	Wear safety glasses and wash your hands after the experiment.

In your write-up, add any more risks you can think of, as well as ways to manage them.

Procedure

1 Heat the clean, dry crucible and lid strongly for 5 minutes over a Bunsen flame. Place the crucible in the desiccator and allow it to cool.

2 Weigh the cooled crucible and lid on an accurate balance. Add 2–3 g of your hydrated crystals and immediately reweigh the crucible.

3 Place the lid on the crucible, leaving a small gap, and heat it for 5 minutes in a low flame. Then heat it strongly for 20 minutes. Allow the crucible to cool in the desiccator and then reweigh it.

4 Heat the crucible strongly for a further 5 minutes, cool it in the desiccator and reweigh it. Repeat until constant weight has been obtained; that is, to within 1–2 mg.

Results

Record your results in a table like the one below.

Mass of crucible and lid	
Mass of crucible, lid and hydrated salt	
Mass after heating of crucible, lid and salt	

Analysis of results

1 Calculate the mass of hydrated compound used (e.g. $MgSO_4.nH_2O$).

2 Calculate the mass of anhydrous compound remaining after heating, and the mass of water removed.

3 Refer to a table of relative atomic masses to calculate the molar mass of compound (e.g. $MgSO_4$ and water H_2O).

4 Calculate the amount of substance of water removed (i.e. $n(H_2O)$).

5 Calculate n(sample); for example, $n(MgSO_4)$, remaining after heating.

6 Determine the ratio of salt to water: $n(MgSO_4):n(H_2O)$ as a simple whole number ratio. Hence, write the empirical formula of hydrated compound in the form $MgSO_4.xH_2O$ and state the degree of hydration of the sample.

Discussion

1 What are the main sources of error in your experiment?
2 If your class analysed different compounds, collate the results for the whole class to determine the water of crystallisation for each compound.
4 Compare the experimentally derived formulas with the known values.
5 Which experimentally derived formulas were further from the known values?
6 Brainstorm reasons why there was a range of discrepancies; for example, did this correlate with the main sources of errors? Did the discrepancy correlate with the number of moles of samples each person had?

QUESTION SET 6.3

Remembering

1 Name the three:
   a strong bonding forces.
   b weak intermolecular forces.
2 Recall what causes dipole–dipole interactions and dispersion forces.
3 What are the common groups that exhibit hydrogen bonds?

Understanding

4 Explain why you should not put full water bottles in the freezer.
5 Rank the three intermolecular forces from weakest to strongest. Explain how polarity influences this order.
6 What are the two factors influencing a molecule's polar character?

Applying

7 If water did not exhibit hydrogen bonding, at what temperature should water boil?
8 Explain why H_2S, a heavier molecule with more electrons, has a lower boiling point than water.

Analysing

9 Water has the second highest specific heat capacity of all known substances. It takes a lot of energy to heat 1 gram of water by 1 degree. This is due to the strength of the intermolecular forces within water. Liquid ammonia has the highest specific heat. Explain why ammonia has a higher specific heat capacity than water.
10 Oxygen levels in water are higher near the surface. Oxygen is more soluble in colder water. These two facts do not contradict each other. Why?

Reflecting

11 Geckos have utilised dispersion forces to allow them to support their full weight with just one toe. If scientists could replicate the extraordinary ability of the gecko to attach and detach to smooth surfaces, explain how that knowledge could benefit our society.

6.6 Solutes

Water is considered to be the universal solvent. This is due to its polarity. Solutes are substances that dissolve in a solvent. The solution that is formed from a solute dissolving in a solvent is homogeneous (uniform), as the particles are too small to be seen.

Substances that readily dissolve in water are called **hydrophilic** ('water loving'). These include ionic compounds and polar substances such as some polar organic substances. Substances that don't dissolve in water are **hydrophobic** ('water hating'). These tend to be gases or non-polar substances.

Substances may dissolve in water if they are polar or charged. The solute may form an ion–dipole, dipole–dipole or hydrogen bond with water.

When a salt dissolves in water, ion–dipole bonds are formed. The positive part of the water molecule is attracted to the anion of the salt. More water molecules surround the anion until it is perfectly hydrated. The negative part of water is attracted to the cation and a similar process occurs until the ionic bonds within the salt crystal are broken and the salt has dissolved.

Water is polar and can dissolve polar and charged substances by forming ion–dipole, dipole–dipole or hydrogen bonds with the substance.

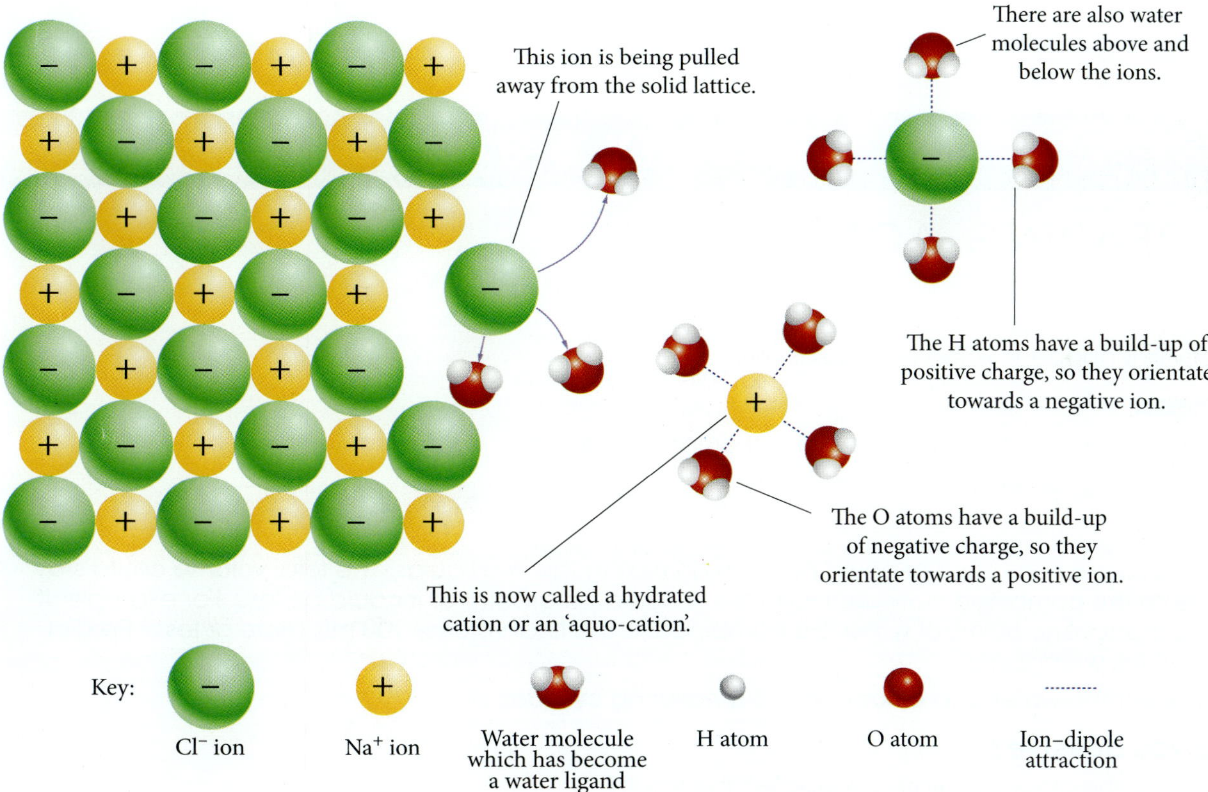

▲ **Figure C6.22**
A simplified model of the solution process for a sodium chloride crystal in water

Organic compounds as solutes

To revise how oil and water don't mix, refer to Context 3, 'Water, the vital substance', page 65.

Most organic molecules are non-polar and hydrophobic. When polar groups are attached to the carbon chain, the hydrophobic nature of organic molecules decreases. Groups that give polarity are –OH, C=O and –NH.

If a large proportion of these polar groups is present, then organic substances such as glucose, amino acids, vitamins, enzymes and hormones may dissolve in water. As you discovered earlier, ethanol dissolves in water because it has a polar end due to the OH group. Because it also has a non-polar end due to the CH_3, ethanol will also dissolve in organic solvents. Larger alcohol molecules that have longer carbon chains are not able to dissolve in water because the hydrophobic carbon chain predominates.

Non-polar substances dissolve in non-polar solvents but do not dissolve in polar solvents. Polar substances dissolve in polar solvents but do not dissolve in non-polar solvents.

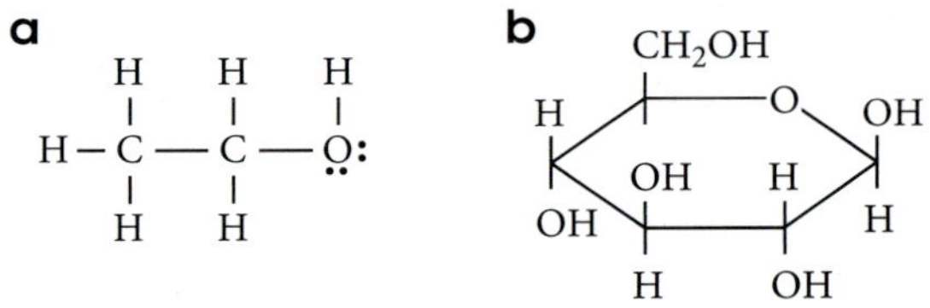

Figure C6.23 ▶ a) Ethanol and b) glucose molecules are polar due to the presence of polar regions.

ACTIVITY 6.3

DISAPPEARING ACT

Aim

To discover what happens when two liquids mix

You will need

- 2 accurately measured identical volumes of ethanol and water
- Large measuring cylinder

What to do

1. There are three options of what could happen on mixing the two liquids. The total volume could stay the same as the combined individual volumes, it could be greater or it could be less. For example, if 50 mL of ethanol and 50 mL of water are combined, is the final volume 100 mL, more or less? Predict what you think will happen.
2. Carefully pour the water and ethanol into a measuring cylinder.

What did you observe?

1. How many of the class accurately predicted the result?
2. Would you expect the same result if ethanol was poured first, or poured faster or slower. What causes any bubbles you may have seen? Can you repeat the experiment without bubbles?
3. Was this a chemical reaction? What must occur for it to be a chemical reaction? Did you have any evidence for this?
4. What interaction could occur between the molecules?
5. Summarise what must have happened as the two liquids mixed.

Other factors influencing solubility

For polar substances to dissolve in water, the energy required to separate the particles has to be less than the energy released when the particles are hydrated. The temperature of the solution alters both the rate at which a substance will dissolve and the amount that can dissolve. If a substance reacts with water, then the solubility is higher.

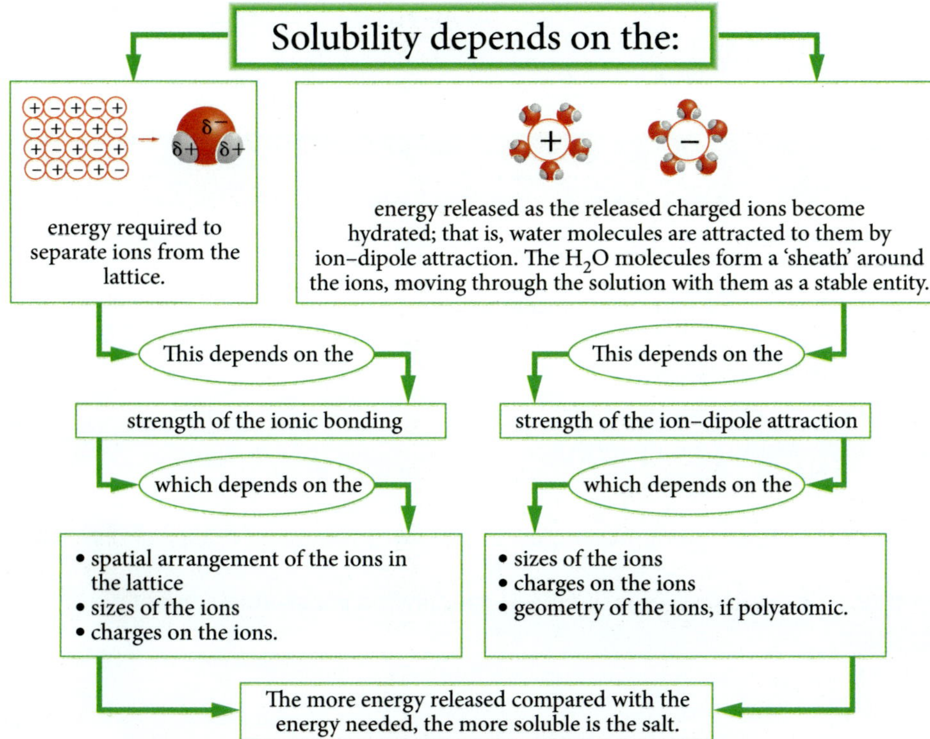

◀ **Figure C6.24** The role of energy factors in controlling the solubility of ionic compounds

Solutions equations (ionic and molecular)

You can tell a lot about what is happening in a reaction from the states in a chemical equation. When salts dissolve, chemists don't generally write the ions separately. For example, dissolved sodium chloride is written as NaCl(aq). The state aqueous (aq) indicates that the ions have **dissociated**. The Na^+ and Cl^- ions are separated, and surrounded by water molecules, as you saw in Figure C6.22. As these are charged ionic particles, the solution can conduct electricity; it is an **electrolytic solution**. Two examples for the dissociation of salts are:

$$NaCl(s) \xrightarrow{H_2O} Na^+(aq) + Cl^-(aq)$$
$$Cu(NO_3)_2(s) \xrightarrow{H_2O} Cu^{2+}(aq) + 2NO_3^-(aq)$$

The equations have to be balanced for charge and number of atoms of each element. The solid has no net charge, so the total combined charge of the dissociated ions has to be zero.

When molecular substances such as carbon dioxide, hydrochloric acid (HCl) or glucose ($C_6H_{12}O_6$) dissolve, the state is aqueous. When acids such as HCl ionise in water, they are written as HCl(aq). This means that the solution contains separate H^+ and Cl^- ions.

$$C_{12}H_{22}O_{11}(s) \xrightarrow{H_2O} C_{12}H_{22}O_{11}(aq)$$
$$HCl(aq) \xrightarrow{H_2O} H^+(aq) + Cl^-(aq)$$

You will learn more about the reactions that occur when substances such as HCl or CO_2 dissolve in water in Chemistry Chapter 7.

After a substance becomes aqueous, further reactions may take place. Carbon dioxide has to dissolve in water before it can react with the water. The reaction with water is written separately. Oxygen has to dissolve for fish and other aquatic organisms to be able to use it.

The ability to dissolve in water may indicate that the substance has polar regions or a that reaction has occurred with water. Typical reactions are ionisation or dissociation reactions.

A solute dissolves when the energy of the bonds it forms with water is lower than the energy of the bonds between water molecules or between the water molecules and ions of the substance being dissolved. The energy difference explains why some polar salts dissolve in water and some don't.

QUESTION SET 6.4

Remembering

1 Draw a diagram that shows sodium chloride (NaCl) dissolving in water.

2 Write the formula for:

- a lithium bromide (LiBr) crystallising as a dihydrate.
- b calcium nitrate ($Ca(NO_3)_2$) crystallising as a tetrahydrate.
- c magnesium nitrate ($Mg(NO_3)$) crystallising as a hexahydrate.

Understanding

3 Hydrogen bromide and methane are small covalent molecules. Explain why methane does not dissolve in water but hydrogen bromide does.

4 Describe what happens when sugar is added to a cup of coffee. Identify the solute and solvent.

5 KBr dissolves in water by dissociation.

- a What is meant by this statement?
- b Write an ionic equation to show what happens.
- c Would this solution be called an electrolyte?

6 Describe what happens on the molecular level when water and ethanol mix. Explain by using the terms 'solute' and 'solvent'. What bonds are broken and what ones are formed? Write an equation that shows this.

Applying

7 Copy and complete the following table.

Bonding type	Solubility in water	Examples
Ionic	Most are soluble	**a**
b	Soluble if hydrogen bonds possible	Ethanol, glucose
	Soluble by reacting with water	HCl, HNO_3, NH_3
	Otherwise insoluble	Dichloromethane
Non-polar molecular	**c**	O_2, I_2
	Most are insoluble	**d**
Very large molecules	Insoluble if highly structured	**e**
	f	Starch, glycogen, enzymes
g	**h**	Diamond, SiO_2
Metals	**i**	**j**
	Unless they react with water	Li, Na, K, Ca, Ba

8 Classify the following as either soluble in water or not soluble. Explain your reasoning.

a HBr
b $MgCl_2$
c CS_2
d $ZnCl_2$
e H_2SO_4
f $SiCl_4$
g Glucose (a sugar, $C_6H_{12}O_6$)
h C_8H_{18}
i NH_2CONH_2

Analysing

9 A student was determining the water of crystallisation of magnesium sulfate. Unfortunately, he couldn't read his initial weights. In his rush to share the results with his prac partners he introduced errors.

His calculations and results are shown below. What are the errors and what was the initial weight for the anhydrous sulfate?

Mass of anhydrous magnesium sulfate = ? Dried weight = 1.65 g

Empirical formula:	$\frac{1.65}{120}$	:	$\frac{3.228}{138}$	convert mass to moles
	0.13749	:	0.02339	simplest ratio
	1		1.7	whole number ratio
	2		3	$2MgSO_4.3H_2O$

Reflecting

10 Enzymes are denatured when their shape changes with high temperatures. However, some organisms can survive at temperature above 60°C. Enzymes in all organisms are made from the same building blocks (amino acids). What differences would be expected in the structure and bonding of these enzymes?

6.7 Measuring solubility

The amount of a substance that dissolves depends upon temperature. The **solubility** of a solute is the maximum amount in grams that can dissolve in 100 g of the solvent at a given temperature.

When adding sugar to a cup of water, there reaches a point at which no more sugar can dissolve at a particular temperature. The extra sugar stays at the bottom of the cup. The solution is said to be saturated with sugar. Before this point was reached, it was an **unsaturated solution**.

If the water is heated, then more sugar will dissolve. If this heated solution is allowed to cool, crystals will start to precipitate again. The temperature at which this happens is the temperature at which the solution is saturated. A solution can become supersaturated. This means that there is more solute dissolved than in a **saturated solution** at the same temperature. If this **supersaturated solution** is bumped, a sugar crystal is added or the side of the glass is scratched, then the extra sugar will precipitate out again. This is because the solution is unstable. Toffee is a supersaturated solution of sugar. To stop the toffee crystallising, a mixture of sugars or the addition of some fats can be used to stabilise the solution.

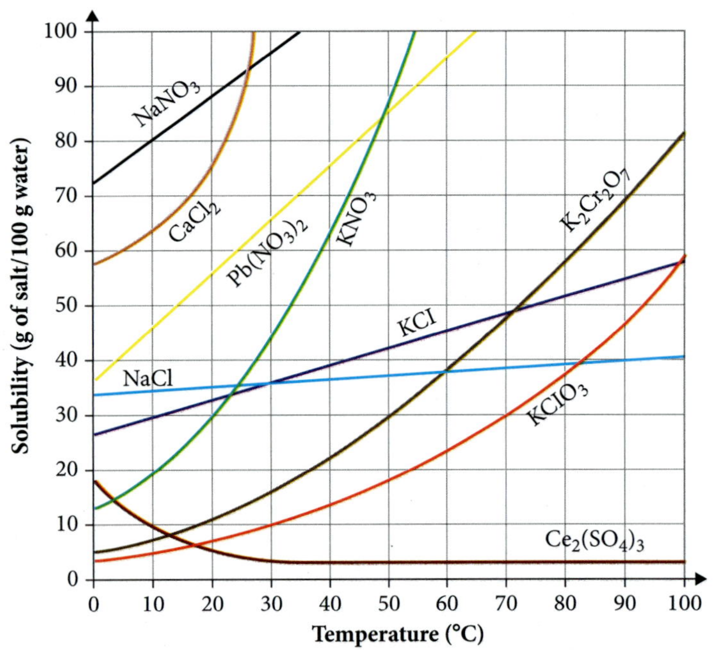

◀ Figure C6.25
A selection of solubility curves for various salts

Solubility curves

A graph showing how much of a solute such as a salt can be dissolved at a particular temperature provides useful information on the solubility of a substance. Generally, as temperature increases more salt will dissolve. This graph is called a **solubility curve**, as shown in Figure C6.25. The amount of solute is measured in grams of solute in 100 grams of water.

WORKED EXAMPLE 6.1

Refer to the solubility curve in Figure C6.25.

a How much potassium dichromate ($K_2Cr_2O_7$) will dissolve at 20°C?

b How much extra will dissolve at 80°C?

Answers	**Logic**
a At 20°C, approximately 12 g of $K_2Cr_2O_7$ will dissolve in 100 g of water.	Find the curve for $K_2Cr_2O_7$ on the graph. Draw a line from 20°C up to the potassium dichromate curve. Then draw a line from this point across to the *y*-axis. This will indicate the amount that will dissolve at 20°C.
b The extra amount that would dissolve at 80°C is 46 g.	Similarly, if the temperature is increased to 80°C, approximately 58 g of the substance will dissolve in 100 g (or 100 mL) of water. 58 − 12 = 46 g This is also the amount that would crystallise out of solution if the temperature of the saturated solution was cooled from 80°C to 20°C.

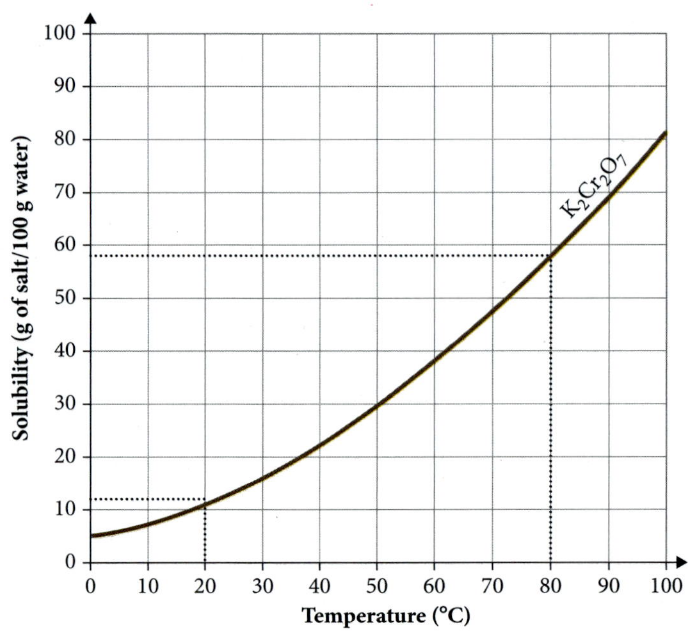

Figure C6.26 ▶
Calculating the amounts of potassium dichromate ($K_2Cr_2O_7$) that can dissolve at 20°C and at 80°C. The difference is how much extra or how much potassium dichromate will crystallise out of the solution on lowering the temperature from 80°C to 20°C.

Try these yourself

a What mass of the following solutes will dissolve in 100 mL of water?

- **i** $K_2Cr_2O_7$ at 50°C
- **ii** NaCl at 100°C
- **iii** $NaNO_3$ at 10°C

b Determine which of $K_2Cr_2O_7$, NaCl and $NaNO_3$ is most soluble in water at 15°C.

c At 30°C, what mass of KCl and NaCl would just dissolve in 100 mL of water. For this mass at this temperature, would the solution be saturated, unsaturated and supersaturated?

The solubility of a substance in a particular solvent is the maximum amount of that substance that can dissolve at a particular temperature.

A saturated solution is one in which no more of a particular solute can dissolve in a given quantity of solvent. A solution with less than this amount is an unsaturated solution.

A supersaturated solution is unstable and has more than the maximum amount of solute dissolved in a given quantity of solvent. It has a higher amount of solute than the saturated solution.

EXPERIMENT 6.2

DERIVING THE SOLUBILITY CURVE OF AMMONIUM CHLORIDE

Aim

To gather data to draw a solubility curve of ammonium chloride

Materials

- 250 mL beaker
- 10 mL measuring cylinder
- Large test tubes
- Bunsen burner
- Tripod
- Retort stand
- Bosshead and clamp
- 0–100°C thermometer
- Stirring rod
- Distilled water
- Balance
- Weighing bottle
- Safety glasses

What are the risks in doing this experiment?	How can you manage these risks to stay safe?
Ammonium chloride is slightly toxic.	Work in a well-ventilated area. Wear safety glasses and wash your hands after the experiment. Dispose of all chemicals according to your teacher's directions. Do not pour them down the sink.
The Bunsen burner will get hot.	Do not use the Bunsen burner if the gas tube is damaged. Ensure long hair is tied back and the flame is away from flammable material. If you burn yourself, place the affected part under cold running water for 10 minutes and inform your teacher.
Broken glassware will cut.	Inspect and discard any chipped or cracked glass wear, no matter how small the damage. Sweep up broken glass with a brush and dustpan; do not use your fingers.

In your write-up, add any more risks you can think of, as well as ways to manage them.

Procedure

1 Carefully weigh out the amount of ammonium chloride assigned to your group.
2 Place the weighed ammonium chloride into a test tube.
3 Add 10 mL of distilled water.
4 Half-fill the beaker with tap water, clamp the test tube securely into the beaker so that it is immersed in the water.
5 Warm the beaker, stirring the contents of the test tube with the stirring rod. Continue until the ammonium chloride has dissolved completely.
6 Allow the test tube to cool. Continue to stir it until tiny flakes of ammonium chloride start to appear. Note the temperature at which this occurs.

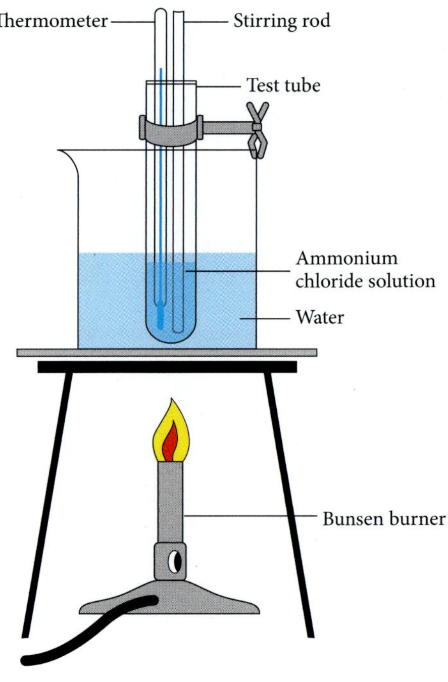

▶ **Figure C6.27**
Experimental set-up for determination of a solubility curve

Results

Copy and complete the following table.

Group	Mass of ammonium chloride (g)	Mass of water (g)	Temperature of recrystallisation (°C) (*x*-axis)	Solubility (g/100 g H_2O) (*y*-axis)
1	4.00	10.00		
2	4.50	10.00		
3	5.00	10.00		
4	5.50	10.00		
5	6.00	10.00		
6	6.50	10.00		

Analysis of results

1 Use the class results to plot a graph of solubility against temperature. Plot the temperature along the horizontal *x*-axis (0–100°C).
2 From the graph, predict the solubility of ammonium chloride at 20°C, 40°C, 60°C and 80°C.

Discussion

1 Describe what happens to the solubility of ammonium chloride as temperature increases.
2 At the point the crystals first appear, the solution is saturated. What does 'saturated' mean?
3 At what temperature would 10 g of ammonium chloride saturate 50 g of water?
4 If a saturated solution of ammonium chloride at 90°C is cooled to 10°C, how many grams of ammonium chloride would crystallise out of 100 g of water?
5 If the theoretical value for the solubility of ammonium chloride at 50°C is 50 g/100 g, what percentage error does your experiment have?

$$\text{Percentage error} = \frac{\text{experimental value} - \text{true value}}{\text{true value}} \times 100$$

6 List possible sources of errors in your experiment.

Taking it further

Describe modifications to the experiment that could improve the accuracy.

Gas solubility curves

The solubility of a gas can also be shown with a solubility curve. The solubility curve for oxygen and carbon dioxide was shown in Context 3 (page 71). As temperature increases, the solubility of non-polar gases generally decreases. Solubilities are generally much lower than for solids and typically the values are shown in parts per million (ppm).

In section 1.9 on page 127, you looked at ppm for a calibration curve of atomic absorption spectroscopy. The unit ppm is very useful when referring to the amount of the solute at very low levels.

Concentration units

Solubility curves use the unit grams of solute per 100 g of water. This can also be called % w/w. If you are weighing the amount of salt added to 100 g of a food, then the % w/w is the easiest to use.

Two other similar units use volumes (% v/v) or a combination of weight and volume (% w/v).

The units you use depend on what you are measuring. For example, you may wish to investigate the alcohol content of drinks. Ethanol is a liquid, so it is easier to measure its volume than its weight. The unit to use is % v/v. The alcoholic content of wine, beer and spirits is measured in volume of ethanol in 100 mL of drink. Note that the volume is 100 mL not 1 litre, which is the standard volume of measurement in chemistry. Having a different unit means that we can use percentage values. This is useful for different-sized containers such as a 750 mL or 1.25 L bottle.

In section 7.2 on page 312, you will learn about another way to measure concentration that uses moles and molarity.

Concentration is a measure of the amount of solute present. There are a variety of units in general use; g/100 g, % v/v (per cent by volume), % w/w (per cent by weight), % v/w, ppm (parts per million, grams of solute in one million grams of solution.)

WORKED EXAMPLE 6.2

a Red wine has an alcohol content of approximately 13 v/v %. In a 750 mL bottle, what amount of alcohol is present?

b A hair sample was found to have a copper level of 41 ppm. It should be less than 2.5 mg % w/w. Is this an acceptable level?

Answers	Logic
a 97.5 mL of ethanol in 750 mL of red wine	Total volume × percentage alcohol $750\text{ mL} \times \frac{13\text{ mL}}{100\text{ mL}} = 97.5\text{ mL}$
b No; 41 mg % w/w is too high.	**1** Convert to an amount in grams. $41\text{ ppm} = 41\text{ g per }10^6\text{ g}$ $= \frac{41 \times 100}{10^6}$ $= \frac{41}{10^4}$ **2** The legal limit has the % w/w as a mg value so convert to milligrams. $= 41 \times 10^{-6} \times 100$ % w/w $41 \times 10^{-6} \times 100 \times 1000 = 4.1$ mg % w/w No, level is too high.

Try these yourself

a 200 mL of a sport drink has 6% w/v of sugar. What amount of sugar is present?

b A 300 mL glass of champagne contains 36 mL of alcohol. What is the concentration (v/v) of alcohol?

c A solution has a cadmium ion level of 500 ppm. How much cadmium would be present in 100 mL of this solution?

It is helpful to remember that ppm = mg L^{-1} = μg g^{-1}

ACTIVITY 6.4

SPORT DRINKS: HOW MUCH SALT CAN YOU DRINK?

Too much salt in our diet can cause health problems. Salt is a common additive in many foods and drinks. Sport drinks are designed to replace salts lost in sweat. If the salt level in the drink is too low, then the drink is hypotonic. If the salt level is too high, then the drink is hypertonic.

Aim

To determine the minimum amount of salt that you can detect in a glass of water.

You will need

- 6 clean glasses
- Table salt (NaCl)
- Electronic balance
- Tap water
- Spoon
- Bucket or waste container
- Marker pens or coloured stickers

All equipment must be food grade and not the general laboratory supplies.

What to do

1 Do not perform this activity in a laboratory because it requires you to taste the samples.

2 Prepare six separate 500 mL water samples that have a range of salt levels from 0 to 5 g of added salt. For a comparison, brackish water has 3 g L^{-1} and sea water 35 g L^{-1}. Ensure that the salt is completely dissolved in each sample. Use marker pens or stickers to distinguish the samples.

3 Pour small amounts of each sample into glasses. Give the samples to your partner in order from no salt to most salt or from high salt to low salt. Make sure that the amounts that they consume are very small, typically less than a few millilitres. Have fresh water to drink in between if required. Do not swallow the sample but spit it into the bucket or waste container for disposal.

What did you discover?

1 Did different people have different detection limits?

2 Do you think that people were influenced by what they observed or the order they were given?

3 Which samples could you drink readily?

4 Convert the amounts in your samples to % w/v and ppm.

5 Why is the unit % w/v better to use than % w/w in this case?

Taking it further

1 Repeat the experiment with sugar. What is the minimum of sugar that can be detected?

2 Research why and how manufacturers chase the 'bliss point' in food.

QUESTION SET 6.5

Remembering

1 When a solvent holds as much of a solute as it normally can at a given temperature, we say the solution is ______________________________. When a solvent has less dissolved solute than it normally can at a given temperature, we say the solution is ______________________________.

Applying

2 Samples of brackish waters were found to have 30 g L^{-1} of salt. How much salt can be obtained from:

a 50 mL?

b 300 mL?

3 Bottles containing 40 mL of brand X, a liquid plant food, contains 3.2 g of seaweed extract. What is the percentage concentrate of seaweed extract in the plant food?

4 The fluoride concentration of drinking water should be less than 1 ppm on a mass to volume basis. A 500 mL sample of water was found to contain 0.3 mg of fluoride ions.

a Calculate the concentration of fluoride ions in the water in % w/v and parts per million.

b Is this below the recommended level?

5 How many grams of potassium nitrate would be needed to saturate 50 g of water at 20°C?

6 Sodium thiosulfate will dissolve in water readily. At 25°C, a saturated solution will have 120 g of $Na_2S_2O_3.5H_2O$ for each 100 g of water.

a You take a test tube and add a crystal of sodium thiosulfate to 2 cm^3 of water. Is it likely to be saturated?

b You take another test tube and add a similar amount of water and then add crystals until no more dissolves. What term describes this?

c In a third test tube, you add 2 cm^3 of crystals and a few drops of water. Why would you not expect all the crystals to dissolve?

d You heat it until the solution is clear and then cool it back to room temperature. What would happen if a crystal were now added to the test tube?

7 A 20 g solution of a pesticide contains 0.5 g of active ingredient. What is the concentration in % w/w and ppm.

8 On the bottle of every sport drink is a label detailing the amount of salts and sugars present, usually in 1 litre of product. Compare sport drinks compositions. If you had to make 750 mL of a sport drink, what and how much would you add to 750 mL of water?

9 A 200 mL sample of tap water was found to contain 0.2 mg of fluoride ions.

a Calculate the concentration of fluoride ions in the water as ppm.

b As the limit should be between 0.7 and 1.2 ppm, does this concentration exceed the recommended limit of 1 ppm?

c The recommended maximum daily intake of fluoride is about 10 mg/day. How many cups would a person need to drink to reach this limit? Assume that water is the only source of fluoride.

Reflecting

10 A gluten-free diet is considered to be the only treatment for people with coeliac disease. Australian manufacturers of gluten-free products must ensure that no detectable amounts of gluten are present. Manufacturers are pushing the regulators to make Australian standards the same as those in Britain and Europe – a maximum of 20 mg of gluten per kilogram. Why is this so important for a person with coeliac disease?

For some examples of the benefits of the practical applications of water, refer to Context 3, 'Water, the vital substance', pages 66 and 69.

6.8 Practical applications

Many practical uses of water rely on its high surface tension and ability to act as a solvent. These applications rely on the surface interaction and intermolecular forces between polar and non-polar substances.

Surfactants: polar and non-polar

Oils and water don't mix; non-polar liquids will not mix with polar solvents. These two substances are **immiscible**; they are unable to be mixed. When a **surfactant** is added to the mixture, the tension between the oil and water is reduced. They appear to dissolve in each other. The result is an **emulsion**, a mixture of the two immiscible substances. The compound that helps form the emulsion is called an emulsifier. Eggs are a good emulsifier for oil and vinegar. The resultant emulsion is called mayonnaise. Milk is considered to be an 'oil in water' emulsion. When whipped sufficiently, it turns into a 'water in oil' emulsion, called butter. The excess water is separated in the process.

ACTIVITY 6.5

EMULSIONS, COLLOIDS AND SUSPENSIONS

When substances are added to a liquid they do not always form a solution. They may form an emulsion, a **colloid** or a **suspension**.

1 What distinguishes these three mixtures?
2 What are the common uses of these mixtures? Give examples that you may see around your home.

Detergents: surfactants that clean

Soaps and detergents are everyday examples of surfactants. One end of a surfactant is hydrophilic and the other is end hydrophobic. Soap has a polar head (typically a sodium ion) attached to a non-polar long hydrocarbon chain tail. Soap forms **micelles**, sub-microscopic clusters in water, which have polar ends on the outside and hydrophobic tails in the middle where they can avoid water. The long tail of the soap dissolves into the grease because of dispersion forces. There is an ion–dipole interaction between the polar head and the water molecules. More and more hydrophobic tails of the soap molecules will tunnel into the grease droplet, eventually incorporating it into the water.

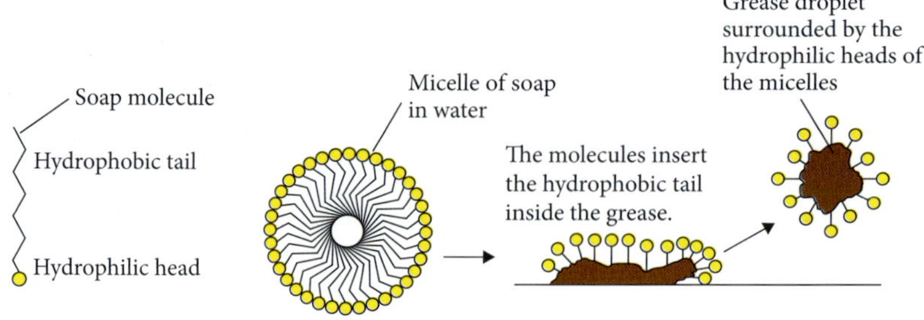

Figure C6.28 ▶ Soap molecules interacting with grease. Over time, more and more soap molecules are incorporated into the grease molecule. Eventually, a small grease droplet is surrounded by the hydrophilic heads of the soap molecules, forming a micelle.

The surface tension of water can be lowered by dissolving surfactants such as soap or detergent. When present, the surfactant breaks the hydrogen bonding between the molecules and the water spreads out onto the surface (wets the surface) rather than remaining in a droplet.

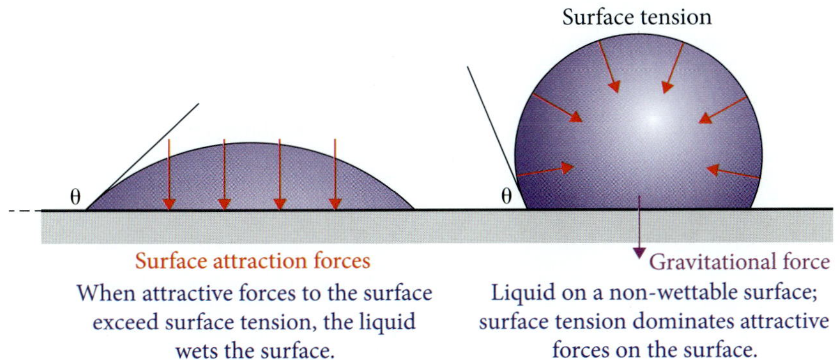

◀ **Figure C6.29**
Wettability: a liquid will wet a surface if there is strong intermolecular attraction between the liquid and the surface. A liquid will bead on the surface if there are stronger cohesive forces within the liquid than the adhesive forces between liquid and the surface. The contact angle (θ) is a way of measuring the wettability.

ACTIVITY 6.6

WETTABILITY

Aim

To determine the difference in bonding of water to various substrates such as plastic, glass and nanotechnology cloth and how this bonding is affected by detergents

You will need

- Dropper.
- Clean glass
- Plastic
- Water-resistant nanotechnology material
- Detergent

What to do

1. Place a few drops of water onto each of the surfaces and notice the shape of the droplet.
2. Add a single drop of detergent to each drop of water.
3. Record what happens.

What did you discover?

Describe what happened when you added the detergent to each drop.

Soap is one of the oldest surfactants. It is made from fatty acids in animal and vegetable oils. The sodium or potassium salts of these fatty acids have the general formula $RCOO^-Na^+$, where R represents a long carbon chain. Sodium stearate is a soap made from the fat of cattle or sheep reacting with a strong base such as NaOH. The **saponification** reaction is:

$$\underset{\text{Fatty acid (oleic acid)}}{C_{17}H_{33}COOH + NaOH} \xrightarrow{\text{heat}} \underset{\text{Base soap}}{R{-}COONa^+} + \underset{\text{Alcohol}}{R'OH}$$

The soap formed has a polar head ($COONa^+$) and a long non-polar hydrocarbon tail (R). The polar head is strongly attracted to the water molecules and the long tail is hydrophobic. The tail will dissolve into the grease or oil. Detergents are used in **hard water** rather than soap, which does not form a lather. Detergents have an ionic head based upon sulfate ions. These sulfate ions do not precipitate the calcium or magnesium ions in hard water as soap does. Detergents are classified according to the charge on their polar head (Figure C6.30). **Anionic** detergents have a negatively charged sulfate ion. **Cationic** detergents have a positively charged end. **Non-ionic detergents** do not have any ionic groups.

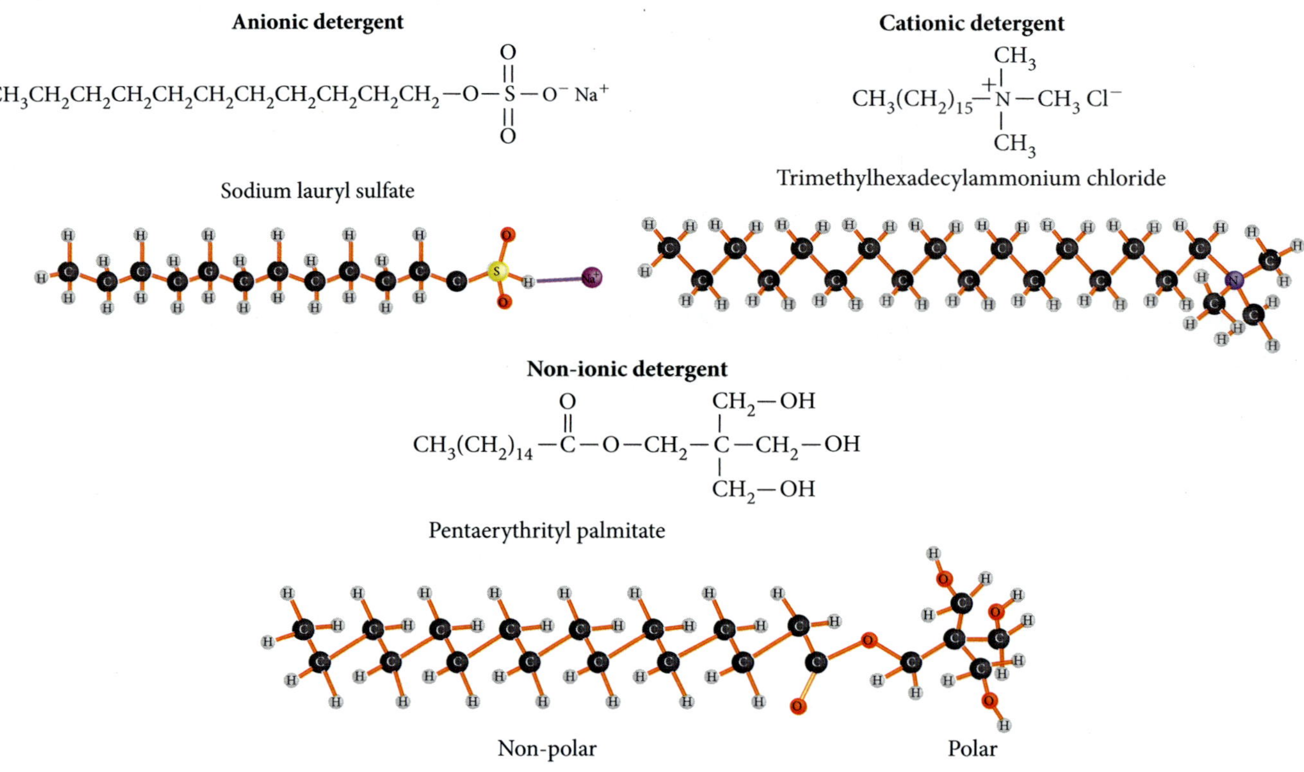

▲ Figure C6.30
The three types of detergent: anionic, cationic and non-ionic. What enables these molecules to dissolve in water?

WOW

No need to bathe or wash again?

Nanotechnology offers many advantages in a world becoming more concerned about water availability. By changing the surface of materials, there may be no need to clean them. Sweat or dirt will just not stick. A surface that beads the water, which then runs off, offers advantages to the builders of high-rise buildings, who may be daunted at the prospect of cleaning windows 40 floors up. Car paint companies are looking at applying nanotechnology to the outside of cars to remove the need to clean the car. Clothing companies are some of the biggest users of nanotechnology.

Surfactants such as detergents interrupt the surface tension of water. Surfactants have both polar and non-polar regions.

6.9 Chromatography

Knowledge of how solutes dissolve in solvents and interact with the surface of materials is the basis of the analytical techniques of chromatography. There are three main techniques: thin-layer chromatography, gas chromatography and high-performance liquid chromatography.

The chemical basis for these techniques is that different substances will **adsorb** onto a surface and **desorb** into a solvent at different rates. This difference is due to many factors of the substance being tested (**analyte**), including:

- the different types of polar groups
- the amount of charged and polar chemical groups present
- its molecular weight
- its geometry
- the positions and numbers of carbon–carbon double bonds.

By altering the solvent (**mobile phase**) and the surface (**stationary phase**), chemists can separate a mixture of solutes.

To see applications of chromatography in forensic testing, refer to Context 1, 'Matter in the universe', page 11.

To see applications of chromatography in blood and water testing, refer to Context 3, 'Water: the vital substance', pages 73 and 74.

Chromatography is the name of the group of techniques that separate substances based upon differential distribution between a stationary phase and a mobile phase.

INVESTIGATION 6.2

CHROMATOGRAPHY OF PLANT PIGMENTS

The plant pigment chlorophyll absorbs the light energy used in photosynthesis. When trying to discover how chlorophyll works, chemists had to first isolate chlorophyll, find out how many different types there are and determine whether all plants had the same pigment.

Aim

To separate plant pigments using chromatography

What will you need?

You will need to design a method to separate the plant pigments in plants by chromatography. Hint: Plant pigments will dissolve best in an ethanol and water mixture. Filter paper or chalk can be used to separate the pigments. Sand can help to break open the cells by grinding the plants. Calcium carbonate can neutralise acid released from the cells as they are crushed.

Complete the risk assessment to show any risks in undertaking this investigation and ways to reduce them. Ask your teacher to check your risk assessment before you proceed.

What are the risks in doing this investigation?	How can you manage these risks to stay safe?
Heating a mixture of ethanol and water will cause the beaker to get hot.	If you burn yourself, place the affected part under cold running water for 10 minutes and inform your teacher.
Ethanol is a toxic and flammable liquid.	Only directly heat solutions that have less than 50 % v/v of ethanol. Keep bottles away from flames and heat sources. Do not come into contact with the liquid or breathe the fumes.
Plants may be poisonous.	
Glassware	
Filter paper	

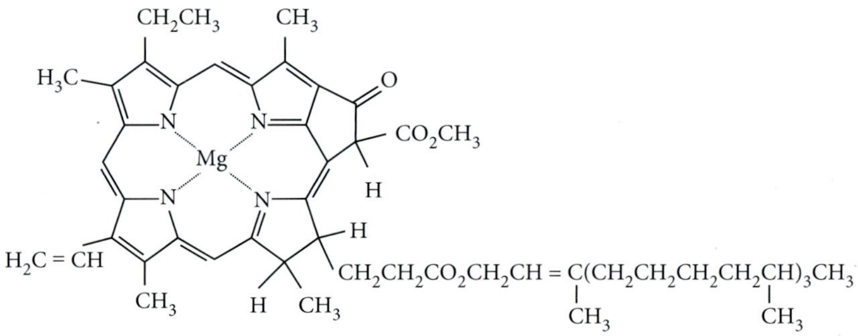

Chlorophyll a

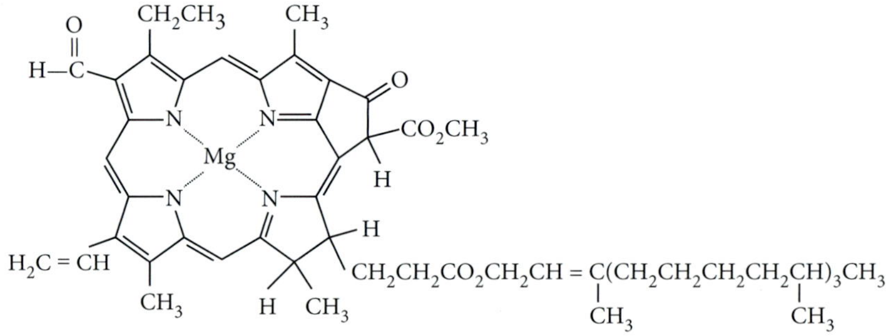

Chlorophyll b

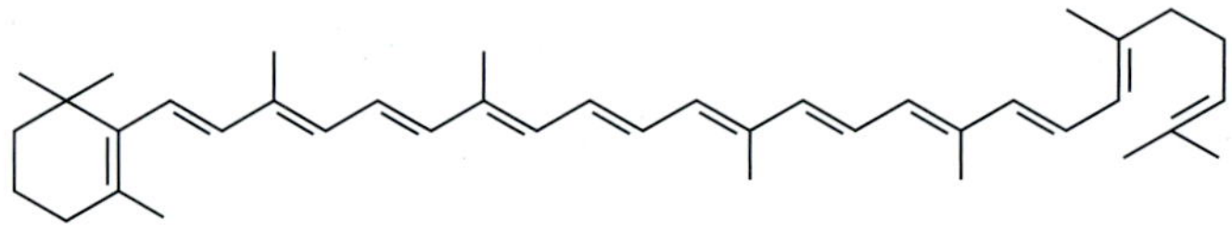

γ-Carotene

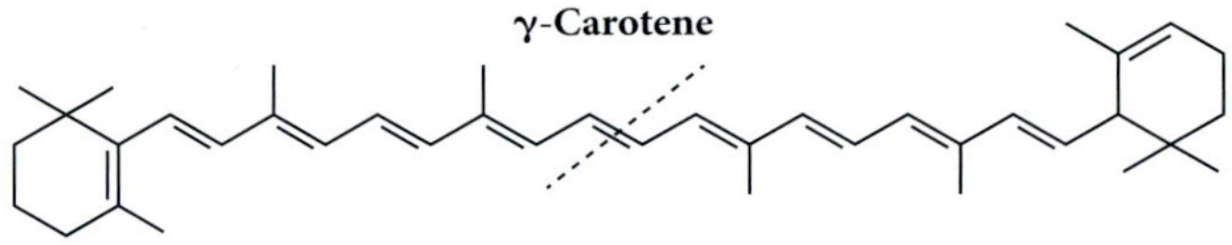

Figure C6.31 ▶ Pigments in plant cells absorb light energy. There are many different pigment molecules, each a different colour. Four are shown here.

Questions to get you started

1 How will you carry out your investigation? Which method will you try first?
2 Will large pieces of plant interfere with your technique? If so, what should you do?
3 How long will you boil the plants for? How will you judge whether you have achieved the maximum extraction?
4 How will you use the filter paper or chalk to separate the pigments from the solvent?
5 How fine will the plant pieces have to be? What factors would increase the amount of pigment released?
6 How will you analyse your results?

After the experiments

1 How well did you manage to extract plant pigments?
2 Did other students succeed in extracting plant pigments? Compare the effectiveness of their method with yours.

Ideas for improvement

1 What were some of the problems encountered? What errors were encountered?
2 How would you refine this method? How will the refined method provide more consistent results?
3 If you suspected that one of the components was chlorophyll, how would you confirm this?

Thin-layer chromatography

Thin-layer chromatography (TLC) is used to separate and analyse a wide variety of molecular mixtures. It is used to analyse the presence of particular drugs and amino acids; TLC is faster and provides more separation than paper chromatography.

The TLC plate typically consists of a 0.1-mm-thick layer of absorbent material bonded to a glass or plastic support. The absorbent material consists of many microscopic plates. These surfaces provide a large surface area for chromatographic separation. This is the stationary phase. The **origin** line is drawn in pencil about 1 cm from the base. This is where the samples are placed.

Separation is achieved by the solvent (mobile phase) moving up the stationary phase. This movement is due to capillary action. The solvent moves through the samples towards the top of the plate. The faint wavy line seen is called the solvent front. The samples will desorb into the solvent and travel up the plate with the solvent. If all components spent the same amount of time in the solvent, then they would all travel at the same rate. They don't because components have differences in chemical structure, which alters the adsorption and desorption rates. The particles most attracted to the stationary phase move the slowest. The particles that are most strongly attracted to the solvent move the fastest.

In TLC, you measure the distance the spot has moved from the origin compared to the distance the solvent has moved from the origin. This ratio is called the retardation factor or $\boldsymbol{R_f}$. The R_f value must be between 0 and 1 and is defined as the distance travelled by the sample from its origin divided by the distance of the solvent front from the sample origin. A substance that does not migrate from the sample origin has an $R_f = 0$, while one that is not adsorbed at all (migrated with the front) has a $R_f = 1$. The R_f value is characteristic for a particular substance with a given absorbent and solvent system. The R_f value cannot be greater than 1.

The separation of the components depends upon the length of time the plate is in contact with the solvent. If this is too short, then there is only a small separation. If this is too long, then all the spots would eventually make it to the top. A longer plate and a reasonable time will provide the best separation. It is not always possible to identify all the components of a sample with one TLC plate. Some samples may have very similar R_f values. Different solvent mixtures and stationary phases can be tried to achieve better separation.

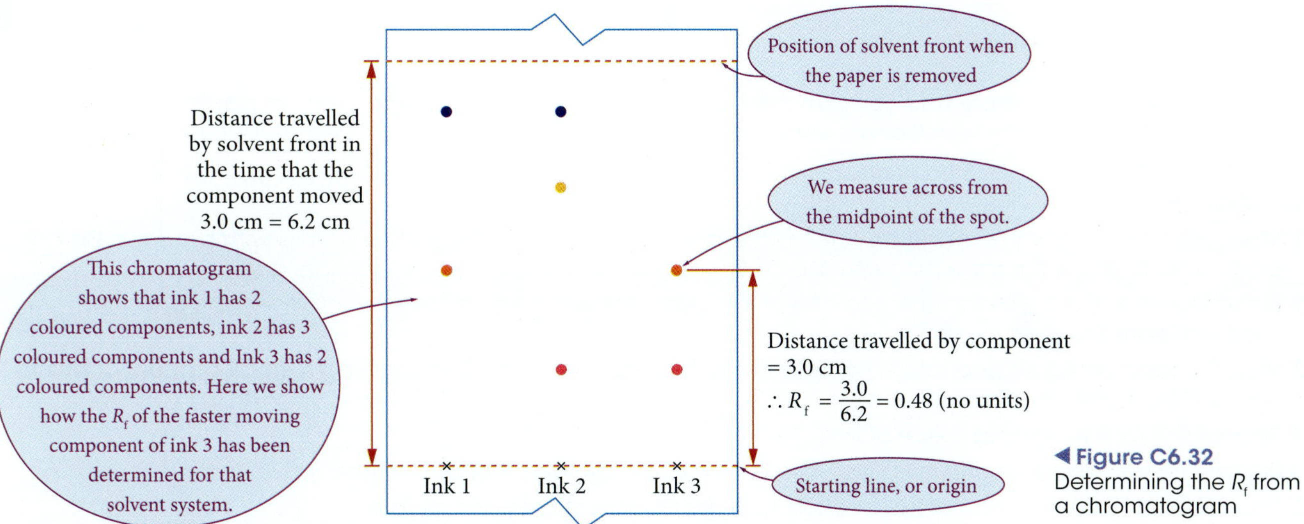

◀ **Figure C6.32** Determining the R_f from a chromatogram

The proportion of time an analyte spends adsorbed compared with desorbed is consistent for that mobile and stationary phase. A material with an R_f of 0.5 will move 2.5 cm when the solvent front has moved 5 cm. If a second plate was allowed to run for longer, then the same sample will move 5 cm when the solvent has moved 10 cm. Therefore, you can consider the samples to be the same when the R_f values are the same.

You can confirm the identity of an unknown compound by running a pure sample of the compound (a standard) under the same conditions. If the substances are the same, then the R_f values should be the same.

EXPERIMENT 6.3

THIN-LAYER CHROMATOGRAPHY

Thin-layer chromatography (TLC) is an invaluable technique used in chemistry and biochemistry for the separation and analysis of a wide variety of molecular mixtures.

In this experiment, two different solvent systems will be used to separate a mixture of dyes on a TLC plate. The solvent should be allowed to move up about half the distance of the plate. The dye molecules contain different types and amounts of charged and polar chemical groups. They also have different molecular weights, geometry and positions and numbers of carbon–carbon double bonds.

After the experiment is completed and the plate is partially dry, a faint wavy line can be observed at the last location of the leading edge of the solvent in the absorbent stationary phase.

Aim

To use TLC to separate mixtures of dyes to determine their composition

Materials

- 2 TLC plates (5 cm × 10 cm)
- 5 different-coloured food dyes
- Distilled water
- Methylated spirits
- 2 × 250 mL beakers
- 5 capillary tubes
- Pencil

What are the risks in doing this experiment?	How can you manage these risks to stay safe?
Methylated spirits is a toxic, flammable liquid.	Keep bottles away from flame and heat sources. Do not come into contact with the liquid or breathe the fumes.
Glassware	

In your write-up, add any more risks you can think of, as well as ways to manage them. In particular, expand on the two risks listed to identify specific risks involved with each. Ask your teacher to check your risk assessment before you proceed.

Procedure

1. Measure 1 cm up from an edge on each plate and lightly draw a straight line across with a pencil. Do not press hard or you will scrape the cellulose off its support. This line will be the sample origin.
2. Place the end of a capillary tube into one of the dyes. Allow the liquid to move into the tube and then remove it from the dye.
3. Hold the tube vertically and touch the end of the tube to the pencil line, allowing a small amount of dye to be adsorbed onto the film.
4. Repeat this for the second piece of film.
5. Repeat steps 2 and 3 for different dyes. Use a new capillary tube for each different dye.
6. Pour distilled water into one of the beakers to a depth of approximately 5 mm.
7. Place the bottom edge of one of the plates in the beaker. Make sure the water level is below the pencil line.
8. Repeat steps 6 and 7 with methylated spirits.
9. Allow the plates to stand in the solvent for approximately 10 minutes.
10. Withdraw the plate, place it on a piece of paper towel and, using a pencil, mark the solvent front.
11. Measure the distance from the pencil line to the solvent front for each solvent and record this measurement.
12. Measure the distances each of the dye components has travelled from the origin in each solvent and record this in a results table.

Results

Copy and complete the following results table.

Dye	Distance for distilled water	R_f	Distance for methylated spirits	R_f

Analysis of results

Calculate the R_f values for each of the dyes and their components by dividing the distance travelled by the dye component by the distance travelled by the solvent.

Discussion

1 Compare the R_f values of the dyes in both solvents.

2 Which dye was adsorbed most strongly in:

  a water?

  b methylated spirits?

3 Which solvent was the best for separating the dyes?

4 TLC can be one of several methods used to help identify an unknown compound. Explain how this might be accomplished using the dyes as an example.

Gas chromatography

To revise calibration curves, refer to Chemistry section 1.9 on page 127.

Gas chromatography (GC) is a separation technique for small organic molecules that can withstand relatively high temperatures. Blood alcohol levels are measured by gas chromatography.

The gas chromatograph consists of a gas bottle, an oven, a column, a detector and a recorder. The sample is injected into the oven where the gas pushes the sample into the long, thin column. In the column, smaller particles and those that adsorb onto the stationary phase the least leave the column first. Larger particles and those that adsorb more readily take longer to leave the column.

The time the sample takes to **elute** (leave the column) is characteristic for the substance. This time is called the **R_t**, the retention time. The R_t value is used to identify the component, as R_f is in TLC and paper chromatography.

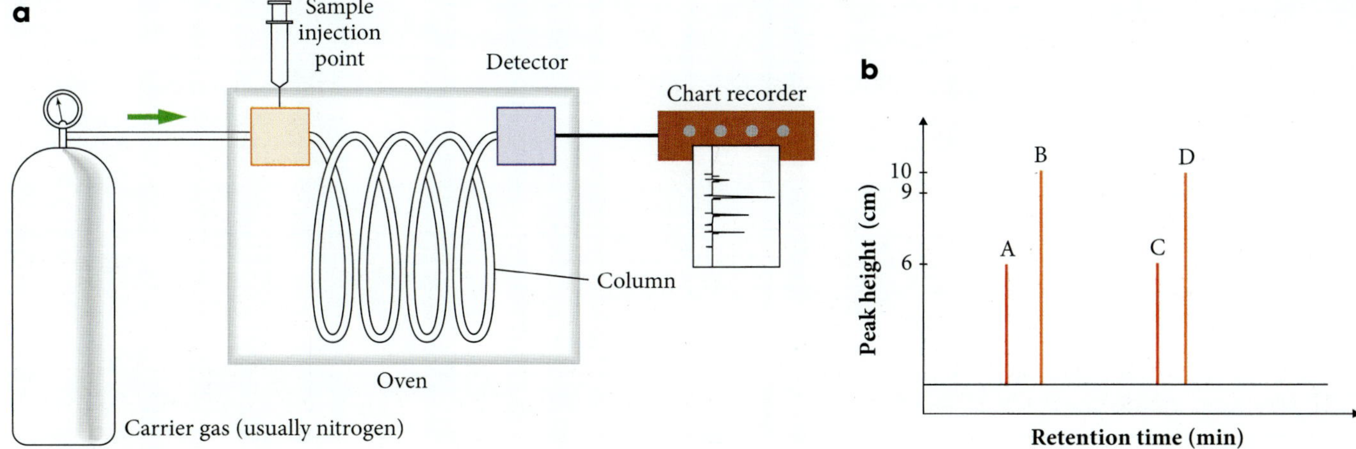

▲ Figure C6.33
A schematic diagram of: a) a gas chromatograph and b) a hypothetical gas chromatogram

LEARN MORE ABOUT HPLC

Watch this video to learn more about HPLC.

This technique can also measure the amount of the analyte. If a wine is thought to contain 12% alcohol, then a series of standards with alcohol ranging from 5% to 20% is run through the gas chromatograph and the peak areas are measured. A calibration curve is drawn and the unknown value is read off. It is important that the unknown value lies within the range of the standard values.

The R_t value will be the same for each standard. What is different is the height of the peak. The peak area will alter in proportion to the amount of the substance present. A calibration curve is plotted and the unknown wine's peak area is used to determine the amount of alcohol present.

High-performance liquid chromatography

High-performance liquid chromatography (HPLC) is used for larger organic molecules. It can be used for substances that are unstable to heat because there is no oven. The stationary column is shorter than in the gas chromatograph. The mobile phase is a liquid not gas. A pump is used because pressure is required to move the liquid through the column. As with GC, R_t values are used to identify compounds, standards are run to confirm the identity (same R_t) and a calibration curve is used to determine the amount of the substance present.

Figure C6.34 ▶ A schematic diagram of: a) a high-performance liquid chromatograph and b) a high-performance liquid chromatogram of pigments extracted from a *Dianella* plant

a

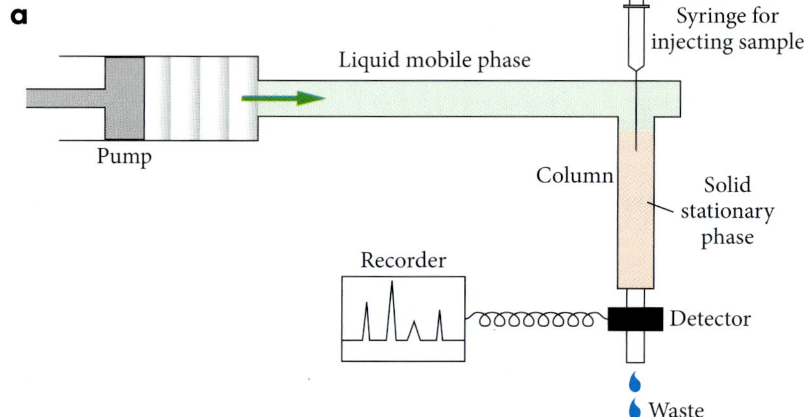

b

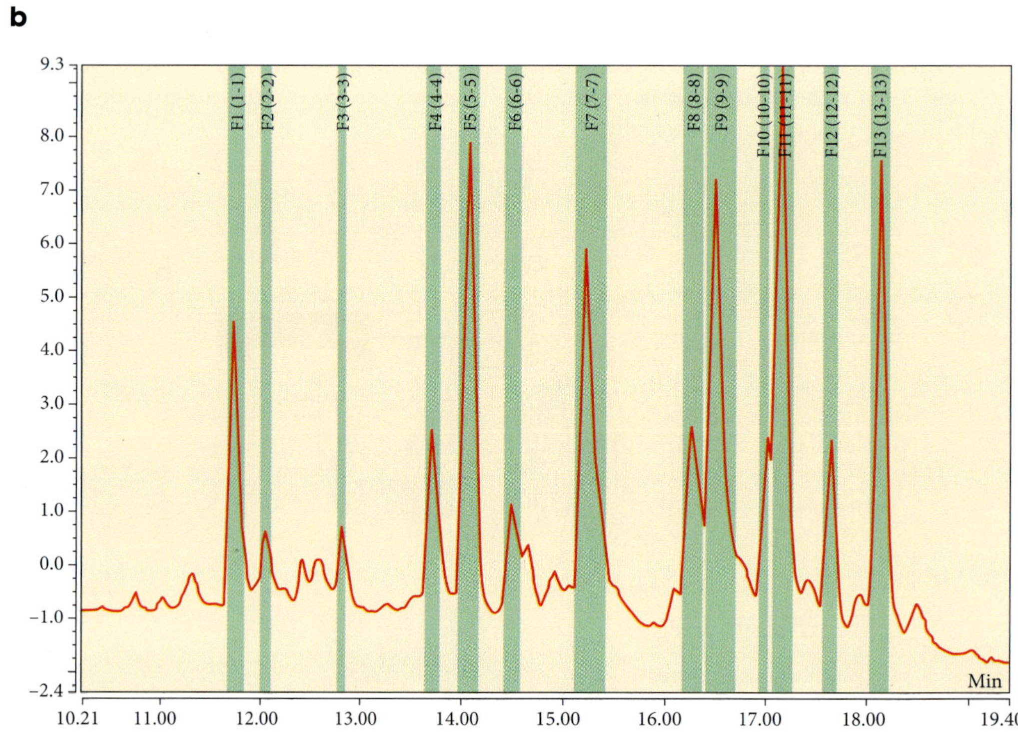

Table C6.3 Summary of chromatography techniques

Technique	Basis	Stationary phase	Mobile phase	Measures	Uses
Gas chromatography	Carrier gas is mobile phase. Sample is injected into oven. The gas moves the vaporised particles through a fine long column. The smallest solutes that adsorb the least, elute first. R_t depends on temperature, length of column and flow rate as well as chemical structure	Long, thin column	Inert gas such as N_2, CO_2 or He	R_t indicates what the sample is. Area under peak indicates amount of substance present. Standards are used to confirm identity (qualitative) and to prepare a calibration curve for calculations of amounts (quantitative analysis)	Small, heat-stable organic compounds such as ethanol
High-performance liquid chromatography	Similar to GC but mobile phase is a liquid. Unlike the gas in GC, the liquid plays an important role in adsorption and desorption. Needs a pump to push the liquid through the denser packed column	Short column (particle size of stationary phase is extremely small, allowing better separation)	Water–methanol or water–acetonitrile mixes	R_t indicates what the sample is. Area under peak indicates amount of the substance present. Standards are used to confirm identity (qualitative) and to prepare a calibration curve for calculations of amounts (quantitative analysis)	Larger organic compounds
Thin-layer chromatography	Solvent moves over stationary phase by capillary action. Components adsorb and desorb at different rates due to chemical structure and ability to form intermolecular bonds to the stationary phase and mobile phase.	Fine powder on glass or plastic	Solvent may be organic or water mixture	R_f is characteristic for each sample at the same conditions. Standards are used to confirm identity (qualitative)	Organic mixtures such as plant pigments and drugs

QUESTION SET 6.6

Remembering

1 List the differences between soap and a detergent. Describe why detergents work with hard water but soaps don't.

2 Figure C6.35 represents a thin-layer chromatogram of drug samples confiscated by the police. Included on the plate are commercial drugs, labelled X, Y and Z.

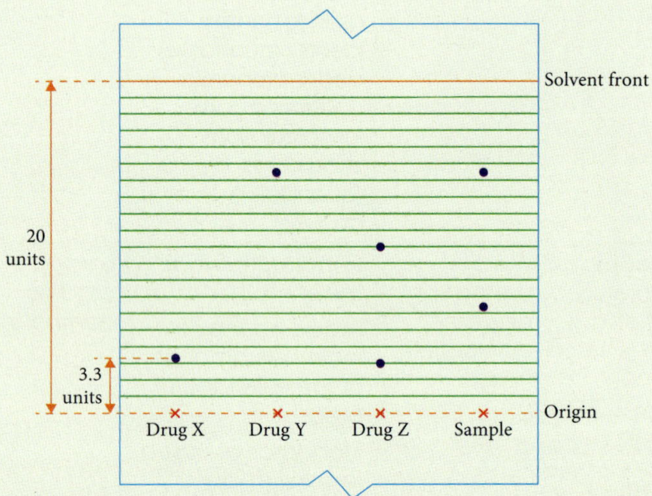

◀ **Figure C6.35**

a Why do chemists include commercial drugs on the same plate as the samples from the police?

b Does the confiscated sample contain any of the commercial drugs? Justify your answer using the data provided.

3 Describe the basic structure of a gas chromatograph.

4 How are compounds identified in GC and HPLC?

Understanding

5 Explain how a gas chromatogram is produced.

6 Figure C6.36 shows the gas chromatogram for a sample of common pesticides. Predict which of compounds 1–5 is the smallest? Explain your reasoning.

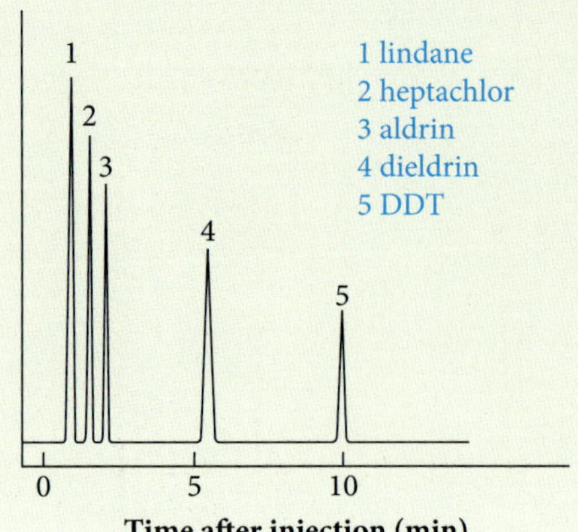

Figure C6.36 ◀
Gas chromatograms for some common pesticides

Applying

7 A mixture of pesticides was run through a gas chromatograph. The trace showed a very large peak at 4 minutes after injecting the sample, and smaller peaks between 1 and 3 minutes. What do you conclude about the composition of the sample?

8 Honeys vary in their composition. Honey consists generally of the disaccharides maltose and sucrose, and the monosaccharides glucose and fructose.

HPLC is commonly used to analyse honey and to find out which sugars are present and in what percentage. The chromatogram in Figure C6.37 was obtained when a sample of Australian honey was tested.

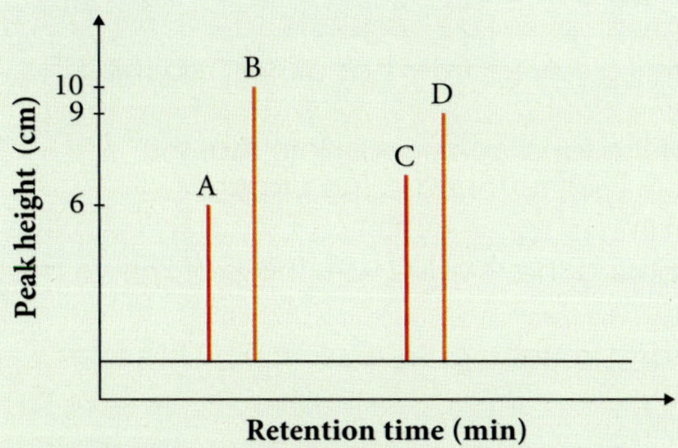

Figure C6.37

a Which molecule spent most time in the stationary phase? Explain how you determined this.

b What is the percentage of C in the mixture?

c Why was HPLC used in preference to GC in this analysis?

d It is suspected that peak C is fructose. How can this be verified?

9 Table C6.5 gives the retention times and peak areas for a series of ethanol standards and a sample of champagne when they were run through a gas chromatograph.

Table C6.5

% Ethanol (w/v)	Relative peak area	R_t
8.4	4 343 500	0.90
9.8	5 820 300	0.89
10.6	6 858 600	0.90
12.2	8 117 900	0.89
Champagne sample	5 324 750	0.90

a Construct a calibration curve for the ethanol and use it to determine the percentage of ethanol in the champagne.

b If the density of ethanol is 0.785 g mL^{-1}, express your answer as % v/v.

c Explain why you need a calibration graph and not just peak areas.

d Explain why this calibration curve would not be suitable to find the alcohol content of a low alcohol wine that typically has a value of around 7% v/v.

10 A student's notebook was messy and it was hard to read the practical instructions on how to separate the individual dye components of food colouring. Not only was the order strange, but there were errors. The instructions were as follows.

- **I** Using a gel pen, draw a reference line 2 cm from the bottom of a strip of chromatography paper.
- **II** Put more drops of the food colouring over the first to make a concentrated spot that is at least 15 mm in diameter so that it has adequate contact with the paper and can be easily seen.
- **III** Allow the drop to dry.
- **IV** Carefully place a drop of food colouring on the line.
- **V** Place solvent in the bottom of a tall jar.
- **VI** Hang the strip of chromatography paper in the jar, making sure that the spot of colouring is completely covered by the solvent.
- **VII** Allow the paper to remain in contact with the solvent until a solvent front has almost reached the top of the paper.
- **VIII** Observe the separation of the coloured components of the food colouring. Note that the component that adsorbs most strongly to the paper pulls itself furthest up the paper.
- **IX** Calculate an R_f value for each spot on the chromatogram.
- **X** Identify the components of the food colouring by comparing the R_f value with those of known dyes.
  - **a** Identify the correct order for the instructions (I–X).
  - **b** Identify three fundamental errors in the instructions and in each case explain how the error should be corrected.
  - **c** Explain how R_f values are calculated.

Reflecting

11 What challenges and opportunities have arisen from the development of HPLC and GC methods in the detection of drugs and other banned substances in sport?

CHAPTER CHECKLIST

You should know:

- molecules can be represented by electron dot formulas and valence structures
- VSEPR theory can be used to help determine the structure of molecules and some ion complexes
- polarity is due to the presence of a permanent dipole in molecules and asymmetry in the molecule
- intermolecular forces of dispersion, dipole–dipole and hydrogen bonding are present in molecules
- dipole–dipole interactions occur when there is an unequal sharing of electron density between different elements
- hydrogen bonding is a special case of dipole–dipole interactions. It is due to the uneven sharing of electrons between hydrogen and a strongly electronegative element (N, O and F)
- dispersion forces exist between all molecules and non-metal atoms
- hydrogen bonding in water is responsible for the unique properties of water
- water is the universal solvent
- surfactants and detergents affect surface tension
- the interactions between molecules are the basis for practical applications such as the chromatography techniques TLC, GC and HPLC
- chromatography techniques (HPLC, GC and TLC) rely on absorption onto a stationary phase and desorption back into a mobile phase
- by running a standard under the same conditions, calculation of the R_f and R_t can identify the species present
- the use of calibration curves of the amount under the peak at a particular R_t can be used to determine the amount present in a sample.

You should be able to:

- predict whether a molecule will have dipole–dipole, hydrogen bonding or dispersion forces between the molecules
- link the intermolecular bonding present to properties such as solubility
- draw electron dot formulas and valence structures of compounds and simple molecules
- determine the shape of the compounds and simple molecules using the VSEPR model
- determine whether a solution is saturated, unsaturated or supersaturated by using a solubility curve
- determine the concentration of solutes in the units ppm, % w/w, % w/v and % v/v
- interpret chromatograms, including the use of R_f and R_t for particular components
- use calibration curves to determine the amount of a component in a chromatograph
- know which instrument (HPLC or GC) to choose for a particular application.

CHAPTER GLOSSARY

adhesive force the electrostatic force of attraction between unlike particles as displayed between water molecules and glass

adsorb to be attracted to the surface of a material; the opposite of desorb

analyte the substance being analysed

anhydrous contains no water

anionic negatively charged; for example, Cl^-

bonding pair a pair of electrons shared by two atoms to form a covalent bond

capillary action the process in which a liquid is drawn up a narrow tube

cationic positively charged; for example, Na^+

cohesive force the electrostatic force of attraction between like particles, which causes water molecules to stick to each other

colloid a mixture in which tiny clusters of particles are dispersed through another substance; colloids do not settle out due to gravity and the particles are too small to be filtered

desorb the action of the substance moving from the stationary phase into the mobile phase

dipole a permanent build-up of negative charge at one end, and positive charge at another end, of a covalent bond or molecule

dipole–dipole force the attraction between molecules with permanent dipoles

dissociated when ionic salts in a solvent dissolve, ions of opposite charge no longer associate with each other and move freely

dispersion force the weak attractive force between atoms and molecules caused by an instantaneous temporary change in dipole moment, arising from the movement of orbiting electrons

electrolytic solution a solution that can conduct electricity due to the presence of ions

electromagnetic attraction the attraction of oppositely charged particles

electronegativity the ability of an atom to attract electrons

elute movement of a mobile phase through a column

emulsion a mixture of two substances that would not normally mix

hard water water that contains high levels of Ca^{2+} and Mg^{2+} ions, which interferes with the action of soaps

heteroatomic a molecule containing more than one type of element

hydrogen bond the intermolecular attraction between hydrogen in –OH, –NH or –FH and the lone pairs on nitrogen, oxygen or fluorine on an adjacent molecule

hydrophilic 'water-loving', a particle with polar regions that bond to water

hydrophobic 'water-hating', a particle with mostly non-polar regions that do not bond with water

immiscible liquids that do not mix

intermolecular forces forces between individual particles (molecules, ions, atoms) of a substance

ion–dipole attraction between the charge on one ion and a dipole in another molecule

ligand an ion or molecule attached to a metal atom (or larger molecule) by dipole bonding

lone pair a pair of valence electrons that are not involved in bonding

meniscus the shape of the surface of a liquid near the edge of a container

micelle small spherical cluster of surfactant molecules suspended in a liquid

mobile phase the phase that carries solutes through the stationary phase

non-ionic detergent a surfactant with a non-charged hydrophilic part

origin in chromatography, the line that samples are placed on; the distance the solvent moves is measured from here

Pauli exclusion principle an orbital can contain 0, 1 or 2 electrons

polar a molecule with a separation of charge; one end positive, another region negative

polar covalent bond a covalent bond that has a separation of charge

R_f retardation factor; the distance a solute moves compared to the distance the solvent has moved

R_t retention time; the time it takes for a solute to elute from the column in GC and HPLC

saponification the production of soap by mixing fats with strong alkali

saturated solution a solution that has the most solute possible at a particular temperature

solubility the amount of solute that can dissolve in 100 g of water (solvent)

solubility curve a graph showing how solubility changes with temperature

stationary phase the substance that is fixed in place during chromatography, to which the solute adsorbs

structural formula a chemical formula that shows how atoms are linked

supersaturated solution an unstable solution that has more solute than possible at that temperature; it is formed from dissolving solute into the solution at a higher temperature and then allowing it to cool. The extra solute is still in solution but can readily be crystallised out

surface tension the force that arises from the attraction of the surface molecules to the bulk of the material

surfactant a chemical that lowers the surface tension of a liquid

suspension a cloudy, heterogeneous mixture containing solid particles that will eventually settle out

unsaturated solution a solution into which more solute can dissolve

valence structure a diagram of the arrangement of atoms around a central atom

vapour pressure pressure exerted down by a vapour onto its liquid form

CHAPTER REVIEW QUESTIONS

Remembering

1 What are the two types of bonding in liquid propane?

2 Draw a diagram showing the bonding in ice.

3 A mixture of two substances is separated by paper chromatography. The substance with the greater R_f value:

A moves further and is more strongly adsorbed on the paper.

B moves further and is less strongly adsorbed on the paper.

C moves less and is more strongly adsorbed on the paper.

D moves less and is less strongly adsorbed on the paper.

4 Draw the electron dot formula for the following and indicate the number of bonding and non-bonding pairs of electrons.

a NO_2^-

b CO

c CH_4

5 Draw valence structures for the following molecules and indicate whether they are polar or non-polar.

a CO_2

b F_2O

c CH_4

d PF_3

6 Ethanol (CH_3CH_2OH) and glucose ($C_6H_{12}O_6$) can dissolve in water. Write the chemical equations that represent this.

7 Describe what happens when an ionic substance such as potassium iodide dissolves in water.

8 Formamide is a liquid with a very high surface tension, which wets glass. If the end of a thin glass capillary is dipped into a beaker of formamide, what would be observed?

Understanding

9 In a mixture of alcohols, each component can be identified by HPLC. The alcohol with the longest retention time will be the one that:

A is the most soluble in the mobile phase.

B has the smallest molar mass.

C will emerge first from the column.

D adsorbs most strongly on the stationary phase.

10 Explain the existence of the property of the surface tension of a liquid in terms of the interactions between the molecules.

11 You are asked to perform a simple experiment to determine which has the greater surface tension, water or diesel.

a Describe the method you would use.

b Describe the results you would expect to observe.

c Provide an explanation for these results.

12 Explain why a needle can float on water in a Petri dish until a drop of detergent is added.

13 Calculate the total volume of alcohol in a small bottle (375 mL) of wine when the alcohol content is 8.5% v/v.

14 Hydrogen peroxide is a strong bleach used for hair and whitening teeth. Would you expect it to be soluble in water? Explain with the use diagrams the forces that might operate between the molecules of peroxide and water.

15 Use Figure C6.25 on page 280 to determine the mass of potassium nitrate that will dissolve in:

a 100 g of water to form a saturated solution at 20°C.

b 50 g of water to form a supersaturated solution at 30°C.

c 80 g of water to form a saturated solution at 10°C.

16 Serotonin ($C_{10}H_{12}N_2O$) is a compound that conducts nerve impulses in the brain and muscles. A sample of spinal fluid was found to contain a serotonin concentration of 1.5 ng L^{-1}. How much serotonin is there in 1 mL of spinal fluid?

Applying

17 Which one of the following is a property of a 'water in oil' emulsion?

A It feels greasy to the touch.

B It mixes readily with water.

C Its electrical conductivity is higher than that of an 'oil in water' emulsion.

D It can be coloured by the addition of a water-soluble dye.

18 Describe the trend in boiling points of the noble gases and halogens in Table C6.2. How does this relate to intermolecular bonding?

19 Figure C6.25 on page 280 shows the solubility curve of potassium chloride, KCl.

a Use diagrams to model how KCl would dissolve in water.

b Write a balanced equation for the dissolution of KCl in water.

c How much KCl would just dissolve in 50 mL of water at 80°C?

d If the solution in part **c** was cooled to 10°C, would it be saturated, unsaturated or supersaturated? Justify your answer.

e What temperature would you need to heat 100 g of water so that 40 g of KCl would dissolve completely? If you cooled this to 10°C, how much KCl would precipitate out of solution?

20 A commercial liquid medication is known to contain 400 mg of magnesium hydroxide per 10.0 mL of the medication. Express this as % w/v and ppm. (Assume that 1 mL of medication = 1 gram.)

21 Explain why different solutes travel through a gas chromatograph at different rates.

Analysing

22 Steroids do not circulate in the blood system by themselves. Infer the polarity of steroid hormones. Dopamine can travel through the blood system easily. What does this indicate about the polarity of dopamine? Which of these is more likely to be injected directly into the veins?

23 Compare and contrast ammonia (NH_3) and methane (CH_4). Highlight the bonding so that you can explain to another student how to determine which one is soluble in water.

24 Reorganise the data in Table C6.1 on page 262 to clearly show the relationship between the generic formula and the lone pair of electrons.

Reflecting

25 Theories such as VSEPR have enabled scientists to develop a greater understanding of molecular structures. Representations have changed from valence structures to three-dimensional representations of electrons as charged clouds. Explain how theories and models generally develop over time, using this specific theory.

CHAPTER 7 REACTIONS IN AQUEOUS SOLUTIONS

By the end of this chapter you will have covered the following material.

Science Understanding

- The concentration of a solution is defined as the amount of solute divided by the amount of solution; this can be represented in a variety of ways including by the number of moles of the solute per litre of solution ($mol\,L^{-1}$) and the mass of the solute per litre of solution ($g\,L^{-1}$) (ACSCH063)
- The presence of specific ions in solutions can be identified using analytical techniques based on chemical reactions, including precipitation and acid-base reactions (ACSCH064)
- The pH scale is used to compare the levels of acidity or alkalinity of aqueous solutions; the pH is dependent on the concentration of hydrogen ions in the solution (ACSCH066)
- Patterns of the reactions of acids and bases (for example, reactions of acids with bases, metals and carbonates) allow products to be predicted from known reactants (ACSCH067) AC

Figure C7.1 ▲ Water is the most common solvent for natural solutions.

7.1 Aqueous solutions of ionic compounds

As you learnt in Chapter 2, solutions are defined as homogeneous mixtures. The most commonly occurring natural solutions have water as the solvent and a wide variety of different solutes, including solids, liquids and gases. Oceans are solutions of different compounds, many of which are ionic, dissolved in water. The chemical reactions that are responsible for maintaining all living things occur in water solutions. Solutions of substances dissolved in water are called **aqueous solutions**. The term 'aqueous' comes from the Latin *aqua*, meaning water.

In Chapter 6, you looked at how the nature of the solute and solvent and temperature affects the solubility of different substances. You also looked at the processes that occur when a substance dissolves, and different measures for the amount of substance dissolved.

When ionic substances dissolve in water, they dissociate. This means they separate into their ions, which are then able to move freely and independently of each other through the solution.

For example, when sodium chloride (NaCl) dissolves in water, it separates into Na^+ ions and Cl^- ions. This can be represented as:

$$NaCl(s) \xrightarrow{aq} Na^+(aq) + Cl^-(aq)$$

Although most ionic compounds are soluble in water, they do not all dissolve to the same extent. The **solubility** of a substance is the degree to which it dissolves and is temperature dependent. The solubility of a solute is the maximum amount in grams that can dissolve in 100 g of the solvent at a given temperature. The standard temperature for measuring solubility is 25°C.

Chemists use the terms 'soluble', 'insoluble' and 'sparingly soluble' to categorise the solubility of different compounds. 'Soluble' means that a compound dissolves to more than $10\,g\,L^{-1}$ (or 1 g/100 mL), 'insoluble' means that it dissolves to less than $1\,g\,L^{-1}$ and 'sparingly soluble' means that it dissolves between the range $1\,g\,L^{-1}$ to $10\,g\,L^{-1}$.

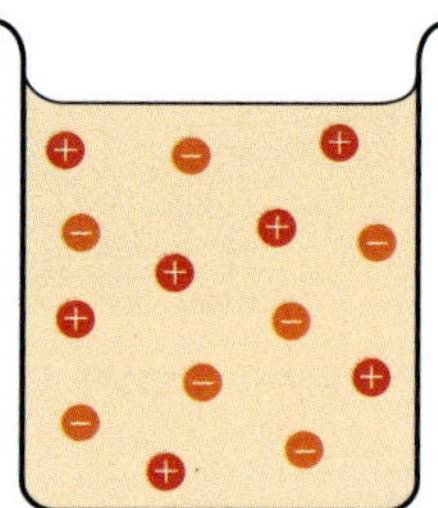

Figure C7.2 ▶ A solution of an ionic substance

Precipitation reactions

When solutions of certain ionic compounds are mixed, they sometimes react to produce a solid called a **precipitate**. This type of reaction is called a **precipitation reaction**.

A precipitate is a solid produced by the reaction between two solutions.

For example, when sodium chloride solution is added to silver nitrate solution, a white solid of silver chloride is formed. When the solution of Na^+ and Cl^- ions is mixed with the solution of Ag^+ and NO_3^- ions, the result is a solution containing all the ions moving randomly throughout (i.e. a precipitate is formed). Remember that each of the ions is surrounded by water molecules. The Ag^+ and Cl^- ions are more strongly attracted to each other than they are to the water, so they cluster together in groups and form a solid, which precipitates out of solution, as shown in Figures C7.3 and C7.4.

▶ **Figure C7.3**
A silver chloride precipitate

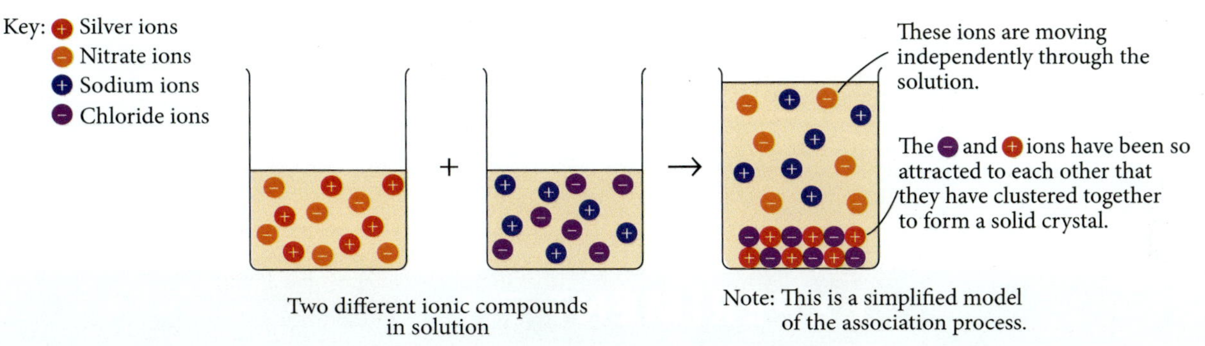

▲ **Figure C7.4**
Explaining the formation of silver chloride precipitate

The Na^+ and NO_3^- ions are not involved in the reaction and remain in the solution as ions. Ions that are not involved in a reaction are called **spectator ions**.

The equation for this reaction can be written in a number of different ways, which may include or omit the spectator ions.

$$Na^+(aq) + Cl^-(aq) + Ag^+(aq) + NO_3^-(aq) \rightarrow AgCl(s) + Na^+(aq) + NO_3^-(aq) \quad (1)$$

$$NaCl(aq) + AgNO_3(aq) \rightarrow AgCl(s) + NaNO_3(aq) \quad (2)$$

$$Ag^+(aq) + Cl^-(aq) \rightarrow AgCl(s) \quad (3)$$

PRECIPITATION ANIMATION

View this animation and predict when a precipitate will form.

Equation 1 is a **complete ionic equation**. It shows all the ions present in the solution. Equation 2 is an overall equation. It shows all the reactants and products as actual neutral compounds. Equation 3 is the **net ionic equation**. It shows only the reacting ions. Removing the spectator ions from the complete ionic equation produces the net ionic equation. This equation most accurately represents the reaction taking place, so is most commonly used to represent precipitation reactions.

Predicting precipitation

By systematically mixing solutions that contain known cations and anions, chemists have experimentally determined which compounds are soluble and which form a precipitate (are insoluble). These results have been summarised into a table of solubility data that can then be used to predict whether or not precipitation will occur when two known solutions are mixed. Sometimes, when only a small amount of precipitate forms, it does not settle to the bottom but stays in suspension and makes the mixture look cloudy.

Figure C7.5 ▲
Solutions and suspensions

EXPERIMENT 7.1

INVESTIGATING PRECIPITATION REACTIONS

Solubility rules can be formulated from experimental results. In this experiment, you will determine which mixtures of solutions form precipitates and use these results to construct solubility rules. These can then be compared with theoretical data.

Aim

To compare the solubility of a range of ionic substances through precipitation reactions

Materials

- Cations (all 0.1 mol L^{-1}) in dropper bottles: sodium nitrate ($NaNO_3$), silver nitrate ($AgNO_3$), lead(II) nitrate ($Pb(NO_3)_2$), ammonium nitrate (NH_4NO_3), magnesium sulfate ($MgSO_4$), iron(II) sulfate ($FeSO_4$), copper(II) sulfate ($CuSO_4$), zinc sulfate ($ZnSO_4$), aluminium sulfate ($Al_2(SO_4)_3$), barium chloride ($BaCl_2$), calcium chloride ($CaCl_2$), iron(III) chloride ($FeCl_3$)
- Anions (all 0.1 mol L^{-1}) in dropper bottles: potassium nitrate (KNO_3), potassium chloride (KCl), sodium sulfate (Na_2SO_4), sodium carbonate (Na_2CO_3), sodium acetate ($NaCH_3COO$), sodium hydroxide (NaOH)
- Distilled water
- 6 test tubes and test-tube rack
- Paper towel
- Test-tube cleaner

What are the risks in doing this experiment?	How can you manage these risks to stay safe?
Lead and barium salts are poisonous.	Wear safety glasses and avoid contact with skin. If contact occurs, wash thoroughly with soap and water.
Lead, barium, silver and copper salts are harmful to the environment.	Do not pour solutions containing these salts down the drain. Dispose of as directed by your teacher.

In your write-up, add any more risks you can think of, as well as ways to manage them. In particular, expand on the two risks listed to identify specific risks involved with each of them.

Procedure

1 Put a few drops of each anion into each of six test tubes. Take care not to mix droppers.
2 Add a few drops of the first cation to each. Record your observations.
3 Clean all test tubes thoroughly because incorrect results can arise from contamination of solutions.
4 Repeat step 2 for the next cation. Continue until all cations have been tested.

Results

Draw up a table similar to the one below to record your observations. For no reaction, record 'NR'; for a precipitate, record 'P' and the colour of the precipitate.

▲ **Figure C7.6**
Experimental set-up

Anion	Cation					
	NO_3^- nitrate	Cl^- chloride	SO_4^{2-} sulfate	CO_3^{2-} carbonate	CH_3COO^- ethanoate	OH^- hydroxide
Na^+						
Ag^+						
Pb^{2+}						
NH_4^+						
Mg^{2+}						
Fe^{2+}						
Cu^{2+}						
Zn^{2+}						
Al^{3+}						
Ba^{2+}						
Ca^{2+}						
Fe^{3+}						

Analysing the results

Summarise your results into statements such as 'All nitrates are . . .'

Discussion

1 a Compare the results of the experiment with the solubility rules in Table C7.1.
  b What, if any, were the discrepancies between your results and Table C7.1? Suggest a reason for these discrepancies and a way of resolving them.
2 Write a balanced net ionic equation for each of the reactions in which a precipitate was obtained.

Conclusion

Summarise the solubility rules according to your results.

Table C7.1 summarises the solubility of common ionic compounds. The table applies to the cations in groups 1 and 2, NH_4^+, Al^{3+}, Cr^{3+}, Mn^{2+}, Fe^{2+}, Fe^{3+}, Cu^{2+}, Co^{2+}, Ni^{2+}, Ag^+, Zn^{2+}, Sn^{2+}, Cd^{2+} and Pb^{2+}. The table is organised in terms of the solubility of various anions. An additional useful generalisation is that all group 1 and NH_4^+ salts are soluble.

Table C7.1 Solubility of common ionic substances

Soluble anions	Exceptions
NO_3^-	None
CH_3COO^-	Ag^+ slightly soluble
Cl^-	Ag^+ insoluble; Pb^{2+} slightly soluble
Br^-	Ag^+ insoluble; Pb^{2+} slightly soluble
I^-	Ag^+, Pb^{2+} insoluble
SO_4^{2-}	Ba^{2+}, Pb^{2+}, Sr^{2+} insoluble; Ag^+, Ca^{2+} slightly soluble
Insoluble anions	**Exceptions**
OH^-	Group 1, NH_4^+, Ba^{2+}, Sr^{2+} soluble; Ca^{2+} slightly soluble
O^{2-}	Group 1, NH_4^+, Ba^{2+}, Sr^{2+}, Ca^{2+} soluble
S^{2-}	Groups 1 and 2, NH_4^+ soluble
CO_3^{2-}	Group 1, NH_4^+ soluble
SO_3^{2-}	Group 1, NH_4^+ soluble
PO_4^{3-}	Group 1, NH_4^+ soluble

WOW

Cyanide (CN^-)

Cyanide (CN^-) has long been used as a poison. In many crime novels, cyanide is chosen as the poison and cyanide 'suicide pills' were kept on hand by spies in the event of their capture. Leading Nazi Hermann Goering resorted to these after being tried at Nuremberg and found guilty of war crimes after World War II.

Applying solubility data

Chemists are often asked to identify substances such as cations and anions that may have contaminated food, soil and water. Precipitation reactions are useful for finding out whether a particular anion or cation is present; for example, chemists could use precipitation reactions to test for lead or cadmium in water supplies. Precipitation reactions can also be used to remove unwanted substances from water.

The data in Table C7.1 can be used to predict whether or not a precipitate will form when particular solutions are mixed.

Figure C7.7 ▶ Chemists working in an analytical laboratory

Thinkstock/Horsche

WORKED EXAMPLE 7.1

a What precipitate, if any, will form when aqueous solutions of magnesium iodide and silver nitrate are mixed? If a precipitate forms, write the neutral species and net ionic equations for the reaction.

b Select a reagent that could be used to precipitate the cation in $BaBr_2$, explaining your selection.

Answers	Logic
a Precipitate is AgI. Neutral species equation: $MgI_2(aq) + 2AgNO_3(aq) \rightarrow 2AgI(s) + Mg(NO_3)_2(aq)$ Net ionic equation: $Ag^+(aq) + I^-(aq) \rightarrow AgI(s)$	**1** Write the formula of the compounds present. Magnesium iodide (MgI_2) Silver nitrate ($AgNO_3$) **2** Predict the products by swapping the partners of the reacting compounds and check that the formulas are correct. $MgI_2 + AgNO_3 \rightarrow AgI + Mg(NO_3)_2$ **3** Use Table C7.1 to see if either of the products is insoluble. All nitrates are soluble, all iodides are soluble except Ag^+ and Pb^{2+} so AgI will precipitate **4** Write the required equations, checking they are balanced. Neutral species equation is: $MgI_2(aq) + 2AgNO_3(aq) \rightarrow 2AgI(s) + Mg(NO_3)_2(aq)$ Net ionic equation is: $Ag^+(aq) + I^-(aq) \rightarrow AgI(s)$
b Na_2CO_3 can be used to precipitate Ba^{2+} from a solution of $BaBr_2$.	**1** Identify the ions present and one to be precipitated. Ba^{2+} and Br^- are the ions present in solution. Ba^{2+} is the ion to be precipitated. **2** Use Table C7.1 to identify an anion which could be used to precipitate the identified cation. From Table C7.1, anions that will produce an insoluble compound with Ba^{2+} are S^{2-}, CO_3^{2-}, SO_3^{2-} and PO_4^{3-}. **3** Choose one of the anions and identify a cation to form a compound. Check that the cation does not react with the anion from the original compound. Generally it is best to choose a cation from group 1 or NH_4^+ because compounds of these are soluble. It would be incorrect to choose Ag^+ and Pb^{2+} as they precipitate with Br^-. **4** Identify the reagent and write the formula. Na_2CO_3 can be used to precipitate Ba^{2+} from a solution of $BaBr_2$.

Try these yourself

a When solutions of potassium sulfide and cobalt acetate are mixed, which salt, if any, will precipitate? If a precipitate does form, write the neutral species and net ionic equations.

b Name and give the formula of the precipitate (if any) that forms when each of the following pairs of substances is mixed.

- **i** Copper(II) chloride and potassium hydroxide
- **ii** Sodium carbonate and ammonium sulfate
- **iii** Lead nitrate and copper(II) sulfate

c What solution would you mix with calcium chloride to obtain a precipitate of the cation?

INVESTIGATION 7.1

IDENTIFYING IONIC COMPOUNDS

The labels on the bottles of five different chemicals have fallen off. The bottles have been labelled A, B, C, D and E. The chemicals are known to be ammonium hydroxide, potassium iodide, sodium hydroxide, silver nitrate and zinc sulfate. You have been asked to identify what is in each bottle.

What is your aim?

Think about what you want to find out in this investigation.
Hint: Look at the all the possible pairings of the chemicals listed and use solubility rules to predict what precipitates will actually form.

What will you need?

Think about the equipment you will need. Will you need any additional chemicals to confirm your results? Be specific about quantities of chemicals and don't forget to include basic materials like distilled water and paper towel.

What are the risks?

Construct a table similar to the one below. Identify specific risks involved in the investigation and ways that you will manage the risks to avoid injuries or damage to equipment. Ask your teacher to check your risk assessment before you proceed.

What are the risks in doing this investigation?	How can you manage these risks to stay safe?

How will you carry out the investigation?

Hint: You need to use solubility rules. Write your method in a logical sequence of steps. Don't forget to be specific and include quantities.

What results will you collect?

Refer back to your aim. Refer to the results of Experiment 7.1 for possible table structure and use those results to help identify products.

What have you found?

Consider what your results show. Look at theoretical predictions you made about possible precipitates and compare your results with these. Are there any differences or inconclusive results? Do you need to repeat any tests? Do you need to use an alternative solution for confirmation?

What do you conclude?

Reflect on what you found. Could you identify each of the solutions?

How could you improve the effectiveness of the identification process?

QUESTION SET 7.1

Remembering

1 Define 'precipitate'.

2 Explain what is meant by 'spectator ion'.

Understanding

3 Which of the following substances are soluble in water?

a Sodium hydroxide

b Aluminium oxide

c Copper(II) sulfate

d Silver carbonate

e Lithium chloride

4 Name a different pair of substances (using suitable amounts) that you could dissolve in the one sample of water to produce a solution identical with that containing:

a potassium chloride and zinc nitrate.

b ammonium sulfate and sodium carbonate.

c magnesium bromide and copper(II) sulfate.

Applying

5 Write overall equations for the reactions (if any) that occur when solutions of the following pairs of substances are mixed. If there is no reaction, write 'NR'.

a Sodium chloride and silver nitrate

b Copper(II) sulfate and potassium hydroxide

c Nickel chloride and potassium sulfate

d Sodium carbonate and iron(II) sulfate

e Zinc nitrate and ammonium sulfide

f Potassium carbonate and calcium chloride

6 Solutions of what substances would you mix to prepare the following compounds by precipitation? Write equations for your reactions.

a Magnesium carbonate

b Lead sulfate

c Silver bromide

Analysing

7 Separate portions of an unknown solution A formed precipitates when potassium carbonate and sodium hydroxide solutions were added to them, but did not give precipitates when solutions of sodium chloride or ammonium sulfate were added. Solution A could have contained:

A Pb^{2+}, Zn^{2+}, Cu^{2+}

B Mg^{2+}, Al^{3+}, Zn^{2+}

C Mg^{2+}, Ba^{2+}, $A1^{3+}$

D Cu^{2+}, Zn^{2+}, Ag^{+}

8 Describe tests to determine whether a solution contained:

a lead nitrate or barium nitrate.

b copper(II) sulfate or iron(II) sulfate.

7.2 Determining concentration

Aqueous solutions can contain many different dissolved substances. The amount of dissolved solute is regularly monitored in many areas of chemistry, including wine production, water quality and swimming pools. In many of these areas, it is important to know the **concentration** of the solute in the solution.

> **The concentration of a solution is the amount of solute present in a specified amount of solution.**

Figure C7.8 ▶ Monitoring water quality involves a range of measurements.

iStockphoto/BartCo

MOLARITY

Use this simulation to learn more about the relationship between moles of substance and volume of solution.

In Chapter 6, you learnt about some different measures of concentration (% w/w, % v/w, % w/v and ppm), which could be used to compare quantities of different substances in solution. Different ways of expressing concentration are used because each has its advantages in different situations.

Molarity

When calculating quantities in chemical reactions, the basic unit is the mole. Therefore, it is convenient to have a measure of concentration in terms of moles. This measure of concentration is called **molarity**. The molarity of a solution is defined as the number of moles of solute per litre of solution.

The concentration of a particular solution is given by:

$$\text{Concentration (mol L}^{-1}\text{)} = \frac{\text{number of moles of solute (mol)}}{\text{volume of solution (L)}}$$

$$c = \frac{n}{V}$$

This formula can also be rearranged to find n and V.

This is also referred to as molar concentration and given the symbol M. In your chemistry experiments, such as Experiment 7.1, you may use a solution that is labelled as 0.1 M sodium chloride, for example. This means 0.1 moles of sodium chloride per litre of solution.

Sometimes, chemists may need to know the mass of solute in a solution so units of g L^{-1} may also be used. In Chapter 5, you learnt the relationship between moles and mass:

$$n = \frac{m}{M}$$

Worked example 7.2 shows how moles, concentration and mass are calculated.

CONCENTRATION

Use this simulation to learn more about factors that cause concentration to change.

Be careful not to confuse M for molar mass in grams with M for molarity, which is used for solutions.

WORKED EXAMPLE 7.2

a Calculate the concentration of copper(II) sulfate in a solution if 250 mL of the solution contains 0.01 mol of copper(II) sulfate.

b What is this concentration in $g\,L^{-1}$?

c Calculate the concentration of a solution prepared by dissolving 13.5 g of sodium hydroxide in 100 mL of solution.

d What mass of silver nitrate is needed to prepare 250 mL of a 0.120 $mol\,L^{-1}$ solution of silver nitrate?

Answers

a Concentration is 0.040 $mol\,L^{-1}$.

Logic

1 Write the relationship.

$$c=\frac{n}{V}$$

2 Identify the quantities and check units.

$c = ?$, $n = 0.01$ mol, $V = 250$ mL $= 0.25$ L

3 Substitute to find c.

$$c=\frac{0.01}{0.25} = 0.040\ mol\,L^{-1}$$

b Concentration is 6.4 $g\,L^{-1}$.

Convert moles to grams using $n=\frac{m}{M}$.

$n = 0.040$ mol, $m = ?$, $M(CuSO_4) = 63.55 + 32.06 + 4 \times 16.0 = 159.6\ g\,mol^{-1}$

$$0.040 = \frac{m}{159.6}$$

$$m = 0.040 \times 159.6 = 6.4\ g$$

Concentration is 6.4 $g\,L^{-1}$

c Concentration is 3.37 $mol\,L^{-1}$.

1 Write the relationships.

$$c=\frac{n}{V}, n=\frac{m}{M}$$

2 Identify the quantities and check units.

$c = ?\ mol\,L^{-1}$, $n = ?$, $V = 100$ mL $= 0.100$ L, $m = 13.5$ g

3 Calculate M and substitute to find n.

$M(NaOH) = 40\ g\,mol^{-1}$

$$n=\frac{13.5}{40}=0.3375\ mol$$

4 Substitute to find c.

$$c=\frac{0.3375}{0.100}\ mol\,L^{-1}$$

$$c = 3.38\ mol\,L^{-1}$$

d Mass is 5.09 g.

1 Write the relationships.

$$c = \frac{n}{V}, n = \frac{m}{M}$$

2 Identify the quantities and check units.

$c = 0.120\ mol\,L^{-1}$, $n = ?$, $V = 250$ mL $= 0.25$ L, $m = ?$

3 Substitute to find n.

$$0.120 = \frac{n}{0.25}\ mol\,L^{-1}$$

$$n = 0.120 \times 0.25 = 0.03\ mol$$

4 Calculate M and substitute to find m.

$M(AgNO_3) = 169.9\,g\,mol^{-1}$

$$0.03 = \frac{m}{169.0}$$

$m = 0.03 \times 169.9 = 5.09\,g$

Try these yourself

a Calculate the molarity of:

i 0.20 mol of zinc iodide dissolved in 1.5 L of solution.

ii 0.04 mol of magnesium chloride dissolved 450 mL of solution.

b Calculate the concentration of the following solutions made by dissolving a known mass in a given volume.

i 8 g potassium chloride dissolved in 250 mL of solution

ii 1.46 g iron(II) sulfate dissolved in 100 mL of solution

iii 7.5 g silver nitrate dissolved in 500 mL of solution

c What mass of solute is needed to make the following solutions?

i 2.00 L of a 0.200 $mo\,L^{-1}$ solution of sodium iodide

ii 250 mL of a 0.01 $mol\,L^{-1}$ solution of potassium phosphate

iii 500 mL of a 0.0075 $mol\,L^{-1}$ solution of barium nitrate

Laboratory preparation of a solution

To prepare a solution of known concentration, chemists must use precise apparatus. The volume of the final solution must be as accurate as possible, so a **volumetric flask** is used because these are designed to allow precise volume determinations.

The steps for preparing a solution of known concentration are as follows.

1 The required amount of solute is weighed and then transferred to a volumetric flask of required volume.

2 The beaker that contained the solute is rinsed with distilled water and the rinse water is transferred to the volumetric flask.

3 More water is added to dissolve the solute and then up to the calibration mark.

4 The stopper is placed in the flask, which is inverted several times to ensure through mixing of the solution.

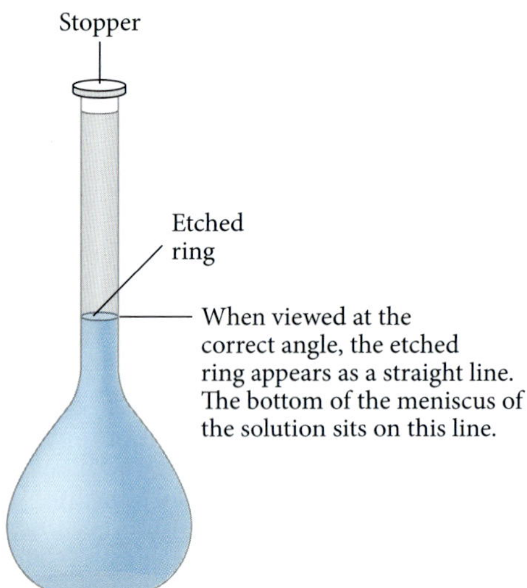

Figure C7.9 ▲
A solution prepared in a volumetric flask

Calculating reacting quantities

Now that a relationship between moles and concentration has been developed, we can extend calculations involving chemical reactions to include those involving solutions.

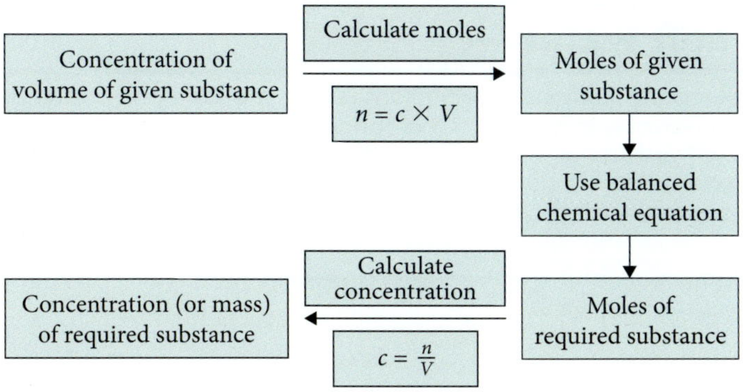

Concentration of volume of given substance → Calculate moles ($n = c \times V$) → Moles of given substance → Use balanced chemical equation → Moles of required substance → Calculate concentration ($c = \frac{n}{V}$) → Concentration (or mass) of required substance

◀ **Figure C7.10** Calculating reaction quantities

WORKED EXAMPLE 7.3

What mass of barium sulfate will be formed when excess sodium sulfate is added to 50 mL of a 0.30 mol L^{-1} solution of barium chloride?

Answer

Mass is 3.5 g.

Logic

1 Write a balanced equation.

$Na_2SO_4(aq) + BaCl_2(aq) \rightarrow BaSO_4(s) + 2NaCl(aq)$

2 Calculate the number of moles of $BaCl_2$.

$c = \frac{n}{V}$

$c = 0.30\ \text{mol L}^{-1}$, $V = 50\ \text{mL} = 0.050\ \text{L}$

$0.30 = \frac{n}{0.050}$

$n = 0.30 \times 0.050 = 0.015\ \text{mol}$

3 Use the balanced equation to calculate the number of moles of $BaSO_4$.

$n(BaSO_4) = n(BaCl_2) = 0.015\ \text{mol}$

4 Calculate the mass of $BaSO_4$.

$n = \frac{m}{M}$

$n = 0.015\ \text{mol}$, $m = ?$, $M(BaSO_4) = 233\ \text{g mol}^{-1}$

$0.015 = \frac{m}{233}$

$m = 0.015 \times 233 = 3.5\ \text{g}$

Try these yourself

a What mass of copper(II) hydroxide is formed when excess copper(II) sulfate is added to 100 mL of a 0.450 mol L^{-1} solution of sodium hydroxide?

b Calculate the mass of magnesium carbonate formed when excess sodium carbonate is added to 75 mL of a 0.08 mol L^{-1} solution of magnesium chloride.

c If excess NaOH is added to 10 mL of a 0.500 mol L^{-1} solution of zinc chloride, calculate the mass of $Zn(OH)_2$ produced.

CONCENTRATION CALCULATIONS

To practise more problems involving concentration calculations, complete this molarity worksheet.

WOW

Hard or soft water

The presence of magnesium and calcium ions affects the taste and usefulness of water. These ions occur naturally in water and the concentration varies with the source and contributes to the hardness of water. Water is described as 'hard' if it is difficult to form a lather or suds when washing. This is because the calcium and magnesium ions react with soaps and detergents to form an insoluble scum, making them ineffective for cleaning.

QUESTION SET 7.2

Remembering

1 What is the relationship between molarity and concentration?

Understanding

2 Calculate the concentration of the following solutions in
 i $mol\,L^{-1}$
 ii $g\,L^{-1}$
 a 2.00 L of solution containing 1.50 moL of potassium bromide
 b 250 mL of solution containing 0.0025 moL of sodium iodide

3 Calculate the number of moles of solute that is needed to make the volume of the solution in the following.
 a 100 mL of a 0.40 $mol\,L^{-1}$ solution of barium nitrate
 b 500 mL of a 0.15 $mol\,L^{-1}$ solution of sodium sulfate
 c 2.00 L of a 0.032 $mol\,L^{-1}$ solution of copper(II) chloride

Applying

4 What mass of solute is required for each of Questions **3a–c**?

5 Calculate the molar concentration of the following solutions.
 a 199 g $NiBr_2$ dissolved in 5.00 L of solution
 b 0.059 g KF dissolved in 227 mL of solution
 c 23 g ethanol (C_2H_5OH) in 750 mL of solution

6 What mass of solute is needed to make the following solutions?
 a 25.0 mL of a 0.02 $mol\,L^{-1}$ solution of sodium carbonate
 b 32.0 mL of a 0.10 $mol\,L^{-1}$ solution of copper(II) sulfate

7 Calculate the volume of a 0.025 $mol\,L^{-1}$ solution that contains 1.0×10^{-3} mol of magnesium nitrate.

Analysing

8 What mass of silver iodide will be formed when excess sodium iodide is added to 25 mL of a 0.15 $mol\,L^{-1}$ solution of silver nitrate?

9 To measure the concentration of calcium ions in a water sample, an analyst took 500 mL of the sample and added sodium carbonate until no further precipitation occurred. The precipitate was filtered off, dried and weighed. It was found to have a mass of 1.72 g.
 a What was the molarity of calcium ions in the original solution?
 b How many grams of calcium ions were there per litre in the original solution?
 c Water is considered to be 'hard' if it contains more than 201–300 $mg\,L^{-1}$ calcium carbonate. Would this water sample be classified as 'hard'?

7.3 Dilutions

In order to reduce transport and packaging costs, many products such as cordials, detergents, laboratory chemicals and pesticides are sold in a concentrated form. Water is added when they are ready for use. The amount of water added to cordials depends on personal preference, but for pesticides there are instructions on the packaging for making solutions of specific concentrations. This is to ensure the final mixture is suitable for its intended purpose. The process of adding water to a solution to make it less concentrated is called **dilution**.

Dilution is an important laboratory technique used to produce solutions of required concentration for testing and analysis procedures. The dilution process must be as accurate as possible, so the concentration of the required diluted solution is correct.

The usual equipment to ensure accurate measures of volume are pipettes and volumetric flasks similar to those shown in Figure C7.11.

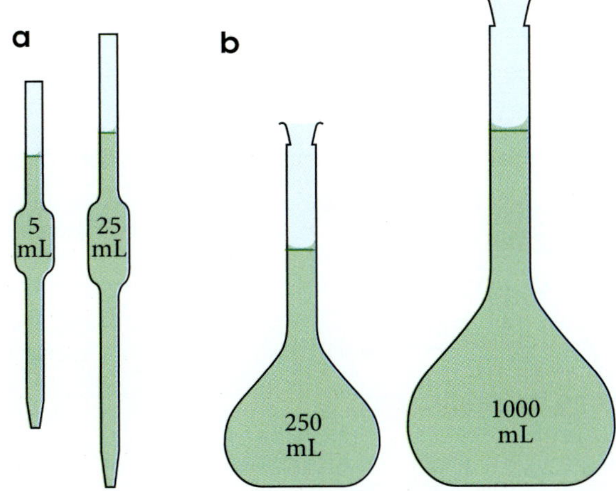

Figure C7.11 a) Pipettes and b) volumetric flasks are used in the laboratory when accurate volumes are required.

Never add water to a concentrated acid solution because the heat generated by the reaction could cause the concentrated acid to boil and spit out of the container. It can also cause the glass container to break. The correct way to dilute a concentrated acid is to add small quantities of the acid to large quantities of water so the heat generated is absorbed by the water.

The dilution procedure involves:

1 using a pipette to measure a definite volume of a solution of known concentration
2 transferring this volume to a volumetric flask
3 adding sufficient water to the volumetric flask to make up the solution to the calibration mark.

To calculate the concentration of the diluted solution, we use the fact that the number of moles (particles) in the diluted solution is the same as in the original volume of solution.

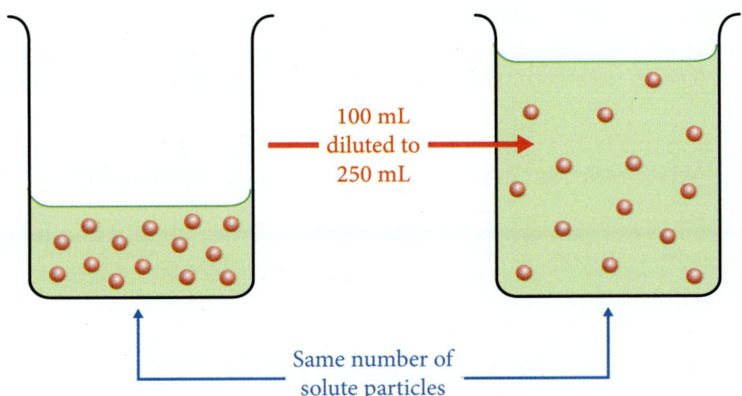

Figure C7.12 Dilution: the same number of moles in a larger volume of solvent

DILUTIONS

To practise more problems involving dilutions, complete this dilutions worksheet.

Because the number of moles of solute remains constant, we can use the following formula to relate concentration and volume of the original solution to that of the diluted solution. This relationship is called the dilution formula:

$$c_1V_1 = c_2V_2$$

where c_1 and c_2 are the initial and final concentrations of the solution in the same units, and V_1 and V_2 are the initial and final volumes of the solution, in the same units.

WORKED EXAMPLE 7.4

a What volume of 12 mol L^{-1} hydrochloric acid is required to prepare 500 mL of a 0.25 mol L^{-1} solution?

b What is the concentration of a solution made by diluting 20 mL of a 5.0 mol L^{-1} solution to 1.00 L?

Answers	Logic
a The volume is 10.4 mL.	**1** Write the relationship. $c_1V_1 = c_2V_2$ **2** Identify the quantities and check units for consistency. $c_1 = 12\ \text{mol L}^{-1}$, $V_1 = ?$, $c_2 = 0.25\ \text{mol L}^{-1}$, $V_2 = 500\ \text{mL}$ **3** Substitute. $12 \times V_1 = 0.25 \times 500$ $V_1 = \frac{0.25 \times 500}{12} = 10.4\ \text{mL}$
b The concentration is 0.10 mol L^{-1}.	**1** Write the relationship. $c_1V_1 = c_2V_2$ **2** Identify the quantities and check units for consistency. $c_1 = 5.0\ \text{mol L}^{-1}$, $V_1 = 20\ \text{mL}$, $c_2 = ?$, $V_2 = 1.00\ \text{L} = 1000\ \text{mL}$ **3** Substitute. $5.0 \times 20 = ? \times 1000$ $c_2 = \frac{5.0 \times 20}{1000} = 0.10\ \text{mol L}^{-1}$

Try these yourself

a What volume of solution is needed to dilute 18 mol L^{-1} sulfuric acid to prepare 2.5 L of 0.1 mol L^{-1} sulfuric acid?

b **i** A student was preparing for an experiment and diluted 10 mL of 5 mol L^{-1} copper(II) nitrate solution to 200 mL. What is the concentration of the final solution?

ii After performing the dilution the student realised they had made a mistake because they actually required 100 mL of a 0.10 mol L^{-1} solution. How much of the solution they diluted in part **i** should they use to obtain the required solution?

c A concentrated solution (10.0 mL) of sodium hydroxide was mixed with 55 mL of water to produce a solution of concentration 0.48 mol L^{-1}. What was the concentration of the original solution?

EXPERIMENT 7.2

PREPARING AND DILUTING SOLUTIONS

Colorimetry is an analytical technique used to determine the concentration of an unknown solution by comparing its colour with those of solutions of known concentration. In this experiment, you will prepare solutions of known concentration and use these to determine the concentration of a solution of unknown concentration.

Aim

To prepare solutions of known concentration and use these to determine the unknown concentration of a solution

Materials

- Approx. 0.1 g potassium permanganate ($KMnO_4$)
- Solution of unknown concentration
- 250 mL beaker
- 25 mL pipette
- 4 × 250 mL volumetric flasks
- Stirring rod
- Filter funnel
- Electronic balance
- Water
- Paper towel

What are the risks in doing this experiment?	How can you manage these risks to stay safe?
Potassium permanganate can stain skin and clothing.	Wear protective clothing and avoid contact.
Potassium permanganate can be irritating to the skin and eyes.	Wear safety glasses and avoid contact with the chemical.

In your write-up, add any more risks you can think of, as well as ways to manage them. In particular, expand on the two risks listed to identify specific risks involved with each of them.

Procedure

1. Weigh out approximately 0.1 g of potassium permanganate in one of the beakers. (Make sure you know the mass you are using accurately.)
2. Add just enough water to dissolve the crystals of potassium permanganate and transfer the solution to a 250 mL volumetric flask.
3. Fill the flask to the 250 mL mark with water and mix thoroughly.
4. Draw up a table like the one in the results section and complete it for this first solution.
5. Pipette 25 mL of the first solution into a clean volumetric flask and fill to 250 mL.
6. Complete the table for this second solution.
7. Repeat steps 5 and 6 using the second and third solutions so you have successive dilutions.
8. Compare the colour of your solution of unknown concentration to the colours of your diluted solutions.

Results

Solution	Relative colour intensity	Initial volume	Final volume	Mass $KMnO_4$	Moles $KMnO_4$	Concentration ($mol\ L^{-1}$)
1						
2						
3						
4						

Analysing the results

1 Complete the results table.
2 What effect does dilution have on:
   a colour intensity?
   b concentration of the solution?
   c mass of the solute in the diluted solution compared to the original mass?

Discussion

1 a What is the concentration of the unknown solution?
   b How easy was it to determine this?
   c What could be done to more accurately determine the unknown concentration?
2 How would you prepare a solution twice as concentrated as solution 2?

7.4 Acids and bases

Our environment is naturally acidic. Rain is slightly acidic as are many foods and drinks. Soft drink contains carbonic acid, citrus fruits contain citric acid and many foods contain vitamin C, which is ascorbic acid. Acids are also important components of many of the compounds that make up living things. Proteins are long chains of amino acids and our stomach produces hydrochloric acid to aid with digestion of food.

Bases are also part of our natural environment, although they are not as commonly recognised as acids. Ammonia is a base found in volcanic gases and produced by rotting plant and animal matter. Caffeine, found in coffee and tea, is a base as is nicotine, which is in tobacco.

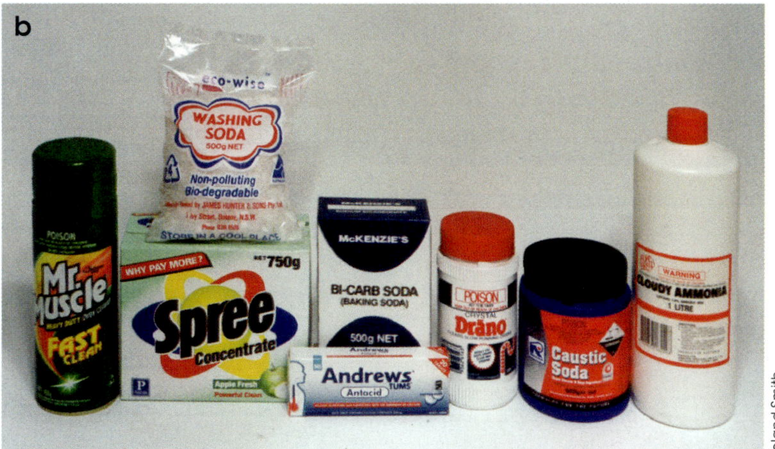

Figure C7.13 ▲
Common substances containing a) acids and b) bases

Definitions of acids and bases

In your Years 9 and 10 science course, you probably studied acids and bases and learnt practical definitions of these substances. Practical definitions are ones based on the common properties of a substance. Acids and bases are families of substances so members of these families can be recognised by their common properties.

Common properties of acids include:

- a sour taste
- sting or burn the skin
- conduct electricity in solution
- turn blue litmus red.

Common properties of bases include:

- a slippery feel in aqueous solution
- a bitter taste
- conduct electricity in solution (not all bases are soluble)
- turn red litmus blue.

WOW

Acid rain

Acid rain is rain that has become acidic because gases from the atmosphere have dissolved in it. About 70% of acid rain comes from sulfur dioxide (SO_2), which dissolves into the water to form sulfuric acid. The rest comes from various oxides of nitrogen (mainly NO_2 and NO_3, collectively called NO_x).

While the effect of acid rain on forests, lakes and rivers has been well publicised, the effects on people have not been as well noted. Many toxic metals are held in the ground in compounds. However, acid rain can break down some of these compounds, freeing the metals and washing them into water sources such as rivers. In Sweden, nearly 10000 lakes now have such high mercury concentrations that people are advised not to eat fish caught in them.

The effects of acids and bases on vegetable dyes such as litmus was initially used to classify substances as acids and bases.

While practical definitions are useful in the laboratory, chemists attempt to define substances by what they are as well as by what they do. The definition of an acid has changed several times over the last 250 years. French scientist Antoine Lavoisier (1743–94) was one of the first to try to establish what an acid is. In the late 19th century, Swedish chemist Svante Arrhenius attempted to explain acids and bases in terms of the particles they produced in aqueous solution. He proposed that acids were substances that ionised in solution to produce hydrogen ions and that acids were strong if they ionised almost completely and weak if the ionised only slightly. Arrhenius defined a base as a substance that in solution produced hydroxide ions. However, not all bases dissolve in water so his definition has been modified to include insoluble bases.

The definition of a base is quite general. Not all oxides are basic. Most oxides formed between a non-metal and oxygen (for example CO_2 and SO_3) are acidic or neutral, while those formed between a metal and oxygen (for example CaO and MgO) are basic or **amphoteric**.

You will learn about acids and bases in more detail in Units 3 & 4.

For present purposes, the following simple definition of acids and bases will be used.

An acid is a substance that produces hydrogen ions (H^+) in aqueous solution. More strictly, the hydrogen ion attaches to a water molecule and forms hydronium ions (H_3O^+).

A **base** is a substance that either contains the oxide ion (O^{2-}) or hydroxide ion (OH^-) or produces hydroxide ion in aqueous solution. A soluble base is called an **alkali**.

Common acids are hydrochloric acid (HCl), sulfuric acid (H_2SO_4) and nitric acid (HNO_3). When these acids are mixed with water to form acidic solutions, the molecules dissociate:

$$HCl \rightarrow H^+ + Cl^-$$
$$H_2SO_4 \rightarrow 2H^+ + SO_4^{2-}$$

Common soluble bases (alkalis) are sodium hydroxide (NaOH), barium hydroxide ($Ba(OH)_2$), potassium oxide (K_2O) and ammonia (NH_3). These substances either contain, or in aqueous solution produce, the hydroxide ion (OH^-). For example:

$$NaOH \rightarrow Na^+ + OH^-$$
$$K_2O + H_2O \rightarrow 2K^+ + 2OH^-$$
$$NH_3 + H_2O \rightarrow NH_4^+ + OH^-$$

Acid–base reactions

Acids and bases react to form compounds called salts and water. The name given to this reaction is **neutralisation**.

If the right amount of acid and base react, the resulting solution will be **neutral** (neither acidic nor basic).

The generalised reaction given above can be used to predict the products of most reactions between an acid and base.

A typical reaction between acids and bases is the reaction between a hydroxide and an acid:

$$\text{Acid} + \text{hydroxide} \rightarrow \text{salt} + \text{water}$$

For example:

$$HCl(aq) + NaOH(aq) \rightarrow NaCl(aq) + H_2O(l)$$

It is usual to write these reactions using the overall equation. The Na^+ and Cl^- ions are spectator ions so the net ionic equation is: $H^+(aq) + OH^-(aq) \rightarrow H_2O(l)$

WOW

Neutralisation reactions

Neutralisation reactions are important in our bodies and in our lives. For example, if too much hydrochloric acid is produced in our stomachs, then we could end up with heart burn or indigestion. This problem is commonly fixed by taking an antacid, which contains a base such as magnesium hydroxide or aluminium hydroxide and neutralises the excess acid.

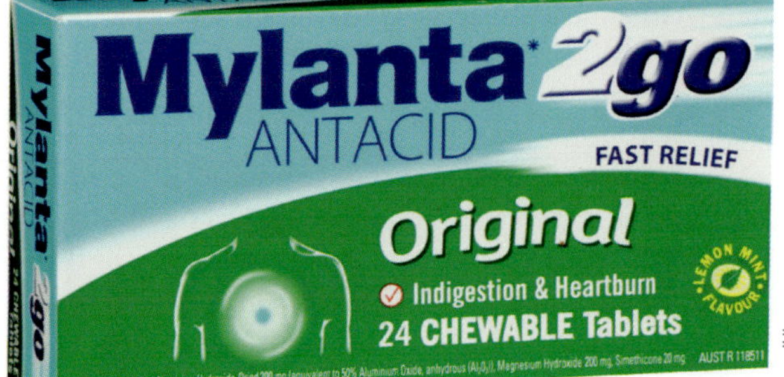

Figure C7.14 ▶ Antacids contain a base, which neutralises excess acid in the stomach.

ANTACIDS

Visit this website to find out more about the chemistry of how antacids work.

Neutralisation reactions can be written as the following generalised form:

$$\text{Acid} + \text{base} \rightarrow \text{salt} + \text{water}$$

Another typical reaction is that between a basic or amphoteric oxide and an acid:

$$\text{Acid} + \text{oxide} \rightarrow \text{salt} + \text{water}$$

For example:

$$2HCl(aq) + MgO(s) \rightarrow MgCl_2(aq) + H_2O(l)$$

Another common but not typical acid–base reaction worth noting is the reaction between an acid and ammonia. This reaction does not adhere to the generalisation given above because the H^+ ion attaches to the NH_3 molecule (not H_2O) to produce $NH_4^+(aq)$.

For example:

$$HCl(aq) + NH_3(aq) \rightarrow NH_4Cl(aq)$$

While ammonia is considered to be a weak base due the production of OH^- ions when it reacts with water,

$$NH_3(g) + H_2O(l) \rightarrow NH^{4+} + OH^-(aq)$$

the ammonium ion is considered to be a weak acid. This means it will react with a base to produce a salt and water. However, ammonia gas will also be produced in the reaction as seen in the following example:

$$NaOH(aq) + NH_4Cl(aq) \rightarrow NaCl(aq) + NH_3(g) + H_2O(l)$$

Ammonium salts such as ammonium sulfate or ammonium nitrate are commonly used as fertilisers. This is one reason why a farmer wouldn't treat a field with an ammonium compound at the same time as using an alkali (a soluble base) such as lime. The ammonium salt will react with the base to give off ammonia, which will escape into the atmosphere as a gas, and so not be available to the plants.

QUESTION SET 7.3

Remembering

1 Define 'acid' and 'base'.

2 What is meant by neutralisation? Give an example.

3 What is the difference between the hydrogen and hydronium ions?

Understanding

4 Write the dissociation reaction for the following acids in water.
  - a Nitric acid (HNO_3)
  - b Hydrochloric acid (HCl)
  - c Phosphoric acid (H_3PO_4)

5 a Use Table C7.1 to determine which of the following oxides and hydroxides are soluble: $Ca(OH)_2$, CuO, NH_4OH, Na_2O.
  - b Write the equation for the reaction of each with water.

Applying

6 Write balanced overall equations for the following reactions.
  - a Nitric acid with potassium hydroxide
  - b Sulfuric acid with zinc oxide
  - c Hydrofluoric acid with magnesium hydroxide
  - d Phosphoric acid with ammonia
  - e Acetic acid with sodium oxide
  - f Calcium oxide with ammonium nitrate

ACIDS AND BASES

Visit this online lab, which has many different resources showing properties and reactions of acids and bases as well as pH and indicators

Analysing

7 a Describe a method of preparing a solution of known concentration of the salt Na_2SO_4.

b How could the salt be obtained from the solution?

c What needs to be taken into account in the preparation to ensure only this salt is obtained?

7.5 Indicators and pH

Litmus is a mixture of water-soluble dyes extracted from lichens. Acids and bases change the colour of litmus. Litmus turns red in acidic solutions, and blue in basic solutions. Litmus is called an acid–base **indicator**.

Indicators are commonly used to measure acidity and alkalinity in soil and water, such as swimming pools, as well as monitor waste from industries, laboratories and sewage treatment plants.

An indicator can also be used to provide information on the degree of acidity or alkalinity of a solution. However, because chemists need more precise measures of the degree of acidity and alkalinity of solutions, they have developed a scale to measure the concentration of acid in a solution. This scale is called the **pH scale**.

©CSIRO

Figure C7.15 ▲
Checking the approximate pH of soil using an indicator colour chart

The pH scale

The term 'pH' stands for 'hydrogen power'. This name came about because the scale is based on the concentration of hydrogen ions in solution. Danish biochemist Soren Sorensen devised the scale to make writing the hydrogen ion concentration in solution more manageable.

The pH scale is based on the concentration of hydrogen ions in solution. It provides a measure of the acidity or alkalinity in aqueous solutions.

The pH scale generally ranges from 0 to 14. The lower the pH, the more acidic a solution is; the higher the pH, the more basic the solution is. Acids have a pH of less than 7. Bases have a pH greater than 7. A substance that has a pH equal to 7 is neutral.

Table C7.2 shows the relationship between pH and hydrogen ion concentration.

Exponential notation is a better way of representing the numbers involved in the concentration of H^+.

Table C7.2 pH and hydrogen ion concentration

pH	0	1	2	3	4	5	6	7	8	9	10	11	12	13	14
H^+ concentration (mol L^{-1})	1	10^{-1} or 0.1	10^{-2}	10^{-3}	10^{-4}	10^{-5}	10^{-6}	10^{-7}	10^{-8}	10^{-9}	10^{-10}	10^{-11}	10^{-12}	10^{-13}	10^{-14}

Notice the relationship between the pH value and the superscript number in the hydrogen ion concentration. They are the same number because pH is based on the hydrogen ion concentration of a solution.

An important aspect of the pH scale is that a change in 1 unit on the scale is equivalent to a 10 times change in the hydrogen ion concentration. For example, a change from pH 1 to 2 is equivalent to a change in the hydrogen ion concentration from 0.1 to 0.01 mol L^{-1}.

Although the pH scale is based on the hydrogen ion concentration, it can also be used to determine the alkalinity and therefore the hydroxide ion (OH^-) concentration. A pH of 7 represents a neutral solution and, as you learnt earlier, neutralisation occurs when an acid and base (alkali) react to produce a salt and water. Therefore, if equal concentrations of H^+ and OH^- ions were to react, the resultant solution should be neutral. At pH 7, the concentration of H^+ is 10^{-7} mol L^{-1} so the concentration of OH^- must also be 10^{-7} mol L^{-1}.

Neutralisation occurs when an acid and base (alkali) react to produce a salt and water. While this statement is correct, there are other factors related to the salt produced that will affect the pH of the final solution. You will learn more about this in Units 3 & 4.

Table C7.3 includes the concentration of OH^- ions compared to pH.

Table C7.3 pH and OH^- ion concentration

pH	0	1	2	3	4	5	6	7	8	9	10	11	12	13	14
H^+ concentration (mol L^{-1})	1 or 10^0	10^{-1} or 0.1	10^{-2}	10^{-3}	10^{-4}	10^{-5}	10^{-6}	10^{-7}	10^{-8}	10^{-9}	10^{-10}	10^{-11}	10^{-12}	10^{-13}	10^{-14}
OH^- concentration (mol L^{-1})	10^{-14}	10^{-13}	10^{-12}	10^{-11}	10^{-10}	10^{-9}	10^{-8}	10^{-7}	10^{-6}	10^{-5}	10^{-4}	10^{-3}	10^{-2}	10^{-1}	10^0

You should notice that as the H^+ ion concentration decreases, the OH^- ion concentration increases. You should also notice that the concentration superscripts always add to 14.

In Units 3 & 4, you will learn more about this relationship and apply it to solve problems related to unknown concentrations of acids and bases.

Measuring pH

There are many natural and synthetic indicators that are used to indicate whether a solution is acidic or alkaline as well as the degree of acidity. Some synthetic indicators are methyl violet, methyl orange, litmus, bromothymol blue and phenolphthalein. Their colour changes are shown in Figure C7.16. The actual range of acidity or alkalinity over which an indicator changes colour varies from one indicator to another.

Chemists have developed a **universal indicator** that produces several colour changes depending on the pH of the solution. It is a mixture of several indicators.

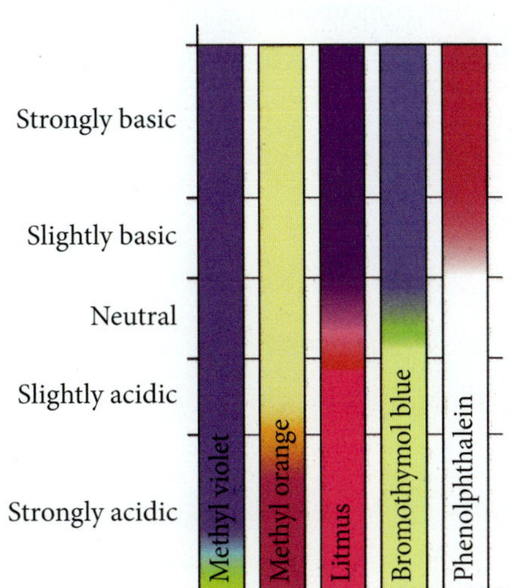

◀ **Figure C7.16** Colour changes in a range of indicators

pH SCALE AND CONCENTRATIONS

Visit this website to run a virtual lab that shows the relationship between the pH scale and acid and base concentrations.

Table C7.4 pH of some common substances

Colour of universal indicator	$[H_3O^+]$	pH	Substance	
	10	–1	Concentrated hydrochloric acid	ACID
	1	0	Car battery acid 1 mol l^{-1} hydrochloric acid	
	10^{-1}	1	0.1 mol L^{-1} hydrochloric acid	
	10^{-2}	2	Stomach acid	
	10^{-3}	3	Vinegar Lemon juice	
	10^{-4}	4	Soft drinks Soda water	
	10^{-5}	5	Wine Black coffee	
	10^{-6}	6	Rain water Milk, saliva	
	10^{-7}	7	Very pure water	NEUTRAL
	10^{-8}	8	Blood Sea water	ALKALINE
	10^{-9}	9	Bore water Baking soda solution	
	10^{-10}	10	Toilet soap	
	10^{-11}	11	Laundry detergents	
	10^{-12}	12	Household ammonia Dishwashing machine powders	
	10^{-13}	13	Chlorine bleach solutions	
	10^{-14}	14	Oven cleaners 1.0 mol L^{-1} sodium hydroxide	

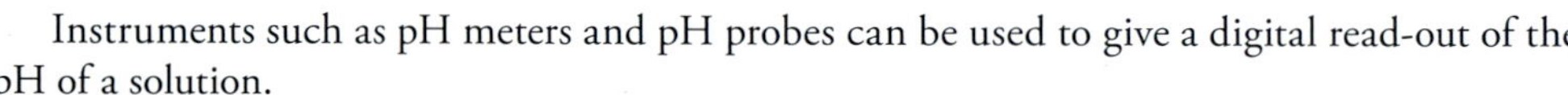

Instruments such as pH meters and pH probes can be used to give a digital read-out of the pH of a solution.

Figure C7.17 ▼ A pH meter

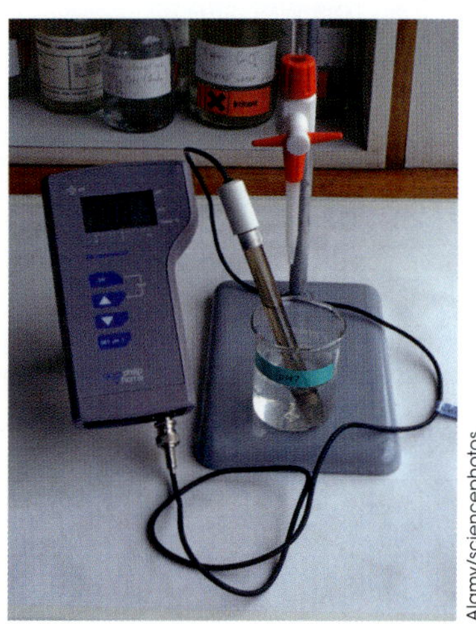

Alamy/sciencephotos

EXPERIMENT 7.3

MEASURING pH

By measuring the pH of various substances, you can determine how acidic and basic they are. In this investigation, you will be able to determine the actual pH of these substances by using universal indicator and a pH meter. An advantage of using a pH meter is that it does not involve adding any substances to the solution and usually gives a more accurate pH provided the meter is calibrated correctly.

Aim

To measure and compare the pH of various substances

Materials

- Universal indicator solution and pH colour chart
- Distilled water
- Various household substances, such as vinegar, lemon juice, dish washer detergent, shampoo, antacid tablet, baking soda, vitamin C, bleach, washing detergent, fabric conditioner, soda water, salt, sugar
- Petri dishes (one for each soil sample)
- Soil samples including potting mix
- Barium sulfate ($BaSO_4$) powder
- Samples of water from swimming pools and local waterways
- Test tubes and test-tube rack
- Spatula
- 1 mol L^{-1} and 0.0001 mol L^{-1} hydrochloric acid (HCl)
- 1 mol L^{-1} and 0.0001 mol L^{-1} sodium hydroxide (NaOH)
- pH meter

What are the risks in doing this experiment?	How can you manage risks to stay safe?
Household chemicals can burn the skin.	Wear safety glasses and protective clothing. Take care when handling the chemicals and tell your teacher if there is a spill.
Food products can become contaminated in the laboratory.	Do not eat or drink any food used in the laboratory.
Soil samples may contain harmful micro-organisms.	Do not touch the soil sample. Work in a well-ventilated area. Wash your hands before eating.

In your write-up, add any more risks you can think of, as well as ways to manage them. In particular, expand on the two risks listed to identify specific risks involved with each of them.

Procedure

Part A: Testing household substances

1 Set up test tubes in the rack. Place a few millilitres (enough to measure with the pH meter) of the substance to be tested in a test tube. (You may need to add distilled water if the substance is a solid.)
2 Test the pH with the pH meter.
3 Rinse the pH meter with distilled water.
4 Add one drop of universal indicator solution.
5 Record the colour and pH.
6 Repeat with all other household substances.

Part B: Testing water samples

1 If the samples contain a lot of suspended sediment, filter or allow them to settle before testing.
2 Repeat the procedure used in part A to test the samples.

Part C: Testing soil samples

1 Place approximately half a spatula of soil in a Petri dish.
2 Add a few drops of indicator to make a paste and stir it.

3 Lightly sprinkle a thin layer of barium sulfate powder over the soil paste. Do not stir.
4 Allow a few minutes for the powder to absorb the indicator. Record the colour and pH values.
5 Repeat steps 1–4 with other soil samples.

Results

Draw up a data table to record the colour, pH from universal indicator and pH from the meter for each substance. Include another column for hydrogen ion concentration.

Analysis of results

1 Use Table C7.2 to determine the hydrogen ion concentration of each solution. You may need to record a range for this value if the reading was not a whole number.
2 Make a list of substances from the most acidic to the most basic.
3 Record which substances, if any, were neutral.

Discussion

1 Identify any significant differences between the pH values obtained using universal indicator and the pH meter. Suggest possible reasons for these differences.
2 Suggest a procedure for measuring the pH of the soil samples using a pH meter.
3 Identify any noticeable difference in the pH of the water samples and suggest a reason for this difference.
4 Compare the hydrogen ion concentration of the known samples of HCl used with the pH reading obtained from the meter. Suggest what a difference in these readings could imply.
5 Reflect on possible errors which could have occurred and suggest what effect these may have had on the results.

Conclusion

Identify any generalisations related to the acidity and alkalinity of the substances tested.

REVISION

Take this quiz to check your understanding of acids, bases and pH.

Calculations involving acids and bases

You have already learnt about the relationship between concentration, moles and volume of a solution. You also learnt about dilutions. We will now apply this information to acids, bases and pH.

WORKED EXAMPLE 7.5

a How many moles of calcium oxide would be needed to neutralise 100 mL of a 0.400 mol L^{-1} solution of HCl?
b What mass of calcium oxide would need to added to the HCl solution?
c What is the pH of the final solution?
d If a solution of HNO_3 with a pH of 3 was diluted by a factor of 100, what is the pH of the final solution?

Answers	Logic
a 0.0200 mol are needed.	1 Write a balanced equation. $2HCl(aq) + CaO(s) \rightarrow CaCl_2(aq) + H_2O(l)$ 2 Calculate the number of moles of HCl using $C = \frac{n}{V}$. $c = 0.400\ \text{mol L}^{-1}$, $n = ?$, $V = 100\ \text{mL} = 0.100\ \text{L}$ $0.400 = \frac{n}{0.100}$ $n(\text{HCl}) = 0.400 \times 0.100 = 0.0400\ \text{mol}$

3 Use the balanced equation to calculate the number of moles of CaO needed to react with the HCl.

$$n(\text{CaO}) = \frac{1}{2}\, n(\text{HCl})$$

$$n(\text{CaO}) = \frac{0.0400}{2} = 0.0200 \text{ mol}$$

b The mass is 1.12 g

4 Calculate the mass of CaO.

$$n = \frac{m}{M}$$

$n = 0.0200$ mol, $m = ?$, $M(\text{CaO}) = 56.08$ g mol^{-1}

$$0.0200 = \frac{m}{56.08}$$

$$m = 0.0200 \times 56.08 = 1.12 \text{ g}$$

c The pH is 7.

5 Determine the pH.

Since the reaction involved using an exact quantity of CaO to neutralise the HCl, the final solution should be neutral. Therefore, the pH should be 7.

d The final pH is 5.

Recall the relationship between pH and concentration. A change by a factor of 10 in concentration is equivalent to a change in 1 unit in the pH scale.

Therefore, a change in concentration by a factor of 100 means a change in pH by two units.

The solution is being diluted so the pH will increase and the pH of the final solution will be 5.

Try these yourself

a How many grams of KOH are needed to neutralise 500 mL of 2 mol L^{-1} H_2SO_4?

b How many grams of $Mg(OH)_2$ are needed to neutralise 1.00 L of 0.25 mol L^{-1} HNO_3?

c If a solution of HCl with a pH of 1 is diluted by a factor of 1000, what will the pH of the diluted solution be?

d A 500 mL solution of NaOH of pH 9 was left to evaporate until the resulting volume was 50 mL. What is the pH of the resultant solution?

QUESTION SET 7.4

Remembering

1 a Define 'pH'.
  b Explain what information the pH scale provides.

2 a What is an indicator?
  b What is the purpose of an indicator?

3 True or false? If a statement is false, rewrite it so it is correct.
  a An acid tastes sour while a base tastes bitter.
  b A soluble base is called an acid.
  c A substance that is neither acid nor base is negative.
  d Bases have the power to neutralise acids.
  e An acid turns litmus blue while a base turns litmus red.

Understanding

4 What is the relationship between concentration of an acid, concentration of a base and pH?

5 An ant bite contains an acid while a wasp sting contains a base. Choose which of vinegar or ammonia would be best to treat each type of sting and explain why.

6 Consider the pH ranges 1–3, 4–6, 7, 8–10 and 11–13.
Which pH range would apply to each of the following solutions?
   a Concentrated hydrochloric acid
   b Distilled water
   c Dilute ammonia solution
   d Oven cleaner

7 If a 100 mL container of orange juice concentrate had a pH of 5 and the instructions on the container were to add the concentrate to 900 mL of water, what would be the pH of the diluted solution?

Applying

Use Figure C7.16 to answer the following question.

8 Which of the following statements is definitely correct?
   A A substance that turns phenolphthalein pink is more acidic than one that turns methyl orange pink.
   B A substance that turns methyl orange yellow is less acidic than one that turns methyl orange pink.
   C A substance that turns phenolphthalein colourless is neutral.
   D A substance that turns litmus red is more acidic than one that turns bromothymol blue a blue colour.

9 Which of the statements in Question **8** could possibly be correct, depending on the pH of the solution. Give the pH range that would make these statements correct.

Analysing

10 An aquarium was at the correct pH when a young child poured a 500 mL bottle of vinegar into it. The label on the bottle showed the vinegar in the bottle to be a 0.83 mol L^{-1} solution of acetic (CH_3COOH) acid.
   a How had the pH of the aquarium changed as a result of this action?
   b How many grams of sodium hydroxide would need to be added to the aquarium to neutralise excess acid?

Science Photo Library/Martyn F. Chillmaid

Figure C7.18 ▲
The reaction between magnesium metal and dilute hydrochloric acid

7.6 Reactions of acids

Acids undergo a number of different types of reactions. As with acid–base reactions, once the pattern of the reaction is known, products for most reactions of these types can be predicted.

Reactions between acid and a metal

Acids react with metals. When rain becomes acidic due to pollutants in the air, it reacts with metals, causing them to corrode more quickly. Not all metals are reactive, but for those that are the reaction can be represented by the general equation:

$$\text{Acid} + \text{metal} \rightarrow \text{a salt} + \text{hydrogen gas}$$

The following reactions are examples of reactions between an acid and magnesium metal.

$$2HCl(aq) + Mg(s) \rightarrow MgCl_2(aq) + H_2(g)$$
$$H_2SO_4(aq) + Mg(s) \rightarrow MgSO_4(aq) + H_2(g)$$

This reaction could also be called a corrosion reaction because the metal is eaten away.

Reaction between an acid and a carbonate or hydrogen carbonate

Many products such as bread and sponge cakes have a honeycomb structure, which contains bubbles. These bubbles are formed by carbon dioxide gas causing the mixture to rise during cooking. Carbon dioxide is formed by adding a carbonate (CO_3^{2-}) or hydrogen carbonate (HCO_3^-) and an acid to the cooking mixture.

When an acid and carbonate react, they produce a salt, carbon dioxide and water.

This can be represented by the general equation:

$$\text{Acid} + \text{carbonate} \rightarrow \text{a salt} + \text{carbon dioxide} + \text{water}$$

The following reactions are examples of reactions between an acid and calcium carbonate.

$$2HCl(aq) + CaCO_3(s) \rightarrow CaCl_2(aq) + H_2O(l) + CO_2(g)$$
$$H_2SO_4(aq) + CaCO_3(s) \rightarrow CaSO_4(aq) + H_2O(l) + CO_2(g)$$

When an acid and a hydrogen carbonate react, they will also produce a salt, water and carbon dioxide according to general equation:

$$\text{Acid} + \text{hydrogen carbonate} \rightarrow \text{a salt} + \text{carbon dioxide} + \text{water}$$

The following reactions are examples of reactions between an acid and sodium hydrogen carbonate.

$$HCl(aq) + NaHCO_3(s) \rightarrow NaCl(aq) + H_2O(l) + CO_2(g)$$
$$H_2SO_4(aq) + 2NaHCO_3(s) \rightarrow Na_2SO_4(aq) + 2H_2O(l) + 2CO_2(g)$$

There is a pattern to all these reactions. Knowing the pattern means the products can be predicted.

Mark Fergus Photography

Figure C7.19 ◀ Marble chips (calcium carbonate) react vigorously with hydrochloric acid.

QUESTION SET 7.5

Remembering

1 Complete the following general equations for the reactions of acids.

a Acid + base →

b Acid + metal →

c Acid + carbonate →

Understanding

2 Using the general reactions as a guide, write balanced equations for the following reactants.

a $H_2SO_4(aq) + NaOH(aq) \rightarrow$

b $CH_3COOH(aq) + Ca(OH)_2(aq) \rightarrow$

c $HNO_3(aq) + Mg(s) \rightarrow$

d $HCl(aq) + Zn(s) \rightarrow$

e $HF(aq) + Ca(HCO_3)_2(s) \rightarrow$

f $HNO_3(aq) + CuCO_3(s) \rightarrow$

7.7 Identifying unknown ions in solution

In Experiment 7.1, you used your understanding to identify ions in unknown solutions. There are several common tests that can be used to identify particular ions in solution. The experimental results are then compared with known test results and this allows ions to be identified.

Table C7.5 Tests used to identify the four selected anions

Anion	Test
Carbonate (CO_3^{2-})	1 Solution has a pH between 8 and 11 (pH paper suffices). 2 Addition of dilute HNO_3 produces bubbles of colourless gas (CO_2).[A]
Sulfate (SO_4^{2-})	1 Addition of $Ba(NO_3)_2$ to an acidified sample of the solution produces a thick white precipitate. 2 Acidification and addition of $Pb(NO_3)_2$ produces a white precipitate.
Phosphate (PO_4^{3-})	1 Addition of ammonia followed by $Ba(NO_3)_2$ produces a white precipitate. 2 Addition of Mg^{2+} in an ammonia/ammonium nitrate buffer produces a white precipitate, $Mg(NH_4)PO_4$. 3 Acidification with HNO_3 followed by addition of ammonium molybdate solution ($(NH_4)_2MoO_4$) produces a yellow precipitate; warming the mixture for a few minutes may be necessary.
Chloride (Cl^-)	1 Addition of $AgNO_3$ to an acidified sample produces a white precipitate,[B] which dissolves in ammonia solution and darkens in sunlight.

[A] Any strong acid would do, but for analysing mixtures we do not want to introduce any Cl^- or SO_4^{2-}.

[B] In non-acidic solutions, silver nitrate also produces precipitates with carbonate and phosphate (and with sulfate at all pH values if sulfate concentration is moderately high), so this test alone does not prove the presence of chloride: it is also necessary to prove the absence of sulfate.

Table C7.6 Precipitates formed between cations and anions

Cation[A]	Anion					
	OH^-	Cl^-	SO_4^{2-}	CO_3^{2-}[B] in alkaline solution	PO_4^{3-}	
					Solution pH < 2	Solution pH > 6
Ba^{2+}	No	No	Yes	Yes	No	Yes
Pb^{2+}	Yes	Yes[C]	Yes	Yes	No	Yes
Ag^+	Yes[D]	Yes	Yes[C]	Yes	No	Yes
Cu^{2+}	Yes	No	No	Yes	No	Yes

[A] In acidic or alkaline solution.

[B] Cannot have carbonate in acid solution: it decomposes to $CO_2(g)$.

[C] Provided concentration of Cl^- or SO_4^{2-} is not too low, say > 0.05 mol L^{-1}.

[D] Precipitate is brown Ag_2O.

In Experiment 7.4, you will use the information provided in Tables C7.5 and C7.6, along with what you have learnt, to identify some unknown anions. A similar process could be used to identify cations.

In addition, cations can be identified by flame tests, which you conducted in Experiment 1.1 on page 123.

Table C7.7 Tests used to identify cations

Cation	Tests
Ag^+	With Cl^- forms a white precipitate that dissolves in ammonia solution
Pb^{2+}	With Cl^- forms a white precipitate (if solution not too dilute, say >0.05 mol L^{-1} and this precipitate does not dissolve in ammonia solution With I^- forms a yellow precipitate
Ba^{2+}	With SO_4^{2-} forms a white precipitate Gives a pale green flame colour No precipitate with OH^- or F^- (compare Ca^{2+})
Ca^{2+}	With SO_4^{2-} forms a white precipitate (if solution not too dilute, say >0.05 mol L^{-1}) With F^- forms a white precipitate Gives a brick-red flame colour
Zn^{2+}	With hydroxide forms a white precipitate This precipitate dissolves in both excess OH^- solution and ammonia solution
Cu^{2+}	With OH^- forms a blue precipitate This precipitate dissolves in NH_3 to form a deep blue solution Gives a blue-green flame colour
Fe^{2+}	With OH^- forms a green or white precipitate, which may turn brown Decolorises acidified dilute potassium permanganate solution
Fe^{3+}	With OH^- forms a brown precipitate With thiocyanate (SCN^-) forms a deep red solution

EXPERIMENT 7.4

IDENTIFYING UNKNOWN ANIONS

In this experiment you will conduct a series of tests to identify specified anions in a solution that contains only one anion. Many of the reactions that occur in this experiment are quite complicated, so you are not expected to write equations for them. The main purpose of this experiment is to understand the processes used.

Aim

To carry out a series of chemical reactions to devise chemical tests for identifying the phosphate, sulfate, carbonate and chloride anions (PO_4^{3-}, SO_4^{2-}, CO_3^{2-} and Cl^-) in solution when these are the only ions that could be present

Materials

- Dropper bottles containing 0.1 mol L^{-1} of test solutions $Pb(NO_3)_2$, $Ba(NO_3)_2$, HNO_3, $AgNO_3$ and NaOH
- Dropper bottles containing 0.1 mol L^{-1} of anion solutions Na_2SO_4, Na_2CO_3, Na_3PO_4 and NaCl
- 4 test tubes and test-tube rack
- Distilled water
- Water bath

What are the risks in doing this experiment?	How can you manage risks to stay safe?
Concentrated NaOH and HNO_3 are corrosive.	Wear safety glasses and protective clothing. Do not allow contact with skin or clothes. If contact occurs, wash with large amounts of water for 10–15 minutes.
Some metal salts (silver, barium and lead) are poisonous. Silver nitrate will leave a brown stain on clothing and skin.	Avoid contact with skin and clothes. Wash hands thoroughly after use. Dispose of the chemicals as directed by your teacher.

In your write-up, add any more risks you can think of, as well as ways to manage them. In particular, expand on the two risks listed to identify specific risks involved with each of them.

Procedure

1 Add 10 drops of each of the anion solutions to each of four test tubes.
2 Add 10 drops of HNO_3 to each of the test tubes and warm them gently.
3 Record the results.
4 Use these results to describe a test for one particular anion.
5 Thoroughly clean all test tubes with distilled water between tests.
6 For the remaining three anions, put 10 drops in each of three test tubes, add 5 drops of HNO_3 and then add 5 drops of Ag^+.
7 Record the results in a table like the one in the results section. Write 'NP' if no precipitate forms and 'ppt' if precipitate forms and give its colour.
8 Repeat steps 6 and 7 with Pb^{2+} and then Ba^{2+}.
9 After testing the Ba^{2+}, add 10 drops of NaOH to each of the test tubes and record any changes.

Extension

1 Use the results from the tests to identify the anion (one per solution) in each of three unknown solutions.
2 Try to identify an unknown solution containing two anions.

Results

Test solution/anion	H^+	Ag^+	Pb^{2+}	Ba^{2+}	Ba^{2+} and OH^-
CO_3^{2-}					
Cl^-					
SO_4^{2-}					
PO_4^{3-}					

Analysis of results

For each test and result below, decide which anion would be the most appropriate fit to these results.

Anion	Test and results
	Gives a ppt with acidified Ag^+ but not with Ba^{2+}
	Gives a ppt with acidified Ba^{2+}
	Gives a ppt with Ba^{2+} in alkaline solution but not acid solution
	Produces bubbles of gas with addition of dilute HNO_3

Discussion

1 Is the lead ion useful in identifying anions? Explain why or why not.

2 When identifying anions in a solution in which two anions may be present, it is necessary to destroy any CO_3^{2-} before further testing for other anions can be carried out. How might this be done? (Consider the tests in this experiment.) Why is this necessary?

Conclusion

How effective was your identification?

WATER QUALITY

Visit this virtual lab in which you will use analysis to investigate water quality.

CHAPTER CHECKLIST

You should know:

- the different types of equations that can be used to show precipitation reactions
- concentration is a measure of the amount of solute in a given volume of solution
- different units of concentration
- how dilution affects concentration
- the relationship between the hydrogen and hydronium ion
- the general equation for the neutralisation reaction between an acid and base
- the purpose of an indicator
- the relationship between pH, hydrogen ion and hydroxide ion concentrations
- the general reactions for acid and metal, acid and carbonate, acid and hydrogen carbonate
- the definition of precipitate, acid and base
- the definitition of pH and pH scale.

You should be able to:

- define precipitate, acid and base
- define pH and the pH scale
- use solubility tables to predict the formation of a precipitate
- write balanced reactions showing precipitation
- use experimental data to identify unknown ions
- calculate concentration
- convert between moles, volume, concentration and mass
- calculate mass of reactants and products in a chemical reaction given the amount of one substance in the reaction
- calculate concentrations and volumes for dilutions
- identify an acid and a base
- predict the products of acid–base reactions
- measure pH and use data to determine pH
- predict products of reactions
- use analytical techniques to identify specific ions.

CHAPTER GLOSSARY

acid a substance that produces hydrogen ions (H^+) in aqueous solution

alkali a soluble base

amphoteric a substance that can act as either an acid or a base depending on the other reactants

aqueous solutions solutions in which water is the solvent

base a substance that either contains the oxide (O^{2-}) or hydroxide ion (OH^-) or produces the hydroxide ion in aqueous solution

complete ionic equations equations that show all the ions present in solution

concentration the amount of solute dissolved in a given volume of solvent

dilution the process of adding water to a solution to reduce its concentration

hydronium ion (H_3O^+) the ion produced when a hydrogen ion (H^+) attaches to a water molecule

indicator a substance that changes colour in solution depending on whether the solution is acidic or basic

molarity the number of moles of solute per litre of solution

net ionic equations an equation that only contains the ions that take part in the chemical reaction; that is, spectator ions are excluded

neutral neither acidic nor basic, pH of 7

neutralisation a reaction between an acid and a base to form a salt and water

pH scale a scale that measures the acidity and alkalinity of a solution, based on the concentration of hydrogen ions in a solution

precipitate a solid produced by reaction between two clear solutions

precipitation reaction a reaction in which a precipitate is produced

solubility the degree to which a substance dissolves

spectator ion an ion that is present in the same form on each side of the equation

universal indicator a mixture of different indicators that displays a range of colours when added to solutions of different pH

volumetric flask a piece of glass laboratory equipment used to prepare solutions of precise concentration

CHAPTER REVIEW QUESTIONS

Remembering

1 Write the definition of:
  a precipitate.
  b acid.
  c base.
  d pH.
  e concentration.

2 What is a spectator ion?

3 What is the relationship between the hydrogen ion and the hydronium ions?

4 What is the relationship between pH, hydrogen ion concentration and hydroxide ion concentration?

Understanding

5 Use the solubility table to determine whether or not a precipitation reaction will occur when the following reactants are mixed. Where a reaction occurs, write the balanced overall equation and net ionic equation for the reaction.
  a Potassium chloride and zinc nitrate
  b Ammonium bromide and lead nitrate
  c Ammonium sulfide and magnesium acetate
  d Strontium chloride and zinc sulfate

6 What solutions would you mix to produce each of the following precipitates?
  a Lead sulfate
  b Iron(II) sulfide
  c Magnesium hydroxide

7 How many moles of solute do you need to make the following solutions?
  a 2.00 L of 1.50 mol L^{-1} sodium chloride
  b 3.5 L of 0.20 mol L^{-1} potassium hydroxide

8 Calculate the molarity of the solutions made by dissolving the following amounts of solute in water and making the volume up to the stated value.
  a 5.0 mol nitric acid in 2.0 L
  b 2.5 mol sodium hydroxide in 0.50 L

9 What mass of the indicated substance do you need to weigh out to make the following solutions?
  a Barium hydroxide to make 0.5 L of 0.060 mol L^{-1} hydroxide ion solution
  b Sulfuric acid to make 250 mL of 0.330 mol L^{-1} hydrogen ion solution

10 The volume of solution in column A was diluted to the volume in column B. Calculate the molarity of the diluted solution.

	Column A	Column B
a	50 mL of 0.242 mol L^{-1} hydrochloric acid	500 mL
b	25 mL of 0.152 mol L^{-1} sulfuric acid	2.0 L
c	10 mL of 0.114 mol L^{-1} sodium hydroxide	250 mL

11 What volume of solution in Column A in the following table is needed to prepare the solution in Column B?

	Column A	Column B
a	0.282 mol L^{-1} hydrochloric acid	250 mL of 0.0113 mol L^{-1}
b	2.42 mol L^{-1} sulfuric acid	2.0 L of 0.121 mol L^{-1}
c	0.318 mol L^{-1} sodium hydroxide	1.0 L of 0.300 mol L^{-1}

12 What colour would you expect to see in a piece of litmus paper to which the following have been added?

a 1 drop of $Ca(OH)_2(aq)$

b 1 drop of $HF(aq)$

c 1 drop of $NaNO_3(aq)$

13 Milk has a pH of 6.5. This means that milk is:

A strongly acidic.

B slightly acidic.

C slightly basic.

D strongly basic.

14 If tomato juice has a pH of 5 and orange juice has a pH of 3, then:

A orange juice is twice as acidic as tomato juice.

B tomato juice is twice as acidic as orange juice.

C orange juice is 100 times more acidic than tomato juice.

D tomato juice is 100 times more acidic than orange juice.

15 Using the general reactions as a guide, write balanced equations for the following.

a $HNO_3(aq) + KOH(aq) \rightarrow$

b $HCl(aq) + Al(OH)_3(aq) \rightarrow$

c $H_2SO_4(aq) + Fe(s) \rightarrow$

d $CH_3COOH(aq) + MgCO_3(s) \rightarrow$

e $H_3PO_4(aq) + NaHCO_3(s) \rightarrow$

Applying

16 A solution was thought to contain either lead nitrate or calcium nitrate. Describe what tests could be carried out to correctly identify the substance.

17 A solution was known to contain either sodium carbonate or sodium chloride. Samples of the solution gave bubbles of gas with HNO_3 but no precipitate with $AgNO_3$. Which anion is present? Explain.

18 A solution was thought to contain either barium chloride or calcium chloride. Samples of the solution gave a white precipitate with Na_2SO_4 and a pale green flame colour. Which of the ions is present? Explain.

19 If a 1 L container of HCl had a pH of 3 and instructions on the container were to dilute 100 mL to 1 L before use, then what would be the pH of the solution being used?

20 To determine the solubility of calcium hydroxide ($Ca(OH)_2$) a chemist took 25 mL of a saturated calcium hydroxide solution and found that it reacted completely with 8.13 mL of 0.102 mol L^{-1} hydrochloric acid. Calculate the:

a molarity of the calcium hydroxide solution.

b solubility of calcium hydroxide in grams per litre.

21 In 16.4 mL of 0.117 mol L^{-1} magnesium chloride solution, how many moles are there of:

a magnesium chloride?

b magnesium ions?

c chloride ions?

Analysing

22 A student was given four solutions and told each contained one of the following cations: Pb^{2+}, Cu^{2+}, Ba^{2+}, Ca^{2+}, Fe^{2+} or Fe^{3+}. The results of a series of tests that were conducted are given in the following table, where NP stands for no precipitate and ppt stands for precipitate .

Reagent added	A	B	C	D
KI	Yellow ppt	NP	NP	NP
H_2SO_4	ppt	NP	ppt	ppt
NaOH	ppt	Brown ppt	NP	ppt

Use the given results to identify possible cations for A–D. If there is not enough information, suggest other possible tests and their expected results that could be used to identify any still unknown solutions.

23 A solution was made by dissolving 22.22 g of calcium chloride in enough water to make a 500 mL solution. This was then reacted with 0.5 mol L^{-1} sulfuric acid to produce insoluble calcium sulfate. Calculate the:

a volume of sulfuric acid needed to react

b mass of calcium sulfate produced

24 Consider the following table of different indicators showing colour change over different pH ranges.

Indicator	Colour change				
	Highly acidic	Slightly acidic	Neutral	Slightly alkaline	Highly alkaline
Methyl orange	Red	Yellow	Yellow	Yellow	Yellow
Bromothymol blue	Yellow	Yellow		Blue	Blue
Litmus	Red	Red	Purple	Blue	Blue
Phenolphthalein	Colourless	Colourless	Colourless	Colourless	Red

a Solution A is red in methyl orange, while solution B is red in phenolphthalein. Which is the more acidic?

b Four different solutions were tested with different indicators. Which of the solutions could be neutral?

i Colourless in phenolphthalein

ii Red in litmus

iii Yellow in methyl orange

iv Blue in bromothymol blue

25 The following table shows the most suitable pH ranges for the growth of some common plants.

Flower	pH range	Crop	pH range
Azalea	4.5–5.5	Barley	6.0–8.0
Calendula	6.0–7.5	Clover	5.5–7.0
Hibiscus	6.0–7.0	Wheat	5.5–7.5
Daffodil	6.0–6.5	Cotton	5.0–6.0
Sweet pea	7.0–8.0		

The pH of a number of soils was tested and the results are shown below.

Soil	pH
A	4.0
B	5.0
C	6.0
D	7.5
E	8.5

a Which soil was most basic?

b In which soil(s) would clover probably grow best?

c Which crop is most likely to grow best in soil A? Why?

d What would need to be added to soil B to make it suitable for growing barley?

e Hydrangea flowers change colour depending on the soil pH. They have blue flowers in acidic soil and pink flowers in basic soil. If some were planted in soil B, then what coloured flowers would they produce?

Reflecting

26 A chain of coffee shops started giving away brightly coloured drinking mugs. It was thought that lead compounds had been used in the paint and that this dissolved in acidic drinks such as soft drinks and fruit juices.

Design an experiment to determine:

a whether lead is present in the paint.

b the quantity of lead dissolved in 100 mL of an acidic drink.

CHAPTER 8
GASES

By the end of this chapter you will have covered the following material.

Science Understanding

- The behaviour of gases, including the qualitative relationships between pressure, temperature and volume, can be explained using kinetic theory (ACSCH060)

8.1 Introducing gases

We live at the bottom of a layer of gases called the atmosphere and we breathe air from the atmosphere into our bodies every few seconds. Life could not exist without the oxygen and carbon dioxide that make up part of our atmosphere. However, gases often go unnoticed except when they are at their most destructive in cyclones and tornadoes. Every day, we use gases for many purposes, including adding fizz to soft drinks, liquefied petroleum gas (LPG) to fuel motor vehicles and natural gas for cooking and heating.

Figure C8.1 ▶ The atmosphere as seen from space

iStockphoto/Baoshan Zhang

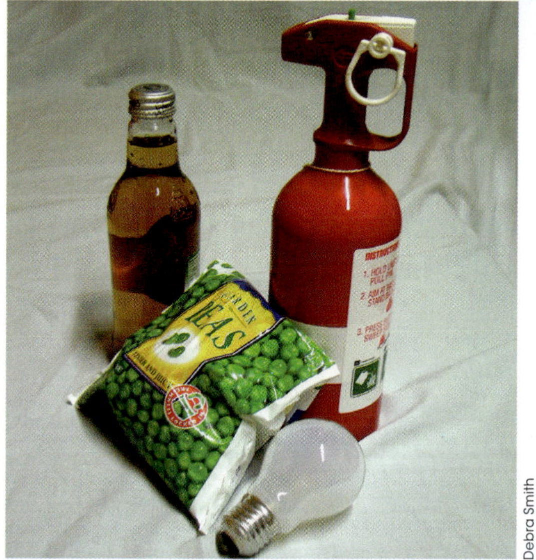

Figure C8.2 ▶ Some products that use components of air.

Debra Smith

USING AIR

Watch this video showing how gases are removed from the air, used in food packaging and transported.

Air

Air is a mixture of many substances. The main four are the gases nitrogen, oxygen, argon and carbon dioxide. The amounts of these are remarkably constant throughout the atmosphere. Air also contains water vapour, but its percentage varies considerably from close to 0% in dry desert air to 5–6% in air above a tropical rainforest. Scientists usually refer to the composition of dry air (Table C8.1) on page 343.

Table C8.1 Major and minor components of dry air

Major constituent	Amount (% v/v)	Minor constituent	Amount (ppm)
Nitrogen	78.08	Neon	18
Oxygen	20.95	Helium	5
Argon	0.93	Methane	1.5
Carbon dioxide	0.035	Krypton	1
		Hydrogen	0.5
		Dinitrogen oxide	0.2
		Carbon monoxide	0.1
		Ozone	0.02
		Nitrogen oxide and nitrogen dioxide	<0.01
		Ammonia	<0.01
		Sulfur dioxide and dihydrogen sulfide	<0.002

The concentrations of some substances are measured in parts per million (ppm) by volume. This measure is often used when concentrations involved are very small; for example, 350 ppm corresponds to a percentage concentration of 0.0350%. Parts per million can also be used for weight/weight (w/w) comparisons.

In Table C8.1 the composition of the air is based on volume, so 100 L of air contains approximately 78 L nitrogen (N_2), 21 L oxygen (O_2) and 1 L argon (Ar). An alternative way to represent composition is in terms of relative amounts of particles or moles, but because the volume of a gas sample varies with temperature and pressure, if gas volumes are to be compared, then they must be measured at the same temperature and pressure

Gas pressure

Gases consist of particles in constant random motion and the gas particles are a long way apart. When the gas particles collide with something, they exert a **pressure**. Gas pressure is a measure of the force per unit area. The International System of Units (SI) unit of gas pressure is the **pascal (Pa)** and is equivalent to a force of 1 newton per square metre.

Dreamstime/Sdbower

◀**Figure C8.3** Hot air ballooning is possible because the hot air inside the balloon is less dense than the outside air, allowing the balloon to rise.

Figure C8.4 ▶ Pressure is exerted by gas particles when they collide with the walls of a container.

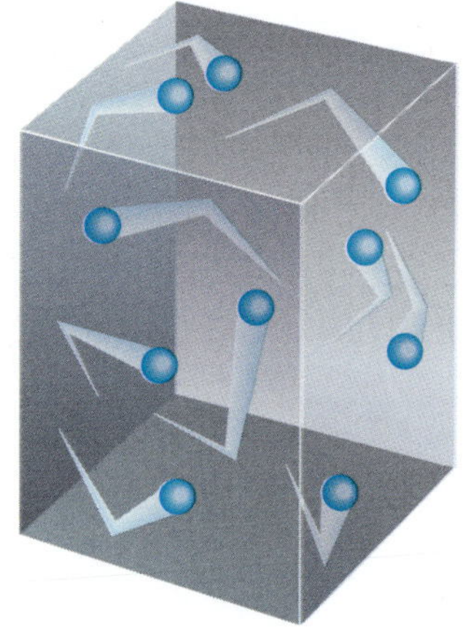

EFFECT OF AIR PRESSURE

Watch this video, which demonstrates the effect of air pressure.

The gases in the atmosphere exert pressure on every exposed surface. The pressure due to the atmosphere is called **atmospheric pressure**. This pressure varies from place to place and time to time. At sea level, atmospheric pressure is like having a kilogram tub of margarine pushing on every square centimetre of every part of the surface of your body all the time but we don't usually notice it is even there.

In order to make communication easier, scientists agreed on a standard of pressure to represent the average air pressure at sea level. This unit of standard atmospheric pressure is 1.10325×10^5 Pa or 101.325 kPa (kilopascals) at sea level. Another common unit for pressure is the atmosphere, which is defined as:

$$1 \text{ atmosphere (atm)} = 101.3 \text{ kPa}$$

Atmospheric pressure varies with height above sea level, temperature and weather conditions, so it is not always 1.00 atm. Meteorologists generally mention high- and low-pressure systems in weather reports and use the hectopascal (1 ha = 100 Pa) or millibar (100 Pa) when discussing pressure. Chemists and physicists generally use the units pascal (or kilopascal) and atmosphere (Table C8.2).

Table C8.2 Commonly used units of gas pressure

Name of unit	Symbol	Comparison
Pascal	Pa	Standard unit, 1 newton per square metre
Kilopascal	kPa	1000 Pa
Atmosphere	atm	101.3 kPa
Hectopascal	hPa	100 Pa
Millimetres of mercury	mmHg	760 mmHg = 1 atm

WOW

Barometer

Italian scientist Evangelista Torricelli is credited with inventing the mercury barometer in 1643 to measure atmospheric pressure. He created a 1-metre-long tube that was sealed at one end, filled it with mercury and set it vertically in a basin of mercury. The column of mercury in the tube fell to a height of 76 cm and this height was originally used as a standard measure of atmospheric pressure. The obsolete unit of pressure, the torr, which equals 1 mmHg, is named after him.

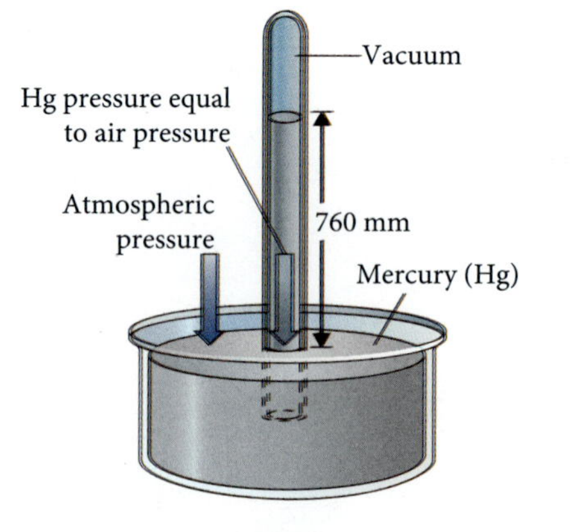

Figure C8.5 ▶ A barometer is used to measure atmospheric pressure.

Figure C8.6 shows how to convert between different units of pressure.

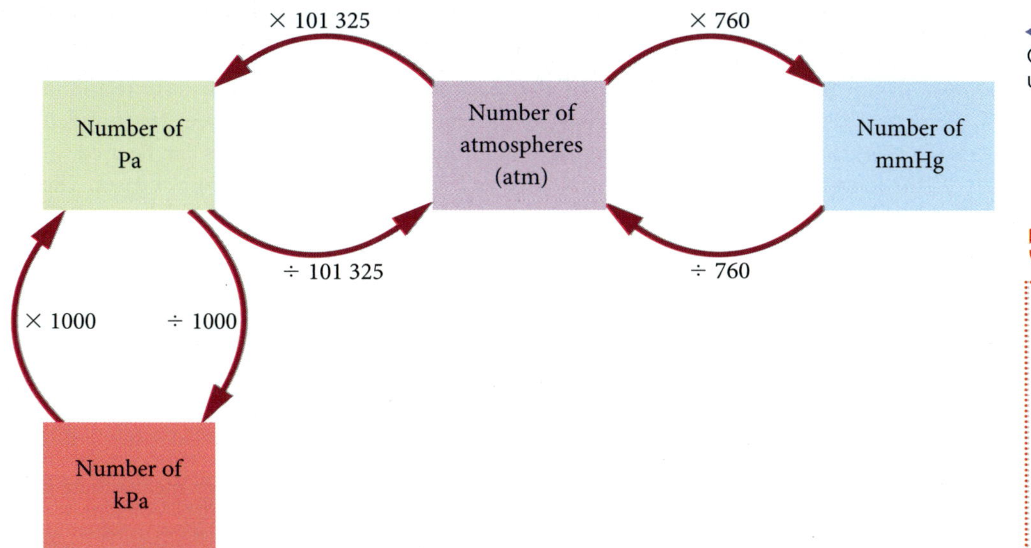

◀**Figure C8.6** Converting pressure units

PRESSURE UNITS DEFINED

Watch this video, which describes different units and explains how to convert between units.

QUESTION SET 8.1

Remembering

1 Explain what causes gas pressure.

2 Name two standard units of pressure.

Understanding

3 Use Table C8.1 to answer the following.
  - a Draw a bar graph to show the composition of the main gases of the atmosphere.
  - b How much argon is in a 100 L of air?

4 Explain what causes air pressure to vary.

5 Convert the following pressures to atmospheres.
  - a 75.5 kPa
  - b 6.83 Pa
  - c 652 mmHg

6 Convert the following pressures to kPa.
  - a 1.03 atm
  - b 0.45 atm
  - c 820 mmHg

Applying

7 Decide whether the following objects would work if there were no atmosphere. Explain why or why not.
  - a Vacuum cleaner
  - b Siphon
  - c Match
  - d Magnet

To see how low pressure and amount of oxygen affect us, refer to Context 4, 'Making reactions work for us', page 92.

8.2 Kinetic theory of gases

The pressure exerted by gases is one of the many similar physical properties they exhibit. Experimental observations by chemists that different gases behave in a similar manner led to the development of mathematical relationships linking volume, mass, pressure and temperature. These relationships are based on the notion of an **ideal gas**; that is, one that obeys the relationships perfectly. While no real gas behaves in a completely ideal manner, most gases are considered to being close enough to ideal under normal conditions that the relationships can be used.

The **kinetic theory of gases** explains much of the physical behaviour of gases. The term 'kinetic' comes from the Greek *kinetikos*, which means 'moving', so many of the properties of gases are related to their motion.

The kinetic theory of gases proposes that:

1 Gases consist of molecules (except the noble gases, which consist of atoms) that move in continual random straight-line motion.
2 The average distance between gas molecules is very large compared to the size of the molecule.
3 Intermolecular forces between molecules are negligible.
4 All collisions of gas molecules are perfectly **elastic collisions**, which means there is no net energy loss during these collisions.
5 Pressure is due to collisions of the molecules with the walls of the container.
6 Temperature is a measure of the average kinetic energy of the molecules.

The kinetic theory of gases is used to explain the physical behaviour of gases.

It is important to remember that this theory uses the model of an ideal gas. Gases, such as hydrogen and helium, which have small, light particles come close to 'ideal' under conditions of low pressure and high temperature. Chemists apply kinetic theory to real gases under normal conditions.

Considering temperature

The kinetic theory states that temperature is a measure of average kinetic energy. Although the molecules of a sample of gas have the same average kinetic energy, the individual molecules move at various speeds. Some move faster and some move slower than the average. Collisions between particles change individual speeds.

However, at a given temperature, molecules of all gases, no matter what size, shape or mass, have the same average kinetic energy. Kinetic energy is given by the formula:

$$\text{Kinetic energy} = \frac{1}{2}mv^2$$

where m is the mass and v is the speed.

Because kinetic energy is determined by both mass and speed, heavier gases will have lower speeds than lighter gases at the same temperature, as shown in Figure C8.7.

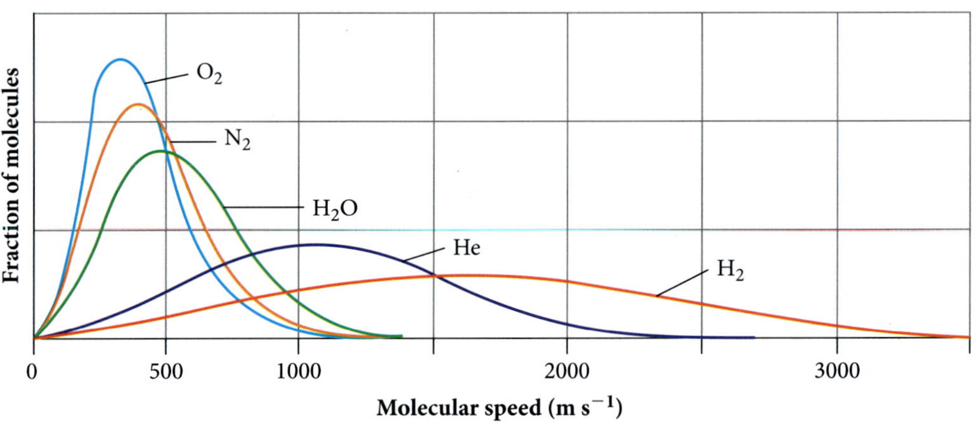

Figure C8.7 ▶ Distribution of speeds for different gases at the same temperature

As the temperature increases, the kinetic energy of the particles increases, and vice versa. Although the gas particles have a range of speeds at any given temperature, the average speed of the particles of a gas increases as temperature increases. Figure C8.8 shows how.

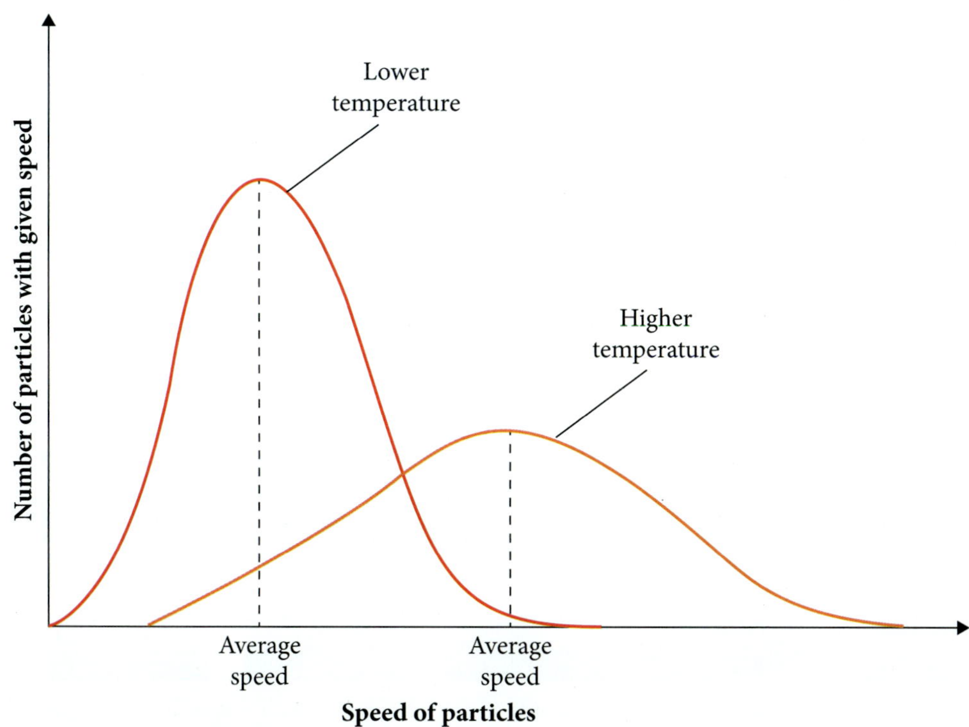

◀Figure C8.8 Average speed of particles at two different temperatures

Therefore, decreasing temperature will result in the particles slowing down. Theoretically, at low enough temperatures, the particles will eventually stop moving. The theoretical lowest temperature possible is known as **absolute zero** (0 K), which is equivalent to −273.15°C.

Absolute (kelvin) temperature scale

The **absolute temperature** is a measure of the average kinetic energy of its molecules. It was developed by British scientist William Thomson, who was also known as Lord Kelvin. He used experimental data and concluded from relationships between volume and temperature that there had to be a lower limit to temperature and that this limit was −273°C.

Lord Kelvin set up a new temperature scale and made the lowest number on the scale 0, which is also called absolute zero. The scale has the same sized units as the Celsius scale. Because he was the first person to recognise the importance of absolute zero, the unit on the scale is called the kelvin (K). The units do not use the degree sign. Figure C8.9 shows how to convert between Celsius and absolute temperature scales.

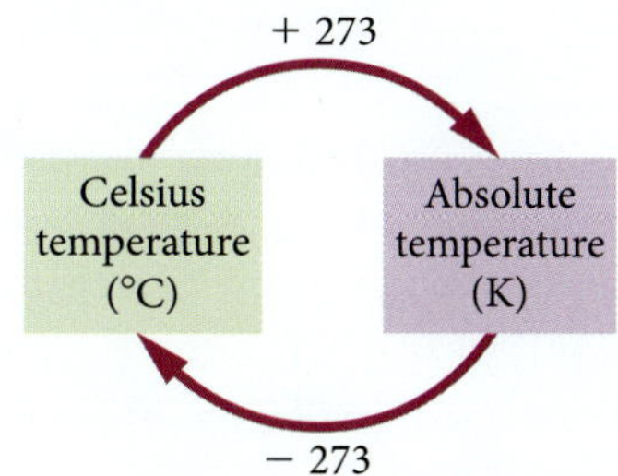

◀Figure C8.9 Converting temperature units

WOW

A new state of matter

In 1924–25 Satyendra Bose and Albert Einstein predicted that a new state of matter would appear as a gas approached absolute zero. This state was called the Bose–Einstein condensate. Seventy years later, in 1995, Eric Cornell and Carl Wieman produced the first Bose–Einstein condensate using a gas of rubidium atoms cooled to 1.7×10^{-7} K.

ABSOLUTE ZERO

Watch this interview of scientists who almost achieved absolute zero.

Applying the kinetic theory

The kinetic molecular theory was developed using the findings of many different scientists. It can be used to explain many properties of gases.

- Gases **diffuse**, or spread, because gas particles have rapid, random motion and a lot of empty space between them so they quickly spread and mix.
- Gases can be easily compressed because there is a lot of empty space between particles.
- Gases spread to fill a container because there are negligible forces between particles.
- Gases exert pressure because the particles move rapidly and collide with the surface of any container or object.
- Gases have low density because there is a lot of empty space between them.

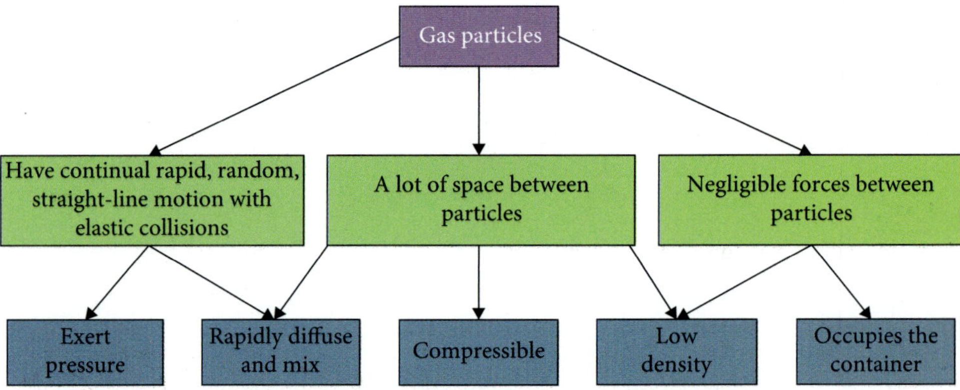

Figure C8.10 ▶ The kinetic molecular theory of gases – a schematic diagram

QUESTION SET 8.2

Remembering

1 List the proposals of the kinetic theory.

2 What is meant by elastic collisions between particles?

3 What is absolute zero?

Understanding

4 Use the kinetic theory to explain why gases:

- a can be compressed.
- b can mix rapidly.
- c have low density.

5 Explain why an increase in the temperature of a gas-filled balloon will cause it to burst.

6 What would happen to the volume of a gas-filled balloon if gas collisions were not elastic?

7 Explain why the particles of hydrogen gas move faster than those of oxygen gas at the same temperature when they have the same kinetic energy.

8 What is the difference between an ideal gas and a real gas?

9 Convert the following temperatures.

- a 237°C to K
- b −25°C to K
- c 120 K to °C
- d 345 K to °C

Applying

10 On a hot day, a metal can is half-filled with petrol and sealed. The next morning, after a cold night, it is observed that the can is partially crushed. Explain why this could happen.

11 Why do real gases behave more ideally at high temperature and low pressure?

12 At a temperature of 25°C, molecules of which of the following gases would move the fastest?

A N_2

B F_2

C CO_2

Analysing

13 What is wrong with saying the gas bottle is half-full?

14 Suppose you have two containers, one filled with ammonia gas (NH_3) and the other filled with chlorine gas (Cl_2). If you were the same distance from both and they were opened at the same time, which one would you expect to smell first? Why?

8.3 Gas laws

You have already learnt that there are relationships between pressure, volume and temperature of a gas. When a balloon is heated, it gets bigger, thus maintaining a constant pressure. If a sealed container is exposed to extreme heat, such as that from a fire, the gas pressure can build up and the container may explode.

Alamy/Tengku Mohd Yusof

◀**Figure C8.11**
A gas cylinder after an explosion

Standard conditions

In order to compare changes in conditions that involve gases, scientists use a set of standard conditions for temperature and pressure. These are defined in Table C8.2.

Table C8.2 Standard conditions for temperature and pressure

Standard conditions	Temperature (°C)	Pressure (kPa)
Standard temperature and pressure (STP)	0	100
Standard laboratory conditions (SLC)	25	100

Boyle's law

More than 300 years ago, Robert Boyle was the first chemist to perform quantitative experiments on gases. By measuring the pressures and volumes of many different gases, he observed that, at a constant temperature, the volume of a given mass of gas is inversely proportional to the pressure. This is known as **Boyle's law**.

In mathematical terms, Boyle's law can be expressed as $V \propto 1/P$, which means the larger the applied pressure, the smaller the volume.

Another way of stating Boyle's law is $PV = k$, where k is a constant. For a particular sample of gas at a constant temperature, the product of the pressure and volume is a constant value.

BOYLE'S LAW

View the NASA simulation of Boyle's law.

Boyle's law states that at a constant temperature, the volume of a given mass of gas is inversely proportional to the applied pressure.

A useful alternative form of Boyle's law is:

$$P_1V_1 = P_2V_2$$

where P_1 is the initial volume
V_1 is the initial volume
P_2 is the final volume
V_2 is the final volume.

Note that any units for pressure and volume can be used, as long as they are consistent.

In Experiment 8.1, you can observe Boyle's law in action.

WORKED EXAMPLE 8.1

A sample of gas collected in a 242 mL container has a pressure of 87.6 kPa. What would the volume of this gas be at 101.3 kPa? (Assume temperature remains constant.)

Answer

The volume would be 209 mL.

Logic

1 Write the relationship:

$P_1V_1 = P_2V_2$

2 Determine the quantities:

$P_1 = 87.6$ kPa, $P_2 = 101.3$ kPa

$V_1 = 242$ mL, $V_2 = ?$

3 Check the units are consistent.

4 Substitute in the formula:

$87.6 \times 242 = 101.3 \times V_2$

$$V_2 = \frac{87.6 \times 242}{101.3} = 209 \text{ mL}$$

Try these yourself

Assume temperature is constant.

a If 4.42 L of gas was collected at a pressure of 94.2 kPa, what volume would it occupy at 1 atm pressure?

b A 210 mL sample of nitrogen gas at 1.0 atm is compressed until the final volume is 150 mL. What is the final pressure?

c When the pressure is 522 kPa, a gas has a volume of 55 mL. What is its volume at atmospheric pressure?

d A sample of hydrogen gas at 73.3 kPa in a 2.5 L flask is expanded into a total volume of 7.2 L. Calculate the final pressure.

EXPERIMENT 8.1

PRESSURE AND VOLUME

Cartesian divers are simple but fascinating toys that have been around for centuries. They consist of an object suspended near the top of a flexible, liquid-filled container. Cartesian divers are useful in demonstrating the relationship between pressure and volume.

Aim

To investigate pressure–volume relationships using a Cartesian diver

Materials

- 1 L clear, flexible screw-cap bottle with lid
- Clear plastic dropper
- Plasticine or Blu-Tack
- Water

Procedure

1 Totally fill the bottle with water.
2 Cut the plastic dropper about 1 cm below the bulb.
3 Half-fill the dropper bulb with water and weight it with some plasticine or Blu-Tack around the bottom, being careful not to block the hole.
4 Place the weighted dropper bulb in the bottle of water so it floats near the top. (You may need to adjust the amount of weight.)
5 Screw the top on the bottle and squeeze the container.

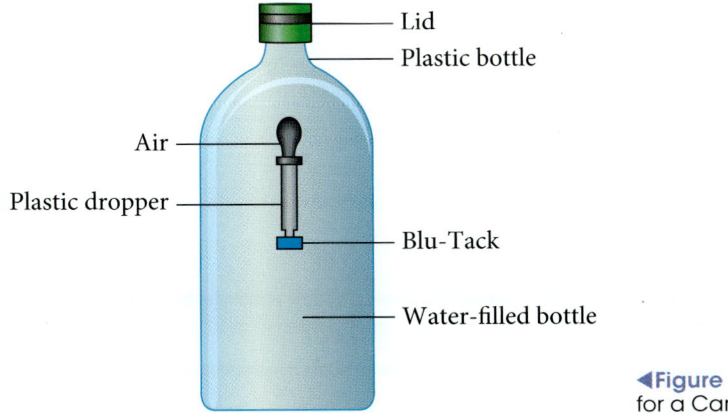

◀**Figure C8.12** Experimental set-up for a Cartesian diver

Results

Describe what happens to the Cartesian diver when the bottle is squeezed and released.

Discussion

Explain what is happening in terms of Boyle's law and densities.

Charles' law

About 100 years after Boyle investigated the relationship between pressure and volume, French physicist Jacques Charles studied the relationship between temperature and volume using gases. The results of his many experiments led him to determine that for a fixed quantity of gas at a constant pressure, the volume increases linearly with temperature. This relationship is called **Charles' law**.

When experimental temperature and volume data are plotted, a linear relationship emerges, as can be seen in Figure C8.13.

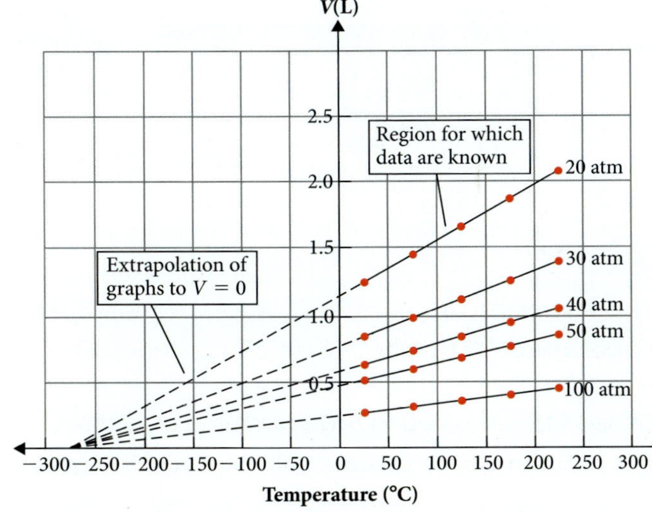

◀**Figure C8.13** The graph of volume for 1 mole of hydrogen gas at different temperatures

Extrapolation of the data leads to a temperature of −273°C, which is absolute zero. So Charles's experiments led to the development of a new temperature scale. If the data is plotted using the absolute temperature scale, all lines will pass through the origin of the graph and therefore temperature (K) is directly proportional to volume. Therefore, in calculations involving gases, temperature must be given in kelvin.

Charles' law states that at constant pressure, the volume of a fixed quantity of gas is proportional to its absolute (kelvin) temperature.

In mathematical terms, Charles' law can be expressed as $V \propto T$ or:

$$\frac{V}{T} = \text{constant}$$

An alternative, more practical, form of this equation is:

$$\frac{V_1}{T_1} = \frac{V_2}{T_2}$$

where V_1 is the initial volume
T_1 is the initial temperature (K)
V_2 is the final volume
T_2 is the final temperature (K).

Note that any unit for volume can be used, as long as it is used consistently, but temperature must be in kelvin.

WORKED EXAMPLE 8.2

A 225 mL volume of gas is collected at 58°C. What volume would this sample occupy at 25°C?

Answer

The volume would be 202 mL.

Logic

1 Write the relationship:

$$\frac{V_1}{T_1} = \frac{V_2}{T_2}$$

2 Determine the quantities and convert temperature to kelvin by adding 273:

$V_1 = 225$ mL, $V_2 = ?$

$T_1 = 58°C = 331$ K, $T_2 = 25°C = 298$ K

3 Check the units are consistent.

4 Substitute in the formula:

$$\frac{225}{331} = \frac{V_2}{298}$$

$$V_2 = \frac{225 \times 298}{331} = 202 \text{ mL}$$

Try these yourself

Assume constant pressure.

a A sample of oxygen has a volume of 3.50 mL at 20°C. What volume would it occupy at 80°C?

b If a sample of gas had a volume of 1.7 L at 25°C and 1 atm pressure, what:

  i would its volume be at 100°C?

  ii temperature would it need to be cooled to have a volume of 0.80 L?

c A gas occupies a volume of 60.0 mL at 36°C. What volume would the gas occupy at 0°C?

Combined gas law

When dealing with gases, we usually find that temperature and pressure change at the same time. By combining Boyle's law and Charles' law, a new relationship called the **combined gas law** is produced. This gives a relationship between volume, pressure and absolute temperature for a given mass of gas. It is most useful in the following form:

$$\frac{P_1V_1}{T_1} = \frac{P_2V_2}{T_2}$$

where P_1, V_1 and T_1 represent initial pressure, volume and temperature (K), and P_2, V_2 and T_2 represent final pressure, volume and temperature (K).

The temperature must be in measure in kelvin and the units for volume and pressure must be consistent.

GAS LAWS

Use these simulations of Boyle's and Charles' laws.

WOW

Boiling eggs

When a hen lays an egg, the egg is warm. As it cools, the insides contract and a pocket of air is formed. Over time, the porous shell allows carbon dioxide and water to diffuse out and air to enter to equalise the pressure. When the egg is placed in rapidly boiling water, the air quickly expands and causes the shell to crack.

WORKED EXAMPLE 8.3

A sample of gas at a pressure of 105 kPa has a volume of 2.50 L at a temperature of 25°C. What pressure is needed to compress it to 1.0 L at 20°C?

Answer

The pressure would be 258 kPa.

Logic

1 Write the relationship:

$$\frac{P_1V_1}{T_1} = \frac{P_2V_2}{T_2}$$

2 Determine the quantities and convert temperature to kelvin by adding 273:

$P_1 = 105$ kPa, $P_2 = ?$

$V_1 = 2.50$ L, $V_2 = 1.0$ L

$T_1 = 25°C = 298$ K, $T_2 = 20°C = 293$ K

3 Check the units are consistent.

4 Substitute for values in the formula:

$$\frac{105 \times 2.50}{298} = \frac{P_2 \times 1.0}{293}$$

$$P_2 = \frac{105 \times 2.50 \times 293}{298 \times 1.0}$$

$P_2 = 258$ kPa

Try these yourself

a A sample of oxygen at 80°C and 135 kPa pressure has a volume of 202 mL. What will the volume of the sample be at 0°C and 101.3 kPa?

b The volume of 1 mole of gas at 25°C and a pressure of 101.3 kPa is 24.5 L. Calculate the molar volume:

i at the top of Mt Everest: temperature –30°C, pressure 30 kPa.

ii on the surface of Mars: temperature 27°C, pressure 0.80 kPa.

c A 20.0 L car tyre is inflated to a pressure of 200 kPa at 15°C. After an hour's driving, the tyre temperature has risen to 40°C. If the tyre has expanded to a volume of 24 L, what is the tyre pressure?

QUESTION SET 8.3

Remembering

1 a What does STP stand for?

b What are standard laboratory conditions?

2 State:

a Boyle's law.

b Charles' law.

Understanding

3 Explain why car owners need to add more air to their tyres in winter and release air from them in summer.

4 A 4 L capacity balloon is two-thirds full. If the kelvin temperature of the gas in the balloon doubles, what will happen?

Applying

5 a A sample of hydrogen gas has a volume of 6.20 L at a pressure of 1.05 atm. What is the volume of the sample if it is compressed until the pressure is 3.00 atm? (Assume temperature remains constant.)

b A motorbike tyre with a volume of 24 L is filled with air to a pressure of 2.3 atm. What volume does this air occupy at 1 atm pressure at the same temperature?

6 a A sample of argon gas has a volume of 450 mL at 100°C. At what temperature will it have a volume of 200 mL if the pressure does not change?

b A sample of gas is heated until its temperature doubles from 25°C to 50°C. If its final volume is 100 mL, what was the original volume if the pressure remained constant?

7 A sample of helium occupies a volume of 10.0 L at 25°C and a pressure of 1.2 atm. What volume will it occupy at STP?

8 The temperature of a closed vessel containing air at a pressure of 103 kPa is raised from 15°C to 200°C. What is the pressure inside the vessel if its volume increases by 10% as the temperature is raised?

Analysing

9 A metal cylinder has a safety valve that opens at a pressure of 100 atm. It is to be filled with nitrogen gas heated to 300°C. What is the maximum pressure to which it can be filled at 25°C?

10 When a cyclone forms, the wind rushes in to fill an area of low pressure. A cyclone approaches and the air pressure falls from 101.3 to 79.4 kPa.

a If the volume of your living room was 430 m^3 initially what volume would the air in the room change to so pressure was equalised inside and outside?

b Why is it advised that windows be left slightly ajar as a storm approaches?

8.4 Avogadro's hypothesis and molar volumes

MORE GAS LAWS

Visit this website for more practice problems using the gas laws

In all the examples considered so far, the amount of gas has been kept constant while temperature, pressure and volume have been varied. Adding or removing gas will affect pressure, volume and temperature. Pumping more air into a bicycle tyre will increase the pressure within the tyre. Releasing gas from an inflated balloon causes the volume to decrease and releasing gas from an inflated tyre causes the valve to become quite cold.

In 1811, Amadeo Avogadro attempted to show how the volume of a gas was related to the amount of gas present. He proposed that equal volumes of gas contain equal numbers of particles at the same temperature. This became known as **Avogadro's hypothesis**.

Avogadro's hypothesis states that equal volumes of any gas, measured at the same temperature and pressure, contain the same number of particles.

You learnt about the Avogadro constant in Chapter 5.

This means that no matter what the gas is, the total number of particles would be the same under the same conditions. So, provided temperature and pressure are the same, 1 L of hydrogen has the same number of particles as 1 L of helium, or 1 L of oxygen and so on. Also, if the number of particles were doubled, then the volume would double.

Molar volume

A mole is defined as a fixed number of particles (Avogadro's constant, 6.02×10^{23}), so it follows that equal volumes of different gases, measured at the same temperature and pressure contain the same number of moles. This is called the **molar volume of a gas.**

- At standard temperature and pressure (0°C and 100 kPa) the molar volume of a gas is 22.71 L.
- At standard laboratory conditions (25°C and 100 kPa) the molar volume of a gas is 24.79 L.

The relationship between number of moles and volume of a gas can be expressed as:

$$\text{Number of moles of gas} = \frac{\text{volume of gas at STP}}{\text{volume of one mole of gas at STP}}$$

$$n = \frac{V}{22.4}$$

where n is the number of moles and V is the volume of gas (L) at STP.

WOW

Icy cylinders

Condensation and even ice can form on the outside of a gas cylinder as it is being used. As the gas is used and the pressure in the cylinder decreases, stored liquid petroleum gas (LPG) 'boils' to a gas vapour, absorbing heat energy from the steel walls of the container

WORKED EXAMPLE 8.4

a How many moles of carbon dioxide gas are present in 350 mL of gas at STP?

b What is the mass of carbon dioxide in part **a**?

c What volume is occupied by 160 g of helium at STP?

Answers

a There is 0.0154 mol.

Logic

1 Write the relationship.

$$n = \frac{V}{22.71}\text{ mol}$$

2 Determine the quantities and check the units.

$n = ?,\ V = 350\text{ mL} = 0.350\text{ L}$

3 Substitute in the formula.

$$n = \frac{0.350}{22.71}$$

$n = 0.0154\text{ mol}$

b The mass is 0.678 g.

1 Write the relationship.

$$n = \frac{m}{M}$$

2 Determine the quantities.

$n = 0.0154\text{ mol}$

$M(CO_2) = 12 + 2 \times 16 = 44\text{ g mol}^{-1}$

3 Substitute:

$$0.0154 = \frac{m}{44}$$

$m = 0.0154 \times 44 = 0.678\text{ g}$

c The volume is 908 L.

1 Write the relationship.

$$n = \frac{V}{22.71}\text{ mol}$$

2 Determine the quantities and check the units.

$n = ?,\ V = ?,\ m = 160\text{ g}$

3 Calculate the number of moles.

$$n = \frac{m}{M}$$

$m = 160\text{ g},\ M(\text{He}) = 4.00\text{ g mol}^{-1}$

$$n = \frac{160}{4.00} = 40\text{ mol}$$

4 Calculate the volume.

$$n = \frac{V}{22.71}$$

$$40 = \frac{V}{22.71}$$

$V = 40 \times 22.71 = 908\text{ L}$

Try these yourself

a What volume container would you need to store 0.095 mol of nitrogen at STP?

b i How many moles are present in 0.450 L of oxygen gas at STP?

ii What is the mass of oxygen in part **i**?

c i What volume will 10 g of hydrogen gas occupy at STP?

ii What volume will 10 g carbon dioxide occupy at STP?

iii Why are these volumes different when the mass is the same?

The general gas equation

All the gas laws studied so far have required that at least one of the conditions of pressure, temperature, volume and quantity of gas remain constant. In reality, when dealing with gases in situations such as scuba diving and using anaesthetics, all the conditions can vary. However, if Boyle's and Charles's laws are combined with Avogadro's hypothesis, the resulting relationship, called the **general gas equation**, includes all conditions as variables.

The general gas equation is also known as the ideal gas equation and the universal gas equation. It provides a relationship between pressure, volume, temperature and the amount of gas:

$$PV = nRT$$

where P is pressure in kPa

V is volume in L

n is total number of moles of gas present

R is the **universal gas constant** = 8.31 J K^{-1} mol^{-1}

T is temperature in K.

The general gas equation, $PV = nRT$, gives the relationship between pressure, temperature, volume and number of moles of a gas.

It is important to use the correct value of R in the general gas equation. One way of reducing mistakes is to always include the unit and the number when recording the value of a physical quantity.

Regardless of the chemical formula and properties of different gases, the general gas equation describes the physical properties of all gases. The equation is called the *ideal* gas equation because real gases deviate slightly from it, particularly at high pressures and low temperatures, because the intermolecular forces between real gases come into play as the particles slow down and move closer together.

The value of R, the universal gas constant, is dependent on units. The value given above is based on an SI unit for pressure. Sometimes, it is more convenient to use pressure in atmospheres. When the unit for pressure is atmospheres, the value of R is 0.0820 L atm K^{-1} mol^{-1}. Table C8.3 shows the units and values for R.

Table C8.3 Units and values for the universal gas constant, R

Condition	R = 8.314 J K^{-1} mol^{-1}	R = 0.082 L atm K^{-1} mol^{-1}
Pressure	kPa	atm
Volume	L	L
Temperature	K	K
Amount of gas	mole	mole

WORKED EXAMPLE 8.5

What is the mass of carbon dioxide in a sample that occupies 1.50 L at 20°C and 125 kPa?

Answer

The mass is 3.39 g.

Logic

1 Write the relationship.

$PV = nRT$

2 Determine the quantities and check the units.

$m = ?, P = 125$ kPa, $V = 1.50$ L, $n = ?, T = 20°C = 293$ K

3 Decide on the appropriate value for R based on the units for P.

$R = 8.314$ J K^{-1} mol^{-1}

4 Substitute in the equation.

$125 \times 1.50 = n \times 8.314 \times 293$

$$n = \frac{125 \times 1.50}{8.314 \times 293}$$
$$= 0.077 \text{ mol}$$

5 Calculate mass from moles.

$$n = \frac{m}{M}$$

$M(CO_2) = 44$ g

$$0.077 = \frac{m}{44}$$

$m = 0.077 \times 44 = 3.39$ g

Try these yourself

a What is the volume of 0.34 mol of H_2S gas at 100°C and 75.0 kPa pressure?

b A sample of oxygen gas is collected and found to occupy a volume of 527 mL at 1.05 atm and 20°C. What is the mass of the sample?

c What is the pressure of 4.00 g of He in a container of volume 7.5 L at 30°C?

EXPERIMENT 8.2

MEASURING THE MOLAR MASS OF BUTANE

Most disposable cigarette lighters are filled with butane (C_4H_{10}). Butane exists as a gas at room temperature (boiling point, −0.5°C), but is compressed to a liquid in lighters. When the pressure is released (the lighter valve is opened), the butane vaporises and is ignited by a flint in the lighter.

In this experiment, you will determine the molar mass of butane experimentally and then compare this with the theoretical value. The mass of liquid butane released from the lighter will be the same as the mass of gas produced, but the volume will be different.

Aim

To determine the molar mass of butane experimentally and compare this value with the theoretical value

Materials

- Pneumatic trough or large plastic tub
- Disposable butane lighter with flint removed
- Thermometer
- 250 mL measuring cylinder
- Retort stand, clamp and boss head
- Barometer
- Water
- Electronic balance
- Paper towel

What are the risks in doing this experiment?	How can you manage these risks to stay safe?
Butane is highly flammable.	Ensure the flint is removed from the lighter and there are no open flames in the laboratory.
Butane gas can be irritating.	Work in a well-ventilated area and avoid breathing the gas.

In your write-up, add any more risks you can think of, as well as ways to manage them. In particular, expand on the two risks listed to identify specific risks involved with each of them.

Procedure

1 Two-thirds fill the pneumatic trough or large plastic tub with water.
2 Wipe the lighter clean and dry, weigh it and record the mass.
3 Fill the 100 mL measuring cylinder with water.
4 Submerge the cylinder in the tub and gradually turn it upside down.
5 Ensure no air bubbles are present. If bubbles are present, then repeat step 4.
6 Lift the measuring cylinder just high enough above the base of the tub so the lighter fits underneath. Clamp the cylinder in position (see Figure C8.14).
7 Hold the lighter directly under the measuring cylinder and press the lever on the lighter. Ensure all the gas is bubbling into the cylinder.

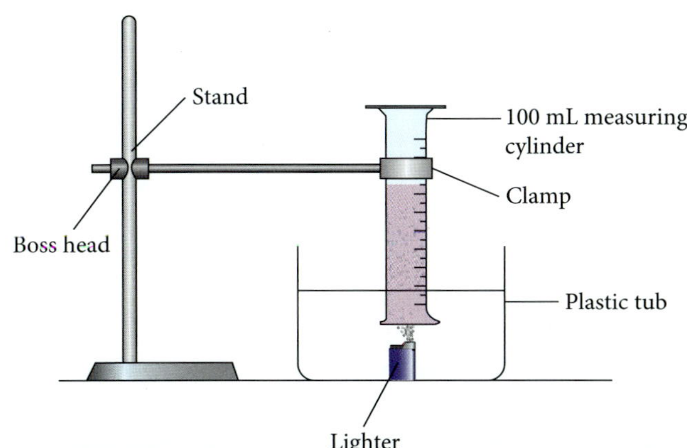

▲Figure C8.14
Experimental set-up

8 Continue to hold down the lever until the water levels inside and outside the measuring cylinder are about the same. Remove the lighter.
9 Adjust the height of the measuring cylinder so the water levels inside and outside are the same. Record the volume of gas in the measuring cylinder.
10 Thoroughly dry the lighter with paper towel. Be careful not to press the lever.
11 Reweigh the lighter and record the new mass.
12 Measure and record the temperature of the water in the tub.
13 Measure and record the barometric pressure.
14 If time allows, repeat steps 2–11.

Results

Record your results in a table like the one below.

Initial mass of lighter	
Final mass of lighter	
Volume of gas	
Temperature of water	
Barometric pressure	

Analysing the results

1 Calculate the mass of butane released.
2 Calculate the number of moles of butane.
3 Use Table C8.4 on page 360 to help determine the vapour pressure of water at the experimental temperature.
4 Calculate the pressure of butane by subtracting the vapour pressure of water from the atmospheric pressure reading.
5 Calculate the experimental molar mass of butane.

▶

Table C8.4 Vapour pressure of water at various temperatures

Temperature (°C)	Pressure (atm)	Temperature (°C)	Pressure (atm)
18	0.0204	25	0.0312
19	0.0217	26	0.0332
20	0.0230	27	0.0351
21	0.0245	28	0.0372
22	0.0261	29	0.0395
23	0.0276	30	0.0419
24	0.0295	31	0.0443

Discussion

1 Compare the experimental molar mass with the reference molar mass.
2 Identify possible sources of experimental error and describe how these would affect the result.
3 Why did the vapour pressure of water have to be subtracted to determine the pressure of butane?
4 Suggest any improvements you might make to the experiment to make it more accurate.

QUESTION SET 8.4

Remembering

1 State Avogadro's hypothesis.
2 a Define 'molar volume'.
  b State the general gas equation.

Understanding

3 a If *R* is the general gas constant, explain why it has different values.
  b Why can't the general gas equation be applied at high pressure and low temperature?

Applying

4 What volume would a mole of gas occupy at 500°C and 0.80 atm?
5 A 5.0 L sample of gas exerts a pressure of 20 kPa at a temperature of 300 K. How many moles of gas are present?
6 Calculate the pressure exerted by 2.59 g of carbon monoxide in an 8.0 L flask at 30°C.
7 What volume is occupied by 0.200 mol of oxygen gas at 20°C and a pressure of 0.97 atm?
8 At STP, the mass of 1.00 L of a gas is 1.89 g. What mass of gas will occupy 1.00 L at 200°C and a pressure of 1.25 atm?
9 A sample of noble gas had a mass of 0.20 g and exerted a pressure of 0.98 atm in a container of volume 0.13 L at 27°C. Which noble gas is it?

Analysing

10 You have been given three 5.0 L balloons. One is filled with hydrogen, one with carbon dioxide and one with air. The average molar mass of air is 29.0 g mol^{-1} and the temperature and pressure of all balloons are 27°C and 1.0 atm. Calculate the mass of gas in each balloon and use this to justify why a balloon filled with hydrogen rises while one filled with carbon dioxide sinks.

8.5 Calculating gases in chemical reactions

In Chapter 5, mass calculations based on amounts of reactants and products were considered. The calculations needed to use the mole relationships given by coefficients in a balanced chemical equation in order to determine amounts of reactants and products. This relationship can now be extended to include volumes of gases.

GENERAL GAS LAW

Visit this website to practise more problems applying the general gas law.

PVT

Use this simulation to investigate the relationships between pressure, volume and temperature.

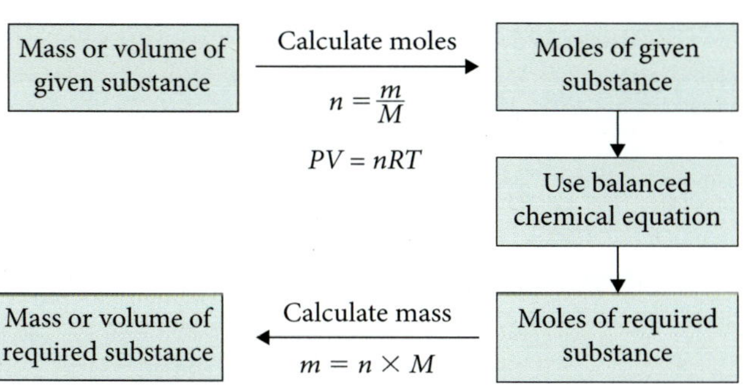

◀Figure C8.15 Mass–volume relationships, where *m* is mass in grams, *n* is number of moles and *M* is molar mass

WORKED EXAMPLE 8.6

a What volume of hydrogen gas would be produced at STP when 0.325 g of zinc reacts with dilute hydrochloric acid?

b Sulfuric acid is formed when sulfur dioxide and water react with oxygen in the presence of a catalyst. If 3.20 g of SO_2 react, what volume of O_2 at 200 kPa and 50°C is necessary for complete reaction?

Answers

a The volume is 113 mL.

Logic

1 Write a balanced equation.

$Zn(s) + 2HCl(aq) \rightarrow ZnCl_2(aq) + H_2(g)$

2 Calculate the number of moles of Zn using the relationship $n = m/M$.

$m = 0.325$ g, $n = ?$, $M(Zn) = 65.4$ g mol^{-1}

$n = 0.325/65.4 = 4.97 \times 10^{-3}$ mol

3 Use the balanced equation to calculate the number of moles of H_2.

From the equation, 1 mol of Zn produces 1 mol of H_2

$n(H_2) = 4.97 \times 10^{-3}$ mol

4 Use molar volume at STP to calculate the volume of H_2.

$$n = \frac{V}{22.71}$$

$$4.97 \times 10^{-3} = \frac{V}{22.71}$$

$V = 4.97 \times 10^{-3} \times 22.71 = 0.113$ L $= 113$ mL

b The volume is 0.336 L

1 Write a balanced equation.

$2SO_2(g) + O_2(g) + 2H_2O(l) \rightarrow 2H_2SO_4(l)$

2 Calculate the number of moles SO_2 using the relationship $n = m/M$.

$m = 3.20$ g, $n = ?$, M can be calculated

$M = 32.1 + 2 \times 16 = 64.1$ g mol^{-1}

$$n = \frac{3.20}{64.1} = 0.0499 \text{ mol}$$

▶

3 Use the balanced chemical equation to calculate the number of moles of O_2 and H_2SO_4.

From the equation, 2 moles of SO_2 reacts with 1 mole of O_2.

$$n(O_2) = \frac{1}{2} \times n(SO_2)$$

$$n(O_2) = \frac{1}{2} \times 0.0499 = 0.025 \text{ mol of } O_2$$

4 Determine the volume using $PV = nRT$.

$P = 200$ kPa, $V = ?$, $n = 0.025$ mol, $T = 50°C = 323$ K, $R = 8.314$ J K^{-1} mol^{-1}

$200 \times V = 0.025 \times 8.314 \times 323$

$$V = \frac{0.025 \times 8.314 \times 323}{200} = 0.336 \text{ L}$$

Try these yourself

a A mass of 4.86 g of magnesium is ignited in nitrogen dioxide. The products of the reaction are magnesium oxide and nitrogen gas. Calculate the volume of nitrogen gas produced at 273 K and 101.3 kPa.

b The complete combustion of carbon in oxygen produces carbon dioxide. Calculate the volume of oxygen that would react at SLC with 10.0 g of carbon and the volume of carbon dioxide produced under the same conditions.

c What mass of potassium chlorate must be heated to give 120 mL of oxygen at 1.3 atm and 32°C? The equation for the reaction is:

$2KClO_3(s) \rightarrow 2KCl(s) + 3O_2(g)$

d Hydrogen peroxide (H_2O_2) decomposes to give water and oxygen gas. How many grams of hydrogen peroxide are needed to produce 100 L of oxygen at SLC?

e 10.01 g of calcium carbonate is decomposed by heat to produce calcium oxide solid and carbon dioxide gas. What volume of carbon dioxide gas is formed at 20°C and 110 kPa?

CHAPTER CHECKLIST

You should know:

- gas pressure is caused by collisions of gas molecules with the walls of a container
- the main units of gas pressure are pascal (Pa) and atmosphere (atm)
- the kinetic theory of gases explains the physical properties of gases in terms of the motion of their molecules and proposes that:
    1. gases consist of molecules (except the noble gases, which consist of atoms) that move in continual random straight-line motion
    2. the average distance between gas molecules is very large compared to the size of the molecules
    3. intermolecular forces between molecules are negligible
    4. all collisions of gas molecules are perfectly elastic, which means no net energy loss occurs
    5. pressure is due to collisions of the molecules with the walls of the container
    6. temperature is a measure of the average kinetic energy of the molecules
- absolute temperature is measured in kelvin according to the absolute temperature scale, which has the same-sized divisions as the Celsius scale but an end point of zero, called absolute zero
- Boyle's law states that at a constant temperature, the volume of a given mass of gas is inversely proportional to the applied pressure
- Charles' law states that at constant pressure, the volume of a fixed quantity of gas is proportional to its absolute (kelvin) temperature
- the combined gas law states a relationship between initial and final volume, pressure and absolute temperature for a given volume of gas
- Avogadro's hypothesis states that equal volumes of any gas, measured at the same temperature and pressure, contain the same number of particles
- the general gas equation, $PV = nRT$, gives the relationship between pressure, temperature, volume and number of moles of a gas.

You should be able to:

- convert between absolute and Celsius temperature scales
- select and use appropriate mathematical representations to solve problems, including using the mole concept and general gas equation to calculate the mass of chemicals and/or volume of a gas (at standard temperature and pressure) involved in a chemical reaction.

CHAPTER GLOSSARY

absolute temperature temperature measured in kelvin according to the absolute temperature scale, which has the same sized divisions as the Celsius scale but an end point of zero

absolute zero lowest possible temperature (0 K or −273.15°C)

atmospheric pressure pressure due to the atmosphere; at sea level has an average value of 1 atm

Avogadro's hypothesis equal volumes of any gas, measured at the same temperature and pressure, contain the same number of particles

Boyle's law a constant temperature, the volume of a given mass of gas is inversely proportional to the applied pressure

Charles' law at constant pressure, the volume of a fixed quantity of gas is proportional to its absolute (kelvin) temperature

combined gas law a relationship between initial and final volume, pressure and absolute temperature for a given mass of gas

diffuse the spreading out of a gas to fill a space

elastic collision a collision in which there is no net loss of energy

general gas equation an equation that gives the relationship between pressure, temperature, volume and number of moles of a gas as $PV = nRT$

ideal gas a gas that perfectly obeys all the proposals of the kinetic theory of gases and the gas laws

kinetic theory of gases a theory that explains the physical properties of gases in terms of the motion of their molecules

molar volume of a gas the volume occupied by 1 mole of a gas at known temperature and pressure

pascal (Pa) the SI unit of gas pressure, equivalent to a force of 1 newton per square metre

pressure the force exerted by a gas or gases due to the collisions of the particles with the walls of a container; expressed as force per unit area

universal gas constant (R) a constant value used in the general gas equation, equal to 8.314 J K^{-1} mol^{-1} or 0.082 L atm K^{-1} mol^{-1}

CHAPTER REVIEW QUESTIONS

Remembering

1 a What is the difference between a real gas and an ideal gas?

b When does this difference become significant?

2 To convert °C to K you:

A subtract 273.

B add 373.

C add 273.

D it depends on the pressure.

3 Which of the following best expresses Boyle's law?

A $V \propto \frac{1}{P}$

B $T \propto V$

C $P \propto V$

D $V \propto \frac{1}{T}$

4 Gases are more easily compressed than liquids and solids. The reason for this behaviour is that:

A gas molecules move with much greater velocities than the molecules of liquids and solids, permitting gases to adjust more rapidly to a change in volume.

B gas molecules undergo elastic collisions with the walls of a container and elastic substances are easily compressed.

C the average distance between gas molecules is much greater than that between particles in liquids or solids, so the volume may be reduced more easily.

D attractive forces between gas molecules are much smaller than those between particles in solids and liquids, and these small forces can be readily overcome during compression.

5 Which of the following is *not* one of the assumptions of the kinetic theory of gases?

A All collisions of gas molecules are non-elastic. This means energy is lost during collisions.

B Gases consist of tiny particles called molecules, except for noble gases, which consist of atoms.

C The molecules of a gas exert negligible attractive or repulsive forces on each other.

D The average distance between molecules in a gas is very large compared with the size of each gas molecule.

Understanding

6 A gas fills a 100 mL cylinder fitted with a piston at a particular temperature and pressure. If the volume of the gas is halved by pushing in the piston, and at the same time the absolute temperature is doubled, the pressure of the gas will:

A increase four times.

B be doubled.

C be unchanged.

D be halved.

7 Which of the following statements about the properties of gases is correct?

A The solubility of a gas in a liquid increases as the liquid is heated.

B One mole of any gas at SLC occupies 24.5 L.

C Larger gas molecules diffuse more quickly than smaller molecules.

D The thermal energy of a gas is given by its temperature.

8 Decide whether each of the following statements is true or false. For those that are incorrect, write the correct statement.

a Halving the number of particles in a given volume of gas decreases the pressure by one-half if the temperature is kept constant.

b At absolute zero, the volume occupied by a gas would be zero.

c At constant temperature and pressure, the volume of a gas decreases as the number of moles of the gas increase.

d According to the combined gas law, $T_2 = P_1 \times \frac{V_2 \times T_1}{P_2 \times V_1}$.

9 A sample of methane has a volume of 4.60 L at 27°C. What volume would it occupy at 120°C if the pressure remained constant?

Applying

10 As a man dives into the sea, the pressure of gas in his lungs changes from 100 000 Pa to 150 000 Pa. If his lungs initially held 6 L of gas, the volume of his lungs would become (assume temperature remains constant):

A 3 L.

B 4 L.

C 6 L.

D 9 L.

11 Ammonia was prepared by the following reaction:

$N_2(g) + 3H_2(g) \rightarrow 2NH_3(g)$

If 1 L of nitrogen is reacted with excess hydrogen, how many litres of ammonia will be produced?

A 2 L

B 3 L

C 1 L

D 0.5 L

12 What is the final pressure of gas (at a constant temperature) if 50 mL at 100 kPa expands to 60 mL?

13 An industrial plant manufacturing polythene (polyethylene) stores ethene (ethylene) gas (C_2H_4) in a vessel with a capacity of 2 L and capable of withstanding a pressure of 1500 kPa. What is the maximum mass of ethene that could be stored in the vessel at a temperature of 42°C?

14 350.0 mL of a gas at 26°C and 105.6 kPa has a mass of 0.35 g. Calculate the molar mass of the gas.

15 Carbon monoxide burns in oxygen according to the equation:

$2CO(g) + O_2(g) \rightarrow 2CO_2(g)$

Calculate the volume of oxygen required at 25°C and 1.00 atm for the combustion of 28 g of carbon monoxide.

16 Ammonia gas may be prepared in the laboratory by heating a paste of ammonium chloride and calcium hydroxide according to the reaction:

$2NH_4Cl(s) + Ca(OH)_2(s) \rightarrow 2NH_3(g) + 2H_2O(g) + CaCl_2(s)$

If 107 g of ammonium chloride reacted fully with excess $Ca(OH)_2$, what volume of ammonia is produced at 0°C and 1.0 atm pressure?

Analysing

17 An experiment was performed to determine the effect of pressure on the volume of a fixed mass of ammonia at 25°C. The following readings were obtained.

P (atm)	0.1000	0.2000	0.4000	2.000	8.000
V (L)	244.5	122.2	61.02	12.17	2.925

Plot these results on graph paper.

Ammonia deviates from an ideal gas at higher pressures. On the same graph, plot the expected volume for all the pressures. (Assume the first pair of readings are correct.)

18 Nitroglycerine has a density of 1.59 g mL^{-1}. It explodes to form several gases according to the reaction:

$4C_3H_5O_9N_3(s) \rightarrow 12CO_2(g) + O_2(g) + 6N_2(g) + 10H_2O(g)$

A sealed 1.00 mL container is filled with nitroglycerine and detonated. Assuming standard temperature and that the container would not break upon detonation, calculate the pressure inside the container in atmosphere.

Reflecting

19 A car with typical petrol consumption of about 14 km L^{-1} will produce about 0.16 kg of carbon dioxide per kilometre.

a If the typical annual distance travelled is 20 000 km, how much carbon dioxide will it produce in a year?

b There are about 1 billion (10^9) motor vehicles in the world. How many kilograms of carbon dioxide do they produce per year?

This is a lot of carbon dioxide but how does it compare with the amount of CO_2 in the atmosphere? The volume of CO_2 in the atmosphere is approximately 1.8×10^{18} L.

c Assuming the temperature is 0°C, calculate the mass of CO_2 in the atmosphere.

d Using your answer from part **b**, calculate the fraction of CO_2 cars contribute to the atmosphere.

e If a fully grown tree can convert about 300 kg of CO_2 a year to oxygen, how many trees would there need to be to offset the effects of motor vehicle emissions of CO_2?

CHAPTER 9
REACTION RATES

By the end of this chapter you will have covered the following material.

Science Understanding

- Varying the conditions present during chemical reactions can affect the rate of the reaction and in some cases the identity of the products (ACSCH068)
- The rate of chemical reactions can be quantified by measuring the rate of formation of products or the depletion of reactants (ACSCH069)
- Collision theory can be used to explain and predict the effect of concentration, temperature, pressure and surface area on the rate of chemical reactions by considering the structure of the reactants and the energy of particles (ACSCH070)
- The activation energy is the minimum energy required for a chemical reaction to occur and is related to the strength of the existing chemical bonds; the magnitude of the activation energy influences the rate of a chemical reaction (ACSCH071)
- Energy profile diagrams can be used to represent the enthalpy changes and activation energy associated with a chemical reaction (ACSCH072)
- Catalysts, including enzymes and metal nanoparticles, affect the rate of certain reactions by providing an alternative reaction pathway with a reduced activation energy, hence increasing the proportion of collisions that lead to a chemical change (ACSCH073)

AC

9.1 The rate of a reaction

Just as some people run faster than others, some chemical reactions go faster than others. As scientists' knowledge of what happens during a reaction has deepened, they have been able to apply this knowledge to control the speed, or rate, of a reaction to suit our purposes.

EXPERIMENT 9.1

OBSERVING RATES OF REACTIONS

Aim

To observe the rates at which different reactions occur

Materials

- Zinc granule
- 10 mL of 1 mol L^{-1} hydrochloric acid (HCl)
- 2 mL of 1 mol L^{-1} copper(II) nitrate ($Cu(NO_3)_2$)
- 2 mL of 1 mol L^{-1} sodium carbonate (Na_2CO_3)
- 2 test tubes

What are the risks in doing this experiment?	How can you manage these risks to stay safe?
Chemicals could splash into your eyes.	Wear safety glasses at all times.
1 mol L^{-1} HCl is corrosive to skin and clothing.	Wear safety glasses and protective clothing. Take care when pouring and clean up spills immediately. If hydrochloric acid is spilt on your skin, wash the affected area with plenty of water and notify your teacher.

In your write-up, add any more risks you can think of, as well as ways to manage them.

Procedure

1 Place the zinc granule in a test tube.
2 Cover the zinc with hydrochloric acid.
3 Record your observations for the reaction.
4 Place 2 mL of copper(II) nitrate in the second test tube.
5 Add 2 mL of sodium carbonate to the second test tube.
6 Record your observations of the second reaction.

Results

Present your observations in a simple table.

Analysis of results

1 How could you tell that a reaction had occurred in each of the test tubes?
2 Which reaction was quicker? How did you know?
3 How could you measure how quickly each reaction occurred?

Discussion

1 What other reactions do you know that occur quickly?
2 What other reactions do you know that occur slowly?

What is a rate?

Rate is how quickly one quantity changes compared to another quantity. The rate at which a car moves is how many kilometres the car travels in an hour. The rate at which someone loses weight would be the weight loss (often in kilograms) per week. Another common rate is rate of pay, which describes how much someone is paid per hour. In terms of reactions, the rate is how quickly a reaction occurs.

Alamy/Neil Tyler

▲ **Figure C9.1**
These animals are travelling at different rates. How can you tell?

Science Photo Library/Martyn F. Chillmaid

◀ **Figure C9.2**
Two beakers with the same reactants. Which one is reacting at the faster rate? How do you know?

ACTIVITY 9.1

RATES IN OUR LIVES

Aim

To become aware of rates that affect our lives

What to do

1. Consider what you do in one day.
2. Make a list of anything you use, do or make for which the rate is important. For example, when you make a cup of tea, the rate at which the water heats up affects how long you have to wait for the water to boil.
3. For each item on your list, state whether a fast or slow rate is desirable.
4. If possible, make a note of how you are able to affect the rate of each item.

To see how reaction rates affect our bodies, refer to Context 4, 'Making reactions work for us', page 85.

Figure C9.3 ▲
Which sugar would dissolve the fastest? How could you measure it to accurately compare the three? Try this to see if your prediction is correct.

Measuring the rate of a reaction

To calculate any rate, divide the change in one quantity by the change in the other quantity.

$$\text{Rate} = \frac{\text{change in quantity 1}}{\text{change in quantity 2}}$$

For example, at the 2012 London Olympic Games, Usain Bolt won the 200-metre sprint in 19.26 seconds. Therefore, his rate of running, or speed, was:

$$\text{Rate} = \frac{200}{19.26} = 10.4 \text{ m s}^{-1}$$

To measure the rate of a reaction, you need something measurable to change, such as:

- mass
- colour
- volume
- pH
- concentration.

If you can measure this quantity at different times during the reaction, then you can determine how quickly it is changing during the reaction. For example, in the reaction between solid calcium carbonate ($CaCO_3$) and a solution of hydrochloric acid (HCl), the mass of the solid will decrease as it reacts and forms the products calcium chloride ($CaCl_2$), water (H_2O) and carbon dioxide (CO_2), according to the equation:

$$CaCO_3(s) + 2HCl(aq) \rightarrow CaCl_2(aq) + CO_2(g) + H_2O(l)$$

If you measure the mass of the solid at the start of the reaction and end of the reaction, and measure the time it takes for the reaction to occur, then you can calculate the rate of the reaction. For example if 2.00 g of calcium carbonate takes 48 seconds to completely react, the rate would be:

$$\text{Rate of reaction} = \frac{\text{change in mass}}{\text{change in time}} = \frac{2.00}{48} = 4.17 \times 10^{-2} \text{ g s}^{-1}$$

However, the rate does not have to be for the overall reaction. You can also calculate the rate at various stages of the reaction by taking measurements throughout the reaction time. These may then be used to calculate the rate directly, or you can graph the changes that occur during the reaction.

Consider the reaction between magnesium and hydrochloric acid:

$$Mg(s) + 2HCl(aq) \rightarrow MgCl_2(aq) + H_2(g)$$

The mass of the hydrogen gas produced during the reaction is calculated by the mass lost from a flask. This is shown in Table C9.1.

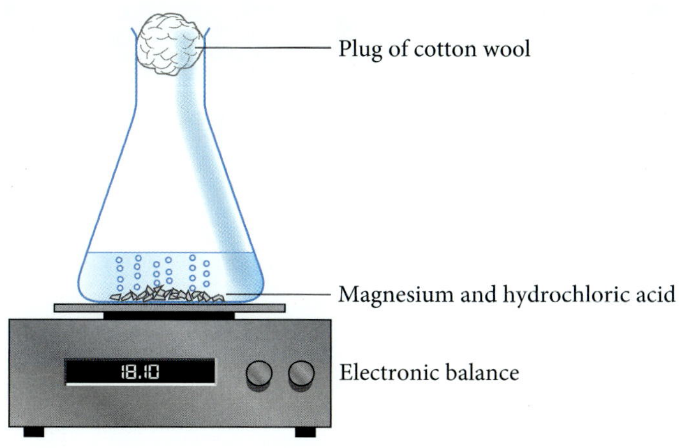

◀ **Figure C9.4** Measuring the mass of hydrogen gas lost during the reaction

Table C9.1 Mass during the reaction between magnesium and hydrochloric acid

Time (s)	Mass of flask and contents (g)	Mass of hydrogen gas released (g)
0	20.00	0.00
10	18.60	1.40
20	18.30	1.70
30	18.20	1.80
40	18.15	1.85
50	18.10	1.90
60	18.06	1.94
70	18.04	1.96
80	18.02	1.98
90	18.01	1.99
100	18.00	2.00
110	18.00	2.00
120	18.00	2.00

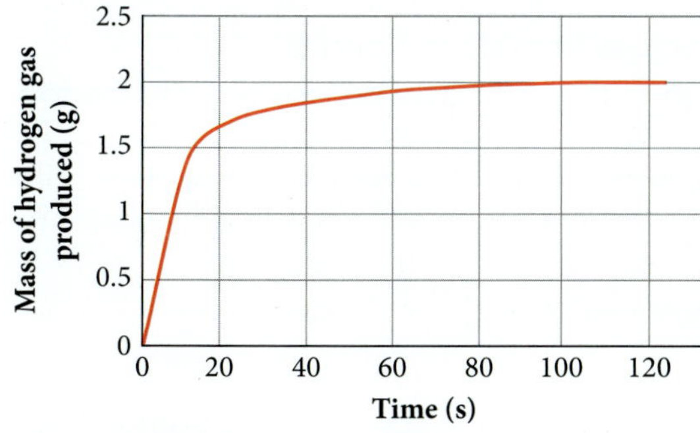

▲ **Figure C9.5**
Mass of hydrogen gas produced

On a graph showing the changes during a reaction, the slope or **gradient** represents the rate of a reaction because it shows how quickly the reactants or products change. The graph in Figure C9.5 shows that rate of the reaction was the greatest at the start of the reaction and gradually slowed to zero at the end of the reaction. This is supported by the raw data, which shows that the changes in mass are greatest at the start of the reaction. You will learn more about why this happens later in this chapter.

GRAPHING THE CHANGES DURING A REACTION

Use this simulation to observe a reaction and the shape of the graphs produced.

EXPERIMENT 9.2

MEASURING THE RATE OF A REACTION

The rate of a reaction is determined by measuring how much reactant is consumed per unit of time, or how much product is made per unit of time.

The reaction between calcium carbonate and hydrochloric acid produces carbon dioxide, water and calcium chloride.

$$CaCO_3(s) + 2HCl(aq) \rightarrow CaCl_2(aq) + CO_2(g) + H_2O(l)$$

By measuring the amount of carbon dioxide produced at different time intervals, you can determine the rate of the reaction.

Aim

To observe and calculate the rate of reaction between calcium carbonate and hydrochloric acid at various stages

Materials

- 2 g of $CaCO_3$ powder
- 250 mL of 0.25 mol L^{-1} HCl
- 100 mL beaker
- 500 mL conical flask
- Spatula
- 250 mL measuring cylinder
- Paper
- Graph paper
- Electronic balance
- Stopwatch
- Calculator

What are the risks in doing this experiment?	How can you manage these risks to stay safe?
Chemicals could splash in your eyes.	Wear safety glasses at all times.
0.25 mol L^{-1} HCl is corrosive to skin and clothing.	Wear safety glasses and protective clothing. Take care when pouring and clean up spills immediately. If hydrochloric acid is spilt on your skin, wash the affected area with plenty of water and notify your teacher.

In your write-up, add any more risks you can think of, as well as ways to manage them.

Procedure

1. Measure 2 g of $CaCO_3$ powder into the clean, dry beaker.
2. Accurately measure 250 mL of 0.25 mol L^{-1} HCl.
3. Place the HCl in a clean conical flask and place this on the electronic balance. Record the mass of the flask and acid.
4. Calculate the mass of the flask, HCl and $CaCO_3$ by adding your values from steps 1 and 3.
5. Place the $CaCO_3$ on a clean piece of paper, and roll the paper so that it can easily be inserted into the top of the flask.
6. Add the $CaCO_3$ to the flask (while it is on the electronic balance) and immediately start the stopwatch and record the mass.
7. Record the mass (and time) as frequently as you can.
8. When the reaction is complete, take the flask off the electronic balance.

Results

Organise your data into an appropriate table, remembering to include a column to calculate the cumulative mass of carbon dioxide gas produced for each time.

Analysis of results

1. What happened to the mass of the flask and its contents? Why did this occur?
2. How can you use the mass of the flask to determine the mass of the CO_2 produced?

3 Calculate the cumulative mass of the CO_2 produced for each time.
4 Graph the mass of CO_2 produced against time.
5 When was the rate of reaction the fastest? How could you know this from the graph?
6 When was the rate of reaction the slowest? How could you know this from the graph?
7 At what time was the reaction complete? How could you know this from the graph?
8 Calculate the average rate of reaction (average rate $= \frac{\text{total change in mass}}{\text{time for mass change}}$).
9 Draw tangents to the graph at three different times and calculate the gradients (or slopes) of the lines. These are the rates of the reaction at those times. Ask your teacher if you do not know how to draw a tangent.

DRAWING A TANGENT TO A CURVE

Visit this website to see how to draw a tangent to a curve.

Discussion

1 What is the difference between qualitative data and quantitative data?
2 During the experiment, how could you observe the rate of the reaction?
3 Did your quantitative data support your qualitative data regarding the rate of the reaction?
4 What aspects of the experiment may have reduced the accuracy of the data?
5 How could you improve the experiment to ensure that the results were more accurate and reliable?

Conclusion

What conclusion can you make about the rate of the reaction at various stages of a reaction?

QUESTION SET 9.1

Remembering

1 Define 'rate'.
2 What data would you record to be able to calculate or compare the rates of different reactions?
3 During a reaction, when is the rate the fastest?

Understanding

4 Describe how to calculate the rate of a reaction.
5 Describe how the slope (or gradient) of a graph indicates the rate.

Applying

6 During a reaction 20 grams of product is formed in 4 minutes. Calculate the rate of the reaction.
7 Copy the graph in Figure C9.6 and draw tangents to the curve at times 0, 1 and 5 to calculate the rate of reactions at each of these times.

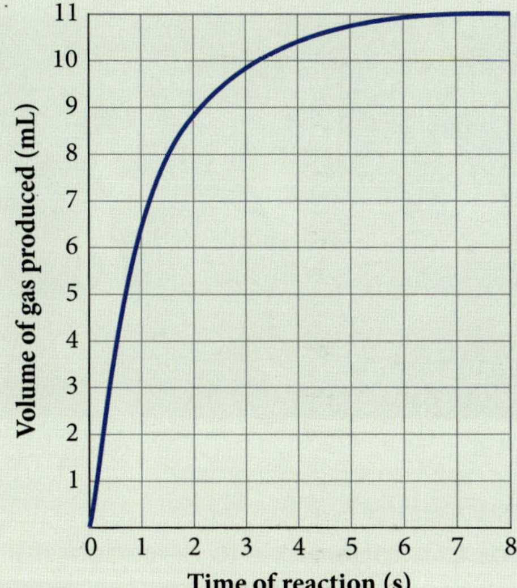

◀ **Figure C9.6** Production of carbon dioxide gas during the decomposition of copper(II) carbonate

9.2 What is needed for a reaction

To be able to understand why some reactions occur faster than others, you must first understand what happens to the particles during a reaction.

ACTIVITY 9.2

MODELLING A REACTION

Aim

To use molecular models to represent a chemical reaction

You will need

- Molecular model kit; for example, Molymod

What to do

1 Using the kit, make two molecules of hydrogen gas (H_2) and one molecule of oxygen gas (O_2) as shown in Figure C9.7.
2 Use these molecules to make two molecules of water (H_2O).
3 Describe what had to happen for the reactants (hydrogen gas and oxygen gas) to become the product (water).

What did you discover?

Based on the reaction you modelled, what has to happen for the reactants in a chemical reaction to become products?

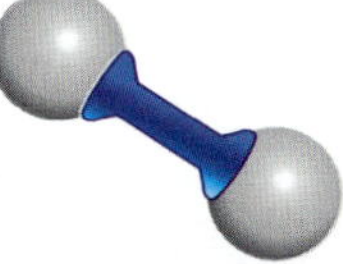

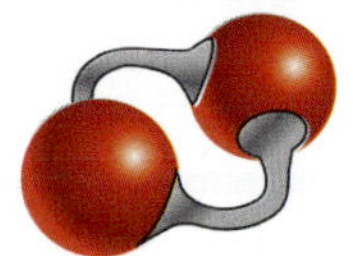

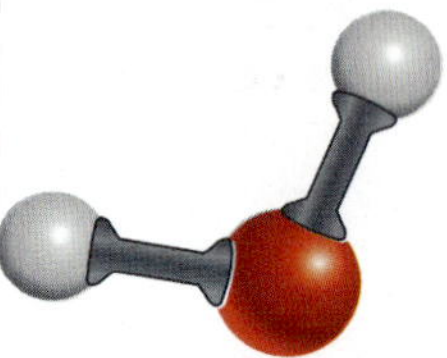

Figure C9.7 ▲
Modelling a chemical reaction

ACTIVITY 9.3

DODGEM CAR CRASH ANALOGY

In this activity you are going to compare a 'successful' dodgem car crash with a chemical reaction.

Aim

To compare the components of a chemical reaction with those of dodgem cars when they crash

What to do

1 In small groups, brainstorm the following ideas.
   - Consider a dodgem car crash.
   - What would make a dodgem car crash 'successful'?
   - Are all dodgem car crashes successful?
   - What has to happen during a dodgem car crash for it to be successful?
2 Write a paragraph describing a dodgem car crash and what has to happen for it to be successful.
3 Save your paragraph to refer to later.

Alamy/Art Directors & TRIP

Figure C9.8 ▲
Is this a successful dodgem car crash?

Consider the reaction between hydrogen gas and oxygen gas that occurs in a hydrogen fuel cell and the 'pop test'.

$$2H_2(g) + O_2(g) \rightarrow 2H_2O(g)$$

For this reaction to occur:

- bonds in hydrogen molecules must break to separate the hydrogen atoms
- bonds in oxygen molecules must break to separate the oxygen atoms
- new bonds must form between the hydrogen and oxygen atoms to make the water molecules.

This reaction requires enough energy to overcome the forces of attraction holding the atoms of the reactants in their original form. So energy is required to break the covalent bonds within the hydrogen molecules and oxygen molecules.

This reaction also requires the appropriate atoms to come into contact with each other so that the new bonds of the products can form. This means that two hydrogen atoms must come in contact with one oxygen atom to be able to form a water molecule.

Collision theory

To understand why reactions occur quickly or slowly, you need to know what happens during a chemical reaction. **Collision theory** is a model that allows us to explain what happens during a chemical reaction and the factors that affect it. This theory states that, for a reaction to occur, the reactants must collide with sufficient energy and in an appropriate orientation. Collisions between reactant particles can be classified as either successful or unsuccessful. A successful collision has sufficient energy and an appropriate orientation to allow old bonds to be broken and new bonds to be formed. An unsuccessful reaction is one in which the energy and/or orientation are not satisfactory. The more successful collisions there are in a given time, the faster the rate of the reaction because more reactants are forming products.

Collision theory states that, for a reaction to occur, the particles must collide with sufficient energy and in the appropriate orientation.

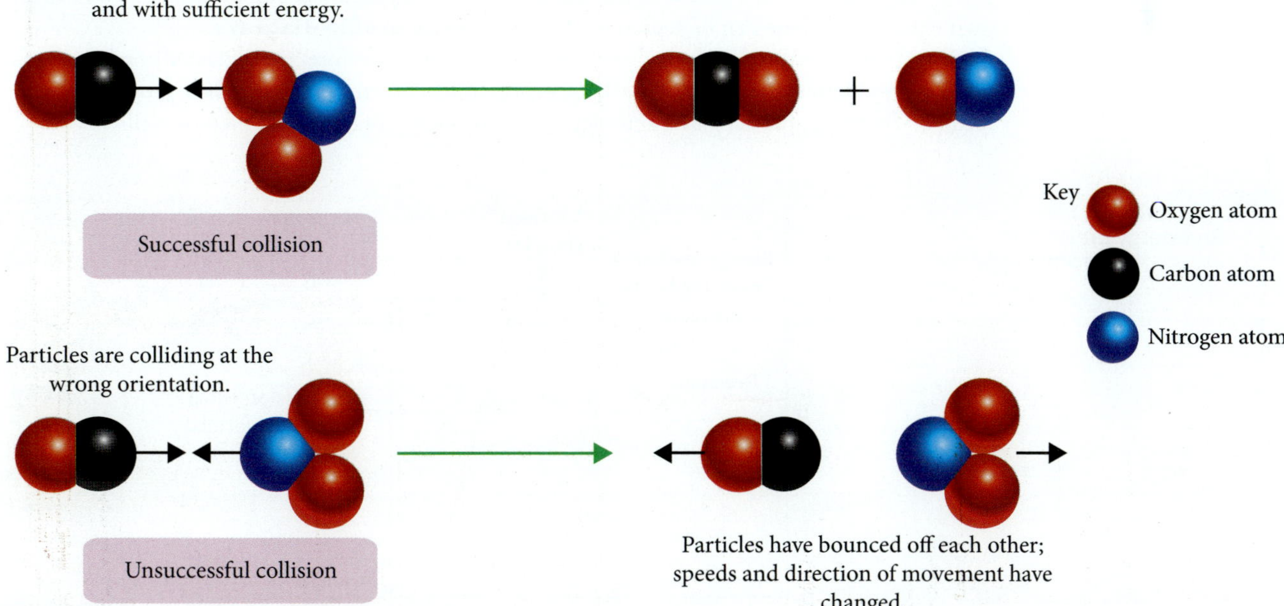

▲ **Figure C9.9**
The correct orientation is needed for a successful collision.

DODGEM CAR CRASH ANALOGY AND COLLISION THEORY

1 Review the dodgem car crash analogy in Activity 9.3.

2 In a dodgem car crash, what would be the equivalent of these parts of the collision theory?
   a The collision
   b Enough energy
   c Appropriate orientation

3 For each part of Question **2**, explain the relevance to a 'successful' dodgem car crash.

During the progress of a reaction, as reactants form the products, there exists a brief moment where the bonds of the reactants have broken, but the bonds of the products are not yet formed. This produces a temporary, highly unstable structure called the **activated complex** or **transition state**. As energy was required to break the bonds of the reactants, the activated complex has more energy than either the reactants or products.

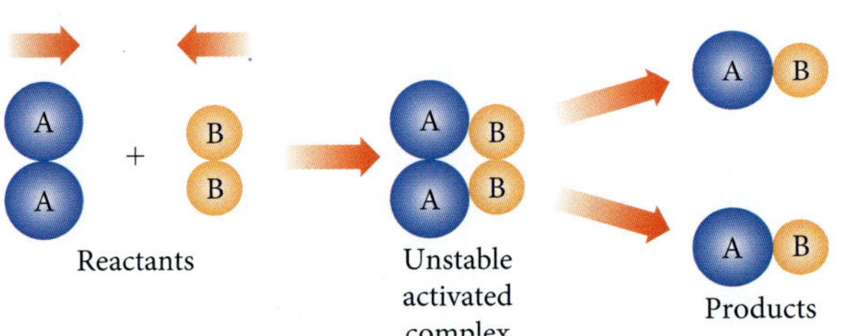

Figure C9.10 ▶ Stages of a reaction ($A_2 + B_2 \rightarrow 2AB$)

The activated complex will quickly form new bonds to produce the products, releasing energy. This means that the atoms that will bond need to be positioned next to each other so they can form the new bonds. This is why the orientation of the colliding particles needs to be correct.

Energy profile diagrams

The total energy possessed by a chemical substance is called **enthalpy** or heat content, and is given the symbol H. In a reaction, the reactants, activated complex and products each have their own enthalpies. These can be represented in an **energy profile diagram**.

Energy is needed to break the bonds of the reactants; therefore, the activated complex has more energy than the reactants. Similarly, energy is lost when the bonds of the products are formed. Therefore, the activated complex has more energy than the products as well.

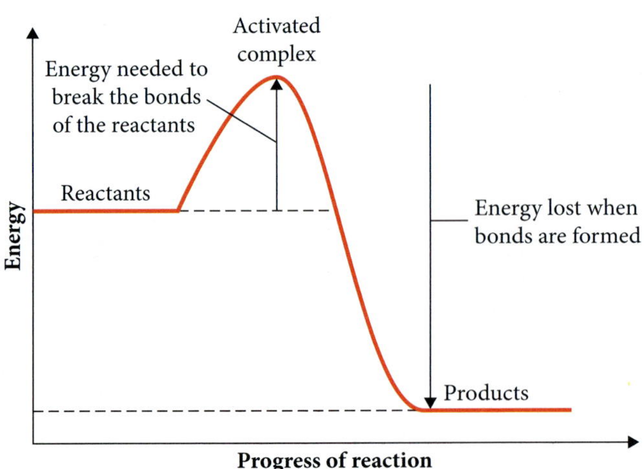

Figure C9.11 ▶ An energy profile diagram

If more energy is lost when the products are formed than was gained to form the activated complex, then the difference in energy is lost from the system to the surroundings. This occurs in an **exothermic reaction**. Conversely, if more energy is gained than is lost, the excess must be gained from the surroundings. This occurs in an **endothermic reaction**.

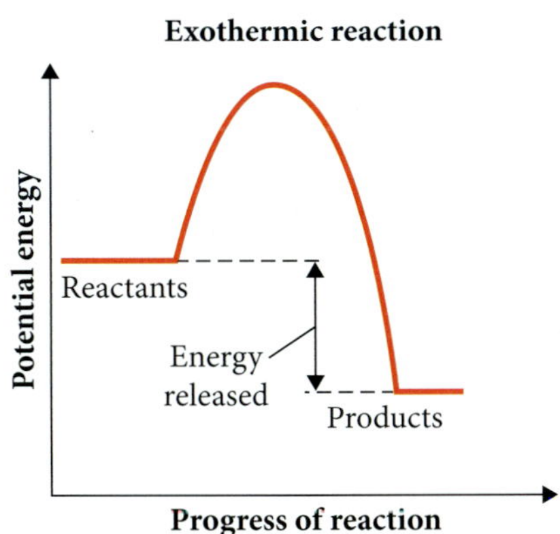

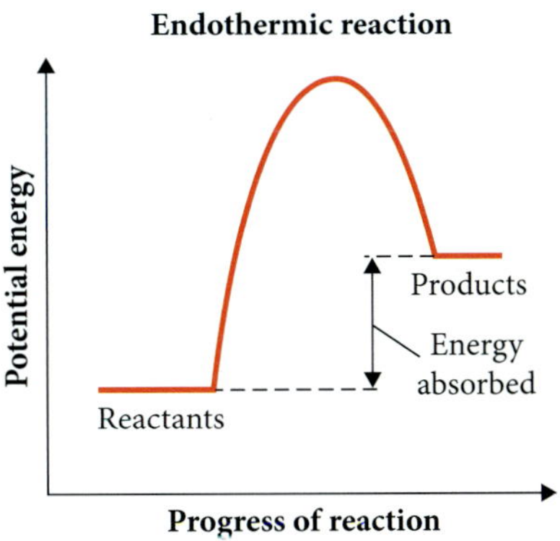

◀ **Figure C9.12** Energy profile diagrams for endothermic and exothermic reactions

To revise exothermic and endothermic reactions, see Chemistry section 4.3 on page 213.

Activation energy

The amount of energy needed to form the activated complex is called the **activation energy** and is often represented as E_a. As this energy is needed for the bonds of the reactants to break, the activation energy is the minimum energy required for a reaction to occur. The activation energy can be determined from the difference in energy between the activated complex and the reactants.

EFFECT OF ENERGY ON COLLISIONS

Use this collision simulation to see what happens if the particles in a reaction have or don't have enough energy.

The activation energy is the minimum energy required for a chemical reaction to occur and influences the rate of a chemical reaction.

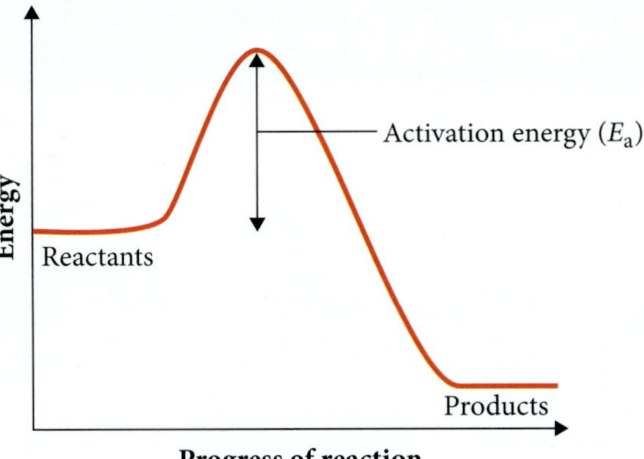

◀ **Figure C9.13** Activation energy

At a given temperature, particles will have different amounts of **kinetic energy**, with the average kinetic energy determining the temperature. This is represented in the **Maxwell–Boltzmann distribution**, which graphically shows the energies of the particles at a particular temperature (Figure C9.14).

From Figure C9.13, you can see that not all particles have the same amount of energy. Only those particles with the activation energy, or greater, will have enough energy to break the bonds of the reactants and have a successful collision. On the graph, the proportion of particles with that amount of energy is represented by the area under the curve above the activation energy (Figure C9.15 on page 378).

EFFECT OF ENERGY AND ORIENTATION ON COLLISIONS

Use this simulation to see the effect of energy and orientation on reactions.

Figure C9.14 ▶
The Maxwell–Boltzmann distribution shows the energy of molecules at a particular temperature.

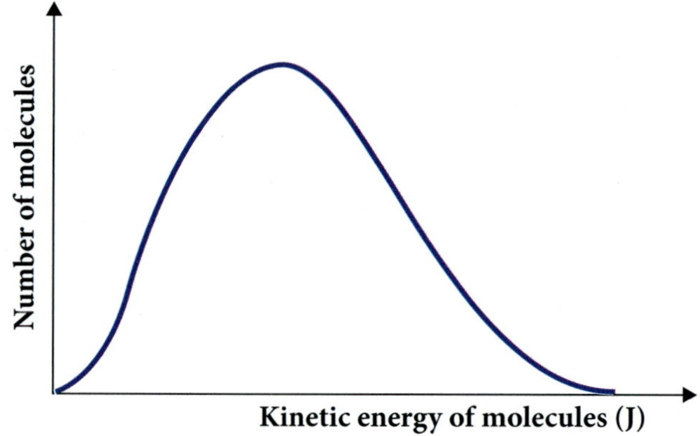

Figure C9.15 ▶
The number of molecules with enough energy for a successful collision

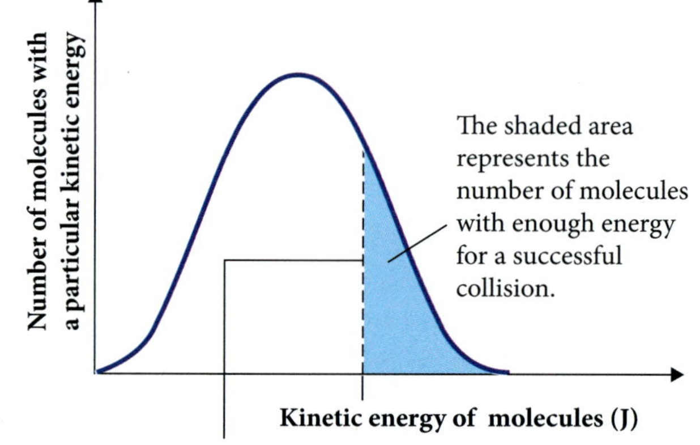

To see applications of collision theory on reactions, refer to Context 4, 'Making reactions work for us', page 85.

QUESTION SET 9.2

Remembering

1 State the collision theory.

2 Define:

  a activation energy.

  b activated complex.

3 Draw a Maxwell–Boltzmann distribution.

Understanding

4 Explain why a successful collision requires:

  a sufficient energy.

  b the correct orientation.

5 What will happen to two particles that collide with energy less than the activation energy?

Applying

6 Draw and label an energy profile diagram for an:

  a endothermic reaction with an activation energy of 1000 J.

  b exothermic reaction with an activation energy of 200 J.

9.3 What affects rate

Collision theory tells us that the rate of a reaction depends on the number of successful collisions between reactant particles in a given time. Therefore, anything that increases the number of successful collisions will increase the rate of the reaction.

The number of successful collisions depends on the:

- total number of collisions
- percentage of collisions that are successful.

Any factor that will increase either of these factors will lead to an increase in the reaction rate. Conversely, anything that decreases these factors will lead to a decrease in the reaction rate.

For example, if there are 1000 collisions between reactants per second and 60% are successful, there will be 600 successful collisions per second. If the number of collisions is increased to 2000 per second, even if only 60% are successful, there will now be 1200 successful collisions per second and therefore the reaction rate has increased. Alternatively, if there are still 1000 collisions, but 80% of them are successful, there are now 800 successful collisions per second and so the reaction rate has still increased.

Table C9.2 Successful collisions

Number of collisions	Percentage of collisions that are successful	Number of successful collisions
1000	60	600
1000	80	800
2000	60	1200
2000	80	1600

INVESTIGATION 9.1

CHANGING THE COLOUR OF POTASSIUM PERMANGANATE

Rhubarb contains oxalic acid, which will react with acidified potassium permanganate, causing it to decolorise. This can be represented by the following equation.

$$2MnO_4^- + 5C_2H_2O_4 + 6H_3O^+ \rightarrow 2Mn^{2+} + 10CO_2 + 14H_2O$$

The time for this change in colour to occur can be measured and used to calculate the rate of the reaction.

The rhubarb can be fresh or frozen rhubarb sticks, or in the form of a puree that can be made by gently boiling rhubarb in water.

Your task is to consider all the factors that may affect the time for the permanganate to decolorise. Choose one of these factors and design an investigation to determine its effect on rate of the reaction.

What is your aim?

Write down an aim for your investigation, stating clearly what you are investigating.

What will you need?

Make a list of all the materials and equipment that you will need to carry out your investigation.

Make sure that you are specific about the type, number and size of all the materials and apparatus needed.

What are the risks?

Construct a table similar to the one on page 380. Identify specific risks involved in the investigation and ways that you will manage the risks to avoid injuries or damage to equipment. Ask your teacher to check your risk assessment before you proceed.

REACTING WITH RHUBARB

Visit this website to find out more about the reaction between oxalic acid in rhubarb and potassium permanganate.

HEALTH AND SAFETY NOTES FOR RHUBARB REACTION

Check your risk assessment for the rhubarb investigation by visiting this website.

What are the risks in doing this investigation?	How can you manage these risks to stay safe?

How will you carry out your investigation?

Clearly write down, in numbered steps, how you are going to carry out your investigation to achieve your aim. Make sure your steps clearly show how you are going to control or measure all your variables.

What results will you collect?

1 What is your independent variable? How will you change this?
2 What is your dependent variable? How will you measure this?
3 Is there other data that you need to measure and record?
4 Will you collect quantitative and/or qualitative results? State what sort of results you will be collecting and when you will collect them.

How will you analyse your results?

1 How will you calculate the rate of reaction from your data?
2 State how you analysed your results and write the analysis here.

What have you found?

What did you find when you analysed your results?

What do you conclude?

1 How can science be used to explain your conclusion?
2 Write a conclusion to your investigation based on your aim and your results.

Ideas for improvement

Did you experience any problems with your materials or method? How would you overcome this if you were to repeat this investigation?

Taking it further

Could you undertake any further investigation to find out more information about this topic? If yes, what would you want to find out, and how would you go about doing that?

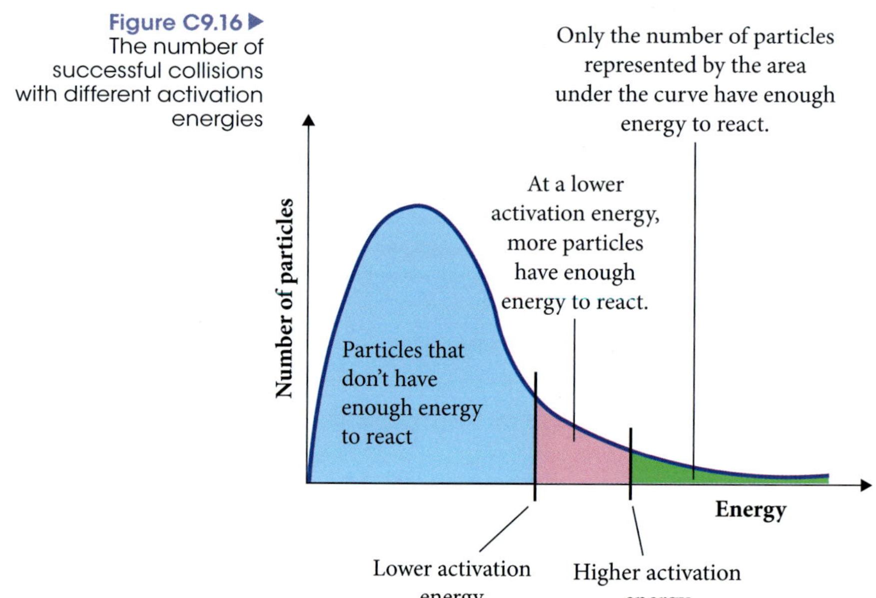

Figure C9.16 ▶ The number of successful collisions with different activation energies

Nature of reactants

The nature of the reactants refers to what the reactants are and what has to happen for them to form the products.

If the bonds of the reactants that need to be broken are weak or few, then they will only require a small amount of energy to break. This means that the activation energy will be low, and a larger number of particles will have enough energy for a successful collision.

The size of the particles may also affect the rate of the reaction. If the particles are larger, then there is a greater chance that they will collide.

ACTIVITY 9.5

DODGEM CAR CRASH ANALOGY AND NATURE OF REACTANTS

What to do

1 Review the dodgem car crash analogy in Activity 9.3.
2 What would be the equivalent of the nature of reactants for your dodgem car crash analogy?
3 How could you increase the 'activation energy' for your dodgem car crash?
4 How could you decrease the 'activation energy' for your dodgem car crash?
5 Are these factors that real car manufacturers are able to consider to make cars safer?

What have you discovered?

Write a paragraph summarising how the nature of reactants will affect the rate of dodgem car crashes.

EFFECT OF THE NATURE OF REACTANTS ON REACTION RATE

View this animation to see the effect of reactant size on reaction rate.

Concentration and pressure

Concentration refers to the number of particles in a given volume. The **pressure** of a gas is determined by the concentration of the gaseous particles, and therefore you can consider the two factors simultaneously. The number of particles in a given volume will determine the number of collisions that occur. The greater the concentration, or pressure, the greater the number of particles and therefore the greater the number of collisions. Even though the percentage of collisions that will be successful will remain the same, there will be more successful collisions in total because there are more collisions. This means that as the concentration, or pressure, of the reactants increases, the reaction rate increases.

EFFECT OF CONCENTRATION ON COLLISIONS

View this animation to compare the number of collisions between acids and a marble chip in different solutions. Which one would have the fastest reaction rate? Choose 'View online', then click the green arrow to start the animation.

Mark Fergus Photography

Figure C9.17 Solutions of different concentrations

EFFECT OF PRESSURE ON REACTION RATE

View this animation of collision theory to see the effect of pressure on the rate of reaction.

ACTIVITY 9.6

SCREAMING GUMMI BEAR

Aim

To observe a fast reaction and investigate the factors that contribute to its speed

What you will need:

- computer with internet access

What to do

1 Watch the video of the experiment at the weblink 'Screaming gummi bear'. This is an example of a very fast reaction. What observations can you make that indicate the rate of this reaction?
2 Research to find the equation for the reaction that is occurring.
3 Research the reaction to find out why it occurs so quickly.
4 Your teacher may demonstrate this experiment to you. How was it similar to the video? How was it different? Why do you think it may have been different? What did you discover?
5 What factors affected the rate of the Screaming gummi bear experiment?

SCREAMING GUMMI BEAR

Watch this video of the screaming gummi bear experiment.

EXPERIMENT 9.3

EFFECT OF CONCENTRATION ON RATE OF REACTION

Sodium thiosulfate reacts with dilute hydrochloric acid to produce sodium chloride, sulfur, sulfur dioxide and water.

$$Na_2S_2O_3 + 2HCl \rightarrow 2NaCl + S + SO_2 + H_2O$$

When the concentration of sulfur is high enough, the solution will become cloudy. The time for this to occur can be measured to indicate the rate of the reaction.

Aim

To investigate the effect of concentration on the rate of the reaction between sodium thiosulfate and dilute hydrochloric acid

Materials

Per trial:

- 45 mL of 0.25 mol L^{-1} sodium thiosulfate ($Na_2S_2O_3$)
- 45 mL of 0.125 mol L^{-1} sodium thiosulfate ($Na_2S_2O_3$)
- 45 mL of 0.0625 mol L^{-1} sodium thiosulfate ($Na_2S_2O_3$)
- 15 mL of 2 mol L^{-1} hydrochloric acid (HCl)
- Stopwatch
- 3 × 100 mL conical flask
- 1 × 50 mL and 1 × 10 mL measuring cylinder
- White paper
- Pen or marker pen

What are the risks in doing this experiment?	How can you manage these risks to stay safe?
Chemicals could splash in your eyes.	Wear safety glasses at all times.
2 mol L^{-1} HCl is corrosive to skin and clothes.	Wear gloves and an apron.
SO_2 gas irritates the nose, throat and airways.	Conduct the experiment in a well-ventilated area and don't breathe in fumes.

In your write-up, add any more risks you can think of, as well as ways to manage them. In particular, expand on the three risks listed to identify specific risks involved with each of them.

Procedure

1 Draw a large cross in the centre of the piece of white paper.
2 Place the conical flask over the cross.
3 Accurately measure 45 mL of 0.0625 mol L^{-1} $Na_2S_2O_3$ in a measuring cylinder.
4 Pour the $Na_2S_2O_3$ into the conical flask.
5 Accurately measure 5 mL of 2 mol L^{-1} HCl in a measuring cylinder.
6 Have a stopwatch ready to start timing.
7 Pour the HCl into the conical flask, starting the stopwatch immediately.
8 Quickly mix the two solutions by carefully swirling the conical flask.
9 Place the conical flask onto the cross and record the time it takes before you can no longer see the cross when looking from above. (Be careful not to breathe in any fumes from the reaction.)
10 Repeat steps 3–9 for the other two solutions.
11 If time and resources allow, repeat steps 3–10 for repeat trials.
12 Place your solutions in the waste beaker, and rinse the conical flask.

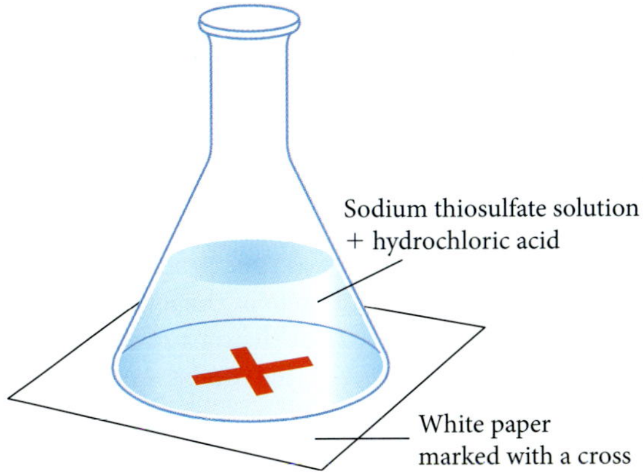

▲ **Figure C9.18**
Reaction between sodium thiosulfate and dilute hydrochloric acid

Results

Enter your results in a results table like the one below.

Concentration of sodium thiosulfate (mol L^{-1})	Time taken for the cross to disappear (s)			
	Trial 1	Trial 2	Trial 3	Average
0.0625				
0.125				
0.250				

Analysis of results

1 Graph the average time taken for the cross to disappear against the concentration of sodium thiosulfate.
2 How is the time taken for the cross to disappear related to the rate of the reaction?
3 What happened to the time taken for the cross to disappear as the concentration of sodium thiosulfate increased?
4 What is the relationship between the concentration of sodium thiosulfate and the rate of the reaction?
5 Does this support the theoretical trend?
6 Were there any anomalies? If yes, why do you think they occurred?

Discussion

1 Use collision theory to explain the relationship between concentration and the rate of the reaction.
2 Where might errors have occurred in the method that would have affected the accuracy of the results?
3 How could the method be improved to increase the accuracy and reliability of the results?

Conclusion

What conclusion can you make about the concentration of reactants and the rate of the reaction?

Taking it further

1 How could you further investigate the effect of concentration on the rate of the reaction in this reaction?
2 How could you further investigate the effect of concentration on the rate of reactions in general?

EFFECT OF CONCENTRATION ON REACTION RATE

View this animation of collision theory to see the effect of concentration on the rate of reaction.

Changing the volume may also change the concentration. In a solution, adding water will dilute the solution, thereby decreasing the concentration. Alternatively, evaporating some of the water will decrease the volume and increase the concentration. In a gas, if the volume of the container is increased, then the concentration of all the gases will decrease. Alternatively, if the volume of the container is decreased, then the concentration of all the gases will increase.

During the course of a reaction, reactants are being used up to form the product. This means that the concentration of the reactants is decreasing. From our understanding of the effect of concentration on the rate of a reaction, it makes sense that the rate of the reaction will decrease as time progresses during the reaction. On a graph of the reactants or products versus time, the rate of the reaction is indicated by the slope of the graph. Therefore, the slope will be steepest at the start of the reaction and will gradually decrease to zero at the completion of the reaction. This can be seen clearly in Figure C9.5 on page 371.

ACTIVITY 9.7

DODGEM CAR CRASH ANALOGY – CONCENTRATION

Aim

To compare the effect of concentration on the rate of a reaction with dodgem car crashes

What to do

1 Review the dodgem car crash analogy in Activity 9.3.
2 In your dodgem car crash analogy what would be the equivalent to concentration?
3 How would changing this concentration equivalent affect the rate of dodgem car crashes?
4 Would it matter if only one 'reactant' or all 'reactants' were changed?

What did you discover?

Write a paragraph summarising how changes in concentration will affect the rate of dodgem car crashes.

Low surface area

High surface area

Figure C9.19 ▲ Dividing a solid will increase the surface area exposed for a reaction.

Surface area

For a reaction to occur, the reactant particles must be able to collide. When one of the reactants is a solid, only the particles on the surface of the solid are available for collision. Once these have reacted, then the particles underneath are exposed and are able to react, and so on.

If the solid is divided or crushed, then the inner particles become exposed and, although the amount of reactant remains the same, the total **surface area** has increased. This means that more of the solid is available for collisions and is able to react. Therefore, there are more collisions and thus the reaction rate will increase.

We naturally use this method of increasing the rate of a reaction every day. It is why we chop wood into small pieces of kindling to start the fire, why we cut potatoes up into small pieces to cook them quickly and why we use granulated sugar instead of sugar cubes in our coffee.

EXPERIMENT 9.4

EFFECT OF SURFACE AREA ON RATE OF REACTION

Calcium carbonate will react with dilute hydrochloric acid according to the following equation:

$$CaCO_3(s) + 2HCl(aq) \rightarrow CaCl_2(aq) + H_2O(l) + CO_2(g)$$

During the reaction, bubbles of carbon dioxide can be seen. Therefore, when there are no more bubbles being produced, the reaction has stopped. The time for this to occur can be measured to compare the rates of reactions.

In this experiment, you will change the surface area of a fixed mass of solid calcium carbonate. By comparing the time taken for the reactions to occur, you will be able to determine the effect of surface area on the rate of the reaction.

Aim

To determine the effect of surface area on the rate of the reaction between calcium carbonate and dilute hydrochloric acid

Materials

- 3 calcium carbonate ($CaCO_3$) chips
- Calcium carbonate ($CaCO_3$) powder
- 50 mL of 2 mol L^{-1} HCl
- Electronic balance
- 10 mL measuring cylinder
- Stopwatch
- 2 large test tubes

What are the risks in doing this experiment?	How can you manage these risks to stay safe?
Chemicals may splash in your eyes.	Wear safety glasses at all times.
2 mol L^{-1} HCl is corrosive to skin and clothes.	Wear gloves and an apron.

In your write-up, add any more risks you can think of, as well as ways to manage them. In particular, expand on the two risks listed to identify specific risks involved with each of them.

Procedure

1. Measure 8 mL of HCl in a measuring cylinder.
2. Weigh and record the mass of three calcium carbonate chips.
3. Place the chips in one test tube.
4. Have a stopwatch ready to start timing.
5. Pour the HCl into the test tube and start the stopwatch.
6. Stop the stopwatch when no more bubbles are produced, and record this time.
7. Carefully weigh the same mass of calcium carbonate powder as that of the calcium carbonate chips.
8. Repeat steps 3–6 for the powder.
9. If time permits, repeat steps 1–8 for a number of trials.
10. Place all the product solutions into the waste beakers as directed by your teacher.

Results

Record your results in a results table like the one below.

Form of calcium carbonate	Time for the reaction to complete (s)			
	Trial 1	Trial 2	Trial 3	Average
Chips				
Powder				

Analysis of results

1 Which form of calcium carbonate took the least time to react?
2 Which form of calcium carbonate has the fastest rate of reaction?
3 Which form of calcium carbonate had the greatest surface area? Explain why.
4 What is the relationship between surface area and the rate of reaction?
5 Do your results support the theory about the effect of surface area on the rate of reaction?

Discussion

1 Was this a fair test? Explain why or why not.
2 Were the results accurate and reliable? Explain why or why not.
3 How could the method be improved to more fairly and accurately test the effect of surface area on the rate of reaction between calcium carbonate and dilute hydrochloric acid?

Conclusion

What conclusion can you make from this experiment?

Taking it further

How could you further investigate the effect of surface area on the rates of reactions?

A similar effect to increasing the surface area is seen when the reactants are stirred. Stirring mixes the reactants together, exposing more of the reactants to each other. This increases the number of collisions and therefore increases the reaction rate. We use this phenomenon when making a cup of tea or coffee. By stirring the sugar, we are able to increase the rate at which it dissolves.

Figure C9.20 ▶ Stirring will have the same effect as increasing the surface area as more particles are exposed to one another.

ACTIVITY 9.8

DODGEM CAR CRASH ANALOGY – SURFACE AREA

Aim

To compare the effect of surface area on the rate of a reaction with dodgem car crashes

What to do

1 Review the dodgem car crash analogy in Activity 9.3.
2 In your dodgem car crash analogy, what would be the equivalent to surface area?
3 How would changing this surface area equivalent affect the rate of dodgem car crashes?

What did you discover?

Write a paragraph summarising how the surface area will affect the rate of dodgem car crashes.

Temperature

The **temperature** of a substance is a measure of the average kinetic energy of the particles. This energy is the energy due to movement and is dependent on the speed, or velocity, of the particles. As the temperature of a substance increases the average kinetic energy also increases, and hence the average speed of the particles also increases.

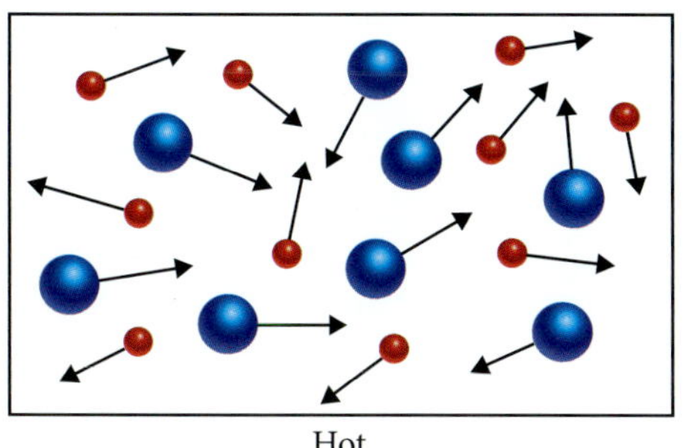
Hot

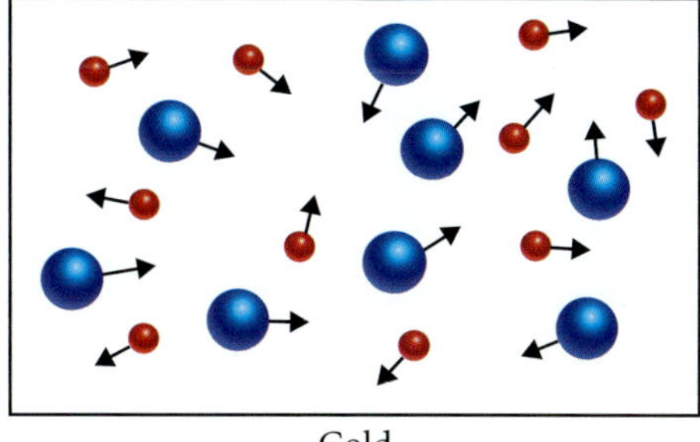
Cold

▲ Figure C9.21
Temperature changes the speed at which particles move (as represented by the length of the arrows).

As the temperature of reactants increases there is a dual effect on the reaction rate.

1 Increased percentage of collisions that are successful

The average kinetic energy of the particles increases. This means that more particles have enough energy to overcome the activation energy, form the activated complex and become products. Therefore, a greater percentage of collisions are successful, increasing the reaction rate.

At a higher temperature the Maxwell–Boltzmann distribution (see Figure C9.22) is pushed to the right. The area under the curve to the right of the activation energy is now greater, representing a greater proportion of particles with enough energy for a successful collision.

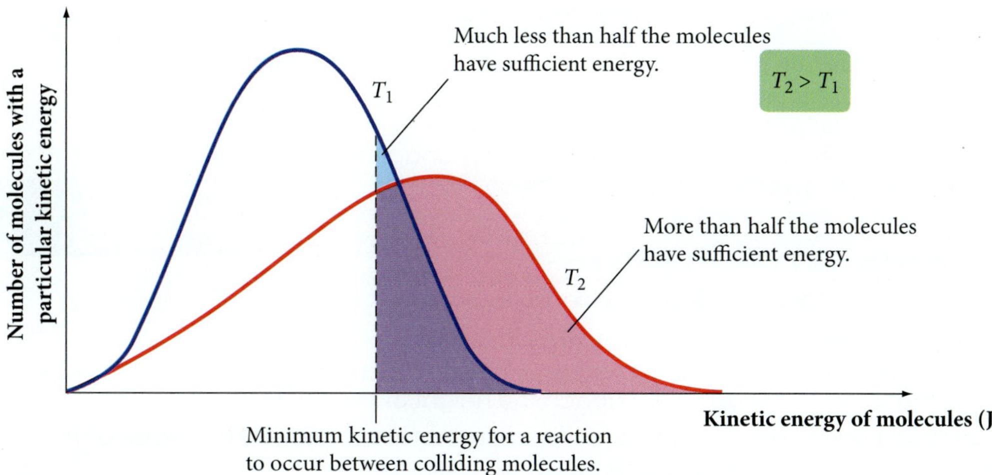

▲ Figure C9.22
The Maxwell-Boltzmann distribution for different temperatures

2 Increased number of collisions

With the increase in temperature, the particles now have a greater average speed. This means that they will collide with each other more frequently, increasing the number of collisions. This will also increase the rate of the reaction. However, this effect is much less significant than the effect of the increased energy.

Kinetic energy can be calculated by using: Kinetic energy = ½ mass × velocity² where velocity is the speed in a straight line. This means that if the kinetic energy of an object increases, then the velocity increases. This also means that if two objects at the same temperature (so have the same kinetic energy) have different masses, then their velocities will also be different. The object with the greater mass will have the lower velocity.

EFFECT OF TEMPERATURE ON REACTION RATE 1

View this animation to see the effect of temperature on reaction rate. Vary the temperature to see how long the reaction takes.

EFFECT OF TEMPERATURE ON REACTION RATE 2

View this animation to see the effect of temperature on reaction rate.

EXPERIMENT 9.5

EFFECT OF TEMPERATURE ON RATE OF REACTION

You have previously seen the reaction between sodium thiosulfate and dilute hydrochloric acid.

$$Na_2S_2O_3 + 2HCl \rightarrow 2NaCl + S + SO_2 + H_2O$$

In this experiment, you will change the temperature of the solutions to investigate the effect of temperature on the rate of the reaction.

Aim

To investigate the effect of temperature on the rate of the reaction between sodium thiosulfate and dilute hydrochloric acid

Materials

Per trial:

- 150 mL of 0.25 M sodium thiosulfate ($Na_2S_2O_3$)
- 15 mL of 2 mol L^{-1} HCl
- Crushed ice
- Hot water
- Stopwatch
- 2 × 1 L beakers (or similar container for ice bath and hot water bath)
- 3 × 100 mL conical flask
- 50 mL measuring cylinder
- 10 mL measuring cylinder
- White paper
- Pen or marker pen
- Thermometer

What are the risks in doing this experiment?	How can you manage these risks to stay safe?
Chemicals may splash in your eyes.	Wear safety glasses at all times.
2 mol L^{-1} HCl is corrosive to skin and clothes.	Wear gloves and an apron.
SO_2 gas irritates the nose, throat and airways.	Conduct the experiment in a well-ventilated area; do not breathe in fumes.

In your write-up, add any more risks you can think of, as well as ways to manage them. In particular, expand on the three risks listed to identify specific risks involved with each of them.

Procedure

1 Draw a large cross in the centre of a blank piece of white paper.
2 Accurately measure 45 mL of the 0.25 mol L^{-1} $Na_2S_2O_3$ in a measuring cylinder.
3 Pour the $Na_2S_2O_3$ into the first conical flask.
4 Place some crushed ice and a small amount of water into the 1 L beaker, and carefully place the conical flask into the ice bath. Ensure that there is enough ice to reach the height of the solution in the flask.
5 Repeat steps 1–4 with the second conical flask, using hot water instead of ice to make a hot water bath.
6 Accurately measure 45 mL of the 0.25 mol L^{-1} $Na_2S_2O_3$ in a measuring cylinder.
7 Pour the $Na_2S_2O_3$ into the third conical flask.
8 Measure and record the temperature of the $Na_2S_2O_3$ in all three conical flasks.
9 Accurately measure 5 mL of the 2 mol L^{-1} HCl in a measuring cylinder.
10 Have a stopwatch ready to start timing.
11 Pour the HCl into the first conical flask, starting the stopwatch immediately.
12 Quickly mix the two solutions by carefully swirling the conical flask.

13 Place the conical flask onto the cross and record the time it takes before you are no longer able to see the cross when looking from above. (Be careful not to breathe in any fumes from the reaction).

14 Repeat steps 9–13 for the other two solutions.

15 If time and resources allow, repeat steps 2–13 for repeat trials.

16 Place your solutions in the waste beaker, and rinse the conical flask.

Results

Record your results in a results table like the one below.

Temperature of sodium thiosulfate (°C)	Time taken for the cross to disappear (s)			
	Trial 1	Trial 2	Trial 3	Average

Analysis of results

1 Graph the average time taken for the cross to disappear against the temperature of the sodium thiosulfate.

2 What happened to the time taken for the cross to disappear as the temperature increased?

3 What is the relationship between temperature and the rate of the reaction?

4 Do your results support the theory about the effect of temperature on the rate of reactions?

Discussion

1 Where might errors have occurred in the experiment and how would these have affected the results?

2 Were your results accurate and reliable? Explain why or why not.

3 How could you improve the experiment if you were to do it again?

Conclusion

What did you find out about the effect of temperature on the rate of the reaction between sodium thiosulfate and dilute hydrochloric acid?

Taking it further

How could you further investigate the effect of temperature on the rate of reactions?

ACTIVITY 9.9

DODGEM CAR CRASH ANALOGY – TEMPERATURE

Aim

To consider the effect of temperature on reaction rates in the dodgem car crash analogy

What to do

1 Review the dodgem car crash analogy in Activity 9.3.

2 In your analogy what would be the equivalent of temperature? (Remember that temperature measures the average kinetic energy.)

3 How would increasing this affect the rate of dodgem car crashes?

4 Is this consistent with the effect of temperature on reaction rates?

What did you discover?

Write a paragraph summarising how the equivalent of temperature will affect the rate of dodgem car crashes.

Catalysts

EFFECT OF A CATALYST ON THE DECOMPOSITION OF HYDROGEN PEROXIDE

View this animation to see the effect of a catalyst on the rate of decomposition of hydrogen peroxide, which catalyst and the amount of catalyst to add.

A **catalyst** is something that can affect the rate of a reaction without being consumed in the reaction. Positive catalysts increase the rate of the reaction. Negative catalysts decrease the rate of the reaction. Usually, we tend to refer to positive catalysts.

There are various ways that a catalyst can affect a reaction, but all positive catalysts provide an alternative pathway for the reaction, which is easier and therefore has a lower activation energy. A lower activation energy means that more molecules have enough energy for successful collisions. Therefore, there are more successful collisions and therefore a faster rate of reaction.

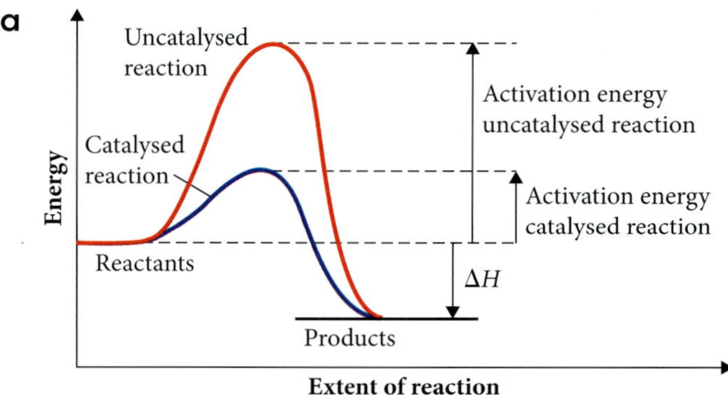

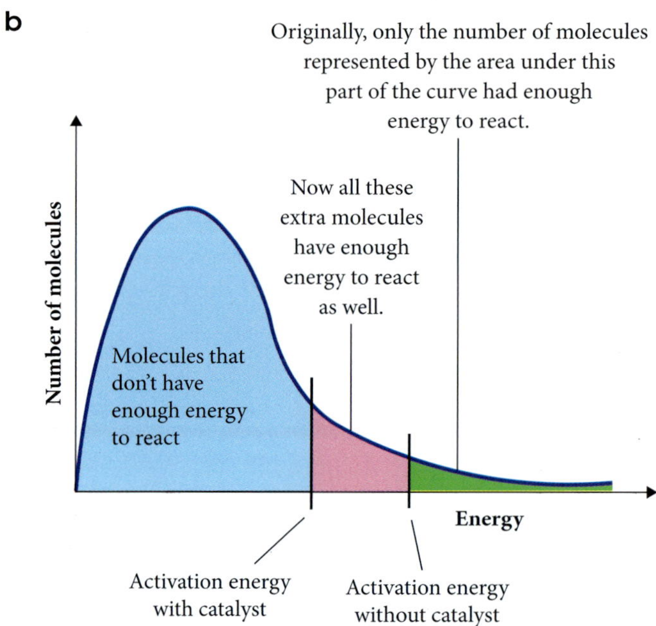

Figure C9.23 ▶ The effect of a catalyst on the number of molecules with sufficient energy for a successful collision: a) energy profile diagram of a catalysed and an uncatalysed reaction; b) Maxwell–Boltzmann distribution for catalysed and uncatalysed reactions

Catalysts affect the rate of certain reactions by providing an alternative reaction pathway with a reduced activation energy.

One type of catalyst that is being used more commonly is **metal nanoparticles**. A nanoparticle is a particle with a diameter less than 100 nm that behaves as a whole unit. Metal nanoparticles have an advantage over other metal catalysts due to their large surface area. This means that a large amount of the reactants can access the catalyst and therefore they can have a greater impact on the reaction rate. Currently, scientists are conducting research into different metal nanoparticles for a number of reactions, including gold and copper nanoparticles as catalysts for fuel cells.

Catalysts in our bodies ensure that the reactions needed for a healthy life are occurring at the appropriate rate. Biological catalysts are called **enzymes** and are usually made of **protein**. Enzymes are affected by temperature and pH. As enzymes have to work within a living organism, it is logical that the optimal temperature and pH is that of the organism. Outside of these conditions, the enzyme is ineffective, and this can often mean that the reaction is not able to occur at the required rate.

For more information about enzymes, refer to Context 4, 'Making reactions work for us', page 85.

Cars are now equipped with catalytic converters. These structures are fitted so that the exhaust fumes pass over a layer of a metal catalyst, usually platinum, rhodium and/or palladium. As the gases pass over the catalyst, the harmful combustion products of nitrogen oxides, volatile organic compounds and carbon monoxide are converted to less harmful substances. Therefore, the catalytic converter is important in reducing pollution from our cars.

EXPERIMENT 9.6

EFFECT OF A CATALYST ON RATE OF REACTION

Oxalic acid reacts with acidified potassium permanganate to produce colourless products. The intense purple of the permanganate disappears as the reaction proceeds. The time for this to occur can be measured and used to compare the rate of reactions.

Manganese(II) sulfate can act as a catalyst for this reaction, increasing the rate of the reaction and therefore decreasing the time taken for the decolorisation to occur.

Aim

To observe the effect of a catalyst on the rate of the reaction between oxalic acid and acidified permanganate

Materials

- 6 mL of 0.02 mol L^{-1} potassium permanganate ($KMnO_4$)
- 50 mL of 2 mol L^{-1} sulfuric acid (H_2SO_4)
- 250 mL of distilled water
- 6 mL of oxalic acid ($H_2C_2O_4$)
- 1 mL of manganese(II) sulfate ($MnSO_4$)
- Stopwatch
- 2 × 250 mL beakers
- 3 × 10 mL, 1 × 25 mL and 1 × 250 mL measuring cylinders
- Stirring rod

What are the risks in doing this experiment?	How can you manage these risks to stay safe?
Chemicals could splash in your eyes.	Wear safety glasses at all times.
2 mol L^{-1} H_2SO_4 is corrosive to skin and clothes.	Wear gloves and an apron.
$KMnO_4$ is an oxidising acid; it can irritate the skin and eyes.	Wear gloves and an apron.
Oxalic acid is harmful to the skin and eyes.	Wear gloves and an apron.

In your write-up, add any more risks you can think of, as well as ways to manage them. In particular, expand on the risks listed to identify specific risks involved with each of them.

Procedure

1 Into the first beaker place 25 mL of 2 mol L^{-1} H_2SO_4, 125 mL of distilled water, 3 mL of 0.02 mol L^{-1} $KMnO_4$ and 3 mL of saturated oxalic acid.
2 Start the stopwatch immediately and record how long it takes for the solution to become colourless.
3 Into the second beaker place 25 mL of 2 mol L^{-1} H_2SO_4, 125 mL of distilled water, 3 mL of 0.02 mol L^{-1} $KMnO_4$, 3 mL of saturated oxalic acid and 1 mL of saturated $MnSO_4$.
4 Start the stopwatch immediately and record how long it takes for the solution to become colourless.
5 Follow your teacher's instructions regarding disposal of chemicals.

Results

Record your results in a table similar to the one below.

Beaker	Time taken for decolorisation (s)
No $MnSO_4$	
With $MnSO_4$	

Analysis of results

1 What was the effect of adding $MnSO_4$ to the reaction mixture?
2 Given that $MnSO_4$ was acting as a catalyst, did your results support the theory about the effect of catalysts on the rate of reactions?

Discussion

How could the experiment be improved to obtain more accurate and reliable data?

Conclusion

What did you determine about the effect of $MnSO_4$ as a catalyst on the reaction between oxalic acid and potassium permanganate?

Taking it further

1 How could you further investigate the effect of catalysts on the rate of reaction?
2 How could you determine if other factors influence the effect of catalysts on the rate of reactions?

ACTIVITY 9.10

DODGEM CAR CRASH ANALOGY – CATALYST

Aim

To consider the effect of a catalyst on reaction rates in the dodgem car crash analogy

What to do

1 Review the dodgem car crash analogy in Activity 9.3.
2 In your analogy, what would be the equivalent of a catalyst? (Remember that a catalyst will make it easier for a collision to be successful because less energy is needed for a successful collision.)
3 How would adding this catalyst affect the rate of dodgem car crashes?
4 Is this consistent with the effect of catalysts on reaction rates?

What did you discover?

Write a paragraph summarising how a catalyst will affect the rate of dodgem car crashes.

QUESTION SET 9.3

Remembering

1 List the factors that affect the rate of a reaction.
2 Define 'catalyst'.
3 Draw a Maxwell-Boltzmann distribution to show the distribution of energy of particles at two different temperatures. Clearly identify which curve is for the higher temperature.

Understanding

4 Explain how crushing a lump of solid will increase its surface area.
5 Compare the average kinetic energy of the particles of substances at 10°C and at 40°C.
6 Explain how the pressure of a gas is equivalent to its concentration.
7 Use collision theory to explain how each of the following factors affects the rate of a reaction.
   a Concentration or pressure
   b Temperature
   c Catalyst
   d Surface area

Applying

8 Ben and Aimee both added a sugar cube to their cups of tea, which were at the same temperature. Ben stirred his tea, while Aimee left her cup to sit. After 1 minute, they each drank their tea. Whose tea would have been sweeter? Justify your answer.

Analysing

9 Five test tubes each contained HCl and magnesium as shown in the following table. Which test tube would contain the fastest reaction? Justify your answer.

Test tube	Concentration of HCl (mol L^{-1})	Temperature (°C)	Size of 2 g of magnesium
1	1	25	Lump
2	2	25	Long strip
3	1	10	Strip cut into small pieces
4	0.5	40	Long strip
5	2	40	Strip cut into small pieces

Reflecting

10 Explain how the development of collision theory has enabled us to understand why reactions occur at different rates.

CHAPTER CHECKLIST

You should know:

- rate tells us how quickly one quantity changes compared to another quantity
- the slope (or gradient) of a graph of a reaction indicates the rate of the reaction
- to measure the rate of a reaction, we measure the changes in reactants or products in a certain time
- we can observe the rate of a reaction by changes in colour, mass, temperature or pH
- collision theory states that, for a reaction to occur, the particles need to collide with sufficient energy and correct orientation
- collision theory allows us to understand what happens during a reaction and, therefore, the reason for the rate of reactions
- activation energy is the amount of energy needed for a successful collision and is the amount of energy required to break the bonds in the reactants
- the correct orientation allows for the appropriate atoms to align so that new bonds can form
- the transition state is the moment when the reactants' bonds have broken but the products' bonds have not yet formed, and the substance is known as the activated complex. This is a very unstable, high-energy state that only lasts for a very short time
- a number of factors affect the number of successful collisions during a reaction, and therefore affect the rate of the reaction
- the concentration or pressure of the reactants affects the number of collisions between reactants. The higher the concentration of the reactants, the more collisions will occur. This means that there will be more successful collision in a given time and a faster rate of reaction

- dividing a solid increases the surface area available for collisions. This means that there are more collisions in total, more successful collisions in a given time and thus a faster rate of reaction
- when temperature increases, the average kinetic energy of particles increases. This means that more particles have enough energy for a successful collision. Therefore, there are more successful collisions in a given time and the reaction rate increases
- when temperature increases, the particles are also moving faster due to the greater average kinetic energy. This means that they will collide more frequently, and the number of successful collisions will be increased and thus the rate of the reaction will be increased. Although this effect is less significant than that due to kinetic energy, it does contribute to the effect on the reaction rate
- a catalyst is a substance that affects the rate of a reaction without being consumed. Most catalysts are positive catalysts because they increase the rate of the reaction by providing an alternative pathway for the reaction, which has a lower activation energy. This means that more particles will have enough energy for a successful collision, increasing the number of successful collisions in a given time and increasing the rate of the reaction.

You should be able to:

- represent the changes in energy during a reaction by an energy profile diagram.

CHAPTER GLOSSARY

activated complex the complex formed when the reactants' bonds are breaking but the products' bonds have not yet formed (see *transition state*)

activation energy the energy required to break the bonds of the reactants and therefore the minimum energy required for a successful collision

catalyst a substance that changes the rate of a reaction without being consumed

collision theory a theory that explains the rate of reaction at a molecular level; it states that, for a reaction, the particles must collide with sufficient energy and the correct orientation

concentration the amount of solute dissolved in a given volume of solvent

endothermic reaction a reaction that absorbs heat from the surroundings because the reactants have less enthalpy than the products

energy profile diagram a diagram showing the relative energies during the progress of a reaction

enthalpy the total energy possessed by a chemical substance at a constant pressure; it is usually expressed as the change in enthalpy

enzyme a protein molecule that catalyses a specific type of reaction by lowering the energy of activation

exothermic reaction a reaction that releases energy to the surroundings because the products have less energy than the reactants

gradient the slope or steepness of a line; it is calculated by dividing the vertical height (rise) by the horizontal distance (run)

kinetic energy the energy of movement

Maxwell-Boltzmann distribution a graph showing the distribution of energies (or velocities) of particles

metal nanoparticle a metal particle with a diameter of less than 100 nm that behaves as a whole unit

pressure the force exerted by a gas or gases due to the collisions of the particles with the walls of a container; expressed as force per unit area

protein a large organic molecule made up of many amino acids joined together

rate how much one quantity changes with respect to another quantity

surface area the total area of all surfaces

temperature a measure of the average kinetic energy of the particles

transition state the temporary, unstable state when the reactants are forming products (see *activated complex*)

CHAPTER REVIEW QUESTIONS

Remembering

1 Define:

a rate.

b surface area.

c concentration.

d pressure.

e catalyst.

2 List the factors that affect the rate of a reaction, and state how each factor achieves this.

3 How can a graph be used to determine the rate at different times during a reaction?

4 Draw a Maxwell-Boltzmann distribution.

5 Copy the energy profile diagram below and label the axes, reactants, products, activation energy and activated complex.

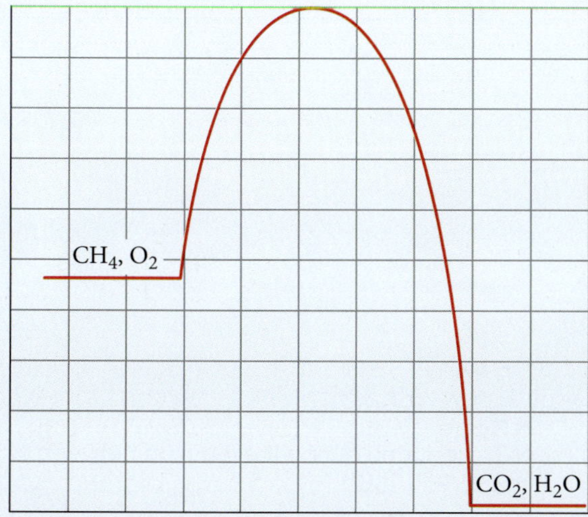

◀ **Figure C9.24**
Energy profile diagram for the combustion of methane

Understanding

6 Explain why particles must collide in order to react.

7 Draw an energy profile diagram for an endothermic reaction in which strong bonds are broken in the reactants. Describe the diagram you have drawn and the reasons for each aspect.

8 Why is the rate of a reaction fastest at the start of the reaction?

9 Explain how catalysts increase the rate of reactions.

10 Why does the activated complex have more energy than the reactant or products?

Applying

11 If someone is paid $525.00 after working 6 hours each day for 5 days, what is their hourly rate of pay?

12 Sketch a graph showing the change in temperature during an exothermic reaction.

13 Sam noticed when she started using cold water to wash her clothes that the washing powder was not dissolving properly. Explain why this might happen.

14 In an experiment, some hydrogen peroxide was added to a reaction flask, which was connected to a gas syringe. A small amount of manganese(II) dioxide was added and a rapid reaction produced bubbles of a gas (oxygen gas), which was collected in the syringe. After 6 minutes, no more gas was produced.

a Sketch a graph for the volume of gas produced in the experiment.

b When the reaction was complete, manganese dioxide was present. What was the purpose of the manganese dioxide?

c Describe one way that you could make the reaction proceed more slowly. How would this affect the volume of gas collected?

15 The graph in Figure C9.25 shows the concentration of nitrogen dioxide (NO_2) gas produced during the reaction between NO and O_2 gas. Sketch the graph you would expect if more NO gas had been initially added to the container.

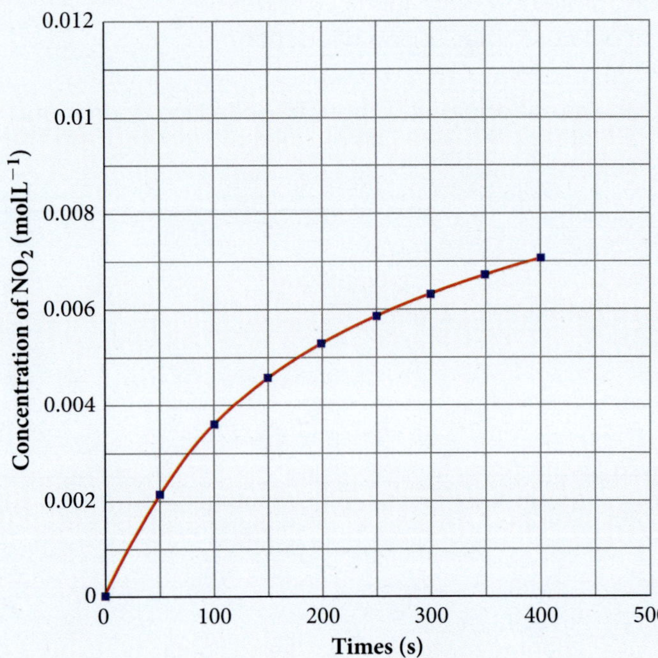

◀ **Figure C9.25**
NO_2 produced during the reaction between NO(g) and O_2(g)

Analysing

16 An analogy of how enzymes work is the comparison with climbing a mountain to get to the other side. In this analogy, the reaction without an enzyme is compared to walking to the very top of the mountain and then back down the other side. The reaction with the enzyme is compared to walking through a tunnel in the mountain to get to the other side. Discuss the merits of this analogy.

Reflecting

17 Use your understanding of collision theory and the factors that affect the rate of a reaction to compare playing darts with a chemical reaction.

18 Draw a concept map to reflect your understanding of collision theory and the factors that affect the rate of a reaction.

CHAPTER 10 SCIENTIFIC INVESTIGATIONS

By the end of this chapter you will have covered the following material.

Science Inquiry Skills

- Identify, research and refine questions for investigation; propose hypotheses; and predict possible outcomes (ACSCH001 AND ACSH040)
- Design investigations, including the procedure/s to be followed, the materials required, and the type and amount of primary and/or secondary data to be collected; conduct risk assessments; and consider research ethics (ACSCH002 AND ACSCH041)
- Conduct investigations, including the use of devices to accurately measure temperature change and mass, safely, competently and methodically for the collection of valid and reliable data (ACSCH003)
- Represent data in meaningful and useful ways, including using appropriate graphic representations and correct units and symbols; organise and process data to identify trends, patterns and relationships; identify sources of random and systematic error and estimate their effect on measurement results; and select, synthesise and use evidence to make and justify conclusions (ACSCH004)
- Communicate to specific audiences and for specific purposes using appropriate language, nomenclature, genres and modes, including scientific reports (ACSCH008 AND ACSCH047)
- Conduct investigations, including measuring pH and the rate of formation of products, identifying the products of reactions, and testing solubilities, safely, competently and methodically for the collection of valid and reliable data (ACSCH042)
- Represent data in meaningful and useful ways, including using appropriate graphic representations and correct units and symbols; organise and process data to identify trends, patterns and relationships; identify sources of random and systematic error; identify anomalous data; estimate the effect of error on measured results; and select, synthesise and use evidence to make and justify conclusions (ACSCH043)

AC

10.1 Introduction

Performing investigations is your chance to experience what doing science is really like. Science is about finding things out through observation and experiment, which is what doing investigations is all about. This is why investigations are central to science, *and* why they are so much fun.

Science Photo Library/Phil Boorman

Figure C10.1 ▲ Students conducting an investigation on the acidity of soft drinks

Sometimes, an important advance in science begins with a casual observation or a lucky accident. This happened in 1799, when Alessandro Volta produced electricity from alternating discs of copper and zinc separated by material soaked in acid. This led to further inquiry, which allowed scientists to understand how a chemical reaction can produce electricity, which has led to the development of the variety of batteries available today.

Scientific investigations can take years to complete and may involve collaboration among many scientists. They may require access to special equipment in Australia or other countries. They may cost a lot of money, sometimes millions of dollars, to complete. Hence, scientists invest time in *planning* investigations before they begin. When scientists apply for grants to carry out investigations, they need to show that they have carefully planned what they will do and how any money provided will be spent. Good planning is crucial to the success of the investigation.

Scientists then make careful *measurements and observations* and record their *results*. They *keep records* of all their experiments. This is a legal requirement. Typically, experimental results need to be kept for 5–7 years. There are also requirements on how and where data are stored.

Once collected, the data need to be *analysed*. There are various ways this is done, but in the physical sciences (physics, chemistry and geology), it almost always involves constructing graphs. Once a relationship is established graphically, a mathematical relationship can be derived.

Finally, the results of the investigation must be *communicated*. Usually this involves publishing a scientific paper in either a journal or conference proceedings. It often includes presenting the results in talks or posters at conferences. If the result is funded by a grant, then a research report must be submitted. If the results are really exciting, then the scientists may write a media release. However the results are communicated, this step must happen for the investigation to be completed.

When *you* perform scientific investigations, you will also need to plan carefully. You will find out about the topic by reading about what other people have done. You will collect data from your own experiments and secondary sources. You will then analyse that data and draw conclusions about what it means. Finally, you will communicate your findings. Each of these steps is outlined below.

10.2 Planning your investigation

There are many things to consider when planning an investigation. You need to think about how much time you will have inside and outside class. You will also need to think about what space and equipment you will need and where you will go if you want to make measurements or observations outside.

You may be working in a group or on your own. Most scientists work in groups. If you can choose who you work with, think about it carefully. It is not always best to work with friends. Think about working with people who have skills that are different from your own.

Finally, and probably the first thing that most students think about, is the topic of the research. You will need to come up with a **research question** or **hypothesis**.

Choosing a research question

Obviously, it is a good idea to investigate something that you find interesting. If you are working in a group, try to find something that everyone in the group finds interesting.

A good way to start is by 'brainstorming' for ideas. This works whether you are working on your own or in a group. Write down as many ideas as you can think of. Don't be critical at this stage. If you are working in a group, get everyone to contribute and accept all contributions uncritically. Write down all ideas.

After you have run out of ideas, it is time to start being critical. Decide which questions or ideas are the most interesting. Think about which of these it is possible to investigate given the time and equipment available. Make a shortlist, but keep the long list too for the moment. Once you have your shortlist, it is time to start refining your ideas.

Alamy/Jim West

▲ **Figure C10.2**
Brainstorm as many ideas as you can in your group.

Researching and refining your question

The next step is to find out what is already known about the ideas on your list. Use the Internet, your textbooks and the library to find out. Make sure you *keep a record* of the information that you find as well as *the sources*. You should start a **logbook** at this stage. You can write in references, or attach printouts to your logbook. This can save you a lot of time later on! Many research students forget to do this when they first start reading about their topic and then have to search all over again.

Good record keeping is important in scientific research, and it begins at this stage of the investigation.

Be critical of what you read. Do not assume that everything you read online or even in books is true. Try to find **reliable** sources of information. Textbooks, websites from universities and government research agencies are usually very reliable. Publications and webpages from professional associations, such as the Royal Australian Chemical Society and equivalent international organisations are also good sources. Blogs and homepages of other students are not usually reliable, although they are useful to give you ideas. Be sceptical about websites that are trying to sell you something. Talk to your teacher about sources of information as well. They will be able to tell you if a website is reliable, and suggest sites that they know are suitable.

Shutterstock.com/Ermolaev Alexander

▲ **Figure C10.3**
Start researching your topic and make sure you keep a record of all your references.

LEARN CHEMISTRY

Visit this website for many ideas for experiments in chemistry.

You may find examples of similar investigations to the one you are thinking of. It is a good idea to look at these, so you can learn from the experience of other researchers. However, in general, it is better not to try to replicate someone else's investigation exactly. If you do decide to replicate someone else's investigation, then you must acknowledge and carefully **reference** their work. See the section on References on page 417. If you do not do so, then it is **plagiarism**. This is a very serious form of academic misconduct. Talk to your teacher about how original your research needs to be, how closely it can be based on someone else's work and the correct way to acknowledge the contributions of others. It is much better to do this at the start than to be accused of cheating later on!

Finally, talk to your teacher about your ideas. They will be able to tell you whether your ideas are likely to be possible given the equipment available. They may have had other students with similar ideas and can make suggestions.

After you have researched your questions and ideas, you should be able to narrow down the shortlist to the one question that you want to tackle. If none of the questions or ideas looks possible (or still interesting), then you need to go back to the long list.

Proposing a research question or hypothesis

Once you have decided on what you will investigate, you need to turn it into a research question or a hypothesis.

A research question is one that can be answered by performing experiments or making observations. A hypothesis is a prediction of the results of an experiment, which can be tested by performing experiments or making observations.

Research questions

A research question may be of the form 'How does the temperature of a solution affect how quickly a reaction occurs?'. The aim of your research is then to answer the question.

You need to frame the question carefully. It needs to be specific enough that it guides the design of the investigation. A specific question rather than a vague one will make the design of your investigation much easier. Asking 'Will a reaction occur more quickly if the temperature is higher?' tells you what you will be varying and what you will be measuring. It also gives a criterion for judging whether you have answered the question.

Figure C10.4 ▶ Your research question will guide your investigation. These students are finding out how temperature affects the rate of a reaction.

Corbis/Frederic Cirou

Asking 'What will make a reaction occur the best?' is not a good question. This question does not say what will be varied, nor does it tell you when you have answered the question. 'Best' is a vague term. What you mean by 'best' may not be what someone else means.

A good research question identifies the **variables** that will be investigated. Usually you will have one **dependent variable** and one **independent variable**. For a lengthy investigation, you may have two or more independent variables. Variables are discussed in more detail on page 402.

Finally, a good research question should be answerable with the time and equipment available.

Hypotheses

A hypothesis is a tentative explanation or prediction, not yet confirmed by experiment, such as 'As the temperature of the sodium thiosulfate solution increases, the rate of the reaction with hydrochloric acid will increase'. Your hypothesis should give a prediction that you can test, ideally quantitatively.

A hypothesis is usually based on an existing **model** or **theory**. It is a prediction of what will happen in a specific situation based on that model. For example, collision theory is used to describe what happens during a chemical reaction. This can be used to develop a hypothesis regarding factors that affect the rate of a reaction, including temperature.

A hypothesis should give you a prediction that you can test by performing an experiment. This means it should at least be **falsifiable** (can be proven false). A good hypothesis should be able to be disproved. However, you will *not* generally be able to claim that you have proved your hypothesis.

If your experiments agree with predictions based on your hypothesis, then you can claim that they support your hypothesis. This *increases your confidence* in your model, but *it does not prove that it is true*. Hence, an aim for an experiment should not start 'To prove …', as it is not possible to actually prove a hypothesis, only to disprove it.

If your experimental results disagree, then you may have disproved your hypothesis. This is *not* a bad thing! Often the most interesting discoveries in science start when a hypothesis based on an existing model is disproved. This means that the model it was based upon is either not a good model, or does not apply to the particular situation. You could then try to work out why the model does not apply, or try to formulate a better model. What to do when your hypothesis is not supported is discussed further in section 10.4 on page 411.

In summary, a good research question is a question that is specific and can be answered by performing experiments and making measurements. A good hypothesis is a statement that predicts the results of an experiment and can be tested using measurements.

Even if your question or hypothesis meets these criteria, do not be surprised if you change or modify it during the course of your investigation. In scientific research, the question you set out to answer is often only a starting point for more questions.

Designing your investigation

Once you have a specific research question or hypothesis, you need to design your investigation. It is fun to start making measurements or observations immediately, but it is also important to spend time learning how to use the equipment, and experimenting to find the best way to set up your investigation. You may also discover that you need different or more equipment. This may save you time later on.

It is also important to keep focused on the purpose of your investigation. At the end of the process, you need good data that answer your question or test your hypothesis. Having a plan helps to ensure that you will take the measurements you need. The longer the investigation, the more important it is that you have a clear plan. There are several things to consider:

- What data will you need to collect?
- What materials and equipment will you need?
- When and where will you collect the data?
- If you are working in a group, who will collect the data?
- Who will be responsible for record keeping?
- How will the data be analysed?

The data that you collect will always include **secondary data**, and will usually include **primary data**. Secondary data are data that have been collected by someone else.

You will already have collected some secondary data when you investigated your research topic to formulate your question or hypothesis. You will probably want to collect more secondary data. If your topic is not one for which you can collect primary data, then you will need to rely on secondary data. Remember that when you collect secondary data, it is important to use reliable, reputable sources.

Primary data are data that you collect yourself. You can collect data by performing experiments. In chemistry, primary data include measurements such as temperature, mass, volume and time. You should also make observations such as colours, smells, sounds and other changes that you can see. You will have had practice at measuring some or all of these things already. You need to decide which variables you will measure and which variables you will control. Consider which variables you can control, and which you cannot.

Consider how you will analyse the data. Will you need access to specific software such as a graphing or statistics package? If so, make sure that you know how to use it. If you are using software to draw graphs, then you need to know how to produce a scatter graph and draw a **line of best fit**. Note that a line of best fit is *not* the same as joining the dots. You should *never join the dots*, even though this is often the default setting in spreadsheet software. You should consult a reference guide or the 'help' menu for your software, or ask your teacher. Graphs are discussed in more detail in section 10.4 on page 411.

Keep a record of your planning. This should go in your logbook. Writing down what you plan to do, and why, will help you stay focused during the investigation. If you are working in a group, then dividing the tasks and keeping a record of what each person agrees to do during the investigation will help to coordinate the project.

Variables and measurements

Variables are anything that may change the results of an investigation or experiment. For each investigation, the variables can be classified as independent, dependent or controlled variables.

The independent variable is what you change. For example, in the investigation to determine the effect of temperature on the rate of a reaction, you would change the temperature. Therefore, the temperature is the independent variable.

The dependent variable is what you are measuring or finding out. In the temperature and reaction rate investigation, you are measuring the reaction rate. Therefore, reaction rate is the dependent variable.

Your hypothesis should clearly show your predicted relationship between the independent and dependent variables. For example, 'As the temperature increases, the reaction rate will increase'.

It is important that everything else that may affect the results is kept constant. Therefore, all other variables must be controlled. Your method should clearly show how you are controlling these variables. For example, you should describe what volume and concentrations of solutions you are using, whether you are stirring the solutions, when timing will start and stop, what size flasks you will use and so on. If these variables change as well as the independent variable, then you would not be able to tell which variable caused the change in the dependent variable. For example, if you had changed the concentration and temperature of the solution, then which one caused the increase in the reaction rate?

In most investigations, there will only be one independent variable. However, in a longer investigation, it is possible to investigate the effect of more than one variable. In this case, it is important to plan carefully so that each set of data has only one variable changed. For example, if you were going to investigate the effect of concentration and temperature on the rate of a reaction, then you would test a variety of concentrations at one particular temperature, then repeat the process at the next temperature and so on. This way you can clearly see the effect of concentration and the effect of temperature separately.

When variables have a numerical value, you make **quantitative measurements**. You measure that numerical value in the appropriate units. For example, you may measure temperature in degrees Celsius, mass in grams or volume in litres.

Continuous variables may take any possible value, usually within some range. Length, time, mass, temperature and volume are examples of continuous variables used in chemistry. A variable that may take only fixed values is called a **discrete** variable. Often these are whole numbers of things that cannot be broken into smaller parts, such as electrons, protons or atomic number.

In some investigations, you may use **qualitative** measurements or data, which are descriptions such as colour, formation of a precipitate or presence of bubbles. For example, a chemical reaction may lead to a colour change. You would usually describe the colour in words, such as 'pink' or 'green', rather than using a number. Sometimes you use a combination of qualitative and quantitative data.

Once you have decided on the variables you will be measuring, you will be able to identify the equipment and other resources you will need.

Identifying the resources required

Now that you have decided on your hypothesis, you should consider what you will need to test your hypothesis. Make a list of all the materials and equipment that you will need.

- What chemicals will you need for your reaction? Be specific and include volumes, masses and concentrations where applicable.
- What equipment will you need to hold the chemicals? Include quantities and sizes or volumes of equipment such as beakers and conical flasks.
- How will you measure the chemicals? Include quantities and sizes or volumes of equipment such as pipettes and measuring cylinders.
- How will you change the independent variable?
- How will you measure the dependent variable?
- Have you chosen equipment that will allow accurate measurements?

▲ **Figure C10.5** Your equipment list should be thorough and detailed to include everything that you will need to conduct your investigation.

You must also consider the safety aspects of your experiment. Look up and print out the Material safety data sheets (MSDS) for the chemicals that you are using and include any safety equipment needed. You should also consider where the experiment will be conducted. Some reactions should only be carried out in a fume cupboard. Other, general, safety equipment such as safety glasses, aprons or labcoats and gloves should also be included in your equipment list.

As you develop your method, you may need to adjust your equipment list to ensure that it is thorough and detailed.

Planning the experimental procedure

The most common problem that students have when doing research is time management. It is important to plan to have enough time to perform the experiments, *and* to analyse them, *and* to report on them. You also need to allow time to learn how to use the equipment if you have not used it before.

Your procedure is what you are going to do in your investigation. This needs to clearly state what will be done at each step in as much detail as possible. As you are planning, consider the following:

- How many variations of your independent variables are you going to test? For example, how many different temperatures will you have?
- What are you using to measure each of the chemicals?

- What will you use to hold the chemicals?
- Will you stir the solution?
- How will you measure your dependent variable?

In your planning, ensure that you only change one thing – the independent variable. Everything else needs to be kept constant, and you need to show this will be achieved – for example, the volume and concentration of solutions will you use, the order chemicals are added, whether the solutions are mixed or not and when a stopwatch will be started and stopped.

Whenever possible, repeat your measurements, so allow time for this. This allows you to check that your measurements are **valid**. Valid results are affected only by a single independent variable. If the results are similar each time, then your results are likely to be valid. If a result is not **reproducible**, it is probably not a valid result. A result is reproducible if you make exactly the same measurement more than once and get the same result, within the limits of experimental uncertainty. If a result is not reproducible, then a variable other than the one you are controlling is affecting its value. If this is the case, you need to determine what this other variable is, and control it if possible.

Corbis/Frederic Cirou

Figure C10.6 ▲ Sometimes experiments just don't work.

Sometimes experiments do not go as planned. This is often due to other factors influencing the experiment. It is a good idea to do a trial run of your experiment to check that your procedure will work. You can then adjust your procedure as required before conducting all of your trial.

Make sure you allow time for analysis. Ideally, do as much analysis as you can while you collect results. If you plot graphs as you take measurements, then you will be able to identify **outliers** early. An outlier is a data point that does not fit the pattern of the rest of the data. If you identify an outlier while you still have access to equipment and space, then you can check the measurement and make sure that you didn't make a mistake.

After you have analysed your results, you need to write your report or communicate your findings in some other form. You need to plan ahead how this will be done. If you are working in a group, decide who will write which part of the report and when. Who will proofread it? Who will be responsible for making sure all the parts fit together?

You may find a timeline useful. A timeline helps keep you on track, and reminds everyone of their responsibilities. If you are working in a group, then get everyone to agree on it.

You can use the following table as a template.

Date and place	What will be done	Who will do it	Outcomes

Risk assessment

You may be required to complete a risk assessment before you begin your investigation. Even if this is not a requirement, it is a good idea to think about it. You need to think about three things:

1. *What are the possible risks* to you, to other people, to the environment or property?
2. *How likely is it* that there will be an injury or damage?
3. If there is an injury or damage to property or environment, *how serious are the consequences* likely to be?

A 'risk matrix', such as Table C10.1, can be used to assess the severity of a risk associated with an investigation. The consequences are listed across the top from negligible to catastrophic.

Negligible may be getting clothes dirty or a very minor injury such as a scratch. Marginal might be a cut from broken glass. Severe could be a more substantial injury or a burn. Catastrophic would be a death or the release of a toxin into the environment. When you are considering the risks involved with chemicals, read the MSDS for each chemical used or produced. (These are readily available on the Internet.)

In general, you need to ensure that your investigation is low risk. You can use a risk matrix either for individual identified risks, or for the investigation overall. If there are multiple experiments, then you would use a risk matrix for each one.

Table C10.1 Risk matrix for assessing the severity of risk

	Consequences			
Likelihood	**Negligible**	**Marginal**	**Severe**	**Catastrophic**
Rare	Low risk	Low risk	Moderate risk	High risk
Unlikely	Low risk	Low risk	High risk	Extreme risk
Possible	Low risk	Moderate risk	Extreme risk	Extreme risk
Likely	Moderate risk	High risk	Extreme risk	Extreme risk
Certain	Moderate risk	High risk	Extreme risk	Extreme risk

Once you have considered the possible risks, you need to think about what you will do about them. What will you do to minimise them, and how will you deal with the consequences if something does happen? This may be as simple as, 'Wear appropriate safety gear'. You can use a risk assessment table like the one shown.

What are the risks in doing this experiment?	How can you manage these risks to stay safe?
2 mol L^{-1} HCl is corrosive to skin and clothes.	Clean up all spills immediately. Wear gloves, safety glasses and an apron.

Ethics

Ethics in research can be controversial. More than one scientist has lost their job for unethical research behaviour. Being ethical in your research has two aspects. The first is about being honest as a scientist. This means recording data accurately, and not ignoring, hiding or changing any data that do not support your hypothesis. It means acknowledging and referencing sources of information, including books, websites, articles and people who have helped you. It means not using other people's ideas or data without their knowledge and permission. Put simply, it is showing integrity or 'doing the right thing'. A good rule is that if you wouldn't want someone to know what you are doing, you probably shouldn't be doing it. It is no different from behaving ethically in any other area of your life.

The other aspect to ethics is treating animals, other people and the environment with care and respect. If your investigation

Science Photo Library/Antonia Reeve

◀ **Figure C10.7** When working with humans, you need to make sure that they are not harmed in any way.

NHMRC

You can find NHMRC human and animal ethics guidelines here.

will be using humans or animals, then you need to make sure you do not harm them, either physically or psychologically. If you are working with animals, then you need to make a strong case for any investigation that harms or could potentially harm them. When scientists want to use humans or animals in their research, they need to be able to show that the benefits to the environment, other animals or humans significantly outweigh the negative effects on the animals or humans used. The National Institute of Health and Medical Research (NHMRC) has guidelines on the ethical use of humans and animals in experimentation.

10.3 Collecting your data

Once you have planned what you are going to do, collected your equipment and set it up, it is time to start collecting data. This is usually the fun part of any investigation. Don't forget you have a question to answer or a hypothesis to test! To complete your research successfully, you need to make sure that you think carefully about what you do and keep good records.

iStockphoto/SteveStone

Figure C10.8 ▶ Make sure you keep an accurate record of what you do as you do it.

Record keeping – your logbook

You will need to keep a record of what you do during your investigation. You do this is in a logbook. Even if you are using data loggers to collect your results and doing your analysis electronically, you should still keep a hardcopy logbook.

Scientists keep a logbook for each project that they work on. It is a record of what they did, why they did it, and what they found out. A logbook is a legal document for a working scientist. If someone's work is questioned, then the logbook acts as important evidence. Every entry in a scientist's logbook is dated, records are kept in indelible form (pen, *not pencil*), and entries may even be signed. Scientists' logbooks include details of experiments, such as methods and results. They include comments and ideas, thoughts about the experiments and analysis. They frequently include printouts of data, photocopies of relevant information, photos and other items. The logbook is the primary source of information when a scientist writes up their work for publication. Sometimes, logbooks are even provided as evidence in court cases; for example, in patent disputes or when a researcher is accused of falsifying data or stealing someone else's results.

Some scientists keep their research records electronically, but most experimental scientists still keep a hardcopy logbook. There are several advantages to a hardcopy logbook over an electronic one. First, electronic records are easy to change, and it is hard to track what was changed, when and by whom. Second, if you are working in a group, then it can be hard to keep track of who has the most recent version of the file(s). Third, files can be easily deleted or corrupted. It takes much more care and discipline to maintain a good electronic logbook than a good hardcopy. Remember that the purpose of a logbook is to record and maintain evidence of what you did. Electronic evidence is not as reliable as a signed hardcopy document.

You should talk to your teacher about what form of logbook records they require you to keep.

If you are working in a group, then you will need to decide whether to keep one logbook for the entire group, or one each. If you will all be working in the same places at the same times, then one for the whole group is best. If you will be in different places, such as doing field observations, then you will need one each. Your teacher may also require each of you to keep your own logbook for assessment, or for authentication purposes.

Your logbook is a detailed record of *what you did* and *what you found out* during your investigation. Make an entry in the logbook *every* time you work on your investigation. At the start of each session you should record the date and the names of all the people with whom you are working at the time.

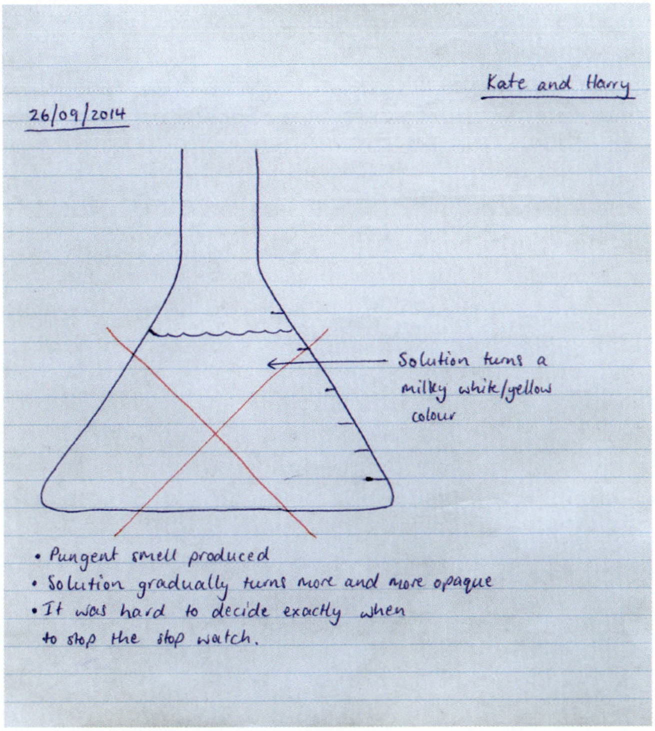

Write down what you do as you do it. It is easy to forget what you did if you do not write it down immediately. An accurate record is important if you need to repeat any measurements or if you get unexpected results.

Write down the names, model and serial numbers of any equipment used.

Include large, clear diagrams of any experimental set-up. Label all the parts or pieces of equipment. You can also include photos of experiments. Diagrams in your logbook do not need to be neat, but you must be able to understand the diagrams later on.

Record the results of *all* measurements *immediately and directly into your logbook, in pen. Never* record data onto bits of scrap paper instead of your logbook. Results must be recorded in indelible form. This means using a pen. Never write your results in pencil. Never use liquid paper or scribble over anything in your logbook. If you want to cross something out, just put a line through it. It is also a good idea to make a note explaining why it was crossed out.

◀ **Figure C10.9**
A page from a student's logbook

A good logbook contains:

- notes taken during the planning of your investigation
- a record of when, where and how you carried out each experiment
- diagrams showing the experimental set-ups, circuit diagrams, etc.
- all your raw results
- all your derived results, analysis and graphs
- all the ideas you had while planning, carrying out experiments and analysing data
- printouts, file names and locations of any data not written directly into the logbook.

It is not a neat record, but it is a *complete* record.

Performing experiments

If you have planned carefully and learnt how to use the equipment, then your experiments should go smoothly.

As stated above, *always* record results immediately, with the correct units and with their **uncertainty**. The raw data should *always* be recorded directly into the logbook unless it is recorded using data loggers connected to a computer. In this case, a printout of the data should be attached to the logbook, and the file name and location recorded.

Make sure that you measure and record everything you will need for your analysis. This will include data such as temperature, mass or volume of chemicals, concentrations of any solutions, and any observations you make. It is much better to measure something and then discover that you didn't need to than to start your analysis and realise that you didn't measure something that you do need.

Always record the units with their uncertainty along with the measurements as you go. Do not try to add them in later.

If you are going to collect multiple data points, then it is a good idea to draw a table to record them in. Label the columns in the table with the name and units of the variables. Do not put the units in the table cells. If you know that the uncertainty in all your measurements is the same, then you can record this in the heading cell at the top of the column as well. Otherwise, each data entry should have its uncertainty recorded in the cell with it. For more explanation of recording uncertainty, see the following section.

Remember to start your analysis while you are collecting your data. As discussed in 'Planning the experimental procedure' on page 403, you will have the opportunity to repeat measurements. If you make a mistake, put a line through the mistake, write in the new data, and make a comment in your logbook. Do not scribble out or remove mistakes, they may turn out to be useful.

If you have not made a mistake, then plotting and analysing as you go allows you to spot something interesting early on. You then have a choice between revising your hypothesis or question to follow this new discovery, or continuing with your plan. Many research projects start with one question and end up answering a completely different one. These are often the most fun, because they involve something new and exciting.

Estimating uncertainties

When you perform experiments, there are typically several sources of uncertainty in your data.

Sources of uncertainty that you need to consider are the:
- **limit of reading of measuring devices**
- **precision of measuring devices**
- **variation of the measurand.**

Limit of reading

For all devices there is an uncertainty due to the limit of reading of the device. The limit of reading is different for **analogue** and digital devices.

Analogue devices include swinging needle multimeters, liquid in glass thermometers and clocks with hands. Analogue devices have continuous scales. For an analogue device, the **limit of reading**, sometimes called the **resolution**, is half the smallest division on the scale. We take it as half the smallest division because you will generally be able to see which division mark that the indicator (e.g., needle, fluid level) is closest to. You may be able to estimate the measurement to one-fifth or even one-tenth of the smallest division if the spacing between divisions is large. However, the limit of reading uncertainty is still half of the smallest division. So, for a liquid in a glass thermometer with a scale marked in °C, the limit of reading is 0.5°C.

Digital devices such as digital multimeters, clocks and thermometers have a scale that gives you a number. It is limited to a specific number of figures, typically three or four, so it is a discrete scale. A digital device has a limit of reading uncertainty of a whole division. So a digital thermometer that reads to whole degrees has an uncertainty of 1°C. For a digital device, limit of reading is *always* a whole division, not a half, because you do not know whether it rounds up or down, or at what point it rounds.

The resolution or limit of reading is the *minimum* uncertainty in any measurement. Usually the uncertainty is greater than this minimum.

Alamy/sciencephotos

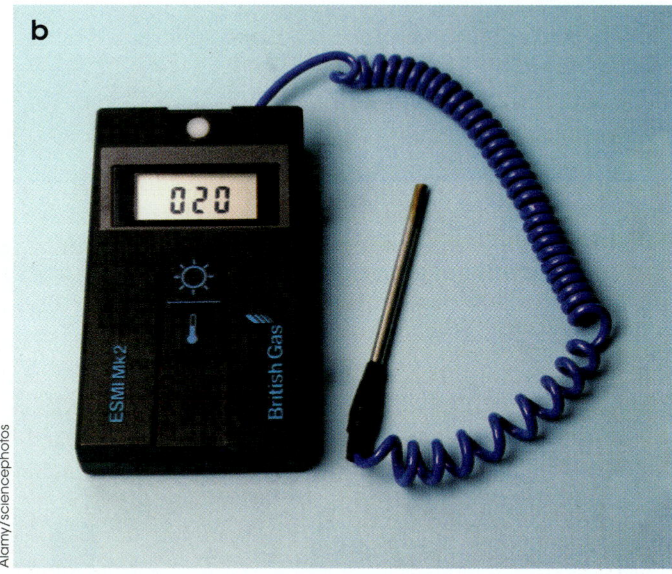

Alamy/sciencephotos

▲ **Figure C10.10**
a) This analogue thermometer has an uncertainty of half the smallest division on the scale.
b) This digital thermometer has an uncertainty of a degree.

Precision of measuring device

The measuring device used will have a **precision**, usually given in the user manual. For example, a multimeter may have a precision of 0.5% on a voltage scale. This means if you measure a potential difference of 12.55 V on this scale, the uncertainty due to the precision of the meter is $0.005 \times 12.55\,\text{V} = 0.06\,\text{V}$. This is greater than the limit of reading uncertainty, which is 0.01 V in this case.

Many students think that digital devices are more precise than analogue devices. This is often not the case. A digital device may be easier for you to read, but this does not mean it is more precise. The uncertainty due to the limited precision of the device is generally greater than the limit of reading.

a

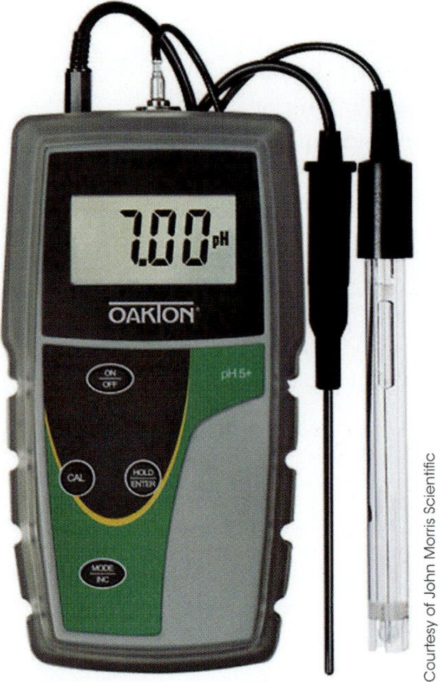

b

Function	Precision
pH range	0.00 to 14.00 pH
Resolution	0.01 pH
Accuracy	+/− 0.01 pH
pH slope range	80 to 120%
No. of calibration points	1 to 3 points (push-button)
Buffer options	pH 4.01, 7.00, 10.01 (USA) pH 4.01, 6.86, 9.18 (NIST) pH 4.10, 6.97 (Pb)
Temperature range	0.0 to 100.0°C
Resolution	0.1°C
Accuracy	± 0.5°C
Temperature comp.	Automatic/Manual (0 to 100°C)

Courtesy of John Morris Scientific

Figure C10.11 ▲
a) A typical pH meter.
b) A page from the user manual giving the precision on various scales

Variation of the measurand

The **measurand** (the quantity being measured) itself may vary. For example, reaction rate is strongly dependent on the temperature, concentration and other factors. Even keeping the conditions as close to identical as possible, it is unlikely that repeat experiments will give you exactly the same results. Making repeat measurements allows you to estimate the size of the variation.

Sometimes you will be able to see how the measurand varies during a measurement by watching a needle move or the readings change on a digital device. Watch and record the maximum and minimum values. The difference between these is the range:

$$\text{Range} = \text{maximum value} - \text{minimum value}$$

The value of the measurand is the average value, or the centre of the range:

$$\text{Measurand} = \text{minimum value} + \tfrac{1}{2}(\text{range})$$
$$= \text{minimum value} + \tfrac{1}{2}(\text{maximum value} - \text{minimum value})$$

The uncertainty in the measurement is half the range:

$$\text{Uncertainty} = \tfrac{1}{2}\text{ range} = \tfrac{1}{2}(\text{maximum value} - \text{minimum value})$$

For example, if you are using an analogue multimeter and you observe that the needle fluctuates between 12.2 V and 12.6 V, then your measurement should be recorded as (12.4 ± 0.2) V. Note that the measurement and uncertainty are together in the brackets, indicating that the unit applies to both the measurement and its uncertainty.

When you take repeat measurements, the best estimate of the measurand is the average value. If you have taken fewer than 10 measurements, then the best estimate of the uncertainty is half the range. If you have more than 10 measurements, then the best estimate of the uncertainty is the standard deviation, given by:

$$\text{Standard deviation} = \left(\frac{\Sigma(x_i - x)^2}{n-1}\right)^{\frac{1}{2}}$$

where x_i is an individual value of the measurand, x is the average value of the measurand and n is the total number of measurements. The sum is over all values of x_i. Most calculators have built-in statistical functions such as standard deviation. Spreadsheet software such as Microsoft Excel also calculates functions such as standard deviation. Remember that repeat measurements means repeating under the same conditions. It is not the same as collecting lots of data points under different conditions.

Random and systematic errors

Sources of uncertainty all give rise to **random errors**. That means that repeated measurements will be randomly spread about the '**true value**', and centred on that value.

You may also have **systematic errors** in your data. These typically occur when there is a calibration error, such as a zero error, in a measuring device. Always check that your equipment reads zero when you expect it to. For example, the pH meter should read 7 (at 25°C) in distilled water. You should also ensure that an electronic balance is tared to zero before weighing the mass of your chemicals. If your devices are not calibrated correctly, then all your readings will be faulty.

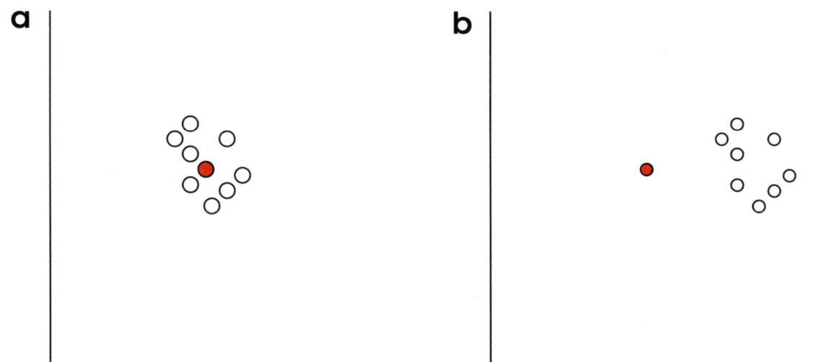

◄ **Figure C10.12** a) Results are clustered about the true value when the errors are random. b) Results are clustered about some other value as a result of systematic errors.

10.4 Analysing your data

GUM

Read more about uncertainties in the GUM (Guide to expressions of Uncertainty in Measurement).

Once you have collected your data, you will need to analyse it. Record all your analyses in your logbook. If this is done on a computer, then record the file name and location and attach a printout of the analyses into your book. Many scientists have logbooks that are bulging with printouts.

The first step is organising your data. This will usually involve tabulating it. You can then graph these data. Plotting graphs is a useful way to begin the analysis of your data. Graphs are a very useful way of representing data so that you can identify trends and relationships. There are many different sorts of graphs that can be used to organise and display data. These are described on page 416.

You will usually need to do some calculations with your data to be able to answer your question or test your hypothesis. Remember to keep units on all quantities, so that any derived values have the correct units. You will also need to calculate uncertainties on any derived quantities.

Organising your data

If you have more than a few data points, then it is a good idea to display them in a table. It is important that you record all of your data in a logical manner. Clearly label each set of data so that you will know what each is when you come to analyse your data. If you have multiple experiments, then you may need a table for each. For example, if you are investigating the effect of temperature and concentration on the rate of a reaction, then you may have a table for each temperature that you tested. Each table would show the rate of the reaction for the different concentrations.

Identifying trends, patterns and relationships

You may be able to see a pattern simply by looking at a list of numbers in a table. However, the most reliable way to identify a pattern in data or a relationship between variables is to plot a graph. If you have a hypothesised equation, then use it to generate a fit on a graph of your data, as described on page 412. Do not substitute your data into your hypothesised equation and try to show that it fits.

A graph should be large and clear. The axes should be labelled with the names of the variables and their units. Choose a scale so that your data take up most of the plot area. This will often mean that the origin is not shown in your graph. Usually there is no reason why it should be.

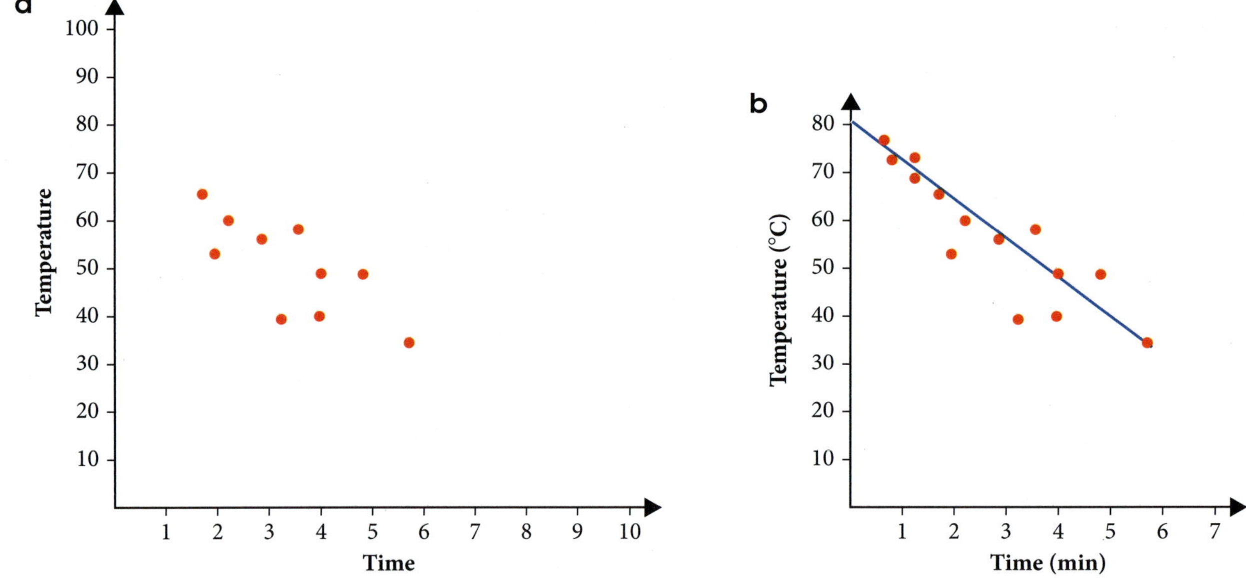

Figure C10.13 ▲
Examples of a) a poor graph and b) a good graph. How many differences can you see?

When you are looking for a relationship between variables, plot a **scatter graph**. This is a graph showing your data as points. *Do not join them up as in a dot-to-dot picture.* Usually the independent variable is plotted on the *x* axis and the dependent variable is plotted on the *y* axis, unless there is a good reason to do otherwise. In the reaction rate example, temperature would be on the *x* axis and time would be on the *y* axis.

To determine a relationship, you need to have enough data points and the range of your data points should be as large as possible. A minimum of six data points is generally considered adequate if the relationship is expected to be linear, but always collect as many as you reasonably can, given the available time.

DATA POINTS

Some helpful advice on deciding the number of data points.

For non-linear relationships, you need more data points than this. Try to collect more data in regions where you expect rapid variation. For example, if you are measuring the pressure of a fixed mass of gas at different volumes as shown in Figure C10.14, then you should expect an inverse relationship and therefore a hyperbola graph. If there are only a limited number of data points, it would be easy to predict the relationship to be linear instead.

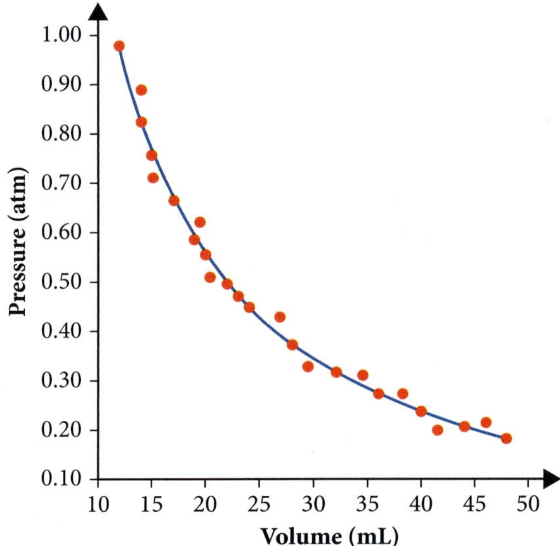

Figure C10.14 ▶
A pressure–volume graph for a gas needs a large number of data points to show the hyperbola shape.

A good graph to start with is simply a graph of the raw data. You will usually be able to tell by looking whether the graph is linear. If it is, then fit a straight line using a graphing package. You can then use a linear regression tool to check how good the straight line fit is. This will give you an R^2 number, which is a measure of 'goodness of fit'. The closer R^2 is to one (or –1), the better the fit. If it is not *very* close to one, then the relationship is not linear. Alternatively, you can calculate the uncertainty in the gradient by using lines of maximum and minimum gradient. If the uncertainty is large, then the relationship may not be linear.

If it is a linear relationship, then finding the equation for the line of best fit may be useful. *Never* force a line of best fit through the origin. Often the intercept gives you useful information. It may even indicate a systematic error, such as a zero error in the calibration of your equipment.

When you plot your raw data you may find that one or two points are outliers. These are points that do not fit the pattern of the rest of the data. These points may be mistakes; for example, they may have been incorrectly recorded or a mistake was made during measurement. They may also be telling you something important. For example, if they occur at extreme values of the independent variable, then it might be that the behaviour of the system is linear in a certain range only. You may choose to ignore outliers when fitting a line to your data, but you should be able to justify why.

When you extend a line of best fit beyond your measured points, this is called **extrapolation**. Any data that you read off a graph outside the range of your data points are extrapolated, and should be viewed with caution. You cannot say for sure that the system continues to behave in the same way beyond the bounds of your data. For example, in an investigation about the effect of enzymes on the rate of reaction at different temperatures, you may find that the rate of the reaction increases for the temperatures that you used. However, if the temperature continues to increase, then reaction rate may decrease.

Reading points, other than data points, from a line of best fit within the region in which you have data is called **interpolation**. You cannot be sure that this is exactly what you would find if you measured that point. However, if your line of best fit really represents the behaviour of the system, then you can use interpolated points in your analysis.

Relationships between variables are often not linear. If you plot your raw data, for instance the volume and pressure of a gas, and it is a curve, *do not draw a straight line through it.* In this case, you need to think a little harder. If your hypothesis predicts the shape of the curve, then try fitting a theoretical curve to your data. If it fits well, then your hypothesis is supported.

If possible, you should **linearise** your data based on your hypothesis. Remember that linear graphs have equations of the form $y = mx + c$. Here y is the variable plotted on the vertical axis, usually the dependent variable. The independent variable x is the variable plotted on the horizontal axis. The gradient is $m = \Delta y/\Delta x$. The constant c is the y intercept.

Sometimes the relationship between variables will be more complicated than a linear relationship. In this case, a graph is still useful but the most you might be able to say is that one variable increases with another, or that there is a peak at a particular position. A graph is still a useful way of identifying trends and patterns, even if you are not able to extract a mathematical relationship from the graph.

▼ **Figure C10.15** You will usually need to analyse your raw data in some way.

Performing calculations with your data

You will usually have to do some calculations with your data as part of your analysis. When you recorded your data, you wrote down the units for all your measurements as well as the uncertainties. You may need to convert the units and the uncertainties to SI units; for example, mL to L. Include the units with all numbers as you do your calculations. In this way, you will make sure you have the correct units on all derived data.

It is good practice *in general*, not just in investigations, to include units at each step in all your calculations.

Interpreting your results

Once you have analysed your results, you need to interpret them. This means being able to either answer your research question or state whether your results support your hypothesis.

If your hypothesis is not supported

It is not enough to simply say 'our hypothesis is wrong'. If the hypothesis is wrong, *what* is wrong with it?

Go through your method, results and analysis. Check that your equipment was correctly calibrated, and that you were using it correctly. Check that data are recorded in the correct units, and that units are correctly carried through all calculations during analysis. Check your analysis carefully. If you are working in a group, get another person to repeat the calculations.

You should also consider what other factors may have affected your results. Were there variables that you weren't able to control? Were there variables that you forgot to control?

It is never good enough to conclude that 'the experiment didn't work'. Either a mistake was made or the model used was not appropriate for the situation. It is your job to work out which.

10.5 Communicating your results

If research is not reported on, then no one else can learn from it. An investigation is not complete until the results have been communicated. Most commonly a report is written.

Writing reports

A report is a formal and carefully structured account of your research. It is based on the data and analysis in your logbook. However, the report is a *summary*. It contains only a small fraction of what appears in the logbook. Your logbook contains all your ideas, rough working and raw data. The report typically contains none of this.

A report consists of several distinct sections, each with a particular purpose. These are:

- Abstract
- Introduction
- Materials
- Method
- Results and analysis
- Discussion
- Conclusion
- Acknowledgements
- References
- Appendices

Reports are always written in the third person and past tense, because they describe what you have done.

Abstract

The abstract is a very short summary of the entire report. It is the most important part, because often it is the only part that people read. Typically an abstract is between 50 and 200 words long. It appears at the start of the report, but is always the last section that you write. Try writing just one sentence to summarise each part of your report.

Introduction

The introduction tells the reader why you did the investigation and what your research question or hypothesis is. This is the place to explain why this research is interesting.

The introduction also provides any background information needed to be able to understand the rest of the report. This is the place to summarise any existing theories and models. You need to do this to justify your hypothesis. You should also summarise any similar investigations. All of this should be correctly referenced, as described in the References section on page 417.

Method

The method describes what you did. It is not a recipe for someone else to follow.

The method summarises what you measured and how you measured it. It also explains, briefly, why you chose a particular method or technique.

Write your method using sentences, not dot points. Remember that these need to be written in past tense – *it is not a recipe*. You are not commanding anyone to do anything. You are telling people what you did. For example, you would write 'The temperature was measured', not 'Record the temperature'.

Include any diagrams needed to make your method clear. The diagrams in your logbook will usually be rough sketches (see Figure C10.16). The figures in your report should be very neat and carefully labelled. Flowcharts can be useful to describe any procedures in which a series of steps was followed. Each included item should have a number and you should refer to it in the text of your report. Position each item close to where it is referred to in the text. You should take the time to learn how to position figures neatly using your word processor software.

▼ Figure C10.16
a) A sketch of a titration from a logbook and
b) a scientific illustration for a report

a

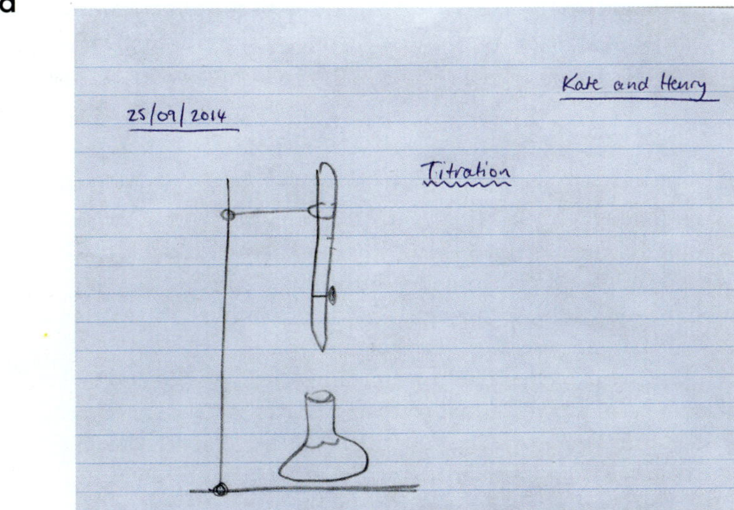

b

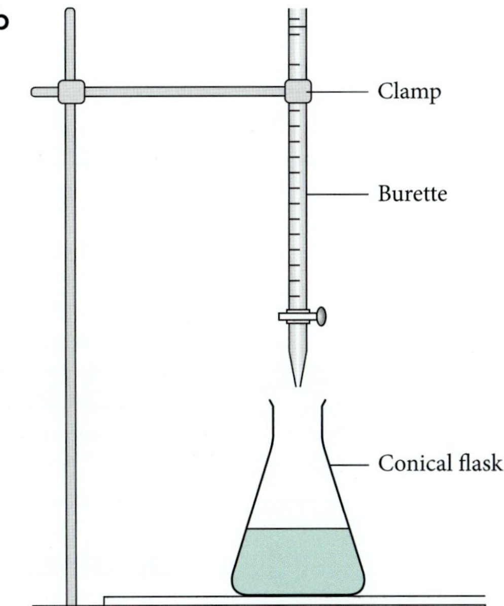

Results and analysis

The results section is a *summary* of your results. It is usually combined with the analysis section, although they may be kept separate.

Avoid including tables of raw data in your report unless they compare the results of a few different experiments. For example, you would include a table showing time taken for the reaction at different temperatures.

If a table has more than a few rows of data, it is better to represent those data in some other way. Usually this will be a graph.

Where appropriate, you should also include a graph to represent your data. However, your graph should be for the averages, not all the raw data. Think about what sort of graph is appropriate. If you want to show a relationship between two variables, then use a scatter plot. Display your data as points and clearly label any lines you have fitted to the data. Always make sure you label your axes, including units. Choose an appropriate scale so that the data take up most of the plot area.

Column and bar charts are useful for comparing two data sets, such as the average time taken for 5 grams of calcium carbonate chips and powder to react. *Do not* use a column or bar chart to try to show a mathematical relationship between variables. Remember to add headings for all figures and tables.

Figure C10.17 gives examples of the two types of graphs.

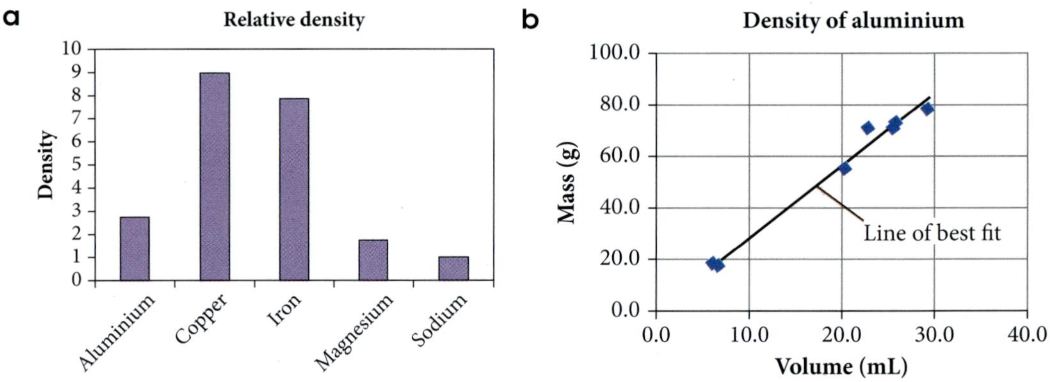

Figure C10.17 ▶ a) A column graph can be used for discrete data. b) A line graph or a scatter plot is used to show a mathematical relationship between continuous data.

a

Relative density

Density

10 9 8 7 6 5 4 3 2 1 0

Aluminium Copper Iron Magnesium Sodium

b

Density of aluminium

Mass (g)

100.0 80.0 60.0 40.0 20.0 0.0

Line of best fit

0.0 10.0 20.0 30.0 40.0

Volume (mL)

Any data and derived results should be given in correct SI units with their uncertainties. If you performed calculations, then show the equations you used. You might want to show one example of your calculation, but do not show more than one if the procedure used is repeated.

Discussion

The discussion should summarise *what your results mean*. If you began with a research question, then give the answer to the question here. If you began with a hypothesis, then state whether your results support your hypothesis or not. If not, explain why. You may not be able to say why, other than that the model was not suitable for the situation being investigated.

If there are any implications of your work, such as what conditions to use for a reaction or which materials to use for a particular purpose, then put them here.

The discussion is also the place to briefly describe any difficulties that you had and make suggestions for improving the process. Remember that you should never say 'the experiment didn't work' if you didn't get the results you expected. You might choose to make some comments on a more appropriate model and possible further work that could be done.

REFERENCING GUIDE

This guide is designed to help you with referencing your sources for assignments.

Conclusion

The conclusion is a *very* brief summary of the results and their implications. Say what you found out and what it means. A conclusion should only be a few sentences long.

Acknowledgements

You should thank anyone who helped you in your investigation. This includes people who supplied equipment or funding, as well as people who gave you good ideas or helped with the analysis. In science, as in other aspects of your life, it is polite to say thank you; however, this is not a necessary section of a report.

References

A reference list details the source of each piece of information used, and is linked to that information in the report.

A reference list details the sources of all information that was used to write the report. Wherever a piece of information or quotation is used in your report, it must be referenced *at that point*. This is typically done by placing either a number in brackets at the point [2], or the author and year of publication (Smith, 2014). The reference list is then provided either in a footnote at the end of the page, or as a single complete list at the end of the report. Referencing must be done in a consistent style. Check with your teacher what style they prefer. There are several good online guides to referencing.

A reference list is *not* the same as a bibliography. A bibliography is a list of sources that are useful to understanding the research. They may or may not have been used by the report authors. You should have a bibliography in your logbook from the planning stage of your investigation. The references will be a subset of these sources.

Appendices

Appendices may be used to provide additional information such as raw data that is not necessary to understanding the report but which might be of interest to some readers. Your teacher might require you to provide raw data in an appendix. Reports do not always have appendices.

Other ways of communicating your results

You may want to present the results of your investigation in another way. Scientists communicate their work in many ways; for example, by presenting a poster or seminar. They may write an article or produce a website. Scientists usually use more than one way, and sometimes several, to communicate about a really interesting investigation.

Look at examples of articles in the scientific and popular media, on websites, posters and so on. This will give you an idea of the different styles used in the different modes. Think about the purpose. Is it to inform, to persuade or both? What sort of language is used?

Think about your audience and use appropriate language and style. A poster is not usually as formal as a report. A website may be more or less formal, depending on your audience.

Posters and websites use a lot of images. Images are usually more appealing than words and numbers, but they need to be relevant. Make sure they communicate the information you want them to.

Keep readability and accessibility in mind if you are creating a poster or website. Posters should use large clear fonts and not have too much text. They should be readable from a few metres away.

Reproduced with permission of Helen Kiriazis (photographer) and Heart News & Views, International Society for Heart Research

Figure C10.18 A poster session is a common way to present scientific findings at a conference.

REFERENCING I-TUTORIAL

This tutorial will help you understand referencing and show you how to avoid plagiarism.

REPORT WRITING EXAMPLE 1

The brochure outlines the key features of a design report.

REPORT WRITING EXAMPLE 2

This online resource guides you through the sections of a typical report.

REPORT WRITING EXAMPLE 3

This online resource will help you write a case study.

WEBSITE ACCESSIBILITY

The Royal Society for the Blind has information on making websites accessible.

Fonts also need to be large enough and clear on websites and digital images should have tags. Follow the weblink 'Website accessibility' for more information on accessibility and web-page design.

However you communicate your work, make sure you know what the message is and who the audience is. Once you have established that, you will be able to let other people know about the interesting things you have discovered in your investigation.

CHAPTER GLOSSARY

analogue a device or scale that gives a continuous measurement; the scale is continuous and may show any value in a range

continuous able to include any value, sometimes within a fixed range; for example, a rainbow is a continuous spectrum

dependent variable the variable that changes as a result of changes to the independent variable

discrete able to include only specific values, not continuous; for example, a line spectrum is a discrete spectrum

extrapolation extension beyond the measured range of data to read or construct new data that have not been measured

falsifiable able to be disproved

hypothesis a tentative prediction, usually based on an existing model or theory. Also a tentative explanation of an observation based on an existing model or theory

independent variable a variable upon which another variable depends

interpolation to read or construct a new data point that has not been measured but is within the range of measured data

limit of reading the minimum uncertainty in a measurement due to the precision with which the scale can be read

line of best fit the line that most accurately fits the data, usually calculated using linear regression

linearise to make linear; to convert into a form that can be described by a straight line

logbook the record of an experiment or investigation kept by the scientist performing the experiments; it is a legal record of the experiments and their results

measurand the quantity being measured

model a representation of a system or phenomena that explains the system or phenomena. A model may be mathematical equations, a computer simulation, a physical object, words or other form

outlier a data point that does not fit the pattern shown by other measured data points

plagiarism presenting someone else's work, including their words or ideas, as your own

precision the variation in repeated measurements, or the uncertainty of a measuring device

primary data data that you have measured or collected yourself

qualitative measurement a measurement with descriptive or non-numerical results

quantitative measurement a measurement with a numerical value

random error a variation that affects a measurement in a random way so that the measurement is as likely to change in any one direction as in any other

reference the source of a specific piece of information or quotation

reliable highly likely to be true; a trustworthy source of information or reproducible data

reproducible giving the same result, within uncertainty, when repeated measurements are made

research question the specific question that a particular experiment or investigation is attempting to answer

resolution the limit of reading or precision of a measuring device

scatter graph a graph or plot showing data points, without a line joining the points, and used to demonstrate or determine a mathematic relationship between variables. The axes are defined by the variables

secondary data data or information that has been collected by someone else

systematic error an error that acts to give a consistent offset in data; for example, a zero error

theory a collection of models and concepts that explain specific systems or phenomena. Scientific theories allow predictions to be made and hence are falsifiable

true value the exact value of a measurand; an idealisation as all measurements have uncertainty and many measurands have values that vary

uncertainty an estimate of the range of values within which the 'true value' of a measurement or derived quantity lies

valid results that are affected only by a single independent variable and hence are reproducible

variable something that can change or be changed, as distinct from a constant, which does not

APPENDICES

Appendix 1: Periodic table

Key

atomic number → 26; symbol of element → Fe; name of element → iron; standard atomic weight → 55.85

- gas at room temperature (dark red)
- liquid at room temperature (green)
- solid at room temperature (black)
- synthetic (does not occur naturally) (blue)

- s block (orange)
- p block (blue)
- d block transition metals (yellow)
- d block lanthanoids and actinoids (pink)

1	2	3	4	5	6	7	8	9	10	11	12	13	14	15	16	17	18
1 H hydrogen [1.007, 1.009]																	2 He helium 4.003
3 Li lithium [6.938, 6.997]	4 Be beryllium 9.012											5 B boron [10.80, 10.83]	6 C carbon [12.00, 12.02]	7 N nitrogen [14.00, 14.01]	8 O oxygen [15.99, 16.00]	9 F fluorine 19.00	10 Ne neon 20.18
11 Na sodium 22.99	12 Mg magnesium [24.30, 24.31]											13 Al aluminium 26.98	14 Si silicon [28.08, 28.09]	15 P phosphorus 30.97	16 S sulfur [32.05, 32.08]	17 Cl chlorine [35.44, 35.46]	18 Ar argon 39.95
19 K potassium 39.10	20 Ca calcium 40.08	21 Sc scandium 44.96	22 Ti titanium 47.87	23 V vanadium 50.94	24 Cr chromium 52.00	25 Mn manganese 54.94	26 Fe iron 55.85	27 Co cobalt 58.93	28 Ni nickel 58.69	29 Cu copper 63.55	30 Zn zinc 65.38(2)	31 Ga gallium 69.72	32 Ge germanium 72.63	33 As arsenic 74.92	34 Se selenium 78.96(3)	35 Br bromine [79.90, 79.91]	36 Kr krypton 83.80
37 Rb rubidium 85.47	38 Sr strontium 87.62	39 Y yttrium 88.91	40 Zr zirconium 91.22	41 Nb niobium 92.91	42 Mo molybdenum 95.96(2)	43 Tc technetium	44 Ru ruthenium 101.1	45 Rh rhodium 102.9	46 Pd palladium 106.4	47 Ag silver 107.9	48 Cd cadmium 112.4	49 In indium 114.8	50 Sn tin 118.7	51 Sb antimony 121.8	52 Te tellurium 127.6	53 I iodine 126.9	54 Xe xenon 131.3
55 Cs caesium 132.9	56 Ba barium 137.3	57–71 lanthanoids	72 Hf hafnium 178.5	73 Ta tantalum 180.9	74 W tungsten 183.8	75 Re rhenium 186.2	76 Os osmium 190.2	77 Ir iridium 192.2	78 Pt platinum 195.1	79 Au gold 197.0	80 Hg mercury 200.6	81 Tl thallium [204.3, 204.4]	82 Pb lead 207.2	83 Bi bismuth 209.0	84 Po polonium	85 At astatine	86 Rn radon
87 Fr francium	88 Ra radium	89–103 actinoids	104 Rf rutherfordium	105 Db dubnium	106 Sg seaborgium	107 Bh bohrium	108 Hs hassium	109 Mt meitnerium	110 Ds darmstadtium	111 Rg roentgenium	112 Cn copernicium		114 Fl flerovium		116 Lv livermorium		

57 La lanthanum 138.9	58 Ce cerium 140.1	59 Pr praseodymium 140.9	60 Nd neodymium 144.2	61 Pm promethium	62 Sm samarium 150.4	63 Eu europium 152.0	64 Gd gadolinium 157.3	65 Tb terbium 158.9	66 Dy dysprosium 162.5	67 Ho holmium 164.9	68 Er erbium 167.3	69 Tm thulium 168.9	70 Yb ytterbium 173.1	71 Lu lutetium 175.0
89 Ac actinium	90 Th thorium 232.0	91 Pa protactinium 231.0	92 U uranium 238.0	93 Np neptunium	94 Pu plutonium	95 Am americium	96 Cm curium	97 Bk berkelium	98 Cf californium	99 Es einsteinium	100 Fm fermium	101 Md mendelevium	102 No nobelium	103 Lr lawrencium

Appendix 2: Relative atomic masses (atomic weights) of common elements

Element	Symbol	Relative atomic mass	Element	Symbol	Relative atomic mass
Aluminium	Al	26.98	Lead	Pb	207.2
Antimony	Sb	121.75	Lithium	Li	6.94
Argon	Ar	39.95	Magnesium	Mg	24.31
Arsenic	As	74.92	Manganese	Mn	54.94
Barium	Ba	137.34	Mercury	Hg	200.59
Beryllium	Be	9.01	Molybdenum	Mo	95.94
Bismuth	Bi	208.98	Neon	Ne	20.18
Boron	B	10.81	Nickel	Ni	58.71
Bromine	Br	79.90	Niobium	Nb	92.91
Cadmium	Cd	112.41	Nitrogen	N	14.01
Caesium	Cs	132.91	Osmium	Os	190.2
Calcium	Ca	40.08	Oxygen	O	16.00
Carbon	C	12.01	Palladium	Pd	106.4
Cerium	Ce	140.12	Phosphorus	P	30.97
Chlorine	Cl	35.45	Platinum	Pt	195.09
Chromium	Cr	52.00	Potassium	K	39.10
Cobalt	Co	58.93	Radium	Ra	226.03
Copper	Cu	63.55	Rhenium	Re	186.2
Fluorine	F	19.00	Rhodium	Rh	102.91
Gallium	Ga	69.72	Rubidium	Rb	85.47
Germanium	Ge	72.59	Ruthenium	Ru	101.07
Gold	Au	196.97	Scandium	Sc	44.96
Hafnium	Hf	178.49	Selenium	Se	78.96
Helium	He	4.00	Silicon	Si	28.09
Hydrogen	H	1.01	Silver	Ag	107.87
Indium	In	114.82	Sodium	Na	22.99
Iodine	I	126.90	Strontium	Sr	87.62
Iridium	Ir	192.22	Sulfur	S	32.06
Iron	Fe	55.85	Tantalum	Ta	180.95
Krypton	Kr	83.80	Technetium	Tc	98.91
Lanthanum	La	138.91	Tellurium	Te	127.60

Element	Symbol	Relative atomic mass	Element	Symbol	Relative atomic mass
Thallium	Tl	204.37	Vanadium	V	50.94
Thorium	Th	232.04	Xenon	Xe	131.30
Tin	Sn	118.69	Yttrium	Y	88.91
Titanium	Ti	47.90	Zinc	Zn	65.38
Tungsten	W	183.85	Zirconium	Zr	91.22
Uranium	U	238.03			

Appendix 3: Guidelines for using significant figures

When we measure quantities on instruments, the last figure in the measurement is usually uncertain. This is because of the in-built uncertainty in the instrument itself, even if we have avoided errors such as parallax error. Thus, if our electronic balance gives a reading of, say, 10.514 g, then we need to be aware that the last figure (4) will be uncertain. It is likely that the true mass is somewhere between 10.512 g and 10.516 g.

Significant figures show how many digits in the reading are meaningful. The last figure is always deemed to be uncertain. By keeping track of the number of significant figures in all the instrumental measurements used to calculate a quantity, we can determine the extent to which our answer correctly represents the accuracy of our instruments. To reflect this accuracy, we always give our answer to the same number of significant figures as the *least accurate* data used. (If we use an even lower number of significant figures than this, then we might as well use less accurate instruments!)

The rules

1. Every non-zero digit is significant. For example, 3.78 and 294 both have 3 significant figures.
2. Every zero in the middle of a reading is significant. For example, the mass reading of 10.514 g has 5 significant figures.
3. Every zero to the right of a reading is significant. For example, 31.20 has 4 significant figures. The exception to this is a number with no decimal point and a trail of zeros, such as 500 mL. This volume may have 1, 2 or 3 significant figures. To avoid this ambiguity, we must be given more information, stated in standard form. For example, if the volume is provided as 5.00×10^2 mL, then we know that it has 3 significant figures. If this is not clarified, we assume that it has the maximum number of significant figures.
4. Every zero before a number is not significant, and only shows the place value. For example, 0.005 only has 1 significant figure and 0.0090 has 2 significant figures. Again, rewriting these numbers in standard form clarifies this. (These numbers would be written as 5×10^{-3} and 9.0×10^{-3} respectively.)

Calculations

Rounding off

For rounding off an answer to a given number of significant figures, we examine the next figure on the right only. If it is 5 or more, then we round up.

Example

If we need to round off 10.9847 to:

- 4 significant figures, then we write 10.98 (8 is the fourth significant figure. The next figure on the right of 8 is 4, so we do not round up.)
- 3 significant figures, then we write 11.0. (9 is the third significant figure. The next figure on the right of 9 is 8, so we round up. Adding 1 to the 9 causes 10.9 to become 11.0.)

Adding or subtracting numbers

The answer cannot have more significant figures after the decimal point (i.e. decimal places) than the least accurate data.

Example

73.251 ← 3 decimal places

+1.4 ← 1 decimal place; therefore, this is the least accurate data

74.651

This answer cannot have more than 1 decimal place, so must be rounded off to 74.7.

Multiplying or dividing numbers

Again, the answer cannot have more significant figures than the least accurate data.

Example

$7.53 \times 6.0958 = 45.9$ (to 3 significant figures)

The least accurate figure (7.53) has 3 significant figures, so we round off the answer to 3 significant figures.

Use of data such as relative atomic masses

Ideally, physical data such as relative atomic masses and molar volumes should be quoted to at least the same number of significant figures as the experimental data.

However, physical data is not taken into account when determining the number of significant figures in the answer – only experimental data such as masses and volumes is considered.

Advice

Data provided in a question or in a test should not be rounded off prior to its use in a calculation. For example, if A_r(H) is given as 1.0 amu, then in a calculation we write 1.0, not 1. Similarly, if a volume is given as 20.00 mL, we write 20.00 in the calculation, not 20 (even though on our calculator we would only enter the number as 20). This makes the accuracy of the reading clear for the purposes of determining the number of significant figures we provide in the answer.

Appendix 4: Common ions, ion colours, flame colours and solubilities

Polyatomic ions

1+	1–	2–	3–
Ammonium NH_4^+	Nitrate NO_3^-	Carbonate CO_3^{2-}	Phosphate PO_4^{3-}
Hydronium H_3O^+	Nitrite NO_2^-	Sulfate SO_4^{2-}	
	Hydrogen carbonate HCO_3^-	Sulfite SO_3^{2-}	
	Hydrogen sulfate HSO_4^-	Hydrogen phosphate HPO_4^{2-}	
	Dihydrogen phosphate $H_2PO_4^-$	Dichromate $Cr_2O_7^{2-}$	
	Hydroxide OH^-	Silicate SiO_3^{2-}	
	Permanganate MnO_4^-		
	Thiocyanate SCN^-		
	Hypochlorite OCl^-		
	Ethanoate (acetate) CH_3COO^-		

Ion colours and flame colours

Ion	Solution colour	Ion	Flame colour
Groups 1, 2, 17	Colourless	Li^+	Bright red
Cr^{2+}	Blue	Na^+	Yellow
Cr^{3+}	Green	K^+	Violet
Co^{2+}	Pink	Ca^{2+}	Yellow-red
Cu^+	Green	Sr^{2+}	Bright red
Cu^{2+}	Blue	Ba^{2+}	Yellow-green
Fe^{2+}	Pale green	Cu^{2+}	Blue (halides); green (others)
Fe^{3+}	Yellow–brown	Pb^{2+}	Light blue-grey
Mn^{2+}	Pale pink	Zn^{2+}	Whitish green
Ni^{2+}	Green		
$Cr_2O_7^{2-}$	Orange		
MnO_4^-	Purple		

Solubility of ionic compounds at standard laboratory conditions (SLC)

Anion	Cl^-, Br^-, I^-	S^{2-}	OH^-	SO_4^{2-}	CO_3^{2-}, PO_4^{3-}, SO_3^{2-}	CH_3COO^-	NO_3^-
High solubility (aq) $\geq$ 0.1 mol L^{-1} (at SLC)	Most	Group 1, group 2 NH_4^+	Group 1, Sr^{2+}, Ba^{2+}, NH_4^+	Most	Group 1, NH_4^+	Most	All
Low solubility (s) < 0.1 mol L^{-1} (at SLC)	Ag^+, Pb^{2+}, Cu^+	Most	Most	Ag^+, Pb^{2+}, Ca^{2+}, Ba^{2+}, Sr^{2+}	Most	Ag^+	None

All group 1 compounds and all ammonium compounds have high solubility in water.

Appendix 5: Common units of measurement

Common values of mass, temperature and pressure

1 tonne (1 t) = 10^6 g = 1 Mg

STP = 0°C and 100 kPa

SLC = 25°C and 100 kPa

0°C = 273.15 K

1 atm = 101.325 kPa = 760 mmHg

Standard values

Avogadro constant $N_A = 6.022 \times 10^{23}$ particles mol^{-1}

Universal gas constant R = 8.31 kPa L mol^{-1} K^{-1}

Molar volume of gases V_M = 22.711 L mol^{-1} at STP; 24.790 L mol^{-1} at SLC

Density of water = 1.00 g cm^{-3} at 25°C

Prefixes for physical quantities

Fraction	Prefix	Symbol	Multiple	Prefix	Symbol
10^{-1} (tenth)	deci	d	10^1 (ten)	deca	da
10^{-2} (hundredth)	centi	c	10^2 (hundred)	hecto	h
10^{-3} (thousandth)	milli	m	10^3 (thousand)	kilo	k
10^{-6} (millionth)	micro	μ	10^6 (million)	mega	M
10^{-9} (billionth)	nano	n	10^9 (billion)	giga	G

NUMERICAL ANSWERS

Chapter 1

Worked example 1.1

a i 17,18,17
  ii 22,26,22
  iii 13,14,13
  iv 56,81,56

b i $^{65}_{30}Zn$
  ii $^{31}_{15}P$
  iii $^{64}_{29}Cu$
  iv $^{207}_{82}Pb$

Question set 1.1

1 Protons +1, neutron 0, electron −1

2 Proton/neutron in nucleus; electrons in electron cloud around nucleus

4 a Fluorine 9,10,9
  b Bromine: 35,45,35

5 a $^{27}_{13}Al$
  b $^{9}_{4}Be$

Worked example 1.2

a 6.92

b 24.32

Question set 1.2

8 36 amu

9 63.62 amu

Question set 1.4

1 18

5 a N: $1s^2\,2s^2\,2p^3$: 2,5
  b P; $1s^2\,2s^2\,2p^6\,3s^2\,3p^3$: 2,8,5
  c H $1s^1$: 1
  d Sc $1s^2\,2s^2\,2p^6\,3s^2\,3p^6\,4s^2\,3d^1$: 2,8,9,2
  e Ge $1s^2\,2s^2\,2p^6\,3s^2\,3p^6\,3d^{10}\,4s^2\,4p^2$: 2,8,18,4

6 a Mg
  b He
  c Fe

Question set 1.5

1 a 2
  b 7
  c 5
  d 1
  e 8

6 Lithium

7 a O
  b Be
  c F
  d C

Question set 1.6

7 Sodium

Worked example 1.3

a i 3.5 ppm
  ii 1.5 ppm

b ii 3.6 ppm
  iii Not safe

Worked example 1.4

a 2

b 35, 37

c 35

d 75, 25

e 35.5 Chlorine

Question set 1.7

3 Mass spectrum

6 b 2.7 mg L^{-1}

7 a 90,91,92,94,96
  c 91.3 Zr Zirconium

Chapter review questions

4 a 6
  b 7
  c 2
  d 6

5 Decrease across; increase down

12 a 2,3 or $1s^2 2s^2 p^1$
  b 2,8,6 or $1s^2 2s^2 2p^6 3s^2 3p^4$
  c 2,1 or $1s^2 2s^1$
  d 2,8,11,2 or $1s^2 2s^2\,2p^6 3s^2 3p^6 3d^3 4s^2$

13 a Oxygen
  b Argon
  c Cobalt

15 a $^{27}_{13}Al$
  b $^{119}_{50}Sn$
  c $^{16}_{8}O$
  d $^{127}_{53}I$
  e $^{85}_{37}Rb$

16 28.1

17 Helium

18 a Zinc

d $12\,mg\,L^{-1}$

e 12 ppm, which exceeds the level on the label.

19 b Five isotopes

c Isotope 44

Chapter 2

Question set 2.1

2 a i Homogeneous

ii Heterogeneous

iii Homogeneous

iv Homogeneous

v Homogeneous

vi Heterogeneous

vii Homogeneous

viii Heterogeneous

ix Heterogeneous

x Homogeneous

b i Element

iii Mixture

iv Mixture

v Pure substance

vii Element

x Element

6 a C

b B

c A

Question set 2.2

2 a Physical

b Physical

c Chemical

d Physical

e Chemical

f Physical

g Physical

h Chemical

3 Precipitate forms; gas is produced; colour changes; temperature changes; insoluble solid disappears

4 a Chemical

b Physical

c Chemical

d Physical

e Chemical

f Chemical

g Physical

h Chemical

5 a Roasting → crushing → washing → baking

b Chemical, physical, physical, chemical

Question set 2.3

5 a 1 = sieving; 2 = separating funnel; 3 evaporation

b Iron, sugar and water

c Fractional distillation

Chapter review questions

3 a Physical

b Physical

c Chemical

d Physical

5 a Solution

b Heterogenous mixture

c Element

d Solution

e Heterogeneous mixture

f Element

6 a Chemical

b Physical

c Chemical

d Chemical

e Physical

9 Solubility; magnetic attraction; evaporation point; state of matter at room temperature

10 a Magnetism; magnet

b Magnetism; magnet

c Boiling point; evaporation

d Density; separating funnel

12 Transparent; heated and reshaped; easily cut; will hold shape

Chapter 3

Worked example 3.1

a 1C : 4Cl

b 1B: 3H

c 1Mg : 1O

d 2Al : 3S

Question set 3.1

6 a 0

b 1

c 5

d 3

e 2

7 a 1 : 4

b 3 : 1

c 1 : 3

d 2 : 3

Question set 3.2

1 Loosely held; delocalised

5 a Good conductor of electricity; ductile

b Hard; high tensile strength

c Good conductor of heat; malleable

d Malleable; lustrous

e Good conductor of electricity; very high melting temperature

Question set 3.3

5 a $1s^2 2s^2 2p^6 3s^2 3p^6 4s^1$:$1s^2 2s^2 2p^6 3s^2 3p^6$ // 2,8,8,1 : 2,8,8

b $1s^2 2s^2 2p^6 3s^2 3p^4$: $1s^2 2s^2 2p^6 3s^2 3p^6$ // 2,8,6 : 2,8,8

c $1s^2 2s^2 2p^3$: $1s^2 2s^2 2p^6$ // 2,5 : 2,8

d $1s^2 2s^2 2p^6 3s^2 3p^6 4s^2$: $1s^2 2s^2 2p^6 3s^2 3p^6$ // 2,8,8,2 : 2,8,8

7 a +2

b −3

9 a Mg^{2+} and O^{2-}

b Ag^+ and Cl^-

c Li^+ and I^-

Worked example 3.3

a NaF

b MgO

c KOH

d $Cu(NO_3)_2$

e $Al_2(SO_4)_3$

Question set 3.4

1 a No overall net charge

b Number cations = number anions

2 a Bromide Br^-

b Sulfide S^{2-}

c Barium Ba^{2+}

d Potassium K^+

e Nitride N^{3-}

4 a Potassium iodide

b Barium chloride

c Calcium hydride

d Lead (IV) oxide

e Sodium phosphate

f Zinc nitrate

5 a Mg_3N

b Al_2O_3

c HgS

d NH_4OH

e Cu_2CO_3

f $Zn_3(PO_4)_2$

6 a +1

b +2

c +1

d +1

e −1

f −1

g −2

Question set 3.5

1 b i Two

ii Four

iii Six

2 Carbon and silicon

3 a 8, 2 and 8 respectively

b Ne, He and Ar

6 a F_2O

b BH_3

Question set 3.6

2 a SO_2

b N_2O_5

c CCl_4

d NF_3

e $SiBr_4$

3 a Dinitrogen oxide

b Nitrogen trichloride

c Sulfur trioxide

d Dihydrogen sulfide

e Dinitrogen tetraoxide

Question set 3.7

1 a C_nH_{2n+2}

b C_nH_{2n}

c C_6H_6

4 C_6H_{12} is an alkene (C_nH_{2n}); $C_6H_{14}$is an alkane (C_nH_{2n+2})

7 a Butane

b Nonane

c 2- pentene or pent-2 -ene

d 1- butene or but-1-ene

e 3-methylhexane

f 4-ethyl- 3-methylheptane

g 3-methyl-2-pentene or 3-methyl pent-2-ene

h 2,3-dimethyl pent-1-ene

10 a $C_5H_{12}(l) + 8O_2(g) \rightarrow 5CO_2(g) + 6H_2O(g)$

b $C_4H_{10}(l) + Cl_2(g) \xrightarrow{\text{UV light}} C_4H_9Cl(l) + HCl(l)$

c $CH_3CHCH_2(g) + H_2O(g) \rightarrow CH_3CH_2CH_2OH(g)$

d $CH_2CH_2(g) + Br_2(g) \rightarrow CH_2BrCH_2Br(l)$

e $C_6H_6(g) + Cl_2(g) \rightarrow C_6H_5Cl(g) + HCl(l)$

f $C_6H_6(l) + CH_3CH_2CH_2Cl(l) \xrightarrow{AlCl_3} C_6H_5CH_2CH_2CH_3(l) + HCl(l)$

Chapter review questions

1 a Covalent, ionic and metallic

8 a Silver oxide

b Aluminium trichloride

c Potassium sulfate

d Ammonium bromide

e Magnesium hydroxide

f Iron(II) carbonate

9 a $MgCl_2$
b $CaCO_3$
c $Cu(NO_3)_2$
d NH_4Cl
e K_2S
f PbO_2

10 a i +3
ii +2
b i −1
ii −3

11 a Phosphorus trichloride
b Sulfur tetrafluoride
c Dinitrogen trioxide
d Nitrogen oxide
e Dichlorine heptoxide

12 a BCl_3
b PI_5
c HBr
d N_2O_4
e SiO_2

14 a Butane
b Hexane
c 1 pentene
d 2 butene or but-2-ene
e 2-pentene or pent-2-ene
f 2,3-dimethyl butane
g 3-ethyl hex-2-ene
h 2,3-dimethyl pent-2-ene or 2,3-dimethyl-2-pentene

16 a 1-hexene
b 3-heptene

17 a i $C_5H_{12}(l) + 8O_2(g) \rightarrow 5CO_2(g) + 6H_2O(g)$
ii $2C_7H_{14}(l) + 21O_2(g) \rightarrow 14CO_2(g) + 14H_2O(g)$
b i $C_3H_8(g) + Br_2(aq) \rightarrow C_3H_7Br(g) + HBr(aq)$
ii $C_8H_{18}(g) + Cl_2(aq) \rightarrow C_8H_{17}Cl(g) + HCl(aq)$

18 a $C_5H_{10}(l) + Cl_2(aq) \rightarrow C_5H_{10}Cl_2(l)$
b $C_3H_6(g) + H_2O(l) \rightarrow C_3H_8O(l)$
c $C_6H_{12}(l) + Br_2(aq) \rightarrow C_6H_{12}Br_2(l)$

19 a A Metallic, B Covalent network, C Covalent molecular, D Ionic, E Covalent network

Chapter 4

Worked example 4.1

a Iron + oxygen → iron oxide

b Octane + oxygen → carbon dioxide + water

Worked example 4.2

a $4Fe(s) + 3O_2(g) \rightarrow 2Fe_2O_3(s)$

b $C_6H_{12}O_6(l) + 6O_2(g) \rightarrow 6H_2O(g) + 6CO_2(g)$

Question set 4.1

1 Represents a chemical reaction

3 a $Fe(s) + 2HCl(aq) \rightarrow FeCl_2(aq) + H_2(g)$
b $2NaHCO_3(s) + H_2SO_4(aq) \rightarrow Na_2SO_4(aq) + 2CO_2(g) + 2H_2O(l)$
c $2CuFeS_2(s) + 2SiO_2(s) + 4O_2(g) \rightarrow Cu_2S(s) + 2FeSiO_3(s) + 3SO_2(g)$
d $2PbS(s) + 3O_2(g) \rightarrow 2PbO(s) + 2SO_2(g)$
e $2C_3H_8(g) + 10O_2(g) \rightarrow 6CO_2(g) + 8H_2O(l)$

4 a No: $2H_2(g) + O_2(g) \rightarrow 2H_2O(l)$
b Yes
c No: $C_6H_{12}O_6(aq) \rightarrow 2C_2H_5OH(aq) + 2CO_2(g)$
d Yes
e Yes

5 a $N_2(g) + 3H_2(g) \rightarrow 2NH_3(g)$
b $Mg(OH)_2(s) + 2HCl(aq) \rightarrow MgCl_2(aq) + 2H_2O(l)$
c $Zn(s) + H_2SO_4(aq) \rightarrow ZnSO_4(aq) + H_2(g)$

Worked example 4.3

a $2Al(s) + 6H^+(aq) \rightarrow 2Al^{3+}(aq) + 3H_2(g)$

b $Cu(s) + 4HNO_3(aq) \rightarrow Cu(NO_3)_2(aq) + 2NO_2(g) + 2H_2O(l)$

Question set 4.2

4 a $Fe(s) + Cu^{2+}(aq) \rightarrow Cu(s) + Fe^{2+}(aq)$
b $CO_3^{2-}(aq) + Mg^{2+}(aq) \rightarrow MgCO_3(s)$

5 a $Ca(s) + H_2O(l) \rightarrow Ca^{2+}(aq) + 2OH^-(aq) + H_2(g)$
b $Mg(s) + 2HCl(aq) \rightarrow MgCl_2(aq) + H_2(g)$
c $Zn(s) + 2H^+(aq) \rightarrow Zn^{2+}(aq) + H_2(g)$
d $Pb^{2+}(aq) + 2I^-(aq) \rightarrow PbI_2(s)$

Worked example 4.4

a $2H_2(g) + O_2(g) \rightarrow 2H_2O(g) + \text{heat}\ \Delta H = -242\ kJ\ mol^{-1}$

b $2NH_3(g) + \text{heat} \rightarrow N_2(g) + 3H_2(g)\ \Delta H = +92\ kJ\ mol^{-1}$

c $CaCO_3(s) \rightarrow CaO(s) + CO_2(g)$ $\Delta H = +180\,kJ\,mol^{-1}$

d $4NH_3(g) + 5O_2(g) \rightarrow 6H_2O(g) + 4NO(g)$ $\Delta H = -905\ kJ\,mol^{-1}$

e $H_2O(l) \rightarrow H_2O(s)$ $\Delta H = -6.02\ kJ\,mol^{-1}$

Worked example 4.5

1 a 298 K

b 269 K

2 a 362°C

b −135°C

Worked example 4.6

a 1476 J

b 39.1 g

Question set 4.3

2 Temperature (°C) = temperature (K) −273

3 $2CO(g) + O_2(g) \rightarrow 2CO_2(g)$ $\Delta H = -566\,kJ\,mol^{-1}$

4 283.2 K

6 70 000 J

Worked example 4.7

33.709 kJ

Question set 4.4

4 a $C_6H_{12}(l) + 9O_2(g) \rightarrow 6CO_2(g) + 6H_2O(g)$

b $2C_2H_6(g) + 5O_2(g) \rightarrow 4CO(g) + 6H_2O(g)$ or $2C_2H_6(g) + 3O_2(g) \rightarrow 4C(s) + 6H_2O(g)$

Chapter review questions

1 a The equation without spectator ions

b Energy is given off

c Heat absorbed or given off in a reaction

d Fuel produced at faster rate than consumed

2 Biodiesel; coal; petroleum; gas

3 a Exothermic

b Endothermic

c Exothermic

d Endothermic

6 a $4FeAsS(s) + 11O_2(g) + 6H_2O(l) \rightarrow 4FeSO_4(aq) + 4H_3AsO_3(aq)$

b $C_3H_8(g) + 5O_2(g) \rightarrow 3CO_2(g) + 4H_2O(g)$ $\Delta H = -2220\,kJ\,mol^{-1}$

7 a $Fe_2O_3(s) + 3H_2(g) \rightarrow 2Fe(s) + 3H_2O(l)$

b $2Al(s) + 3Cl_2(g) \rightarrow 2AlCl_3(s)$

c $Cu(s) + 2AgNO_3(aq) \rightarrow 2Ag(s) + Cu(NO_3)_2$

d $V_2O_5(s) + 5Ca(s) \rightarrow 5CaO(s) + 2V(s)$

e $2H_2O_2(l) \rightarrow 2H_2O(l) + O_2(g)$

10 a 30.44 kJ g^{-1}

Chapter 5

Worked example 5.1

a 124

b 44

c 63

d 180

Worked example 5.2

a 41

b 261

c 106

d 159.6

Question set 5.1

1 a 28

b 34.1

c 141.9

d 342

e 98.1

2 a 74.6

b 62.3

c 310.3

d 187.5

e 132.1

3 N_2 or CO

Worked example 5.3

a C 75%, H 25%

b Na 39%, Cl 61%

c Ca 43.5%, C 26%, N 30.5%

d N 22.6%, H 6.5%, C 19.4%, O 51.6%

Question set 5.2

1 % by mass of each element

2 Analysing an unknown compound

3 a C 12.1%, O 16.2%, Cl 71.7%

b K 44.9%, S 18.4%, O 36.7%

4 41.7%

5 a Chalcocite (Cu_2S)

b Chalcopyrite ($CuFeS_2$)

c Bornite (Cu_5FeS_4)

Worked example 5.4

a 54.8 mol

b 0.02 mol

c 0.011 mol

d 5.0 mol

e 1.5 mol

Worked example 5.5

a 6.02×10^{22} molecules

b 3.61×10^{24} atoms

c 3.01×10^{23} ions

d 4.94×10^{22} atoms

e 1.26×10^{24} molecules

Worked example 5.6

a $K = 3$ mol, $Cl = 3$ mol

b 72 mol

c 2 mol

d 4.52×10^{24} individual atoms

e 6.77×10^{24} individual atoms

Question set 5.3

4 a 6.64 mol
  b 0.017 mol
  c 0.2 mol
  d 0.0137 mol

5 a 1.505×10^{24} molecules
  b 1.99×10^{23} atoms
  c 9.63×10^{23} iron ions
  d 2.11×10^{22} silver nitrate units

6 a 3 mol
  b 3 mol
  c 9 mol
  d 5.42×10^{24}

7 a 498 mol
  b 3984 mol
  c 8964 mol

8 a 10 mol
  b 0.4 mol
  c 45.1 mol

Worked example 5.7

a 159.8 g

b 98.1 g

c 74.1 g

d 80 g

e 46 g

Worked example 5.8

a 0.199 mol

b 0.106 mol

c 0.0073 mol

d 55.1 mol

Worked example 5.9

a 234.6 g

b 1.26 g

c 26.5 g

d 22.8 g

Question set 5.4

2 a 34 g
  b 184.1 g
  c 100.1 g
  d 72 g

3 a 0.14 g
  b 1125 g
  c 0.039 g
  d 148.8 g
  e 96.0 g

4 a 0.56 mol
  b 0.17 mol
  c 0.33 mol
  d 0.097 mol
  e 0.048 mol

5 a 0.00064 mol or 6.4×10^{-4} mol
  b 132 g

6 a $1.00 \times 294 = 294$ g
  b 0.021 mol
  c 72.0 g
  d 8.76×10^{18} molecules

7 a 7.53×10^{22} molecules
  b 1.506×10^{23} atoms

8 a 3556 mol
  b 2.14×10^{27} molecules

9 132 g

10 1.2 g

Worked example 5.10

a i CBr_2
  ii $C_4H_{10}O$
  iii $C_4H_5N_2O$

b $C_{20}H_{50}O_5$

Question set 5.5

2 a $COCl_2$
  b C_2H_5
  c NO_2
  d Na_2PO_3

3 Cl_2O

4 a C_5H_7N
  b $C_{10}H_{14}N_2$

5 PbO_2

6 a (0.236 mol CO_2, 0.314 mol H_2O)
  b $n(CO_2) = n(C) = 0.236$ mol C, $n(H) = 2n(H_2) = 0.628$ mol H
  c 2.83 g C, 0.628 g H
  d 3.782 g
  e CH_3O

7 $x = 6$; $MgBr_2.6H_2O$

8 a 7.85 g
  b 1.00 g
  c Cu_2O
  d To prevent metal reacting with air and oxide reforming

Worked example 5.11

a i 2.25 mol
  ii 4.25 mol

b i 0.75 mol
  ii 1.5 mol
  iii 4 mol

c i 35.75 mol
  ii 22 mol CO_2, 27.5 mol H_2O

Worked example 5.12

a i 14 g

ii 11 g

b i 5.4 g

ii 0.080 g

c i 160 g

ii 141 g

d 32 g

Question set 5.6

6 a 30 mol

b 19.4 g

7 125 g

8 a 34 g

b 47.8 g

9 a 405 kg

b 833 kg

10 No

Chapter review questions

7 Law of conversation of mass

8 a $64\ g\ mol^{-1}$

b $101.1\ g\ mol^{-1}$

c $99.4\ g\ mol^{-1}$

d $60\ g\ mol^{-1}$

9 a 54.1%

b 61.5%

10 a Ni 38%, S 20.7%, O 41.3%

b H 3%, P 32%, O 65%

11 a 5 mol

b 0.17 mol

c 2×10^4 mol

12 a 1.8×10^{23} atoms

b 2.4×10^{24} molecules

c 4.5×10^{24} formula units

14 a 0.05 mol

b 0.01 mol

c 0.05 mol

15 a 175.3 g

b 147 g

c 180 g

d 5.85 g

17 a 2 mol

b 9 mol

c 2.1 mol

18 0.12 mol

19 a 8 mol

b 2 mol

c 2.4×10^{24} particles

20 CuO

22 a 0.785 g

b 2.73 g

23 1.1 kg

24 65.4 g

25 a $FeSO_{11}H_{14}$

b $FeSO_4.7H_2O$

26 50.0%

27 97.7%

Chapter 6

Question set 6.1

1 a 1

b 2

c 5

d 3

e 1

f 6

3 a HCl: one bonding, three non-bonding; NH_3: three bonding, one lone pair; CH_4: four bonding pairs

b HCl: linear; NH_3: Pyramid; CH_4: Tetrahedral

6 a Tetrahedral

b Tetrahedral

c V-shaped oxygen with perpendicular hydrogens

d Linear

e V-shaped

f Linear

g Trigonal planar

h Linear

7 a i $1s^2 2s^2 2p^6 3s^2 3p^3$ or 2,8,5

Question set 6.2

1 Oxygen, nitrogen, fluorine, chlorine

3 a O−H: O more negative $\delta-$, H more positive $\delta+$

Cl−H: Cl $\delta-$, H $\delta+$

C−H: very small difference, C $\delta-$, H $\delta+$

P−H: no difference

N−O: O $\delta-$, N $\delta+$

N−Cl: no difference

N−H: N $\delta-$, H $\delta+$

S−H: S $\delta-$, H $\delta+$

C−S: no difference

F−O: F $\delta-$, O $\delta+$

Cl–O: O δ–, Cl δ+

C–O: O δ–, C δ+

P–O: O δ–, P δ+

C–Cl: Cl δ–, C δ+

b Highly polar: O–H, N–H, Cl–H, C–O, P–O, C–Cl

Some polarity: N–O, N–Cl, S–H, F–O, Cl–O,

Non polar: N–Cl, S–H, C–H, P–H, C–S

4 a ii Tetrahedral

iii No polar bonds

v Non-polar

b ii Tetrahedral

iii Polar bonds present

v Polar

c ii Linear

iii No polar bonds

v Non-polar

d ii Pyramid

iii No polar bonds

v Non-polar

5 a Polar

b Polar

c Polar

d Polar

e Polar

6 Tetrahedral

Question set 6.3

1 a Metallic, ionic and covalent

b Dipole–dipole, H-bonding and dispersion forces

3 NH, OH and FH

5 Dispersion, dipole–dipole, hydrogen

7 Less than –100°C

Question set 6.4

2 a $LiBr.2H_2O$

b $Ca(NO_3)_2.4H_2O$

c $Mg(NO_3).6H_2O$

5 b $KBr(s) \rightarrow K^+(aq) + Br^-(aq)$

c Yes

8 a Soluble

b Soluble

c Not soluble

d Soluble

e Soluble

f Not soluble

g Soluble

h Not soluble

i Soluble

Worked example 6.1

a i 30 g

ii 40 g

iii 80 g

b $NaNO_3$

c 36 g, both saturated

Worked example 6.2

a 12 g

b 12% v/v

c 50 mg

Question set 6.5

1 Saturated; unsaturated

2 a 1.5 g

b 9 g

3 8% w/v

4 a 0.06% w/v

b Yes

5 15 g

6 a No

b Saturated

d Crystals would form

7 2.5% w/w; 25 000 ppm

9 a 1.0 ppm

b No

c 40 cups

Question set 6.6

6 Compound 1

8 a D

b 22%

c Honey decomposes when heated over 40°C

9 b 11.9% v/v

10 a I, IV, III, II, V, VI, VII, VIII, IX, X

Chapter review questions

1 Covalent and dispersion forces

3 B

4 a 3 bonding, 6 non-bonding

b 3 bonding, 2 non-bonding

c 4 bonding, no non-bonding

5 a Non-polar

b Polar

c Non-polar

d Polar

6 $CH_3CH_2OH(s) \xrightarrow{H_2O} CH_3CH_2OH(aq)$;
$C_6H_{12}O_6(s) \xrightarrow{H_2O} C_6H_{12}O_6(aq)$

9 D

13 31.9 ml

15 a 30 g

b 23 g

c 16 g

16 1.5×10^{-12} g

17 A

19 b $KCl(s) \xrightarrow{H_2O} KCl(aq)$

c 26 g of KCl

d Supersaturated

e 10 g

20 40 000 ppm

22 Dopamine

Chapter 7

Worked example 7.1

a $K_2S(aq) + Co(CH_3COO)_2(aq) \rightarrow$
$2CH_3COOK(aq) + CoS(s)$
$S^{2-}(aq) + Co^{2+}(aq) \rightarrow CoS(s)$

b i $Cu(OH)_2$ copper hydroxide

ii None produced

iii $PbSO_4$ lead sulfate

c Solution containing sulfide, carbonate, sulfate or phosphate ions

Question set 7.1

3 a Soluble

b Insoluble

c Soluble

d Insoluble

e Soluble

4 a Potassium nitrate and zinc chloride

b Ammonium carbonate and sodium sulfate

c Magnesium sulfate and copper bromide

5 a $NaCl(aq) + AgNO_3(aq) \rightarrow$
$NaNO_3(aq) + AgCl(s)$

b $CuSO_4(aq) + 2KOH(aq) \rightarrow$
$Cu(OH)_2(s) + K_2SO_4(aq)$

c $NiCl_2(aq) + K_2SO_4(aq) \rightarrow 2KCl(aq) + NiSO_4(s)$

d $Na_2CO_3(aq) + FeSO_4(aq) \rightarrow$
$FeCO_3(s) + Na_2SO_4(aq)$

e $Zn(NO_3)_2(aq) + (NH_4)_2S(aq) \rightarrow$
$2NH_4NO_3(aq) + ZnS(s)$

f $K_2CO_3(aq) + CaCl_2(aq) \rightarrow$
$2KCl(aq) + CaCO_3(s)$

6 a $Na_2CO_3(aq) + MgCl_2(aq) \rightarrow$
$2NaCl(aq) + MgCO_3(s)$

b $Pb(NO_3)_2(aq) + CuSO_4(aq) \rightarrow$
$PbSO_4(s) + Cu(NO_3)_2(aq)$

c $NaBr(aq) + AgNO_3(aq) \rightarrow$
$NaNO_3(aq) + AgBr(s)$

7 B

Worked example 7.2

a i 0.133 M

ii 0.089 M

b i 0.43 M

ii 0.096 M

iii 0.088 M

c i 60.0 g

ii 0.533 g

iii 0.980 g

Worked example 7.3

a 2.20 g

b 0.51 g

c 0.50 g

Question set 7.2

2 a i 0.75 mol L^{-1}

ii 89.25 g L^{-1}

b i 0.0100 mol L^{-1}

ii 1.50 g L^{-1}

3 a 0.040 mol

b 0.075 mol

c 0.064 mol

4 a 10.5 g

b 10.7 g

c 8.6 g

5 a 0.182 M

b 0.0044 M

c 0.67 M

6 a 0.053 g

b 0.51 g

7 0.04 L

8 0.88 g

9 a 0.034 M

b 1.38 g in one litre

c Yes

Worked example 7.4

a 13.9 mL

b i 0.25 M

ii 40 mL

c 3.12 M

Question set 7.3

4 a $HNO_3(aq) \rightarrow H^+(aq) + NO_3^-(aq)$

b $HCl(aq) \rightarrow H^+(aq) + Cl^-(aq)$

c $H_3PO_4(aq) \rightarrow 3H^+(aq) + PO_4^{3-}(aq)$

5 a $Ca(OH)_2$: slightly soluble; CuO: soluble; NH_4OH: soluble; Na_2O: soluble

b $Ca(OH)_2$: $Ca(OH)_2(aq) \rightarrow Ca^{2+}(aq) + 2OH^-(aq)$

CuO: $CuO(aq) \rightarrow Cu^{2+}(aq) + O^{2-}(aq)$

NH_4OH: $NH_4OH(aq) \rightarrow NH^{4+}(aq) + OH^-(aq)$

Na2O: Na2O $\rightarrow 2Na^+(aq) + O^{2-}(aq)$

6 a $HNO_3(aq) + KOH(aq) \rightarrow KNO_3(aq) + H_2O(l)$

b $H_2SO_4(aq) + ZnO(aq) \rightarrow ZnSO_4(aq) + H_2O(l)$

c $2HF(aq) + Mg(OH)_2 (s) \rightarrow MgF_2(aq) + 2H_2O(l)$

d $H_3PO_4(aq) + 3NH_3(aq) \rightarrow (NH_4)_3PO_4(aq)$

e $2CH_3COOH(aq) + Na_2O(aq) \rightarrow 2CH_3COONa(aq) + H_2O(l)$

f $CaO(aq) + 2NH_4NO_3(aq) \rightarrow Ca(NO_3)_2 (aq) + NH_3(aq) + H_2O(l)$

Worked example 7.5

a 112 g

b 7.29 g

c 4

d 10

Question set 7.4

3 a T

b F

c F

d T

e F

5 Vinegar: wasp sting; ammonia: ant bite

6 a pH 1–3

b pH 7

c pH 8–10

d pH 11–13

7 pH = 6

8 D

9 C could be correct between the pH range of 6–7

10 a It would be lower.

b 16.6 g of NaOH

Question set 7.5

1 a Acid + base → salt + water

b Acid + metal → salt + hydrogen

c Acid + carbonate → salt + carbon dioxide + water

2 a $H_2SO_4(aq) + 2NaOH(aq) \rightarrow Na_2SO_4(aq) + 2H_2O(l)$

b $2CH_3COOH(aq) + Ca(OH)_2(aq) \rightarrow (CH_3COO)_2Ca(aq) + 2H_2O(l)$

c $2HNO_3(aq) + Mg(s) \rightarrow Mg(NO_3)_2(aq) + H_2(g)$

d $2HCl(aq) + Zn(s) \rightarrow ZnCl_2(aq) + H_2(g)$

e $2HF(aq) + Ca(HCO_3)_2(s) \rightarrow 2CaF_2(aq) + 2CO_2(g) + 2H_2O(l)$

f $2HNO_3(aq) + CuCO_3(s) \rightarrow Cu(NO_3)_2(aq) + CO_2(g) + H_2O(l)$

Chapter review questions

5 a No

b $2NH_4Br(aq) + Pb(NO_3)_2(aq) \rightarrow 2NH_4NO_3(aq) + PbBr_2(s)$

$Pb^{2+}(aq) + 2Br^-(aq) \rightarrow PbBr_2(s)$

c No

d $SrCl_2(aq) + ZnSO_4(aq) \rightarrow SrSO_4(s) + ZnCl_2(aq)$

$Sr^{2+}(aq) + SO_4^{2-}(aq) \rightarrow SrSO_4(s)$

7 a 3.00 mol

b 0.7 mol

8 a 2.5 M

b 5 M

9 a 2.57 g

b 4.05 g

10 a 0.0242 M

b 0.0019 M

c 0.00456 M

11 a 0.0100 L

12 a Purple

b Red

c No colour change

13 B

14 C

15 a $HNO_3(aq) + KOH(aq) \rightarrow KNO_3(aq) + H_2O(l)$

b $3HCl(aq) + Al(OH)_3(aq) \rightarrow AlCl_3(aq) + 3H_2O(l)$

c $H_2SO_4(aq) + Fe(s) \rightarrow FeSO_4(aq) + H_2(g)$

d $2CH_3COOH(aq) + MgCO_3(s) \rightarrow (CH_3COO)_2Mg(aq) + CO_2(g) + H_2O(l)$

e $H_3PO_4(aq) + 3NaHCO_3(s) \rightarrow Na_3PO_4(aq) + 3CO_2(g) + 3H_2O(l)$

17 Sodium carbonate

18 Barium ions

19 4

20 a 0.0166 M

b 1.23 g per litre

21 a 0.00192 mol

b 0.00192 mol

c 0.00383 mol

22 A: Pb^{2+}

B: Fe^{2+}

C: Ba^{2+}

D: Ca^{2+}

23 a 0.400 L

b 27.24 g

24 a Solution A

b i, iii

25 a E

b C

c Azalea

d Base

e Blue

Chapter 8

Question set 8.1

3 b 0.93 ppm

5 a 0.745 atm

b 0.0000674 atm

c 0.858 atm

6 a 104.4 kPa

b 45.6 kPa

c 109 kPa

7 a No

b Yes

c No

d Yes

Question set 8.2

9 a 510 K

b 248 K

c −153°C

d 72°C

12 A

Worked example 8.1

a 4.11 L

b 1.4 atm

c 283 mL

d 25 kPa

Worked example 8.2

a 4.2 mL

b i 2.1 L

ii 140 K = −132°C

c 53 mL

Worked example 8.3

a 208 mL

b i 67.5 L

ii 3123 L

c 181 kPa

Question set 8.3

1 a Standard temperature and pressure

b 25°C and 1.0 atm

4 The balloon would burst.

5 a 2.2 L

b 55 L

6 a −107°C

b 92 mL

7 11 L

8 154 kPa

9 52 atm

10 a 548 m^3

Worked example 8.4

a 2.13 L

b i 0.02 mol

ii 0.64 g

c i 112 L

ii 5.1 L

Worked example 8.5

a 14 L

b 0.736 g

c 336 kPa

Question set 8.4

2 b $PV = nRT$

4 79.2 L

5 0.040 mol

6 29 KPa

7 5.0 L

8 1.36 g

9 38.6 Ar

Worked example 8.6

a 1.1 L

b 20.4 L

c 0.51 g

d 278 g

e 2.2 L

Chapter review questions

2 C

3 A

4 C

5 A

6 A

7 B

8 a T

b T

c F

d F

9 6.0 L

10 B

11 A

12 83.3 kPa

13 32 g

14 23.5 $g\,mol^{-1}$

15 12.2 L

16 74.4 atm

18 98.5 atm

19 a 3200 kg

b 3.2×10^{12} kg

c 3.5×10^{18} g

d 9×10^{-4}

e 1×10^{10} trees

Chapter 9

Question set 9.1

3 At the start

6 0.08 mL CO_2 per s

7 0: 11 mL CO_2 per s

1: 3.5 mL CO_2 per s

5: 0.25 mL CO_2 per s

Question set 9.3

1 Surface area, temperature, concentration, catalyst

8 Ben

9 Test tube 5

Chapter review questions

2 Concentration, surface area, temperature and catalyst

11 $17.50

14 b It is a catalyst

GLOSSARY

$^{A}_{Z}X$ representation a format that describes the composition of the atom, including the element symbol (*X*), mass number (*A*) and atomic number (*Z*)

absolute temperature temperature measured in kelvin according to the absolute temperature scale, which has the same sized divisions as the Celsius scale but an end point of zero

absolute temperature scale a temperature scale that only has positive values since 0 K is defined as the point where there is no thermal motion for any known species. There is a linear relationship between this scale and the Celsius scale: temperature (°C) = temperature (K) −273

absolute zero lowest possible temperature (0 K or −273.15°C)

absorbance a measurement taken by a machine that compares the light passing into a substance with that exiting and gives a value

acid a substance that produces hydrogen ions, H^+, in aqueous solution

actinoids the period of the periodic table that, with the lanthanoids, make up the f block

activated complex the complex formed when the reactants' bonds are breaking but the products' bonds have not yet formed (see *transition state*)

activation energy the energy required to break the bonds of the reactants and therefore the minimum energy required for a successful collision

active site the section of an enzyme where the substrate(s) binds and undergoes a chemical reaction

addition reaction a reaction that involves the breaking of a multiple bond and the addition of new atoms to the compound

adhesive force the electrostatic force of attraction between unlike particles as displayed between water molecules and glass

adsorb to be attracted to the surface of a material; the opposite of 'desorb'

alkali a soluble base

alkali metals the common name of the elements found in group 1 of the periodic table

alkane a compound of carbon and hydrogen containing only single bonds

alkene a compound of carbon and hydrogen that contains carbon–carbon double bonds

allotrope a different physical form of the same element

alloy a mixture of two or more elements, one of which must be a metal

alpha particle a particle containing two protons and two neutrons, sometimes called a helium nucleus

amino acid a small organic molecule that combines to form proteins

amphoteric a substance that can act as either an acid or a base depending on the other reactants

amu atomic mass unit; the mass of one atom of carbon-12

analogue a device or scale that gives a continuous measurement; the scale is continuous and may show any value in a range

analyte the substance being analysed

anhydrous contains no water

anion a negatively charged atom or groups of atoms; formed by the addition of electron(s)

anionic negatively charged; for example, Cl^-

aqueous solutions solutions in which water is the solvent

aromatic hydrocarbon hydrocarbon containing one or more benzene rings

atmospheric pressure pressure due to the atmosphere; at sea level has an average value of 1 atm

atom the fundamental particle of matter; is composed of protons, neutrons and electrons

atomic absorption spectroscopy an analytical technique for determining the unknown concentration of an element based on the amount of light it absorbs

atomic number (*Z*) the number of protons in an atom

atomic orbital the region of space around an atom that has a specific shape and may contain a maximum of two electrons

atomic orbital diagram a diagram showing the space electrons occupy in one of four orbitals – s, p, d and f

atomic radius the distance from the nucleus to the boundary of the cloud of electrons surrounding it

Avogadro constant (N_A) the number of atoms in exactly 12 grams of the carbon-12 isotope

Avogadro's hypothesis equal volumes of any gas, measured at the same temperature and pressure, contain the same number of particles

base a substance that either contains the oxide (O^{2-}) of hydroxide ion (OH^-), or produces an hydroxide ion in aqueous solution

Big Bang the rapid expansion of matter from a high temperature and density state at the origin of the universe

biochemistry the study of chemicals and their reactions in living organisms

biofuel energy source produced from renewable sources such as crops

black hole a region of space with an extremely high gravitational field from which light and matter cannot escape

bonding pair a pair of electrons shared by two atoms to form a covalent bond

Boyle's law at a constant temperature, the volume of a given mass of gas is inversely proportional to the applied pressure

calibration curve a graph constructed during atomic absorption spectroscopy that plots known concentrations against the absorbance values, used to determine the concentration of an unknown substance

capillary action the process in which a liquid is drawn up a narrow tube

carbohydrate an organic molecule made up of one to many molecules of simple sugars

catalyst a substance that changes the rate of a reaction without being consumed

catalytic cracking the chemical process of breaking longer-chain hydrocarbons into shorter-chain hydrocarbons

cation a positively charged atom or groups of atoms; formed by the loss of electron(s)

cationic positively charged; for example, Na^+

cellular respiration the reactions that occur in cells to convert the energy in nutrients into energy that can be used; many organisms rely on oxygen and glucose to react to produce carbon dioxide, water and energy

Charles' law at constant pressure, the volume of a fixed quantity of gas is proportional to its absolute (kelvin) temperature

chelating ligand a covalent substance whose molecules can form several bonds with each single metal ion

chemical bond the electrostatic force of attraction between the protons and electrons of participating atoms

chemical change a change in the chemical composition of a substance to produce a new substance

chemical property a property of a substance relating to its ability to change to new substances during chemical reactions

chromatography techniques used to separate components of aqueous, liquid or gaseous mixtures

co-enzyme a non-protein chemical that binds to an enzyme and is necessary for the function of the enzyme

cohesive force the electrostatic force of attraction between like particles, which causes water molecules to stick to each other

collision theory a theory that explains the rate of reaction at a molecular level; it states that, for a reaction, the particles must collide with sufficient energy and the correct orientation

colloid a mixture in which tiny clusters of particles are dispersed through another substance; colloids do not settle out due to gravity and the particles are too small to be filtered

combined gas law a relationship between initial and final volume, pressure and absolute temperature for a given mass of gas

combustion a reaction with oxygen to form the oxides of each of the elements present; with adequate oxygen, combustion of a hydrocarbon will produce carbon dioxide and water

complete ionic equations equations that show all the ions present in solution

compound a pure substance composed of more than one type of atom chemically combined in fixed proportions

concentration the amount of solute dissolved in a given volume of solvent

continuous able to include any value, sometimes within a fixed range; for example, a rainbow is a continuous spectrum

corrosion a chemical reaction in which a metal degrades in the presence of oxygen and water to form the oxide of the metal

covalent bond the electrostatic force of attraction between shared negatively charged electrons and positively charged nuclei; results in a molecule or covalent lattice

covalent compound a compound composed of atoms of at least two different non-metals chemically combined in definite proportions

covalent molecular substance a discrete molecule in which the atoms are held together by covalent bonding

covalent network substance a three-dimensional network of covalently bonded atoms

cyroprotectant a chemical that is used to stop biological tissues from freezing

delocalised electron an electron that is detached from its atom and is free to move about within the structure

denature altering the chemical structure so that the original properties are lost

dependent variable the variable that changes as a result of changes to the independent variable

desorb the action of the substance moving from the stationary phase into the mobile phase

detector a device used to measure light or particles, found in both atomic absorption spectroscopy and mass spectrometry processes

diatom a single-celled algae found in water sources

diffuse the spreading out of a gas to fill a space

dilution the process of adding water to a solution to reduce its concentration

dipole a permanent build-up of negative charge at one end, and positive charge at another end, of a covalent bond or molecule

dipole–dipole force the attraction between molecules with permanent dipoles

directional bonding bonding that is in a direct line between adjacent particles; typical of covalent bonding

discrete able to include only specific values, not continuous; for example, a line spectrum is a discrete spectrum

dispersion force the weak attractive force between atoms and molecules caused by an instantaneous temporary change in dipole moment, arising from the movement of orbiting electrons

dissociated when ionic salts in a solvent dissolve, ions of opposite charge no longer associate with each other and move freely

distillate the vapour that condenses back to a liquid and which is collected after distillation

distillation a technique for separating a solution and retrieving the liquid component, based on differences in boiling point

double covalent bond a covalent bond in which two pairs of electrons are shared

El Niño a warm ocean current that develops every few years along the coast of Ecuador and Peru

elastic collision a collision in which there is no net loss of energy

electrolyte a charged solute that allows the solution to conduct electricity

electrolytic solution a solution that can conduct electricity due to the presence of ions

electromagnetic attraction the attraction of oppositely charged particles

electron a negatively charged particle found in energy levels around the nucleus of an atom

electron affinity the ability of an atom in the gaseous state to accept an electron and form a negative ion

electron charge cloud diagram a visual representation of the region of space around a nucleus where an electron might be found

electron configuration the arrangement of electrons around an atom in their energy levels

electron dot formula representation of an atom or ion using dots or crosses to show only valence electrons

electron shell diagram a visual representation of electrons in their energy levels around the nucleus

electronegativity the relative ability of an atom to attract electrons

electrostatic attraction a force that pulls particles together when they have an opposite charge

electrostatic repulsion a force that pushes particles apart when they have an identical charge

electrostatic separation a technique for separating substances based on attraction to an electric field

element a pure substance made up of atoms with the same atomic number

elute movement of a mobile phase through a column

eluting separating the parts of a mixture by using the property that they travel through a solvent at different rates

emission spectrum a pattern of bands produced by the emission of light from a source, separated due to the different wavelengths present

empirical formula the simplest whole number ratio of the elements in a chemical compound

emulsion a mixture of two substances that would not normally mix

endothermic a chemical or physical change where the chemical energy of the products is greater than the chemical energy of the reactants; hence, heat energy is absorbed

endothermic reaction a reaction that absorbs heat from the surroundings because the reactants have less enthalpy than the products

energy level a region of the atom in which electrons of the same energy can be found

energy profile diagram a diagram showing the relative energies during the progress of a reaction

energy shell see *energy level*

enthalpy the total energy possessed by a chemical substance at a constant pressure. It is usually expressed as the change in enthalpy

environmental footprint a measure of the impact of humans on Earth's ecosystems and resources

enzyme a protein molecule that catalyses a specific type of reaction by lowering the energy of activation

eutrophication a process in which an additional amount of nutrients leads to excessive growth of plants including microscopic plant-like phytoplankton

excited state when an electron is in a higher energy level than the ground state due to absorption of energy

exothermic a chemical or physical change where the chemical energy of the products is less than the chemical energy of the reactants; hence, heat energy is released

exothermic reaction a reaction that releases energy to the surroundings because the products have less energy than the reactants

extrapolation extension beyond the measured range of data to read or construct new data that have not been measured

falsifiable able to be disproved

fat an organic molecule that is insoluble in water; many are made up of fatty acids joined to glycerol

filtrate the liquid or solution that passes through the filter paper during filtration

filtration a technique for separating a solid from a liquid or a solution, based on solubility

flashpoint the lowest temperature at which a fuel has enough vapour present to ignite in air

food preservation methods of slowing the degradation or spoiling of food

food spoilage negative changes in food causing changes in taste, smell and feel

fossil fuel an energy source, such as crude oil and coal, produced from the decayed remains of animals and plants

fractional distillation a technique similar to distillation, used to separate liquids that have similar boiling points

general gas equation an equation that gives the relationship between pressure, temperature, volume and number of moles of a gas as $PV = nRT$

gradient the slope or steepness of a line; it is calculated by dividing the vertical height (rise) by the horizontal distance (run)

ground state when all the electrons of an atom are in their lowest possible energy levels

group a vertical column in the periodic table that gives information on number of valence shell electrons and trends between atoms

Haber process an industrial process of ammonia production from nitrogen and hydrogen gases

half-life the time taken for the radioactive emissions of an isotope to fall to half the original value

hard water water that contains high levels of Ca^{2+} and Mg^{2+} ions, which interferes with the action of soaps

heat of combustion heat released when a fuel undergoes complete combustion at standard atmospheric pressure

heteroatomic a molecule containing more than one type of element

heterogeneous having non-uniform composition

homeostasis processes that maintain the stable internal environment of an organism within limits

homogeneous having uniform composition throughout

hydrocarbon a chemical compound composed of only hydrogen and carbon

hydrogen bond the intermolecular attraction between hydrogen in –OH, –NH or –FH and the lone pairs on nitrogen, oxygen or fluorine on an adjacent molecule

hydronium ion (H_3O^+) the ion produced when a hydrogen ion (H^+) attaches to a water molecule

hydrophilic 'water-loving', a particle with polar regions that bond to water

hydrophobic 'water-hating', a particle with mostly non-polar regions that do not bond with water

hypertonic concentration of solutes is higher than another solution

hypothermia the condition of a reduced body temperature (below 35°C), lower than needed for normal metabolism

hypothesis a tentative prediction, usually based on an existing model or theory. Also a tentative explanation of an observation based on an existing model or theory

hypotonic concentration of solutes is lower than another solution

ideal gas a gas that perfectly obeys all the proposals of the kinetic theory of gases and the gas laws

immiscible liquids that do not mix

independent variable a variable upon which another variable depends

indicator a substance that changes colour in solution depending on whether the solution is acidic or basic

intermolecular bond the electrostatic forces of attraction between molecules in close proximity

intermolecular forces forces between individual particles (molecules, ions, atoms) of a substance

interpolation to read or construct a new data point that has not been measured but is within the range of measured data

intramolecular bond the electrostatic force of attraction that holds atoms together within a molecule

ion a charged atom, either positive from losing electrons, or negative from gaining electrons

ion-dipole attraction between the charge on one ion and a dipole in another molecule

ionic bond the electrostatic force of attraction between positively charged cations and negatively charged anions; results in an ionic compound with a three-dimensional lattice structure

ionic compound a compound that is made up of positive ions (cations) and negative ions (anions)

ionic equation an equation that gives information about all species in the reaction; aqueous solutions are written as separate ions; for example, NaCl(aq) is written as Na^+ and Cl^-

ionisation energy the amount of energy needed to remove an electron from a neutral atom when it is a gas

isomers (structural) compounds that have the same molecular formula but different structural formulas

isotonic concentration of solutes is the same in both solutions

isotopes different forms of an element with the same number of protons but different numbers of neutrons

isotopic composition the number and amount of isotopes within a sample of an element

kinetic energy the energy of movement

kinetic theory of gases a theory that explains the physical properties of gases in terms of the motion of their molecules

lanthanoids the period of the periodic table that, with the actinoids, make up the f block

law of conservation of mass the law that states that in a chemical reaction matter is neither created nor destroyed

ligand an ion or molecule attached to a metal atom (or larger molecule) by dipole bonding

limit of reading the minimum uncertainty in a measurement due to the precision with which the scale can be read

line emission spectrum a pattern of lines showing the component wavelengths in light

line of best fit the line that most accurately fits the data, usually calculated using linear regression

linearise to make linear; to convert into a form that can be described by a straight line

logbook the record of an experiment or investigation kept by the scientist performing the experiments; it is a legal record of the experiments and their results

lone pair a pair of valence electrons that are not involved in bonding

magnetic separation a technique for separating substances based on attraction to a magnetic field

main sequence stars that are in a state of equilibrium between gravity and the pressure produced by nuclear fusion

mass number (*A*) the total number of protons and neutrons in an atom

mass spectrometry an analytical method that uses the different masses of particles to measure their relative abundance in a sample

mass spectrum a graph produced during mass spectroscopy that shows the mass and relative abundance of substances present

material a substance with particular qualities or that is used for specific purposes

matter a physical substance; anything that has mass and occupies space

Maxwell–Boltzmann distribution a graph showing the distribution of energies (or velocities) of particles

measurand the quantity being measured

meniscus the shape of the surface of a liquid near the edge of a container

metal nanoparticle a metal particle with a diameter of less than 100 nm that behaves as a whole unit

metallic bond the electrostatic force of attraction that holds metal atoms together; acts between negatively charged delocalised valence electrons and positively charged metal cations

metalloid an element that has properties of both metals and non-metals

micelle small spherical cluster of surfactant molecules suspended in a liquid

micro-organism a very small organism that can be seen only with the use of a microscope such as a bacterium, virus, or protozoan

mixture matter that contains two or more different materials or substances with varying composition

mobile phase the phase that carries solutes through the stationary phase

model a representation of a system or phenomena that explains the system or phenomena. A model may be mathematical equations, a computer simulation, a physical object, words or other form

molar mass (*M*) the mass of one mole of a substance

molar volume of a gas the volume occupied by 1 mole of a gas at known temperature and pressure

molarity the number of moles of solute per litre of solution

mole (*n*) the amount of substance containing about 6.02×10^{23} particles of that substance

monochromator a device used in atomic absorbance spectroscopy to select light of a single wavelength

multiple covalent bond a covalent bond in which more than one pair of electrons are shared

nanomaterial a substance that is made of or incorporate nanosized particles, with unique properties

nanoparticle a particle that is on the nanometre scale (10^{-9} m)

nanotechnology a branch of science dealing with particles in the range of 1–100 nm

net ionic equation an equation that only contains the ions that take part in the chemical reaction; that is, spectator ions are excluded

neutral neither acidic nor basic, pH of 7

neutralisation a reaction between an acid and a base to form a salt and water

neutron a neutral particle found in the nucleus of an atom

neutron star a celestial body composed of neutrons, with high density and strong magnetic field, formed after the death of a massive star

non-bonding or lone pair a valence pair of electrons present in a covalent substance that is not involved in the covalent bond

non-directional bonding bonding that occurs in most directions; typical of metallic and ionic bonding

non-ionic detergent a surfactant with a non-charged hydrophilic part

nucleus a region of the atom containing all the protons and neutrons; it occupies only a small part of the volume of the atom but contains most of the mass

ocean acidification lowering of pH of the ocean due to the reaction of carbon dioxide with water molecules

organic chemistry the branch of chemistry that studies carbon compounds

origin in chromatography, the line that samples are placed on; the distance the solvent moves is measured from here

outlier a data point that does not fit the pattern shown by other measured data points

pascal (Pa) the SI unit of gas pressure, equivalent to a force of 1 newton per square metre

Pauli exclusion principle an orbital can contain 0, 1 or 2 electrons

percentage composition the percentage by mass of each different element in a compound

period a horizontal row of elements in the periodic table that gives information on the number of energy levels occupied by electrons in an atom

periodic table a chart of the elements, arranged in increasing atomic number; it is organised into groups and periods to show trends in the elements

pH scale a scale that measures the acidity and alkalinity of a solution, based on the concentration of hydrogen ions in a solution

physical change a change in the physical properties of a substance but not the chemical composition of the substance

physical property an observable feature of a substance that can be measured without changing the identity of the substance, such as colour, density and hardness

phytoplankton microscopic plant-like organisms found at the surface of oceans, seas and lakes

plagiarism presenting someone else's work, including their words or ideas, as your own

polar a molecule with a separation of charge; one end positive, another region negative

polar covalent bond a covalent bond that has a separation of charge

polyatomic ion an ion that consists of two or more atoms that are strongly bonded by covalent bonding

precipitate a solid produced by reaction between two clear solutions

precipitation reaction a reaction in which a precipitate is produced

precision the variation in repeated measurements, or the uncertainty of a measuring device

pressure the force exerted by a gas or gases due to the collisions of the particles with the walls of a container; expressed as force per unit area

primary data data that you have measured or collected yourself

protein a large organic molecule made up of many amino acids joined together

proteolytic enzyme an enzyme that breaks proteins into smaller lengths

proton a positively charged particle found in the nucleus of an atom

qualitative analysis analysis that identifies an element or substance but does not make numerical measurements about it, such as mass or concentration

qualitative measurement a measurement with descriptive or non-numerical results

quantitative analysis analysis that measures values such as amount, concentration or volume rather than just identifying the substance

quantitative measurement a measurement with a numerical value

radioactive decay spontaneous disintegration of an atom due to instability in the nucleus, releasing particles or electromagnetic radiation

random error a variation that affects a measurement in a random way so that the measurement is as likely to change in any one direction as in any other

rate how much one quantity changes with respect to another quantity

reactivity the likelihood of an element or substance undergoing a chemical reaction

red giant a star in the late stage of its life cycle, not undergoing hydrogen fusion, with high luminosity but low temperature

red supergiant a massive star that is undergoing helium fusion, with low temperatures and extremely large radii

reference the source of a specific piece of information or quotation

relative atomic mass the mean mass of an element that takes into account the isotope masses and the relative abundance on Earth; it is measured against carbon-12

relative formula mass (M_r) the mass of one formula unit of an ionic compound on a scale in which the mass of an atom of the carbon-12 isotope is exactly 12

relative molecular mass (M_r) the mass of one molecule of a substance relative to the mass of an atom of the carbon 12 isotope taken as exactly 12

reliable highly likely to be true; a trustworthy source of information or reproducible data

reproducible giving the same result, within uncertainty, when repeated measurements are made

research question the specific question that a particular experiment or investigation is attempting to answer

residue the solid remaining in the filter paper after filtration

resolution the limit of reading or precision of a measuring device

R_f retardation factor; the distance a solute moves compared to the distance the solvent has moved

R_t retention time; the time it takes for a solute to elute from the column in gas chromatography and high-performance liquid chromatography

saponification the production of soap by mixing fats with strong alkali

saturated compound a compound that contains only carbon–carbon single bonds

saturated solution a solution that has the most solute possible at a particular temperature

scatter graph a graph or plot showing data points, without a line joining the points, and used to demonstrate or determine a mathematic relationship between variables. The axes are defined by the variables.

secondary data data or information that has been collected by someone else

semiconductor a substance that has little to no electrical resistance when cooled to extremely low temperatures

separating funnel equipment used for separating immiscible liquids

sieving a technique for separating a mixture based on particle size

solubility the amount of solute that can dissolve in 100 g of water (solvent)

solubility curve a graph showing how solubility changes with temperature

solute the substance that is dissolved or the smaller component of a solution

solution a homogeneous mixture that is formed when a solute dissolves in a solvent

solvent the substance in which the solute dissolves or the greater part of a solution

species a general term that refers to atoms, ions and molecules

specific heat capacity heat energy required to raise the temperature of 1 gram of a substance by 1 kelvin

spectator ion an ion that is present in the same form on each side of the equation

spectroscope a device used to separate light into its component wavelengths

spectroscopy the branch of chemistry involving absorption and emission of light from substances

stationary phase the substance that is fixed in place during chromatography, to which the solute absorbs

stoichiometry the study of the amounts of substances in a chemical reaction

strong nuclear force an attractive force that exists between particles in the nucleus; it is a short-range force acting only on adjacent particles

structural formula a chemical formula that shows how atoms are linked

subshell a part of an energy level that contains orbitals of the same energy

substance a homogeneous material made up of the same type of particle

substituent an atom or group of atoms that has replaced a hydrogen atom on a hydrocarbon

substrate the chemical that fits into the active site of an enzyme

supernova the explosion of a star at the end of its life cycle, resulting in the ejection of most of the star's mass

supersaturated solution an unstable solution that has more solute than possible at that temperature; it is formed from dissolving solute into the solution at a higher temperature and then allowing it to cool. The extra solute is still in solution but can readily be crystallised out

surface area the total area of all surfaces

surface tension the force that arises from the attraction of the surface molecules to the bulk of the material

surfactant a chemical that lowers the surface tension of a liquid

suspension a cloudy, heterogeneous mixture containing solid particles that will eventually settle out

synthetic element element that does not exist in nature, but has been made in a laboratory

systematic error an error that acts to give a consistent offset in data; for example, a zero error

temperature a measure of the average kinetic energy of the particles

theory a collection of models and concepts that explain specific systems or phenomena. Scientific theories allow predictions to be made and hence are falsifiable.

transition elements elements found between groups 2 and 13 in the periodic table, also known as the d-block elements

transition metals the metals in groups 3–12 of the periodic table

transition state the temporary, unstable state when the reactants are forming products (see *activated complex*)

triple covalent bond a covalent bond in which three pairs of electrons are shared

true value the exact value of a measurand; an idealisation as all measurements have uncertainty and many measurands have values that vary

turbidity cloudiness of water due to suspended microscopic particles. High turbidity reduces how far light penetrates into water

ultraviolet light invisible, high-energy, high-frequency light of wavelengths 10 nm to 400 nm

uncertainty an estimate of the range of values within which the 'true value' of a measurement or derived quantity lies

universal gas constant (*R*) a constant value used in the general gas equation, equal to 8.314 J K^{-1} mol^{-1} or 0.082 L atm K^{-1} mol^{-1}

universal indicator a mixture of different indicators that displays a range of colours when added to solutions of different pH

unsaturated compounds compounds in which there is at least one multiple bond (double or triple) between carbon atoms

unsaturated solution a solution into which more solute can dissolve

valence electron an electron in the valence (outermost) shell of an atom in its ground state

valence shell the outermost shell of an atom that contains electrons

valence structure a diagram of the arrangement of atoms around a central atom

valency the combining power of an atom equal to the number of hydrogen atoms that atom could combine with or displace from a compound (hydrogen has a valency of 1)

valid results that are affected only by a single independent variable and hence are reproducible

vaporisation a technique for separating the liquid component of a solution, based on boiling point

vaporised when a substance is heated so that it turns into its atomic form

vapour pressure pressure exerted down by a vapour onto its liquid form

variable something that can change or be changed, as distinct from a constant, which does not

viscosity the ability of a fluid to resist flow; honey has high viscosity, water has low viscosity

volatile can change from a liquid to a gas (vaporise) easily

volumetric flask a piece of glass laboratory equipment used to prepare solutions of precise concentration

water of crystallisation the water molecules that are bonded inside crystals; there is an exact ratio between the water molecules and either the salt or the polar molecule because of the number of ion–dipole or hydrogen bonds present between the two

wavelength a property of light related to the length of the wave, which can give properties of light such as colour

white dwarf a planet-sized, dense star formed when a small star ceases nuclear fusion

yield the amount of product produced during a reaction

zeolite an aluminosilicate mineral; i.e. contains Al, Si and O

INDEX